Human Factors Engineering and Ergonomics

This textbook comprehensively covers the basic principles and most recent advances regarding visual displays, auditory and tactile displays, and controls; psychophysics; cognitive processes; human–computer interaction, artificial intelligence and artificial life; stress and human performance; occupational accidents and prevention; human group dynamics and complex systems; and anthropometry, workspace, and environmental design. The systems perspective emphasizes nonlinear dynamics for system performance changes and emergent behaviors of complex person–machine systems.

This book:

- Surveys principles of conventional and computer-based machine interaction.
- Assesses the relative effectiveness of accident analysis and prevention strategies.
- Highlights nonlinear dynamics for system performance changes.
- Examines artificial intelligence and complex systems.
- Investigates sources of cognitive workload and fatigue.

The textbook will be a valuable resource for advanced undergraduates and graduate students in diverse fields including ergonomics, human factors, cognitive science, computer science, operations management, and psychology. The textbook brings together core principles of person–machine interaction, accident analysis and prevention strategies, risk analysis and resilience, artificial intelligence, group dynamics, and nonlinear dynamics for an enhanced understanding of complex person–machine systems.

Human Factors Engineering and Ergonomics

A Systems Approach

Third edition

Stephen J. Guastello

CRC Press
Taylor & Francis Group
Boca Raton London New York

Third edition published 2023
by CRC Press
6000 Broken Sound Parkway NW, Suite 300, Boca Raton, FL 33487-2742

and by CRC Press
4 Park Square, Milton Park, Abingdon, Oxon, OX14 4RN

CRC Press is an imprint of Taylor & Francis Group, LLC

© 2023 Stephen J. Guastello

[First edition published by CRC Press 2006]
[Second edition published by CRC Press 2013]

ISBN: 9781032416588 (hbk)
ISBN: 9781032081625 (pbk)
ISBN: 9781003359128 (ebk)

DOI: 10.1201/9781003359128

Typeset in Sabon
by Newgen Publishing UK

Contents

Preface   *xix*
New to this edition   *xxi*
Author biography   *xxiii*

1   Introduction to human factors and ergonomics   **1**

  1.1   Evolution of person–machine systems 1
  1.2   The role of human factors and ergonomics 3
    1.2.1   The person–machine system 3
    1.2.2   The cognitive core 4
    1.2.3   The workspace shell 6
    1.2.4   Computer 7
    1.2.5   Beyond the great outdoors 8
  1.3   Criteria of human factors 8
    1.3.1   Performance criteria 8
    1.3.2   Industry standards 9
    1.3.3   Civil liability 10
      1.3.3.1   Strict liability 10
      1.3.3.2   Contributory negligence 11
      1.3.3.3   Negligence 11
      1.3.3.4   Proving the case 12

2   Elements of human factors analysis   **15**

  2.1   Allocation of function 15
    2.1.1   The benefits of machines 15
    2.1.2   The user population 17
    2.1.3   Experts and novices 18
    2.1.4   Workload issues 19
    2.1.5   Flexible allocation of function 20
    2.1.6   Trust in automation 21

2.2   *Human error and system reliability 23*
    2.2.1   *Error probability 24*
    2.2.2   *Redundancy 25*
2.3   *Usability testing 26*
    2.3.1   *Preparation 26*
    2.3.2   *Iterative laboratory testing 27*
    2.3.3   *Field testing 29*
    2.3.4   *Technical manuals 29*
    2.3.5   *Cost–benefit analysis 30*
    2.3.6   *System reengineering 32*

3   Systems change over time   35

3.1   *Communication, information, and entropy 35*
    3.1.1   *Communication model 35*
    3.1.2   *Quantifying information 36*
    3.1.3   *Entropy 36*
3.2   *How system events change over time 38*
    3.2.1   *Perception of change 38*
    3.2.2   *What is random? 39*
    3.2.3   *Simple attractors 40*
    3.2.4   *Bifurcation 41*
    3.2.5   *Chaos 42*
    3.2.6   *Fractals 46*
    3.2.7   *Self-organization 46*
    3.2.8   *Catastrophes 48*
    3.2.9   *Emergent phenomena 50*
    3.2.10 *Synchronicity 51*
3.3   *Networks 53*
    3.3.1   *Social networks 53*
    3.3.2   *Nonhierarchical structures 54*
    3.3.3   *Centrality 55*
    3.3.4   *Small worlds 56*
3.4   *Simulation strategies for complex systems 58*
    3.4.1   *Classical system simulations 58*
    3.4.2   *Artificial life simulations 59*
        3.4.2.1   *Cellular automata 60*
        3.4.2.2   *Agent-based models 61*
        3.4.2.3   *Genetic algorithms 63*

4   Psychophysics   65

4.1   *Classical psychophysics 65*
    4.1.1   *Threshold concepts 65*
    4.1.2   *Fundamental laws 66*

4.2   Scaling procedures 68
    4.2.1   Psychophysical stimuli 68
    4.2.2   Non-psychophysical stimuli 70
4.3   Signal detection theory 71
    4.3.1   Threshold concepts 72
    4.3.2   Discrimination index 72
    4.3.3   Minimization of errors 73
    4.3.4   ROC curves 74
    4.3.5   Individual differences 76
    4.3.6   Power law 78
    4.3.7   Decisions revisited 78
    4.3.8   Predictive power 79
4.4   Fuzzy signal detection theory 79
4.5   Vigilance 81
4.6   Multidimensional stimuli 83
    4.6.1   Multidimensional scaling 83
    4.6.2   Multidimensional nonlinear psychophysics 85
    4.6.3   Chaos and vigilance 87

5   Visual displays   **91**

5.1   The sense of vision 91
    5.1.1   Visual acuity 91
    5.1.2   Color vision 93
    5.1.3   Color vision abnormalities 95
    5.1.4   Photopic and scotopic functions 96
5.2   Perception 97
    5.2.1   Gestalt principles of form 99
    5.2.2   Depth 101
    5.2.3   Motion 103
5.3   Principles of display design 104
    5.3.1   Types of basic displays 105
    5.3.2   Display design criteria 106
    5.3.3   Visibility 107
    5.3.4   Distinguishability 109
    5.3.5   Interpretability 111
    5.3.6   Completeness 112
    5.3.7   Parallax effect 113
    5.3.8   Color and contrast 113
    5.3.9   Historical displays 114
    5.3.10   Predictive displays 115
    5.3.11   3-D displays 116
    5.3.12   Digital versus analog displays 119

5.4   Heads-up displays 119
    5.4.1   Clutter 121
    5.4.2   Design interactions 122
5.5   Iconic displays 123
    5.5.1   Symbolic versus concrete pictograms 124
    5.5.2   Emergent meaning 125
    5.5.3   Shared meaning 127
5.6   Display panel organization 127
    5.6.1   Linking analysis 128
    5.6.2   Ecological interface design 128
5.7   Signs of importance 130
    5.7.1   Design 130
    5.7.2   Standards 131
    5.7.3   Behavioral impact 133
5.8   Illumination and glare 133
    5.8.1   Illumination 133
    5.8.2   Glare 134

6   **Auditory and tactile displays**   137

6.1   Sense of hearing 137
    6.1.1   Loudness 138
    6.1.2   Pitch 139
    6.1.3   Timbre 141
    6.1.4   Binaural hearing 142
6.2   Nonverbal auditory displays 142
    6.2.1   Types of nonverbal auditory displays 143
    6.2.2   Gestalt laws of perception 144
    6.2.3   Streaming 145
6.3   Classic problems and solutions 146
    6.3.1   Localization 146
    6.3.2   Confusability 146
    6.3.3   Desensitization 148
    6.3.4   Recommendations 148
    6.3.5   Adding redundancy 149
    6.3.6   3-D auditory displays 149
6.4   Auditory icons 150
6.5   Speech displays 153
    6.5.1   Speech spectrograms 154
6.6   Noise 155
    6.6.1   Colors of noise 155
    6.6.2   More signal detection 156
    6.6.3   Hearing loss and noise exposure 156

6.7   Tactile displays 159
        6.7.1   Haptic perception 159
        6.7.2   Knobs 160
        6.7.3   Vibratotactile displays 161
        6.7.4   Tactons 162
        6.7.5   Gloves 162

7   Cognition   165

7.1   Organization of human memory 165
        7.1.1   Short-term memory 165
        7.1.2   Long-term memory 167
        7.1.3   Types of memory 168
        7.1.4   Working memory 169
        7.1.5   Attention and perception 171
        7.1.6   Attention tunneling 172
        7.1.7   Task switching 173
        7.1.8   Fluid intelligence 174
7.2   Types of decisions 176
        7.2.1   Simple binary decisions 176
        7.2.2   Optimizing decisions 176
                  7.2.2.1   Expectancy theory 176
                  7.2.2.2   Prospect theory 178
                  7.2.2.3   Incomplete information 179
        7.2.3   Nonoptimization decisions 179
                  7.2.3.1   Planning 179
                  7.2.3.2   Predicting a future state 180
                  7.2.3.3   Divergent thinking 183
        7.2.4   Production paradox 183
        7.2.5   Troubleshooting 183
                  7.2.5.1   Fault isolation 184
                  7.2.5.2   Template matching 184
                  7.2.5.3   Statistical template matching 184
7.3   Cognitive workload 185
        7.3.1   Channels and stages 186
        7.3.2   Limited capacity theory 186
        7.3.3   Variable capacity theory 188
        7.3.4   Resource competition model 189
        7.3.5   Attentional blink 189
        7.3.6   Multitasking 191
7.4   Measurement of cognitive workload 193
        7.4.1   Behavioral indicators 194
        7.4.2   Subjective indicators 195
        7.4.3   Physiological indicators 196

7.4.4   *Autonomic arousal 198*
7.4.5   *Augmented cognition 200*
7.5   *Automatization of cognitive processes 201*
   7.5.1   *Telegraph operation 201*
   7.5.2   *Controlled processes 202*
   7.5.3   *Recognition-primed decision-making 203*
   7.5.4   *Degrees of freedom principle 204*
7.6   *Dynamic decisions and situation awareness 206*
   7.6.1   *Dynamic decisions 206*
   7.6.2   *Control of dynamical systems 207*
   7.6.3   *Situation awareness 209*
   7.6.4   *Measurement of situation awareness 211*
   7.6.5   *Sensemaking 212*
   7.6.6   *Anticipatory thinking 213*
7.7   *Cognitive analysis of a person–machine system 214*
   7.7.1   *Job descriptions 215*
      7.7.1.1   *Functional job analysis 215*
      7.7.1.2   *Task-based job analysis 217*
      7.7.1.3   *Benchmark jobs 217*
      7.7.1.4   *O*Net online 217*
   7.7.2   *Cognitive task analysis 218*
      7.7.2.1   *Cognitive inventory 219*
      7.7.2.2   *Hierarchy of rules, skills, and knowledge 219*
      7.7.2.3   *Hierarchy of goals 220*
      7.7.2.4   *Ecological task approach 220*
      7.7.2.5   *Illustrative examples 221*
   7.7.3   *Think-aloud technique 224*

8   Psychomotor skill and controls   227

8.1   *Reaction and response time 227*
   8.1.1   *Donders' RT 227*
   8.1.2   *Type of stimuli 228*
   8.1.3   *Stimulus–response compatibility 228*
   8.1.4   *Population stereotypes 229*
8.2   *Learning and skill acquisition 230*
   8.2.1   *Skill acquisition 230*
   8.2.2   *Dynamics of learning processes 231*
   8.2.3   *Speed–accuracy trade-off 233*
   8.2.4   *Taxonomy of psychomotor skills 235*
   8.2.5   *Factor analysis and principal components
        analysis 236*

8.3   Types of manual controls 237
    8.3.1   Multidimensional controls 242
    8.3.2   Size 244
    8.3.3   Shape 246
    8.3.4   Space of controls 247
    8.3.5   Labels 248
    8.3.6   Resistance 248
    8.3.7   Control panels 249
    8.3.8   Voice control 251
8.4   Feedback and control 253
    8.4.1   Open and closed loops 254
    8.4.2   Fitts' law 254
    8.4.3   Motor control 258
        8.4.3.1   Walking 259
        8.4.3.2   Reaching and grasping 260
        8.4.3.3   Aiming 263
8.5   Complexity of control systems 263
    8.5.1   Order of controls 263
    8.5.2   Ashby's law 265
    8.5.3   Chaotic controllers 265
        8.5.3.1   Anticipation 266
        8.5.3.2   Adding instability 267
        8.5.3.3   Periodic entrainment 267
        8.5.3.4   Use of control parameters 268

9   Anthropometry and workspace design   271

9.1   Body measurements 271
    9.1.1   Designing for bodies in motion 273
    9.1.2   Facial anthropometry 275
9.2   Safety and related concerns 275
    9.2.1   Machine guards 275
    9.2.2   Overcrowding and confined spaces 277
9.3   Physical abilities 278
    9.3.1   Strength 278
    9.3.2   Flexibility 280
    9.3.3   Body coordination and equilibrium 280
    9.3.4   Stamina 281
    9.3.5   Lean body mass 281
    9.3.6   Personnel selection and physical abilities simulation 282
    9.3.7   Exoskeletons and exosuits 285
9.4   Some common biomechanical issues 285
    9.4.1   Lifting 286
    9.4.2   Walking surfaces 287

9.4.3   *Seating 288*
9.4.4   *Hand tools 289*
9.4.5   *Carpal tunnel syndrome 289*
9.5 *Computer workstations 290*
9.5.1   *Directory assistance operators 291*
9.5.2   *Technical service representatives 291*
9.5.3   *Workstation experiment 292*
9.5.4   *The near point 292*
9.5.5   *Workstations in health care settings 293*

10 Stress, fatigue, and human performance   **297**

10.1 *Physical stressors 297*
10.1.1 *Toxins 297*
10.1.2 *Extreme temperatures 298*
10.1.3 *Extreme altitudes 300*
10.1.4 *Noise 301*
10.2 *Social stressors 301*
10.2.1 *Crowding and isolation 302*
10.2.2 *Electronic monitoring 303*
10.3 *Speed and load 303*
10.3.1 *Working too slowly 304*
10.3.2 *Signal detection tasks 306*
10.4 *Work schedules 306*
10.4.1 *Circadian rhythm 307*
10.4.2 *Dysregulation 309*
10.5 *Consequences of stress 310*
10.5.1 *Performance 311*
10.5.2 *Health 313*
10.6 *Fatigue 314*
10.6.1 *Physical fatigue 316*
10.6.2 *Cognitive fatigue 317*
10.7 *Catastrophe models for cognitive workload and*
      fatigue 319
10.7.1 *Buckling stress 320*
        10.7.1.1 Physical demands 320
        10.7.1.2 Cognitive workload model 322
10.7.2 *Elasticity–rigidity 323*
        10.7.2.1 Affect constructs 324
        10.7.2.2 Coping constructs 327
        10.7.2.3 Fluid intelligence constructs 327
        10.7.2.4 Conscientiousness constructs 328
        10.7.2.5 Summary of cusp models 328

10.7.3 Fatigue 330
    10.7.3.1 Compensatory abilities 331
    10.7.3.2 Summary of cusp models 333
10.7.4 Multitasking 333
    10.7.4.1 Ordering of tasks 333
    10.7.4.2 Vigilance dual task 336
10.8 Other stress and performance dynamics 337
    10.8.1 Arousal, anxiety, and performance 337
    10.8.2 Levels of performance 340
    10.8.3 Diathesis stress 340

11 Occupational accidents and prevention 343

11.1 Causes of death and injury 343
    11.1.1 Death statistics 344
    11.1.2 Occupational accident trends 344
11.2 Structural risk models 347
    11.2.1 Individual accident proneness 347
    11.2.2 Single-cause models 348
    11.2.3 Multiple-cause models 350
    11.2.4 Domino models or event chains 351
    11.2.5 Factorial models 352
    11.2.6 Process-event sequences 353
    11.2.7 Fault trees 353
    11.2.8 Flow charts and Petri nets 354
    11.2.9 A complex and circular causation network 357
    11.2.10 Cusp catastrophe model of the accident process 359
    11.2.11 Complex dynamics, events, and deviations 361
11.3 Group dynamics, safety climate, and resilience 363
    11.3.1 Group dynamics and complex technologies 364
    11.3.2 Safety climate 365
    11.3.3 Cusp model for safety climate 367
    11.3.4 Safety culture 370
    11.3.5 Swiss cheese model 371
    11.3.6 Resilience engineering 372
11.4 Accident-prevention programs 374
    11.4.1 Personnel selection 377
    11.4.2 Technology interventions 379
    11.4.3 Behavior modification 381
    11.4.4 Poster campaigns 382
    11.4.5 Safety committees 382
    11.4.6 Medical management 383
    11.4.7 Near-miss accident reporting 384
    11.4.8 Comprehensive ergonomics 384

11.4.9   Other management interventions 386
11.4.10 Governmental interventions 387
11.5 Emergency response 388
11.5.1   Hazard perception 388
11.5.2   Time ecologies 390
11.5.3   Situation awareness and sensemaking 392

12 Human–computer interaction 395

12.1 The changing nature of the interface 395
12.2 Controls 397
12.2.1   Keyboards 397
12.2.2   Keypunch machines 399
12.2.3   Numeric keypads 401
12.2.4   Membranes 401
12.2.5   Positioning devices 402
12.2.6   Touchscreens 403
12.2.7   Styli 404
12.2.8   Gestural interfaces 405
12.2.9   Mobile devices 407
12.2.10 Gaze control 408
12.3 Memory enhancements 409
12.3.1   The word processing challenge 409
12.3.2   The desktop computer 410
12.3.3   Menu structures 411
12.3.4   Data storage capacity 412
12.3.5   Clouds 413
12.4 Visual displays 415
12.4.1   Error messages 415
12.4.2   Screen organization 416
12.4.3   Graphical user interfaces 419
12.4.4   Use of color 420
12.4.5   Pop up and wait 421
12.5 Auditory and multimedia displays 422
12.5.1   Speech interfaces 422
12.5.2   Earcons and spearcons 422
12.5.3   Animation and hypermedia 423
12.6 The Internet and the web 424
12.6.1   Origins 425
12.6.2   Search engines 427
12.6.3   Information foraging 430
12.6.4   Navigating the site 432
12.6.5   Web pages 433

12.6.6   Interactive pages 435
12.6.7   Responsive design 436
12.6.8   Extreme graphics 437
12.7 Virtual reality 438
12.7.1   Helmet and wall displays 439
12.7.2   Glove and body controllers 442
12.7.3   Anthropometric issues 442
12.7.4   Haptic perception 443
12.7.5   Training systems 444
12.8 Emotions in human–computer interfaces 445
12.8.1   Stress and anxiety 445
12.8.2   Emotions as information 446

13 Programming, artificial intelligence, and artificial life   451

13.1 The evolution of programs 451
13.1.1   Conceptual levels of programs and systems 452
13.1.2   Conceptual levels of system design 454
13.2 Artificial intelligence and expert systems 455
13.2.1   Some basic principles 455
13.2.1.1 Gödel 455
13.2.1.2 Turing 455
13.2.1.3 Von neumann 457
13.2.1.4 Simon 458
13.2.2   Expert system architecture 458
13.2.2.1 Algorithmic systems 459
13.2.2.2 Rule-based systems 461
13.2.2.3 Chaining strategies 462
13.2.2.4 Classification structures 463
13.2.2.5 Frame-based systems 464
13.2.2.6 Example spaces 466
13.2.2.7 Recursive systems 466
13.2.2.8 Interface requirements 467
13.2.2.9 Smart integrated displays 469
13.3 Artificial life 471
13.3.1   Neural networks 471
13.3.2   Machine learning 473
13.3.3   Autonomous agents 474
13.4 Large-scale integrated databases 476
13.4.1   What is possible? 476
13.4.2   Database construction 477
13.4.3   Data mining 479

13.5 *Validation issues 480*
    13.5.1 *Knowledge base validity 480*
    13.5.2 *Expert knowledge space 481*
    13.5.3 *Extraction of knowledge 482*
    13.5.4 *Decision validity 483*
    13.5.5 *Signal detection technique 483*
    13.5.6 *Meta-interpretive reliability 484*
    13.7.7 *Barnum effect 485*

14 Complex systems   489

14.1 *Complex adaptive systems (CAS) 489*
    14.1.1 *Schemata and agents 490*
    14.1.2 *Agent interaction 492*
    14.1.3 *Problem solving and conflict 492*
    14.1.4 *Irreversibility and emergent order 492*
    14.1.5 *Agent fitness 493*
    14.1.6 *Phase shifts 494*
    14.1.7 *Complexity catastrophes 496*
14.2 *Real-world complexity 496*
    14.2.1 *Simplifying designs 497*
    14.2.2 *Modularity 498*
    14.2.3 *Multiple person–machine systems (PMSs) 499*
    14.2.4 *Revenge effects 500*
    14.2.5 *Complex systems of systems 502*
        14.2.5.1 *NextGen air traffic control 502*
        14.2.5.2 *The smart power grid 503*
        14.2.5.3 *Health care systems 505*
14.3 *Collective intelligence 505*
    14.3.1 *Asynchronous problem solving in*
        e-communication 507
    14.3.2 *Sensemaking and situation awareness 509*
    14.3.3 *Networks 510*
    14.3.4 *Network growth 511*
    14.3.5 *Dynamics of feedback loops 513*
    14.3.6 *Other temporal dynamics 515*
    14.3.7 *Learning organizations 516*
14.4 *Team coordination 517*
    14.4.1 *Implicit learning 517*
    14.4.2 *Shared mental models 518*
    14.4.3 *Role of verbalization 519*
    14.4.4 *Game theory 519*
    14.4.5 *Intersection games 520*

14.4.5.1 Nonverbal communication 521
14.4.5.2 Minimum entropy principle 522
14.4.5.3 Changes in team membership 523
14.4.6 Group size 523
14.4.7 Stag hunt and emergency response 524
14.4.8 Human–robot interaction 527
14.4.9 Human–autonomy teaming 533
14.4.10 Group cognitive workload 534
14.5 Team synchronization 535
14.5.1 Autonomic arousal 536
14.5.2 EEG and team neurodynamics 537
14.6 Safety in complex systems 539
14.6.1 Transportation 540
14.6.2 Information technology 540
14.6.3 Medicine 542
14.6.4 Butterfly effects 543

15 Environmental design   547

15.1 Microenvironments 547
15.1.1 Offices 547
15.1.2 Homes 549
15.1.3 Kitchens 550
15.1.4 Stairs 552
15.2 Macroenvironments 553
15.2.1 Building and facility complexes 553
15.2.2 Facilities management systems 554
15.2.3 Defensible space theory 557
15.2.4 Navigation through facilities 558
15.2.5 Special populations 560
15.2.6 Emergency exits 561
15.2.7 Sick building syndrome 561
15.3 The great outdoors 562
15.3.1 Aesthetics and stress 562
15.3.2 Navigation 562
15.4 Playing in traffic 564
15.4.1 Exposure 565
15.4.2 Driver's age 565
15.4.3 Blood alcohol concentration 566
15.4.4 Seat belts 567
15.4.5 Driving speed 568
15.4.6 Risk homeostasis theory 568
15.4.7 Driver response times 570
15.4.8 Roadway configurations 570

15.4.9   *Lighting and signals 572*
15.4.10 *Driver distractions 573*
15.4.11 *Automated vehicles 574*
15.5 *Outer space 575*
15.5.1   *Brief history 575*
15.5.2   *Personnel selection 580*
15.5.3   *Gravitational forces 581*
15.5.4   *Allocation of function 582*
15.5.5   *Anthropometry 583*
15.5.6   *Vision 585*
15.5.7   *Vestibular sense and motor control 585*
15.5.8   *Sleep 586*
15.5.9   *Space habitats 586*

References   *591*
Index   *667*

Preface

This textbook is the outgrowth of teaching human factors engineering for nearly 40 years to undergraduates. The course is an offering of the psychology department, just as it was decades ago when I was a student myself. The field of human factors psychology (or human factors engineering, engineering psychology, or ergonomics) has changed markedly during that time. Although it still stays true to its original concerns about the person–machine interface, it has expanded to include developments in stress research, accident analysis and prevention, and nonlinear dynamic systems theory (how systems change over time), artificial intelligence, and some aspects of human group dynamics and environmental psychology. Computer technology and artificial intelligence have become more ubiquitous, more so since the previous edition. As systems become more complex, theories evolve to reconcile with the new sources of complexity.

It was a challenge to find a textbook for the class under these conditions of technological change. At first, I found one that seemed just perfect with regard to the breadth and depth of coverage I was looking for. After a few years it only needed a supplementary reading or two to help out, but eventually it went out of print, never to return. The other textbook choices by that time had diverged greatly in how they characterized the scope of the field. One approach concentrated on tables and graphs for otherwise traditional topics. A second approach retrenched into the theories of cognitive psychology and focused less on the practical problems in human factors. Meanwhile, the fast pace of technological change presents some interesting challenges for extracting the fundamental principles of the subject area.

In any case, I present the new scope of the psychology of human–machine interaction. The typical roomful of students that I have in mind is usually composed of upper division students and a few graduate students. The class is typically composed of 50% engineering students of different sorts, 40% psychology students, and 10% sundry others. One implicit goal of the course is for the engineers to think more like psychologists, and the psychologists to think more like engineers. The sundry others usually show signs of thinking like both, and make the class situation more interesting for everyone.

I would like to take this opportunity to thank Joseph J. Jacobsen for helping to arrange some of the photographic opportunities that appear throughout this book.

New to this edition

There are several new developments in this third edition, some of which span multiple chapters. Some highlights:

The allocation of functions between person and machine used to be a fairly straightforward separation of labor. Augmented cognition, automation, and artificial intelligence are blurring this distinction. Machines can now adapt their functions when they detect that the operator is under stress, but how does the machine make that determination? Is automation reducing human workload and stress or just changing it from one form to another? How is the effectiveness of automation related to the human not trusting the machine?

The nonlinear dynamical systems paradigm in person–machine systems continues to expand along with a growing recognition that some complex phenomena are better understood by studying patterns of change over time. Special attention is given to events that emerge or synchronize suddenly, apparently out of nowhere. The principles that are most relevant to human factors have been consolidated in Chapter 3. Specific applications to stress and human performance, occupational accidents and prevention, complex systems, and other topics appear throughout the text.

The design of auditory displays has always been a staple topic in human factors. The latest innovations have been triggered by the cacophony of indistinguishable audio signals and alarms in hospitals and the urgent need to redesign them.

The chapter on cognition has been changed substantially in light of developments that have consolidated in several theoretical areas. Look for dynamic decisions, situation awareness, sensemaking, and physiological indicators of cognitive workload.

There is a growing concern in society about cognitive workload and fatigue as more people find that they have too much work to do and not enough time to do it in. Software and automation, which are not always trustworthy, made work more mentally intensive, particularly during the COVID-19 pandemic when many people were forced to work from their homes. The effects of cognitive workload and fatigue have been historically difficult to separate because they both produce negative effects on performance over time, while other processes that transpire over time serve to improve performance. It is now possible to separate those effects for a wide range of cognitive operations with nonlinear models and experimental designs that capture all the ingredients of the models. Whereas systems have been traditionally designed to minimize the role of individual differences in cognition and personality, and with good reason, individual differences in cognitive responses to systems are now gaining importance.

Artificial intelligence was once the label given to computer programs that emulated human thought processes. The programs are now capable of knowledge discovery that is more independent of human control than ever before and managing and manipulating huge databases that never used to exist. It is left to the reader to ponder how close we are getting to a turning point where the machines will control humans more so than the other way around.

Complex systems applications in human factors and ergonomics are becoming a greater concern as many person–machine systems become connected through networks of interaction. The broadest new developments here include complex *adaptive* systems, human collective intelligence, distributed situation awareness and sensemaking, human–autonomy teams, and how to control your very own swarm of robots or drone aircraft. What could make a system of multiple integrated autonomous systems go out of control? Short answer: Small things at absolutely the wrong time can wreak more havoc than big things at other times.

Author biography

Stephen J. Guastello is a professor at the Department of Psychology, Marquette University, USA. His research interests fall into several areas. The largest concerns nonlinear dynamics (chaos, complexity, and catastrophe theories, and related computational methods) and their applications to work motivation, work performance, and turnover, occupational safety and health, creativity, group dynamics, and cognitive workload and fatigue. His other interests include the role of macroeconomics in organizational behavior, human–computer interface design, artificial intelligence, personality, and organizational behavior. He regularly teaches courses in industrial-organizational psychology, human factors engineering, theories of personality, along with occasional seminars on chaos and complexity in psychology. He is Editor in Chief of the journal *Nonlinear Dynamics, Psychology, and Life Sciences*.

Chapter 1

Introduction to human factors and ergonomics

1.1 EVOLUTION OF PERSON–MACHINE SYSTEMS

Ever since prehistoric times, *Homo sapiens* made tools to enhance their physical capabilities and to channel their physical energy. Eventually they developed tools to enhance mental capabilities and channel mental energy as well. The first tools were handheld devices, such as hammers and cutting edges. The first machines such as wheels, levers, and pulleys channeled physical forces as well as human forces.

The industrial revolution that started in the 17th century coincided with major developments in mechanical physics. Sir Isaac Newton is historically recognized for introducing the concept of a mechanical system. A system has parts that all work together to fulfill an objective. If we know how each of the parts work, we can know how the entire system works. This perspective produced clocks, clockwork-type machinery, and myriad devices that, once the human operator set them in motion, they continued their function until completion. Machines such as the cotton gin and textile weavers started to replace tedious work previously done by people; we recognize the phenomenon as *automation*.

The advent of machine tools in the 19th century made it possible to produce thousands of objects that looked and functioned exactly alike. If a product were to be assembled from many elementary machine-tooled objects, the product could now have interchangeable parts. Machine tools facilitated the development of assembly-line production systems.

Electricity was another landmark from several perspectives. It greatly expanded the number and type of tools and machines that could be made and reduced the required amount of human energy further. Before taking the ubiquity and inevitability of electricity for granted, however, it is important to remember that entire empires were built without it. There are still places in the inhabited world where electrical service is not available all day every day.

Communication devices have changed not only our way of life, but also our concept of who we are and how we are connected to other

DOI: 10.1201/9781003359128-1

people. In primitive times we chiseled a piece of rock or scratched on a piece of papyrus. Writing and writing media were memory-enhancement devices at both the individual and collective levels of experience. Not only could the writers extend their own memories, they could also share them with anyone else who found the writing even after the original writer had died.

Books of memories were produced eventually, but each book had to be copied accurately by hand until the invention of the printing press in the 15th century. Although the lead type for the print was set by hand until the late 19th century, the printing press was a major leap forward for automating the process and disseminating the information. Starting in the same time period, electricity was combined with communication goals to produce the telegraph, telephone, radio, and television; these devices allowed us to send messages over great distances (and eventually the entire world) in relatively short amounts of time. Other technical advances allowed the possibility of recorded sound and its automated production once the record master was developed in 1916. Photography displaced the painter as the sole provider of realistic productions of visual events. Photographs were combined with a hand-cranked machine to produce a movie; eventually a motor replaced the crank. The communication devices were more than tools of communication; they also became media of expression.

Automobiles, trucks, and aircraft that developed throughout the 20th century, allowed us to move our entire bodies around the world in ways never before possible. Space travel has started to liberate us from one particular world, although the other nearby worlds tend to be cold and nasty.

The earliest ancestor of the device we know today as the computer was introduced in the late 1940s. The capabilities of the earliest versions of the machine were confined to rapid and accurate calculations, which were of course highly desirable. Other types of thinking processes were introduced into its programming in the years that followed, along with simplicity of use. Desktop models for individual owners appeared in the 1980s. The 1990s saw larger data storage and memory capacities, faster processing, and lightweight briefcase-sized models. Computer-based technologies are now part of almost every contemporary machine. In some cases a desktop computer acts as the controller for a complex electromechanical system.

Although the foundations of the shift were forming earlier, the prototype device we call "the computer" has become intertwined in every communications system available, short of talking or paper and pencil. Actually, the ubiquity of e-mail, cell phones, and text messaging almost appear to preempt real face-to-face relationships and replace paper-based objects and activities with those that are plugged into an energy source and an information network. Can people do anything anymore without a computer-based device becoming involved? This question extends to simple arithmetic and walking across a street. The information technology industries have tacitly answered, "Not if we can help it," and call what their latest efforts "ubiquitous computing."

Computers differ from conventional machines because they are programmable. Programming allows people to change the function of a machine in ways never before seen after the introduction of machines. Not only does the computer enhance human cognitive processes, it can substitute artificial skill for real skill, alter our concepts of ourselves as social beings, and create artificial psychological realities through particular combinations of programming capabilities; the more advanced combinations are called "virtual reality." All of these developments led Turkle (1984) to refer to the computer as "the second self" to emphasize that this class of machines has affected our concept of who we are more so than any other technological advance. Forty years later, the computer is more of the same.

1.2 THE ROLE OF HUMAN FACTORS AND ERGONOMICS

The complexity of machines is obviously growing, as are the demands on the user to operate them. Machines are only as effective as their operation by humans allows, and thus there are opportunities and challenges that lie at the interface between the person and the machine. The setting in which the machine is used, the physical environment, other tasks occurring simultaneously, and other people and machines have a strong impact on the performance of the person–machine system (PMS).

The person–machine interface has been changing alongside the changes in the ways functions are divided between the person and machine. The tool that was under the continuous direct control of the human hand evolved into a machine. The machines evolved so that the primary point of contact with the human was the interface where information was sent and received. Eventually the boundaries between control and display merged together in the form of point-and-click computer programs and touchscreens. Machines have evolved beyond sending and receiving information; they have been picking up some of the thinking tasks as well. With virtual reality (VR), the interface became an environment that surrounds the humans; in other words, the human is in the machine. There are little machines that carry out their functions from within the human's shirt pockets or from inside their bodies.

1.2.1 The person–machine system

Meister (1977) emphasized the concept of a *system*. A system has elements, which interact to produce a result that cannot be reduced to, or understood as, the simple outcome of one of the system elements. The primary elements, of course, are one person and one machine, but PMSs can vary in size and complexity.

The person and the machine exert a close control over each other within the system. The point of control occurs at the interface, which is most often a control and display panel of some sort. The machine should be capable of carrying out functions automatically once the human has activated it. If not, then we have a tool rather than a machine. Although Meister (1977) conceptualized the PMS as only containing those elements or people that have direct control over the machines, ergonomic science in later years recognized less direct sources of system control that do in fact affect the interactions at the interface. In fact, we have now moved on to networks of PMSs and systems of systems.

1.2.2 The cognitive core

The field of human factors engineering (HFE) started in the 1940s as a joint effort between psychologists and engineers (Chapanis, 1975) to study the interactions that occur at the human–machine interface. They studied practical questions such as, "Why did a seemingly competent cockpit crew fly their aircraft into a mountain?" "How could the chemical tragedy in Bhopal, India, ever happen?" "Why are people not using many of our software features?"

Figure 1.1 depicts the traditional scope of HFE. The human interact closely with the machine. The information provided by the machine to the human is a *display*. The information given to the machine by the human is a *control*. Three groups of cognitive psychology topics that were immediately invoked were perception, psychomotor skill, and principles of cognition. Theories of perception explained how the human assimilated the display. Psychomotor

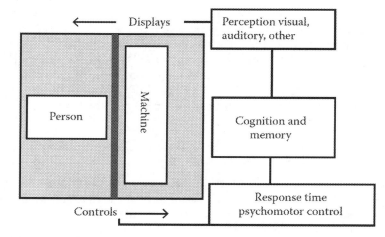

Figure 1.1 Traditional range of HFE topics.

studies explained how the person managed to control the machine effectively. Cognition theories explained what transpired between the human's ears from the moment of display and the moment of control. The overriding objectives were, and continue to be, the optimization of system performance and the minimization of human errors produced by clumsy system designs.

Chapter 2 covers a set of concepts that are pertinent to most human factors initiative. Who is the user of a system? What composes the system? How do we allocate functions between people and machines? When are redundancies in function relevant? What constitutes *trust* in automation? How do we assess the functionality of the system?

Chapter 3 delves into question such as: What is information and how does it flow? How do system events change over time, and how do changing events affect system operation? What propels a system into a state of chaos? Nonlinear dynamical systems theory captures the primary patterns of change of time. Systems also self-organize and synchronize without deliberate human intervention.

Events do not always change smoothly over time or in simple proportion to demands and stressors. Rather, a small change in a system parameter could have a dramatic effect if it occurred at the right place and the right time. Similarly, making a large change in a system at the wrong time could have no discernible or permanent impact on the system's behavior. Nonlinear dynamics also underlie *emergent phenomena:* system states that bear no visible similarity to previous states, and that would appear to have come out of nowhere to the untrained eye.

Chapter 4, "Psychophysics," covers the most basic features of the interactions between people and their real worlds. It explains how people detect a physical stimulus—sound, sight, vibration, and so on—and how to characterize simple decisions that are made on the basis of the presence or absence of stimuli. There are stimulus detection problems that are not nearly so simple, however. Applications to vigilance tasks have always been prevalent, and more so with computer-based technologies—someone has to watch the computer screen and take action as necessary.

Chapter 5 pertains to the processes of visual sensation and perception, and their impact on the usability of visual displays. Chapter 5 pertains to auditory sensation, perception, and displays, and also includes the human interpretation of tactile information. These core topics have a long history in the field, and there are new innovations to keep them exciting. For instance, iconic displays, which started as special property of graphical user interfaces (GUIs), have now become common enough to include as display types for all systems. Auditory displays were a sleepy area of research for a long time, but became prominent in recent years to combat the cacophony of indistinguishable sounds from medical equipment in hospitals.

Chapter 7 delves into the cognitive processes that occur between the perception of a display and the moment of a control action. How did the human make the evaluation and decide what to do? Cognitive workload capacity and measurement have become another expanding area of research; entire conferences have been devoted to it. The principles in this chapter are foundational to Chapter 10 on stress and performance and Chapter 13 on artificial intelligence. Chapter 8 describes controls, human response times, and related issues of human performance when operating machines.

1.2.3 The workspace shell

The subject matter of HFE expanded in the 1980s. The word *ergonomics* arrived from Europe. Initially it denoted the interaction between the human and the nonliving work environment, but today *ergonomics* and *HFE* are interchangeable in meaning. Contemporary ergonomics is now conceptualized as "the design and engineering of human–machine systems for the purpose of enhancing human performance" (Karwowski, 2000, p. 1).

The expanded range of topics integrate well with another classic area in Chapter 9, *anthropometry*. Anthropometry, or human measurement, affected the design of hand tools, workspaces, driver and passenger compartments of vehicles of all types, and requirements for human movement within workspaces. This chapter also covers topics in human strength because of their proximity to other human-measurement concepts and a snapshot of how to set up a personnel selection study for hiring decisions. Two regular topics of concern are the control of back injuries and carpal tunnel syndrome. Both are introduced in Chapter 9, but the broader range of topics in biomechanics is beyond the scope of this text.

Figure 1.2 depicts the scope of HFE as it exists today. The immediate work environment surrounds the person–machine interface, which is in turn surrounded by environments of broader physical scale, known as

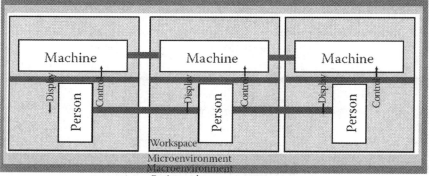

Figure 1.2 An expanded view of HFE.

microenvironments and macroenvironments. Kantowitz and Sorkin (1983) were among the first authors to expand the boundaries of HFE to these broader spectra of physical space.

Environments are sources of many forms of stress. Although the study of some types of stress and human performance is part of classical HFE and nearby areas of work psychology, the major thrust of stress-related theories was not introduced until the 1980s. In a departure from the traditional impression that heat, cold, and noise are the primary stressors, some attention to the social origins of stress are given as well in Chapter 10 along with shift work, cognitive workload, time pressure, and fatigue. Workload and fatigue effects have been historically very difficult to separate because they both occur simultaneously along with other dynamics that improve performance over time. Progress has been made to separate them, however, using a mathematical modeling approach and experimental designs that are sufficiently complex.

We now arrive at the topic of accident analysis and prevention in Chapter 11. Accident analysis and prevention has an independent literature, which has made more use of ergonomics thinking more often in recent years. Traditional HFE appeared to regard the topic as another reason to control human error from the perspectives that it has always taken. Systems thinking, particularly the nonlinear dynamical variety, promoted a refreshed perspective that takes into account the interaction between the person, physical work environment, and social factors that contribute to safe work practices. Although this HFE text is *not* about business management in its usual scope, some aspects of management do play an important role in accident analysis and prevention and ergonomics policy in work organizations. Also, cultures differ with regard to their perceived importance of accident analysis and prevention relative to other economics objectives.

1.2.4 Computers

Human–computer interaction also became an active area of study in the 1980s. Although basic HFE concepts were still relevant to these new applications, computer-based systems raised a number of new questions and issues. The development of computers in the next decades was the result of interplay between developments in programming capability and developments in cognitive science. Chapter 12 covers input-output devices and virtual reality. Chapter 13 explores the fundamentals of artificial intelligence, artificial life, expert systems, and some of the extensive computation components needed to make some applications work properly. The growing range of products containing artificial intelligence components leads to a new issue, which is *trust* in automation.

Chapter 14 examines complex human–machine systems. The concept of a complex adaptive system (CAS) enhances our understanding of work group coordination, human–autonomy teaming, synchronization, and collective intelligence. Some themes in the chapter address questions such as: "Why do attempts to improve or fix a system sometimes backfire and produce a situation that is worse than the previous one?" and "How do events in the system emerge that are not readily explained by events in their elementary parts?"

1.2.5 Beyond the great outdoors

Chapter 15 considers environments that are larger in scale than the immediate work environment. Microenvironments would include private homes, offices containing several workspaces, or factory departments. Macroenvironments are substantially larger and more complex, such as entire factories, housing developments, airports, and even cities and the great outdoors. HFE issues address the organization of the spaces and navigation through them. Automobiles and traffic are two further topics related to the environment. Human factors in outer space comprise the final saga of the text; numerous HFE issues emerged over the decades, and they are becoming more severe as missions to Mars are now underway.

1.3 CRITERIA OF HUMAN FACTORS

If the goal of HFE is to design and engineer PMSs for the purpose of enhancing human performance, there is considerable ground to cover regarding human performance itself and quality of work life, as the foregoing chapter outline suggests. Some best practices have evolved into industry standards that designers would be wise to adopt if they are not legally enforced. There is also the matter of legal responsibility when the consequences of failure are severe enough.

1.3.1 Performance criteria

The primary performance criteria include maximizing output in terms of quantity, minimizing response time, minimizing error, and ensuring safety. The primary criteria are often interrelated, and it is often possible to introduce a design idea that satisfies more than one objective. For example, display designs that can be read more quickly are also the designs that can be read more accurately. On the other hand, gains in speed might also result in higher error rates or compromised safety. Errors can range from trivial, to costly, to deadly, depending on the situation. Thus, it is well-advised to consider multiple aspects of system performance when evaluating a design. HFE specialists often participate in devising training programs to meet the foregoing standards.

Two other criteria that are instrumental to performance objectives are heightening situation awareness and controlling unnecessary cognitive or physical workload. Situation awareness is a mental state of knowing the status of the system, understanding how the facts and procedures work together to produce results, and the ability to anticipate future system states and the results of control actions. Situation awareness is explored in further detail in Chapters 7 and 14. Conscientious operators exert concerted effort to maintaining performance objectives in spite of sometimes large fluctuations in workload demands. Thus, the effects of workload are not always visible in performance measurements, and as a result, workload demands need to be specifically targeted for analysis. Workload and its ramifications are explored in Chapters 7 and 10.

1.3.2 Industry standards

Various industries have at least some standards for equipment design. Human factors issues often fall outside the boundaries of many official regulations, but not in every case (Karwowski, 2006). Standards have different levels of officiousness. A *state of the art* standard represents the most progressive idea that is publicly known. State of the art ideas might not have been thoroughly tested, and could easily be upstaged by an equally experimental new idea before the evaluation of the human factors is reasonably complete. *State of the science* standards reflect scientifically supported ideas that have a reasonable amount of generalizability. The *state of the industry*, however, is often slower than the state of the science to uptake the scientifically sound principles into practical systems. Old equipment that is still functional is not going to be discarded just for the sake of keeping up with trends. New purchase decisions, however, are likely to reflect the more progressive systems designs. Thus within a particular industry one can anticipate a cascade of technologies.

Industry standards reflect the minimum acceptable design and performance standards in many cases. They usually adopt requirements that are least ambiguous in terms of their scientific support for their efficacy in system performance. Although industry standards per se do not have the force of law behind them, they often serve as reference points in civil liability litigation where the goal is to determine whether a design flaw was responsible for an injury or another form of damage. Representative of the industries themselves are usually part of the process of developing regulatory standards.

Regulatory standards have the force of law. As examples, the Federal Communications Commission (FCC) sets standards for the quality of audio and video broadcasts and the technical systems by which the broadcasts will be made. The U.S. Department of Defense (DOD) sets standards for the human factors attributes of equipment purchased for the military. The

standards are strongly influenced by the results of scientific contracts issued by the DOD. Any manufacturer desiring a military contract must meet those standards. Although DOD standards do not pertain to civilian products that are not purchased by the military, their information base might contribute some influential information to a civil litigation.

Another governmental nexus surrounds health and safety. The Occupational Safety and Health Administration (OSHA) regulates safety in the workplace. The majority of its work since its inception in 1971 was restricted to employers' use of equipment and supplies—electrical configurations, chemical handling safety, proper ventilation and exhaust, personal protection equipment, and so forth. OSHA also has standards for exposure to noise and excessive heat, which are two clearly ergonomic matters. In late 2000, Congress passed a law extending OSHA's purview to several other ergonomic matters, but rescinded the law in early 2001. Since that time OSHA has served an informational and advisory role to employers on ergonomics matters. Although its primary function is setting standards, auditing, and enforcement, it now positions itself as a consulting resource to employers who wish to know more and take care of situations themselves. The National Institute of Occupational Safety and Health (NIOSH) is a research branch of OSHA. NIOSH commissions studies of situations and investigates the scientific side of new and emerging issues. NIOSH is currently part of the Center for Disease Control (CDC), which is concerned with public health issues outside the workplace as well as within it.

1.3.3 Civil liability

The 50 states vary somewhat regarding their criterion for proving a case against a manufacturer or other defendant in a civil liability trial, but there are some common concepts nonetheless. They are strict liability, contributory negligence, and negligence.

1.3.3.1 Strict liability

A legal theory of strict liability states that a manufacturer of a product is liable for damages connected to the product because the product is produced for profit. Here one might examine whether a human factors flaw could be detected from the product design, and whether such a flaw could have reasonably induced an error that led to the accident.

Dangerous products such as chainsaws are made and sold regularly of course. One suitable defense for the manufacturer is the obviousness of the danger. Is there any question about the function a chainsaw is supposed to serve and what the blades are for? Some dangers are less obvious, however, and the burden of responsibility is on the manufacturer to provide suitable warnings and instructions for safe operation of the product. Appropriate

warnings constitute another viable form of defense. One might then debate whether the warnings were in fact clear, informative, and relevant to the danger in question.

1.3.3.2 Contributory negligence

In a defense of contributory negligence, the defendant argues that the plaintiff took actions that were irresponsible in some way. One example would be to use the product for a purpose for which it was not intended. A tale from the insurance industry captures the essence of improper use of a product. Several decades ago, a manufacturer of home washing machines designed a model with a glass top. The idea was that the happy housewife could watch the clothes go around without opening the lid. One proud owner, however, decided to dance on the lid of the washing machine. Unfortunately she pounded her foot down too hard, it broke through the glass, and her leg was torn up by the washing mechanism. Although the manufacturer prevailed in this instance, they now had evidence that the glass lid might not be a good idea, and pulled it off the market.

Another group of examples of contributory negligence would charge that the plaintiff ignored a warning. A third group might involve the plaintiff not attending to information on displays, taking an incorrect control action, or taking the action too soon or too late for the purpose. Human factors experts might be called in to determine whether it was reasonable to expect the operator of the device to see, hear, or do something differently that would have prevented the accident. All told, there are reasonable and unreasonable expectations from the human operator. Even a successful legal defense is costly, and designers are advised to idiotproof their products to the extent humanly possible.

1.3.3.3 Negligence

A negligence suit is generally one that is aimed not at the manufacturer, but against a third party that might have provided a service or a piece of equipment for the plaintiff or contrary to the interests of the plaintiff. In these cases the defendant might have operated a piece of equipment improperly, or rented it to the plaintiff in a state of disrepair. The outcome of such cases depends greatly on whether the defendant was providing a service for hire or in some way profiting through the use of the equipment in question.

Another type of negligence suit might occur in an employment context where the employer disregarded safety appointments for the equipment or instituted some other work procedure that compromised safety. Under current law, employees who are injured on the job are entitled to compensation for medical expenses and lost wages from the Workman's Compensation agency. According to the regulations, however, the employer cannot be

sued, although OSHA will investigate and levy fines as necessary. Injured employees who believe their compensation from Workman's Compensation is inadequate, however, might file a suit against the equipment manufacturer for an alleged design flaw; the defendant manufacturer then has the burden of responsibility to shift the blame onto the employer or employee.

1.3.3.4 Proving the case

States differ in their statutes, but Wisconsin serves as a representative example for present purposes. The plaintiff's case consists of two parts: the contributions of the parties to the events that led to the liability, and the monetary valuation of the damages. The jury's task is to determine the percentage of fault associated with each party's negligence or liability in the matter. If the plaintiff is not successful in convincing the jury that the defendant was responsible for at least 50% of the fault, the plaintiff receives no award from the court. If the plaintiff can establish that the defendant's fault is greater than 50%, the award is equal to the amount of damages times the percentage for the defendant's fault. Admittedly, there is some subjectivity as to what constitutes a percentage in these types of decisions.

DISCUSSION QUESTIONS

1. How is a cell phone similar to, or different from, an implant?
2. What is a smart display, compared to a not-so-smart display?
3. Why do attempts to improve or fix a system sometimes backfire and produce a situation that is worse than the previous one?
4. An important theme in engineering is that a technology is neither good nor bad ethically. "Good" and "bad" depend on how it is used. Can you think of some technologies have ethical and unethical uses?
5. What is privacy? Consider both the legal concept and practical experience with PMSs.
6. Consider the following traffic accident. The defendant was driving her car at approximately the speed limit (35 mph) down a two-lane city street. No cars were parked on the driver's side of the street at the time the accident occurred. The car proceeded through a lighted intersection and struck a 12-year-old boy at a point approximately 150 feet past the intersection. The car dragged the boy 41 feet before stopping. Once the car stopped the driver backed up and drove over the boy again. Most of the boy's injuries were to his leg. Damages were assessed at $100,000 for recovery and rehabilitation. What human factors issues should be considered to determine whether the driver was able to stop the car prior to striking the boy? Reconsider your answer to this question after reading a few more chapters.

7. A human operator of an automated vehicle encountered a problem on the road leading her to think that the automated vehicle was going to crash into another car within seconds. When she suddenly turned off the automation to regain manual control, the automated vehicle jumped the median strip and crashed into a different car on the other side of the road. How should the fault of the accident be divided between the manufacturer of the automated vehicle and its operator?

Chapter 2

Elements of human factors analysis

2.1 ALLOCATION OF FUNCTION

The importance of the goals of the system should not be underestimated. Designers should have a clear idea of what the intended users are trying to do, how many varieties of work they need to do, and what types of discretionary controls the users wish to have. It is essential that some of the people contributing to the system design are actually experts at the task themselves.

2.1.1 The benefits of machines

The first design question after establishing the goals of the system and the user population is to determine what functions should be assigned to the machine, and what functions should be assigned to the human. Often a new PMS is designed to replace a strictly human action by a combination of human and machine actions.

Machines are good for tasks or task components that involve repetitious actions that must be produced exactly the same way each time they occur. Machines are also preferred when large mechanical forces are involved. Humans typically get the job of setting the machine into motion, and choosing machine settings for specific outcomes, especially when the system allows some choices and flexibility. Humans are also good at making decisions when imprecise or incomplete information is available. A machine may be able to process information, but can only do so if all the necessary starting information is supplied. Humans also repair the machines, and repair tasks typically require additional tools and machines.

Figure 2.1 depicts one of many automation success stories. Figure 2.1a shows a lathe (1960s design) that is used to make one-of-a-kind machine parts. The operator guides and controls the sequence of cutting motions through completion, a process that may take hours for each rendition of the manufactured object. Figure 2.1b shows the control panel for a contemporary version of a lathe that has the same basic system goals. This time, the operator can program the whole sequence of cutting motions, set

DOI: 10.1201/9781003359128-2

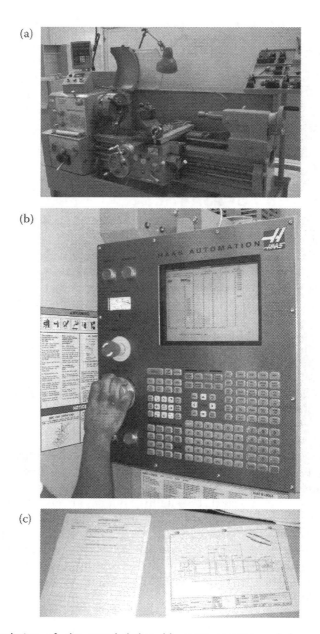

Figure 2.1 Redesign of the metal lathe: (a) semiautomatic and human intensive; (b) controller for a programmable lathe; and (c) blueprint and code sheet.

the zero-point in the cutting chamber (not shown), and then push the start button. The information can be stored to make more copies of the same object if desired and the cutting process is much faster overall. Figure 2.1c shows what the operator still needs to do to translate a blueprint into a sequence of machine codes that will be programmed into the machine.

Both lathes in Figure 2.1 require a substantial amount of skill to operate, although the skill composition is different. Some forms of automation, however, have the effect of deskilling the task overall and are deliberately introduced as a means of cutting some of the labor costs of production (Hammer & Champy, 1996). In those cases, the liability to the overall system is that there might not be anyone around who knows what to do when the automation fails or recognizes the situation while it is happening. Thus training programs that accompany automated systems should teach operators much more than what buttons to push when things go well.

2.1.2 The user population

Satisfactory definitions of system goals depend on a clear knowledge of the user population. The same can be said for virtually every aspect of the interface design. Typical questions the designers should ask are as follows.

Is this system intended for people working in a specific occupation, or is it intended for a broader slice of the general population? Historically, the human factors community has taken the position that the system must work excellently for everyone who is expected to use it. The abilities and skills of occupational users were taken as a given; system designs should not be predicated on assumptions regarding individual differences in basic cognitive abilities, cognitive styles, or personality. If there is too much variability in human performance that can be traced to those types of differences, the system should be redesigned to minimize the role of those differences (Verhaegen, 1993). As consumer products are designed for the broader population along with products intended to interface with wider ranges of industry-specific software and equipment, the principle of one-design-fits-all becomes less realistic. The roles of motivation, personality, and coping strategies become more salient (Baldwin, 2009; Drury, Holmes, Ghylin, & Green, 2009; Guastello, Futch, & Mirabito, 2020; Guastello, Shircel, Malon, & Timm, 2015; Matthews & Campbell, 2009; Szalma, 2009, 2012; Szalma & Taylor 2011; Szymura, 2010). The extreme case is where a product is intended to be specific to individual needs (Hancock, Hancock, & Warm, 2009).

The standard questions about the user population(s) are at least as relevant as they once were. For instance, do the users have normal-range physical capabilities, or do they have specific limitations or handicaps? Some systems are designed as aids for the blind or aging populations or as add-ons to enhance their accessibility. Given that the intended users have normal-range physical and mental abilities, what abilities do they wish to extend through use of the new system?

What is the typical occupational training or educational level of the user? Are the users already skilled in tasks relevant to the new PMS, or are the users seeking to acquire a skill through, for instance, the new virtual reality training system? There are some distinct differences between experts and novices that are explored next.

2.1.3 Experts and novices

The differences between expert and novice system users are typically relative to the tasks within a particular system. Their differences tend to fall into a few, somewhat overlapping, categories. At the level of commands and operations, experts may have developed "population stereotypes" concerning how the system should operate. "I should be able to find the X control [here], and it should allow me to do Y if I select Z." Population stereotypes are considered further in Chapter 8. Designers of relatively new systems should investigate whether new users can be expected to have experience with similar but different PMSs.

Some features of systems could frustrate experts in ways that would not faze novices. As an example, some of the early versions of popular statistics programs did not offer some of the more advanced calculation options that experts might want to use. (Repeated measures analysis of variance was once one of them.) Novices might be content with default settings which might be very reasonable choices in many situations, but ill-advised in others. In the early stages of its development, the system did not have enough capability to meet some of the operators' goals or subgoals. As the capabilities of the statistical software evolved in newer versions, it was a good friend to experts who knew the calculations and how and when to use them; the machine just did the work of making the calculations correctly at lightning speed without having to rope a few different pieces of software together. As novices learned the coding strategy, they could ramp up their deeper knowledge of the procedures without having to make any of the calculations firsthand. As point-and-click interfaces were introduced to replace the users' need to learn how to use the code for operating the system, however, two things happened: The functionality of the software declined and did not recover for at least another decade, and graduate school classes in statistics refocused on to how to use the click-buttons, with less emphasis on knowing how and why. Chemistry students have been known to express a similar frustration: They wanted to know the how and why of chemical reactions, but their instructor only wanted to teach them how to push the buttons on the software. Scenarios such as these lead one to question whether "smarter" technology is making people "dumber" (Hancock, 2014).

Experts and novices also differ in the way they solve problems. The expert is likely to have developed some sense of probable cause-and-effect relationships. Given a problem, they would have some reasonable guesses as to the most likely causes, which aspects of the situation should be checked in what order, and which solution options should be eliminated and which ones are most likely to work. Expertise goes beyond solving problems after they are identified. Experts can anticipate problems and look for clues that signify a critical situation in the making (Vicente, Mumaw, & Roth, 2004). Experts tend to have well-formed mental models of the global system in which specific parts and operations all fit together and operate together. Novices, however, work their way through specific operations toward a full system view. This top-down versus bottom-up approach to solving a problem greatly facilitates problem solving with both regard to speed and accuracy (Hoffman, LaDue, Mogil, Roebber, & Trafton, 2017). Experts also differ in their strategies for solving problems.

2.1.4 Workload issues

As a general rule, a system that is simple and easy to operate is more attractive than a complicated one. The fewer steps needed to complete the operation, the fewer pieces of information one must keep in mind, and the less prior knowledge about the technology needed the better. Workload issues have a clear impact on product usability (Longo, 2018). The measurement of workload is, unfortunately, not straightforward, and something that is again relative to the task environment. The point is expanded further in Chapter 7.

Nonetheless, automation is often effective at reducing mental workload, but there are times when the pre-automation workload is comfortable, and the automation reduces the workload to an uncomfortably low level leading to boredom and reduced attention capacity (Young & Stanton, 2002). Although it is plausible that reducing the demand from one task frees up mental attention for a secondary task, the overall effort and arousal levels can decline to a critical point where human error actually increases if secondary tasks are not available. This U-shaped relationship between arousal or load and errors is considered further in Chapter 10. Meanwhile, too much automation in high-consequence systems can also compromise the operators' levels of situation awareness, which is their understanding of what is happening in the moment, what will happen next, and what actions should be taken if any (Kaber & Endsley, 2004). The received wisdom at present is that intermediate levels of automation are best for meeting the full range of performance objectives—not too much, not too little.

A related issue, however, is that the secondary task might not be a wise choice. There have been numerous initiatives during the last decade to automate aspects of automobile driving, such as monitoring the distance between the following car and the leading car in traffic, or the lateral position of a car in a traffic lane. Although the automatic function appears to work as intended, the mental capacity of the driver, which is already underutilized in normal driving, is diverted to non-driving tasks that are usually regarded as driver distractions (Carsten, Lai, Barnard, Jamson, & Merat, 2012). A practical example of a design-fail shows up in some of the automobiles produced in the last five years with elaborate entertainment options, climate controls, and navigation systems organized around a central graphic display screen. The system pops up a screen at random moments that says (in paraphrase): *WARNING: Taking your eyes of the road to look at this stupid screen can cause accidents.* What is wrong with this picture?

Sometimes the secondary tasks are not really secondary, but embedded in the daily work routines onto which the new system is superimposed. The new system might be the center of the operator's activity or it might pertain to tasks that are done less often or with less priority compared to other tasks. A system might not appear to absorb too much mental workload if it is one that is done in isolation from other tasks, but if other things are going on that require a lot of attention, the secondary system would do well to keep its demands simple and convenient.

Experts can manage their workloads in ways that novices are still learning to do (Vicente et al., 2004). Workload issues are pervasive in human factors as multiple tasks and different demands levels are routinely involved. Some tasks are better done sequentially, while others can be done simultaneously. The expert in any case develops strategies to get better results with the least amount of wasted effort or time.

2.1.5 Flexible allocation of function

Scerbo (2001) distinguished two classes of flexibility in automated systems, adaptable and adaptive. In *adaptable* systems, the operator makes some selection for what functions will fall under user control and which ones will be carried out automatically. There is typically a dialog sequence at the start of an operation where the choices are registered. It is usually possible to change the choices when desired.

In *adaptive* systems, however, the machine makes the choice as to whether to put a function under automation or returning it to manual control. The system senses elevated workload by using electroencephalogram (EEG) signals or other biometric information and goes into automated mode, thereby relieving the operator of some of the demand (Schmorrow & Stanney, 2008). The automated mode turns off when workload returns to normal. The switching between automated and normal modes makes sense if the human performs better than the machine under normal circumstances. One design challenge is to determine the point where the mode switches are most advantageous, which has to be worked out in laboratory studies that resemble the real-world task involvement as closely as possible. Another challenge is to properly alert the operator that the system has gone into automatic mode, which means changing the operator's expectations of what the machine will accomplish during a particular sequence of activities. Changing the operator's mental model of the work environment involves mental switching costs that need to be taken into consideration as well (Reinerman-Jones, Taylor, Sprouse, Barber, & Houston, 2011). Brain-based adaptive systems have made positive impacts on the performance of primary and secondary tasks (Bailey, Scerbo, Freeman, Mikulka, & Scott, 2006).

2.1.6 Trust in automation

The classic contributions of machines have been to reduce the amount of physical force required, to enhance speeded precision in production, and to extend human sensory capabilities. The advent of computer-based systems led to new vistas in automation whereby human thought processes were taken over by the machine. The earliest generation of computer software addressed the need for rapid, accurate, and repeated calculations and led to numerous success stories before artificial intelligence per se was introduced. Other, and often more complicated types of decisions and psychomotor control eventually became targets for automation as well. Machine-centered designers frequently convey the idea that if 80% or more of accidents or unwanted events are the result of human error, all those errors will go away if the human in the system can be automated. Unfortunately, automation might only move the errors around the system rather than eradicate them. Automation can change the nature of the task to make the task more programmable, but in doing so could make the task incompatible with the operator's expectations for proper functioning (Papantonopoulos & Salvendy, 2008) and the means of monitoring it (Wiener & Curry, 1980). Furthermore, there is always a human in the system somewhere. If it is not the operator in the usual sense, it is the programmers who can run away and hide behind licensing agreements that absolve them and their (sketchy) subcontractors from any liability.

Other hidden humans could be contributing to the automation-complacency dilemma as well. Trains are highly automated, although humans can override the automatic process. The automatic process has its limitations, however, as it is not programmed to respond to unusual circumstances such as obstructions or ice on the tracks; the operator would need to override the automation to avoid an incident. Sheridan (2002) recounted a situation where two commuter trains crashed resulting in fatalities and serious injuries because the operator did not override the automation when he saw his train was not slowing as expected when approaching a station. Ice on the tracks foiled the automation. When asked why he did not use the manual override, the operator reported that, according to his understanding, he was working under strict orders from management to maintain automated control (p. 30).

Automation can intervene between the human and system operation at four different stages of information processing: information acquisition, information analysis, decision-making, and action (Parasuraman, Sheridan, & Wickens, 2000; Parasuraman & Wickens, 2008, p. 512). Each of the four types of function intrudes on the operator's decision-making to different extents. Automation that maximizes convenience and organizes information into a form that the operator can use right away is relatively welcome so long as the information is sufficiently accurate. Otherwise the operator needs to invoke new rules for when to believe the information and when to take it with the proverbial grain of salt. Automation that sorts information by priority conditions can also be a work-saver, so long as the system is using the same rules that the human would use. Systems that use decision rules that are not clear to the operator defeat the operator's ability to qualify the results, produce conflict with the operator's better judgment, and are likely to be very frustrating and possibly more trouble than they are worth.

Any of the foregoing sources of inaccuracy in automated process undermine trust in automation and keep the operator oriented toward maintaining system control and delegating less to the automatic process. There have been several new initiatives to introduce new kinds of automation into high stakes conditions with high workload, such as air traffic control. One new system under study was a NextGen air traffic control system that interfaced automated communication systems on aircraft with automated receiver systems in the control tower for hovering and landing routines, thus taking the pilot and human controller out of the loop to some extent. Workload for air traffic control is tightly linked to the number of planes taking off, hovering, and landing, and weather conditions. Controllers who evaluated one particular new system were divided between the novices, who felt they benefitted from the new automation components so that they could attend to the biggest problems that required their judgment, and the experts who thought the overall risks were too high to trust the machine too far (Cho & Histon, 2012). Experts remarked that their workload was not reduced; it just shifted to added demands to figure out which planes had which technologies and how they would interface with the new communication system.

Knowledge of the unreliability of a system is a factor in the operator's response to the incoming information. For example, firefighters worked through a computer simulation of a fire response scenario wherein they received information that changed over time and made decisions similar to their real-world decisions (Omodei, McLennan, & Wearing, 2005). In one experimental condition the participants were informed that the information they were receiving was highly reliable. In another condition the participants were informed that they could not always be sure of the reliability of the information they were receiving. The participants in the high reliability condition spent more time analyzing the information and lost time making optimal responses. Those in the less reliable condition moved more quickly to their own better judgment and ended up making better decisions overall.

In contrast, however, less expert users of automated systems tend to trust the computer program even if it is wrong and a human advisor tells them otherwise (Madhavan & Wiegman, 2007). This *complacency* effect, which is the tendency not to question the automatic system, arises from the user's self-assurance or inexperience with the task, excessive workload, fatigue, or insufficient communication (Bailey et al., 2006; Singh, Molloy, & Parasuraman, 1993). To complicate matters further, if a system has multiple automated components and some of the components are more reliable than others, operators apply a global evaluation to the reliability of the system, and thereby ignore the specific sources of unreliability (Rice & Geels, 2010). This phenomenon can be reasonably traced to a workload problem: How many different evaluations and responses to a system's unreliability can an operator handle and still do the actual work effectively?

Experts have learned the quirks of automated systems and which system responses are more believable than others. Sometimes the presence of an error message is believable, but the more specific nature of the error is not what the error message reports. One learns from repeated exposures to the error message and works through the problem. Sometimes expertise can backfire, however. Wesley and Dau (2017) reported a case of an oil pipeline breakage that resulted in a huge spill. The experts thought the warning signals generated by the system were just another example of the system overstating or overinterpreting the underlying conditions, having seen false alarms too many times. The novices in the group took the warning at face value, however, and it turned out that the novices made the correct call and the experts did not. Unfortunately, the conclusion about who was correct was not made until the damage was done. This episode can be reasonably traced to a questionable alarm system and whether it could be redesigned or reprogrammed to reduce the ambiguity that led to the conflict in interpretation, human error, and ensuring disaster.

2.2 HUMAN ERROR AND SYSTEM RELIABILITY

There are five common types of human error. Errors of *commission* occur when the operator intends to take a necessary action, but selected the wrong action or pushed the wrong button. Errors of *omission* occur when the operator fails to take a needed action ("I just put your engine back together, but I had a couple parts left over. Do you want them for anything?"). *Extraneous acts* are those actions that the operator takes when doing nothing would have been the desirable response ("If it ain't broke, don't fix it.").

Sometimes *when* is as important as *what*. *Sequential errors* are actions that are taken in the wrong order. Sequential errors can be made by one person operating a set of controls. They can also occur between operators doing multiple related tasks; in this case the error would be characterized as a coordination failure. *Timing errors* are those that are taken too soon or too late.

Mode error is the latest arrival to the common types of error. Complex control systems are often designed so that a set of buttons works one way in one mode, but the same buttons take on other functions in a different mode. Engaging the correct mode is a separate control action and another opportunity for error. Mode changes can occur when automated functions turn themselves on and off. In the case of airline pilots, they are usually absorbed in monitoring the actual behaviors of the aircraft such as speed, flight heading, and altitude that they overlook indicators of mode (Sarter, 2008). The source of this type of error is a cognitive operation known as *perceptual cycling*, which is considered in further detail in Chapter 7. It is also an example of automation attempting to reduce workload by condensing the control-display panel but only succeeding in moving the workload around by making up a new monitoring requirement.

The inventory of errors found in one industry probably does not transfer well to other industries (O'Connor, O'Dea, & Melton, 2007). Part of the explanation is that, in a complex system, attempts to minimize a type of error might only have the effect of changing it from one form into another. Thus the *context* of the errors is as important as the errors themselves for defining effective control solutions (Renshaw & Wiggins, 2007, p. 201). For instance, in the context of aviation mishaps involving night vision glasses, Renshaw and Wiggins collected several incident features in their incident reports: organizational factors such as command and training, the importance of risk present in the situation, unpredictability of the situation, time pressure, and workload. In the context of naval underwater diving incidents, O'Connor et al. (2007) developed an extensive list of incident features in the categories of situation awareness, appropriateness of decision-making, communications, stress and fatigue, and supervision or leadership contributions.

2.2.1 Error probability

The human error probability (HEP) is simply the ratio of actual errors to the number of opportunities for error:

$$\text{HEP} = \frac{\text{Number of Errors}}{\text{Opportunities}}. \tag{2.1}$$

Reliability is, therefore, 1 − HEP.

Some examples of HEPs from Adams (1989) would include the following: The HEP for selecting a wrong control from a group of identical controls could range from .001 to .01. The HEP for turning a poorly designed control in the wrong direction could range from .1 to .9. The HEP for operating a sequence of valves 10 times in a row could range from .001 to .05. The HEP for not recognizing a status indicator that is located right under the operator's nose could range from .005 to .05. Some of these HEPs look large at face value whereas others may look very small. If an action is taken 20 times a day or 100 times a week, however, the odds of an error can extrapolate to a very large number of errors.

Equation 2.1 does not take the consequences of an error into consideration. There are times and situations where the error is only a minor annoyance, and there are situations where one error is absolutely one too many.

2.2.2 Redundancy

One way to improve the reliability of a system is to introduce redundancy in the operation. For instance, two or more operators may have to agree on an action before it is taken. In another type of example, two or more indicators from machine displays must be present before the operator takes the action. The effect of redundancy in a system can be stated as:

$$R_{xx} = 1 - (1 - r)^c, \tag{2.2}$$

where R_{xx} is the reliability of the total system, r is the reliability of each system component, and c is the number of redundancies in the system (Kantowitz & Sorkin, 1983, p. 55). Equation 2.2 assumes for convenience that all redundant components have the same individual reliability, but there is no need to assume that real-world system components be equally reliable.

Consider an example that became salient in 2020 during the COVID-19 epidemic. Image a system with two people, either one of whom could have the virus without knowing so. Both people are wearing cloth masks that filter out 60% of particles being exhaled or incoming. The reliability of the two-person system would be

$$R_{xx} = 1 - (.4^2) = .860.$$

Next, supposed one of the two people is wearing a level-2 medical mask with a reliability of .85 under the cloth mask. The reliability of the system improves:

$$R_{xx} = 1 - (.16 \times .15) = .976$$

(assuming that the two masks are firmly connected to prevent aerosol from escaping out the side from between the two masks). If both people are wearing the second mask, the reliability of the system improves a bit more:

$$R_{xx} = 1 - (.16 \times .0225) = .996.$$

The disadvantage of building redundancy into the system is that redundancy slows the system down. It might also require more personnel, which is a labor cost. Hammer and Champy (1996) argued in favor of removing redundancies in work systems as part of their "reengineering of the organization." They did not address the positive effect of redundancy outright, but a wise systems engineer should consider whether some of the redundancies are actually necessary. Sometimes a new automation system can be more reliable than two or more humans in an old system together—or not!

2.3 USABILITY TESTING

Product manufacturers are strongly advised to incorporate HFE principles in the early stages of product development. To paraphrase an old saying, an ounce of problem prevention is worth a pound of retrofit redesign. There are three basic stages to usability analysis and testing: preparation, iterative laboratory testing, and field-testing.

2.3.1 Preparation

Designers in the preparation stage establish the goals of the product. Much of the information that guides them in these decisions is based on marketing analysis, which is recommended but not at the expense of ignoring ergonomic principles. Too often the idea behind the product is attractive and the prototype person–machine interface looks good, but the product is not as functional or usable as it appears. Defining the user populations and allocation of function are also part of the preparation stage; these particular points were covered earlier in this chapter.

Sources of preliminary information for design features would include industry-related publications, surveys of potential users, and focus groups (Ward, 2009). Surveys can take on myriad forms, depending on the questions one includes. Although human factors specialists can sometimes make good use of marketing surveys, and marketing professional would do well to consider the ergonomics of their intended products, human factors specialists are not conducting marketing surveys that ask, "How likely are you to buy this product?" Instead, the objective is to learn about the tasks that comprise various jobs, how often the tasks are performed, and what cognitive operations are required by the person and machine in the current configurations of the activities. This approach to investigation is elaborated further in Chapter 7 regarding cognitive task analysis (CTA).

Focus groups are composed of potential users brought together to discuss a product idea, problems they would like to see solved, and perhaps generate some directions toward solutions. Focus groups are not problem-solving groups in the usual sense of presenting a problem and concluding with a solution that everyone agrees is optimal. Rather the goal of a focus group is to generate the range of design ideas, usability concerns, user preferences, and possible reasons for any preferences expressed.

When in doubt, which is often, designers should visit the potential users at work when possible and ask, "What are they doing? How are they doing it? What do they wish was better or less annoying?" Once again, CTA is helpful for answering questions like these.

2.3.2 Iterative laboratory testing

At this stage the designers make up a prototype interface with the controls and displays organized in a sensible fashion. If there is no clear answer in the HFE literature about the design of a feature for an interface, the best response is to think through some options, make up a few different prototypes, and conduct an experiment. The experiment would contain a small but suitable number of probable users who are given a task to perform using one or more prototypes. Participants would be assigned to prototypes in a randomized or counterbalanced fashion. They would perform a standard task in which one or more aspects of performance are measured. Indicators of speed and accuracy are the common choices in this phase of the research. The ensuing statistical analysis should show that one prototype is better than the others.

The design of the keypad on a touch-tone telephone was the result of using an experimental analysis to determine optimal design (Deininger, 1960). Previously there was no known rule for a good keypad layout, and designers could only guess what the possibilities might be, so they designed several of them. The prototypes are reconstructed in Figure 2.2 (they are not the originals). The candidates were buttons arranged in a circle to emulate

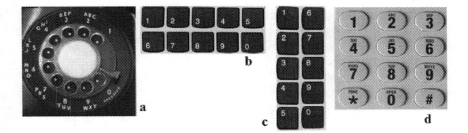

Figure 2.2 Reconstructed prototypes for touch-tone phone keypads: (a) the rotary phone; (b) numeric typewriter keys divided into two rows and horizontally oriented; (c) numeric typewriter keys vertically oriented; and (d) rectangular array.

Figure 2.3 Two examples of the rectangular array for numeric keypads.

a rotary phone, a line of buttons like those on a typewriter cut into two segments, and the now-familiar rectangular arrangement. The research participants were asked to enter phone numbers accurately. The results of the experiment should be obvious by now.

Notice that the arrangement of numbers on a phone is different from what commonly appears on a numeric keypad (Figure 2.3). A device intended for numeric data entry has the 0 and 1 key next to each other. This arrangement was convenient when large amounts of data consisted of 0s and 1s. The operator only needed to twiddle two keys with one hand most of the time. Unlike the telephone, there was no need to associate numbers with letters of the alphabet when keypunching data. The arrangement persists today because of its familiarity to so many users.

There are other issues beyond raw speed and accuracy. If operation speed of the PMS appears slow, there may be a way to revise the workload on the human operator. Speed is relative of course, to particular system requirements. Advantages in speed could be determined simply by comparing the new device with other devices that have similar purposes or designs.

Safety issues may be critical to the survival of the product and the organization selling a device (Williams & Noyes, 2009; Zunjic, 2009). Could someone use the product in an unintended way? Could the operator possibly activate the wrong control to produce a disaster? An investigation into the nature of the operators' errors can produce a great deal of useful information. Safety features could be introduced to prevent injury.

The laboratory testing process is called *iterative* because it may occur more than once before a product design is ready for field-testing in realistic environments. An adaptation of the iterative design and testing process was known as *rapid prototyping* in the software industry a few decades ago. Rapid prototyping is an interactive design process whereby a model is developed, tested with experts or users, and revised until users and experts are satisfied with the results. The rapid prototyping technique is equally applicable to the design of human–computer interfaces, which may themselves be intelligent (Cohen & Howe, 1989; Newell & Card, 1985).

Part of what made the process rapid is that software is often more readily changed than interfaces made from hard materials. For instance, some control-display sequences in computer programs can be designed, improved, and finalized in a matter of days as long as there are no conflicting demands on the programmer's time and the human participants are standing by ready to try the new design. Hard systems, in contrast, may require months to design and test an interface. Thus greater forethought in the initial design will shorten the sequence of iterative tests.

2.3.3 Field testing

Laboratory studies are explicitly designed to vary some variables and control others. The real world is replete with uncontrolled variables and tasks that might not have been anticipated in the laboratory phase.

The field-testing phase is also known as *beta testing* in the software industry. Often enough beta testing is intended to see if the programming actually works; the more basic HFE aspects of control and display design are secondary goals because the main issues should have worked out before reaching this stage of the process. Nonetheless, some controlled monitoring of users and their experiences in their natural work settings should occur at this phase before the final product is released.

The iterative design and field-testing processes are regrettably subject to some abuse. Because of the ease with which user input can be collected and assimilated in software design, we end up with software products being released for sale, and beta tested at the same time. The market then sees a myriad of versions and upgrades leading one to question, "Why couldn't you get it right the first time?" Sometimes it is difficult for the user to distinguish a visible change between an upgrade and a downgrade. Sometimes the changes are not visible at all; one can legitimately question whether the upgrade is actually installing something necessary or useful, or whether it is just an invasive probe.

2.3.4 Technical manuals

The information about user interactions can serve additional purposes beyond system design and upgrades. The user manual is one purpose. Cost–benefit analysis and planning for future redesigns of the system are two others.

Technical manuals of the old school were often written at a reading level that represented several years of education beyond that of the actual user (Chapanis, 1975). One common fault is that the texts often devoted considerable space to explaining how the machine worked from the machine's point of view, with user instructions interspersed at critical points. This format makes the user's experience a frustrating hunt for critical information.

A better format would separate the text into two documents. One document would contain the technical information that people might use if their goal was maintenance and repair. The other document would lay out the big picture so to speak of the machine's organization, identify the controls and displays, and then proceed with operations instructions. Operations instructions should begin with simple and popular tasks, then move on to the more exotic capabilities of the system. Note the emphasis here on two user populations for text documents.

Online help facilities for computer programs have been convenient and popular. They work best when the users know what they want to do and can at least guess at the searchable keywords for finding an instruction. A natural language processor, which the user does not actually see, can translate a user-defined question into one or more possible questions that the system knows how to answer; see Chapter 12 for an elaboration of this point. The format of most searchable help facilities usually does not give a user a good overview, however, of what the system's capabilities are or how they are organized. Separate files that are often called "Tours" or "Getting Started" can fill in this void.

Note, however, that users exercise two broad strategies in their searches for information. One is the directed search for targeted information. The other is exploration to determine what information and new ideas are available.

In spite of the convenience of online help for many purposes, not to mention the cost savings to the manufacturer, there is still a role for a printed technical or user's manual. "Getting Started" means just that—advanced users have probably developed needs for the deeper or more challenging features of the system, which would overwhelm a new user. Furthermore, a user can be in the middle of a job with several screens open and might wish to look something up without having to move windows around or disturb the work layout.

Training *programs* extend well beyond a printed or online manual into a structured learning experience where the user has an opportunity to gain hands-on experience with the system features and capabilities. The program structure is typically organized from simple to complex, such that each new module builds on what was learned in previous modules.

2.3.5 Cost–benefit analysis

As its name implies, the cost–benefit analysis is a mental exercise that potential system users engage to determine whether there is a benefit to adopting a new system. Costs in the concrete financial sense figure heavily into the process. Comparisons may be made between new potential systems or against a present system. Typical questions that might be asked at this stage would be as follows.

Does the new system get the job done faster than the present system? Faster in this context often means delivering a manufacturer's product to the end purchasers faster so that the chain of payments is faster also. Laboratory analyses of control-display sequences often signal reductions in operation time by a matter of seconds. Time savings on this scale might translate into saved labor costs, but in many cases the savings are in user satisfaction, which is a subjective experience and not readily quantifiable in the same manner as gross labor costs.

Does the new system replace a manually driven system or a computer-based system that reuses information that is time consuming to produce? Word processors replaced the conventional typewriter for this basic reason. Computer-driven metal lathes can reduce a task time from 2 weeks to 12 hours in some cases, and reduce the time to make the object a second time to significantly fewer hours because the basic instruction program can be saved. Machine time is expensive too.

What is the training and start-up time? A new system may be ideal for a user or user organization that has never used a version of the system before. The decision to adopt a particular system might be predicated on training time. A user with an overhead in one particular system is now considering whether the downtime for retraining is worth the effort. The potential for substituting higher skill, higher labor cost with lower skill, lower labor cost would factor into the decision.

How many different system goals are served by the new system? In the early days of desktop computers, tasks that involved intensive data analysis favored the mainframe computer over the little desktops by a time savings of 100 to 1. (A colleague of mine timed a representative job of his using a 286-series desktop and a university mainframe at midnight that he could access through a phone line.) The pretty printing that everyone has come to expect from a word processor, however, favored the desktop; mainframes were usually equipped with only rudimentary typographic capabilities. Small businesses could obtain a significant advantage by converting their operation to a desktop that was much less expensive than a mainframe or mainframe service. Businesses with large computational needs had the opposite experience.

Compatibility with other software has become another important design feature. Lange, Oliva, and McDade (2000) compared the market uptake for two products that had similar goals. Market growth for each remained relatively equal for a number of years until one of the manufacturers expanded its offerings in system-compatible products. Other software manufacturers then jumped on the bandwagon by making their products compatible with that manufacturer's items.

2.3.6 System reengineering

Hammer and Champy (1996) extolled the virtues of adopting new technologies that could improve an end user's financial experience by eliminating redundancy and substituting low-skill labor for high-skill labor. The benefit of introducing or retaining redundancy to improve system reliability was mentioned earlier. We have already seen examples of how true experts and novices approach problems differently. The pros and cons of skill and knowledge substitution are discussed further in Chapter 13 in the context of expert systems.

There are positive stories to tell, however. In the home construction industry, there is an expediter whose job it is to coordinate numerous contractors on numerous sites all in the correct sequence. The expediter also needs to inspect finished phases of the work and to go to the sites at critical moments when something unplanned happens. Before the days of cell phones and laptop computers, the expediter had to make a series of phone calls from a hardwired phone in an office and physically drive out to sites where phones did not reach. With the new equipment, however, the expediter's office is now in his truck. Sites can be seen, calls can be made, information files are always handy, delays are reduced, and final products are delivered sooner with less overhead overall.

Sometimes an organization's rate of innovation outpaces the decline of older technologies. As new features of the technology (e.g., a telephone system) are added, new links have to be made among parts of the system. As new features are requested by users, the technology becomes brittle, meaning that a small change requires extensive reprogramming so that the system can stay internally coordinated. At some point the technology reaches a crisis point where it becomes more sensible to rewrite the big program from scratch. The workload demands for doing so can be extensive. At the same time the pressure to change the system from scratch is closely tied to the organization's dwindling knowledge about how the product was initially assembled (Staudenmeyer & Lawless, 1999). Fortunately for one organization with this type of problem, there were a few people remaining in the organization who still remembered how to manipulate the original program code, all 15 million lines of it, after 20 years. The remaining problem was to figure out how to shut down the existing system while the new system was being installed. Staudenmeyer and Lawless named this type of barrier to technological change *sociotechnical embeddedness*.

DISCUSSION QUESTIONS

1. Consider two different examples of the same type of product, such as a handheld computer device or software for a particular application. Imagine that you are looking at two prototypes. Sketch a plan in as much detail as possible for assessing the usability of the devices.

2. Consider a device or piece of equipment that greatly reduced human labor when it was first introduced. How did the allocation of function between the person and machine shift over time?
3. How might differences in the user population affect your product design in Question 1?

Chapter 3

Systems change over time

3.1 COMMUNICATION, INFORMATION, AND ENTROPY

The Shannon and Weaver (1949) communication model was first developed at Bell Labs as a concept for improving the quality of telephone communications. It is applicable to any communications medium, however, with or without the assistance of electronic devices. The communications model was the outgrowth of other important conceptual developments pertaining to the quantification of information.

3.1.1 Communication model

The communication process (Figure 3.1) is a five-stage process. Errors in communication could occur at any stage. The first stage is the sender, who should be speaking the correct words clearly into the phone. In the second stage, the speaker's voice is encoded into electronic impulses that are used by the phone system. In the third stage, the message travels across telephone lines or telephone cells (or other airwaves). Noise can produce signal distortion, and it has been the target of many technological innovations. In the fourth stage, the signal is decoded into sound again at the receiver's telephone. The final stage is the receiver who must interpret what is heard the right way.

There are four important aspects of noise. One aspect is the stray or unwanted signal component that obviates the signals that are intended as the real message. This aspect of noise is considered in Chapter 3 on psychophysics.

The second aspect of noise is the high-volume noise in the receiver's environment. Not only can it obscure signals, it can also lead to hearing loss when operators are exposed to large quantities of it for prolonged periods of time. This aspect of noise is considered in Chapter 5 on auditory displays.

The third aspect of noise is the psychological stress that it can produce. The effects of stress may be health related or performance related, even where the tasks are not primarily auditory. This aspect of noise is considered in Chapter 10 on stress and human performance.

DOI: 10.1201/9781003359128-3

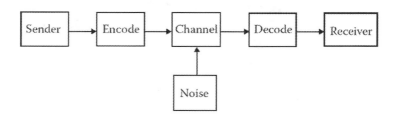

Figure 3.1 The Shannon–Weaver communications model.

The fourth aspect of noise is statistical. Some statisticians use it as a synonym for "error variance" or "measurement error," often with the implication that it could contain more structure, and therefore meaning, than what is captured by a particular statistical or deterministic model.

3.1.2 Quantifying information

A bit of information is the amount of information required to make a binary decision. Given that a system can take on two states, A and B, only one bit of information is required to know whether the state is A or B.

If the system can take on N possible states, we would need $N - 1$ bits of information to know which of the states is taking place. Each bit of information is quantified as 0 or 1. If the bits associated with states 1 through $N - 1$ are quantified as 0, then it follows that state N is taking place.

If we do not have the information that is required to predict the state of the system, then we have uncertainty or *entropy* in its place. Shannon (1948; Ott, Sauer, & Yorke, 1994) defined entropy (H_s) in Equation 3.1, where p_i, is the probability associated with one categorical outcome in a set of r categories:

$$H_s = \sum_{i=1}^{r} p_i \ln(1 / p_i).$$ (3.1)

Log_2 is often used in place of the natural log function (ln). Entropy is greatest when the odds of system states are equal (H_{MAX}), and less than maximum when the *a priori* odds of a state are unequal. Information and entropy add up to H_{MAX}.

3.1.3 Entropy

The construct of entropy has undergone some important changes since it was introduced in the late 19th century. Shannon entropy (Equation 3.1) is actually the third paradigm of the construct to be introduced out of four. In its first historical epoch, the concept of entropy was introduced by Clausius in 1863 in conjunction with relationships among pressure, volume of a container, and temperature along with broader problems in heat transfer. Entropy thus became a concept of *heat loss* or *unavailable energy* (Ben-Naim, 2008), which gave rise to the second law of thermodynamics, which became controvertible a century later. According to the second law, heat loss is inevitable in any system, so that the eventual outcome is "heat death." In this scenario, the system (if not the entire universe) will gravitate toward an equilibrium point where, in essence, nothing more will happen once the heat energy has been expended, and the system is completely disordered.

The second paradigm of entropy was associated with Boltzmann's contributions to statistical mechanics, which concerned the movement of gas molecules throughout a container space. Although it is not possible to follow the movement of every molecule, it is possible to obtain an average value for the molecules in a particular condition. In one extreme we have the condition of absolute zero temperature where molecules of a gas stand still; in the other extreme we have Brownian motion, where molecules are bouncing randomly and completely filling the container's space. This paradigm led to the concept of entropy as being one of *order versus disorder*.

The third paradigm of entropy occurred when Shannon (1948) introduced the quantifiable construct of information. Information, which is not always completely available, is used to predict the motions of the system. Note that in the first three paradigms, entropy is motion that is induced by other conditions affecting the system such as heat, and detected by an outside observer who wants to predict system states. The perspective changes in the fourth paradigm that originated with nonlinear dynamical systems (NDS) theory and the phenomenon of self-organization (Nicolis & Prigogine, 1989). The NDS paradigm contradicts the heat death scenario: Rather than dissipating energy, the system reorganizes itself—without any assistance from outside agents—to produce a more efficient use of its energy. Instead of thinking of information as something the observer uses to predict the state of the system, information in the NDS paradigms is generated by the system as it changes behavior over time. From this perspective, information and entropy are the same quantity. Self-organizing systems are expanded in the next section of this chapter along with other basic NDS constructs.

The metrics for quantifying entropy in the NDS paradigm often build on Equation 3.1 in some way (Guastello & Gregson, 2011). Although the full range of the metrics falls beyond the scope of what human factors and ergonomics has utilized to date, one important feature is worth remembering. Shannon entropy only quantifies the variation in elementary states; it does not account for *patterns of states* over time. For these purposes, constructs such as *approximate entropy*, *sample entropy*, and *topological entropy* are often preferred in nonlinear time series analysis. Symbolic dynamics analyses have been useful for identifying patterns (e.g., in task switching).

3.2 HOW SYSTEM EVENTS CHANGE OVER TIME

3.2.1 Perception of change

One of the critical links between human and animal perception and their ability to survive is their ability to detect change either over spatial location, over time, or both (Gibson, 1979). When a machine system renders information about change in the environment, we can speak of an operator' *situation awareness* (Endsley, 1995) that involves several levels of depth in perception and cognition: (a) Did the operator notice that an event occurred such as a warning light going on or off, or objects entering or moving around the display screen? (b) Did the operator interpret any change in meaning associated with the perceived differences? (c) Did either the change in stimuli or change in meaning signal a change in response? (d) Did any of the foregoing aspects of change also produce a change in the perceived state of the system?

Note that there are two things going on here: One is the more obvious difference in levels of processing. The other is the translation of change at one level of processing into change at another level. The former can be relatively fast, while the other is relatively slow. A critical level of change is required at the lower level to promote a change at the higher level.

Information, as HFE tends to experience it, has different relationships to time. In the simplest relationship, the information is static; the information content remains constant or stable until perhaps a control action has been taken or a noticeable change has occurred to denote a new stable state. Such a stable state might occur if a status light indicates "not ready for action," until such time as the light changes to mean "ready for action." Even here, however, something changes when the light goes on and off.

In other situations, the information is changing over time much more vibrantly. Sometimes the change over time is a deliberate contrivance of a mode of communication; Morse code, which went out of fashion decades ago, is an example. The auditory signals convey all their meaning in terms of the duration of pulses that are heard over time, and the pattern of long and short pulses. Although still interesting in some ways, deliberate time-phased signals are not the central concern here. Rather, the concern is for the changing events for which one must decide whether meaning exists; such decisions require some effort if not also some well-developed intuition.

In conventional thinking, a discernible pattern of system behavior is regarded as a *deterministic* process. If no discernible pattern exists then it is regarded as a *stochastic* process, which is to say a random event. Furthermore, the behavioral sciences have typically treated change as only one type of event. This is where NDS makes its contribution. What might have been interpreted as a stochastic process is often a chaotic deterministic process that only appears random over time. The changes over time can be captured by simple equations. The trick is to determine what the relevant equations might be. In NDS many different patterns of change over time are possible, some of which are more complex than others (Guastello & Liebovitch, 2009).

3.2.2 What is random?

There is a tendency among behavioral scientists to jump to the conclusion that random processes are normally distributed or vice versa. Actually one is neither necessary nor sufficient to conclude the other. Random processes have a rectangular distribution such that all the system's states, or values of a variable are equally likely, and one does not know which state or value will pop up at a given moment. When there are multiple random processes taking place, however, the concatenation of rectangular distributions produces a normal (Gaussian) distribution. Thus a normal distribution can result from many random processes occurring simultaneously. However, it is also possible to predict normally distributed variables from knowledge of other variables that are also normally distributed. Social scientists and others do it all the time, and when they do, it is possible to conclude that the dependent measure is not so random after all because it can be predicted to *some* degree of accuracy.

The overreliance on the normal curve once led to the interpretation of Poisson distributions in accident data as prima facie evidence of accident proneness. Eventually it became clear that it was just another statistical distribution and did not result necessarily from accident prone individuals or deviant situations. Yet accidents can also be predicted with some degree of accuracy (Chapter 11).

NDS functions tend to involve other types of distributions, which are most often the power law, exponential, and complex-exponential distribution. In fact every differentiable process or function has a unique distribution within the exponential family. Thus statistical distributions can tell us something about the processes that produce the data. Because error almost always exists to some extent, prediction is not perfect. Yet knowing that the relationship between variables is a function other than a straight line, together with the particular roles of other variables that go into the function can greatly enhance the prediction of a process (Guastello & Gregson, 2011).

So what is random? The medieval meaning of the word is perhaps the most applicable today: the motion of a horse that the rider cannot predict (Mandelbrot, 1983, p. 201). However, just because the rider cannot predict the motion, it does not mean the horse does not have perfect control (Guastello & Liebovitch, p. 2). Some basic NDS processes are considered next.

3.2.3 Simple attractors

Attractors are spatial structures that characterize the motion of points when they enter the space. Although they exist in different structural varieties, the fixed-point type is widely applicable. Collectively attractors represent patterns of events that were known in the past as "equilibria." In the *fixed-point attractor* (Figure 3.2), a point may enter the space, and once it does, it remains at a fixed point. Other points that are indefinitely close to the epicenter of the attractor are pulled into the attractor space. Points traveling outside the boundaries (or *basin*) of the attractor are not pulled into the attractor region. There are other types of attractors beyond the fixed point, but Figure 3.2 makes the case for the prototypic attractor. Fixed-point attractors can be taken for granted if they are not considered in a context with other more volatile change processes.

Fixed points might not appear especially interesting, but try to move a system out of one. For that challenge we need bifurcations and at least one other stable state to which the system may gravitate. A definition of *stability* would be helpful as well: A system is stable if all the points in the dynamical field are behaving according to the same rule. Thus a fixed-point attractor would be stable, but not everything that is stable is a fixed point.

Oscillators are also structurally stable attractors, and they are often found in biological systems and economics. Human work systems that produce output on a regular basis can be characterized as oscillators also, although there is a tendency for operations managers to view the cycle as a whole and ascribe one output value to the whole work cycle and ignore the internal temporal dynamics. The behavior of oscillators (mostly in engineering contexts) is usually characterized by a regular sine wave function.

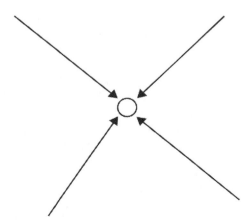

Figure 3.2 Fixed-point attractor.

Auditory signal waves, for example, are often superimposed on each other to make what we hear as complex sounds or multiple tones. Oscillators are commonly decomposed in signal processing using fast Fourier analysis that produces an additive combination of signal functions that are occurring at different frequencies.

The analytic task becomes more challenging, however, when the oscillators interact. According to a theorem by Newhouse, Ruelle, and Takens (1978), three coupled oscillators are minimally sufficient to produce chaos (see below). Although not all combinations of three oscillators will do so, complex systems containing at least three coupled oscillators have the potential for chaotic behavior. Karwowski (2012) noted that because of the growing complexity of PMS in virtually every economic sector, more examples of chaotic behavior should be anticipated.

3.2.4 Bifurcation

A *bifurcation* is a pattern of instability that is observed within a dynamical field. Bifurcations split a dynamical field into local subregions occupied by separate dynamics. One simple well-known bifurcation, the logistic map, changes a behavior pattern from a fixed-point attractor to a chaotic process, as the value of a *control parameter* changes. A control parameter is, in essence, an independent variable that is found in a dynamical system. Unlike the independent variables that are found in ordinary experiments and statistical analysis, control parameters within a system can be different from each other because of their different underlying functions, rather than simply their content. The logistic map is shown in Figure 3.3.

The equation for the logistic map is

$$X_2 = CX_1 (1 - X_1). \tag{3.2}$$

X in Equation 3.2 starts with a value between 0 and 1. It is an iterative process, meaning that we start with value of X_1, use it to calculate X_2, run X_2 through the same equation to produce X_3, and continue for many values of X_i. C is a control parameter. At low values of C, Equation 3.2 produces a series of X_i that denote a fixed-point attractor. When C becomes larger, X goes into oscillations. As C increases further, we start to see period doubling, or oscillations within oscillations. When C exceeds 3.6, the series of X becomes chaotic.

The logistic map has seen some useful applications in ecology and organizational behavior, which fall outside of any HFE applications. There are several other types of bifurcations, however, that have seen several useful applications in HFE.

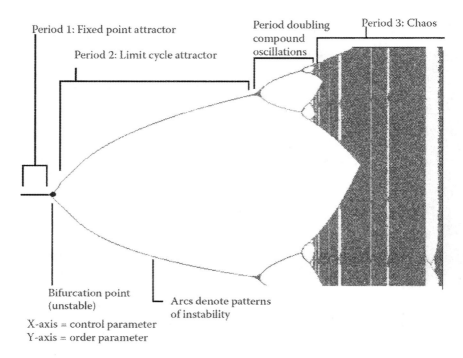

Figure 3.3 Logistic map.

A bifurcation may be as simple as a critical point, like the ones in the logistic map model, or it may be a continuous pattern of instability as found in the catastrophe models. Critical-point bifurcations appear to be inherent in the onset of cognitive automaticity (Chapter 7), or in the speed–accuracy trade-off (Chapter 8). There are also more complicated types that have been useful for explaining sudden movement across discrete system states; this type is explained further below in connection to catastrophe models.

3.2.5 Chaos

Chaos is a series of events that appear to be random in their appearance of values and incidences of occurrence, but are actually accountable by simple deterministic equations. In the ideal case, chaotic time series are nonrepeating, bounded, and sensitive to initial conditions (Kaplan & Glass, 1995; Sprott, 2003). Although chaotic behavior may appear random, chaos is ultimately a deterministic process, hence the initial fascination with the concept in the scientific community (Lorenz, 1963; Dore, 2009) and popular media (Dooley, 2009).

Boundedness means that the values of the observed variable stay within limits. Sensitivity to initial conditions means that if two points start arbitrarily close together, and each point is iterated to the same equation, then after a period of time (number of iterations) the two points become markedly further apart. The reader is encouraged to iterate some numbers through Equation 3.2 where $C = 4.0$ and see what happens. Then try it again with a slightly different value of X_1, and compare the two series. An example of what a pair of trajectories could look like appears in Figure 3.4.

A chaotic time series can result from an underlying attractor, but an attractor structure is not a requirement. In the case of a chaotic or "strange" attractor, the range of values for the observable variable would stay within boundary values, but between the boundary values it would exhibit the type of unpredictability associated with chaos. Ironically, a chaotic attractor is structurally stable, meaning that all points within its boundary are behaving according to the same (unpredictable) rule. Again, it is important to

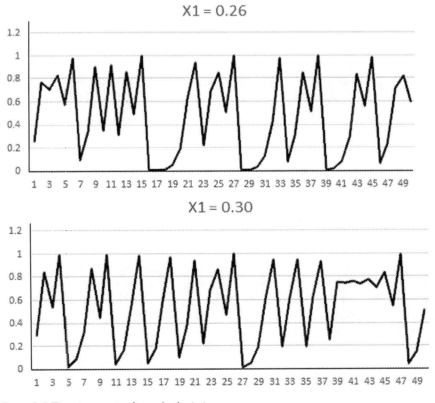

Figure 3.4 Two time series from the logistic map.

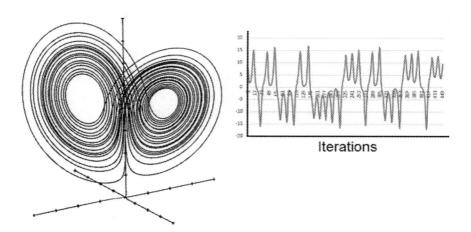

Figure 3.5 The Lorenz attractor and sample time series.

emphasize that "unpredictable" in this context is relative to conventional means of statistical prediction or conventional mental models of prediction that one uses. Within NDS "unpredictable" is better characterized as "non-repeatable" (Guastello & Liebovitch, 2009).

The Lorenz (1963) attractor is one of the more visually appealing chaotic attractors. It consists of three variables that correspond to coordinates on the x, y, and z axes in Figure 3.5:

$$dx/dt = a(y - x); \ dy/dt = -xz + rx - y; \ dz/dt = xy - bz. \tag{3.4}$$

Note that x, y, and z are all functions of each other. There are three control parameters, a, b, and r that affect the orientation of the state space diagram in 3-D space and whether the display of points are shown in the figure or scrunched up into a knotty ball. The discovery of the attractor's property of sensitivity to initial conditions led to the metaphor "Butterfly Effect." Can a butterfly flapping its wings in Texas lead to a typhoon in Japan?

Chaotic behavior in real data is usually not as good-looking. Figure 3.6 depicts chaotic data taken from an experiment in which two people were carrying on a conversation with electrodes attached to their fingers to record electrodermal response (Guastello, Pincus, & Gunderson, 2006); the change in electrical conductivity of the skin is a signal of autonomic nervous system arousal. The three axes of the plot are: electrodermal response from Person A at one point in time, electrodermal response from Person B at the same point in time, and electrodermal response for Person A taken 20 seconds later (vertical axis).

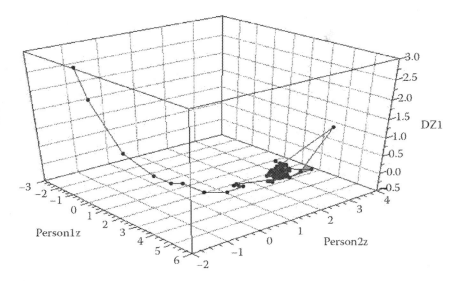

Figure 3.6 Chaotic time series for electrodermal responses from two people engaged in a conversation.

At present there are at least five dozen formally defined chaotic mathematical systems (Sprott, 2003). Testing each and every one would be virtually impossible given that the behaviors in question could result from several possible chaotic systems and the control parameters that are known to drive the formally defined system might not be readily discernible in the real-world database. Thus a simpler yet sufficient analytic strategy is to first determine whether the time series data reflects chaos at all, and at the same time ask a related question: How chaotic is the system? For these purposes two types of metrics have been useful. One consists of entropy statistics that were described briefly earlier. The other is the Lyapunov exponent, which can be calculated directly on relatively noise-free data, or statistically using non-linear regression modeling (Guastello, 2009; Guastello & Gregson, 2011).

The concept nonetheless goes as follows: Chaotic time series exhibit both expansions and contractions over time. Sensitivity to initial conditions is, for the most part, an expansion. The Lyapunov exponent calculates sequential differences in the dependent measure over a short number of observations, then extrapolates those to produce a spectrum of values. In a chaotic system, the largest Lyapunov is a positive value, but there are negatives in the spectrum as well. A value of 0 is a perfect oscillator; slightly positive values indicate aperiodic oscillators, and slightly negative values indicate dampened oscillators. If only negative values are taken from a time series, they signify that the system is a fixed point. There is a relationship between the largest Lyapunov exponent and topological entropy under limiting conditions.

As mentioned earlier, humans must often deal with flows of information that appear irregularly over time often with the help of their display systems. People vary widely in their ability to track and respond to chaotic information, yet some *can* predict values from a chaotic sequence with a respectable degree of accuracy. The nature of this ability has not yet been determined. The widespread limitation has given rise to a class of intelligent displays and controls known as chaotic controllers, which are addressed in Chapter 8. Chaotic functions also show up in Chapter 14 where complex systems are considered in further detail.

3.2.6 Fractals

Another important property of chaotic attractors is that their outer rims, or basins, are fractal structures. *A fractal* is a geometric structure that is characterized by repetition of overall patterns at different levels of scale. Thus if we were to zoom in on a small detail of the fractal, we would see a repetition of the overall shape in the enlarged area. This scaling property of fractals has led to some interesting questions such as whether the fractal nature of a time series of individual biometric data also appears at the level of group collective biometrics (Cooke et al., 2012); sometimes the pattern repeats itself at different levels of scale, and sometimes different patterns appear at different levels (Likens, Amazeen, Stevens, Galloway, & Gorman, 2014).

The dimension of a fractal is fractional, meaning that it does not occupy one of its dimensions completely. Fractals have probably become famous for their pretty geometric designs as much as for their other characteristics. The geometric designs have not shown much direct applicability to HFE or human social behavior. One concept based on fractal theory that has had an impact on real-world applications is the *fractal dimension*. The fractal dimension is related to the Lyapunov exponent under limiting conditions; both have been used as measures of complexity within a time series. In spite of their reticent role in HFE, fractals do connect the concepts of chaos and self-organization, which are more closely related to HFE phenomena.

3.2.7 Self-organization

Self-organization is a dynamical event whereby a system that is in a state of chaos creates its own order by forming feedback loops among its component parts. Although there are several mechanisms of self-organization, the feedback loops that transfer information are their common feature. The feedback itself may be positive or negative. It can be a constant flow, a

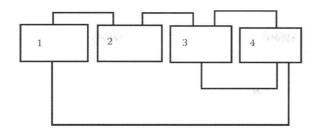

Figure 3.7 A set of feedback loops in a simple self-organized work process.

periodic flow, or even chaotic over time. A general representation of a self-organized system appears in Figure 3.7. Note that both feedback and feed-forward loops are possible. *Positive feedback* loops increase the probability of an event occurring or increase the quantity of a result. *Negative feedback* loops induce dampening effects, and they are often found in situations that benefit from the maintenance of steady states.

The coupling of subsystems through information loops contributes to the complexity of a system. Additionally, the information flows are coupled with local rules by which agents or subsystems interact and produce outcomes.

The spin-glass phenomenon is a simple type of interaction among agents. If molten glass is subjected to high-intensity mixing, the molecules do not homogenize. Rather, molecules group together depending on the similarities of their electron spins; this principle has been studied in the context of living dynamical systems in a few different contexts (Kauffman, 1993; Sulis, 2008).

In more complex systems and interactions, the mixing of agents produces hierarchical structures that serve to reduce entropy or disorder and keep the system apparently stable. The hierarchical structures often take the form of driver–slave relationships among the subsystems, such that the behavioral dynamics of one subsystem drive the dynamics of another (Haken, 1984, 1988). The key point is that the system increases its internal order spontaneously, and reduces its rate of lost potential energy. Within the context of human factors, an agent can be a human, a semi-autonomous machine, or a compound agent consisting of both.

There is a line of thinking that CASs toggle between states of chaos and self-organization (Waldrop, 1992; Dooley, 1997). A chaotic state is one with a high potential for adaptation in many possible directions. A self-organized state is one where the organism has formed a pattern of behavior with lower entropy. Shocks from the environment or from internal sources, however, may propel the organization back into a chaotic state to prepare for an adaptation of a different type (Guastello, 2002). CASs and self-organizing processes are considered further in Chapter 14.

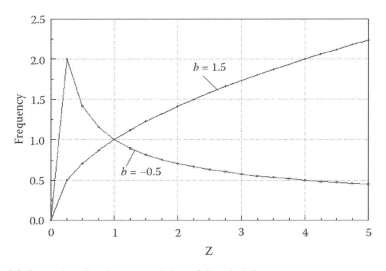

Figure 3.8 Power law distributions with *b* = –0.5 and +1.5.

There are two sets of statistical footprints of a self-organized system. One set is the inverse power law relationship between the values of a variable X and the frequency distribution of X.

$$\text{Freq } (X) = a\, X^b \tag{3.5}$$

Equation 3.5 is saying that X has a power law distribution. If $b < 0$ and a noninteger (which it will be 99% of the time if we confine ourselves to two decimal places), then b is the fractal dimension of X, which in turn denotes that a self-organizing process is taking place (Bak, 1996). An example of an inverse (i.e., $b < 0$) power law distribution appears in Figure 3.8. Beware, however, that the presence of an inverse power law does not prove the existence of a self-organizing process. Functions such as Equation 3.7 often arise from self-organizing processes and are thus a signal to look further for the actual process that could have produced the function.

The second set of footprints for self-organization is the presence of a catastrophe model. Catastrophe models involve fixed-point attractors, multiple-system states, and bifurcations of varying levels of complexity.

3.2.8 Catastrophes

Catastrophes are discontinuous changes of events. According to the classification theorem (Thom, 1975), given a maximum of four control parameters, all discontinuous changes of events can be modeled by one of seven elementary topological models (with qualifications). The models describe change between (or among) qualitatively distinct forms for behavior. The elementary catastrophe models are hierarchical and vary in the complexity of the behavior spectra they encompass. Change in behavior is described by differential equations that represent the structure of the behavior spectrum, or *response surface*. The *cusp* response surface is 3-D and describes changes between two stable states of behavior, which are usually fixed-point attractors. The two attractors are separated by a bifurcation structure (manifold).

Movement of the system around its possible states is governed by two control parameters. The *asymmetry* parameter governs how close the system is to discontinuous change in behavior. The *bifurcation* parameter governs how large the change will be. In contrast, the unstable transients (represented in Figure 3.9 as the cusp point) represent behavioral regions of great instability and indeterminism. Some specific catastrophe models that concern stress and work performance and occupational accidents appear in Chapters 10 and 11.

Not all catastrophe functions are self-organized processes, and not all self-organized processes can be framed as catastrophe models. There are enough occasions for overlap, however, to warrant at least a question of how catastrophe dynamics might produce a self-organized system, or be the result of one. The general thinking here is that a self-organized process is often evidenced by a dramatic shift in the system's behavior, visual appearance, or internal organization. The shift is analogous to phase shifts among gasses, liquids, and solids, which are known to follow the cusp

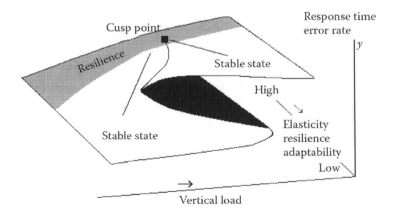

Figure 3.9 Cusp catastrophe model labeled for the cognitive workload and resilience problem.

catastrophe equation literally (Gilmore, 1981). The equation of the cusp response surface for a process that changes over time is:

$$dy/dt = y^3 - by - a \tag{3.6}$$

where y is the dependent measure, b is the bifurcation variable, and a is the asymmetry variable. Methods for the statistical analysis for catastrophe models can be found in Guastello and Gregson (2011).

The cusp model shown in Figure 3.9 is labeled for a problem that we revisit in later chapters. This particular construct of resilience is part of a buckling problem for workload, and draws on the idea of elasticity versus rigidity. The model depicts two stable states of performance; "stable" does not mean "desirable" any more than "catastrophe" means anything but a "sudden jump" from one state to another. There are two control variables involved. As increasing load is put on a system, a rigid system will resist the change in performance very effectively up to a point until it snaps. A more flexible situation will waffle a bit under load, but will not approach the bifurcation where the performance errors suddenly appear. The more flexible state is considered "resilient." The downside of this insight, however, is that "resilient" is not necessarily stable, although it can be made so by first taking the system toward the cusp point, then redirecting it toward locking in the desirable characteristics are needed to cope with the increase in workload.

3.2.9 Emergent phenomena

Emergent phenomena are those that seem to appear out of nowhere, and when they do, they cannot be explained as the simple additive result of the actions of the agents within the system. Something new has taken place. Although there is some debate as to whether all emergent phenomena result from the self-organizing processes outlined above, many of them do so (Goldstein, 2011).

Emergent phenomena should be anticipated whenever multiple PMSs are interacting. They cannot be readily decomposed into more elementary precursors or causes. They often occur suddenly, hence the word "emergency," but suddenness is not a necessary feature of emergence. The earliest concept of emergence dates back to a philosophical work by George Henry Lewes in 1875 (Goldstein, 2011). It crossed into social science in the early 20th century, when Émile Durkheim wanted to identify sociological phenomena that could not be reduced to the psychology of individuals (Sawyer, 2005). The existence of groups, organizations, and social institutions are examples; bilateral interactions among individuals eventually give rise to norms, deeper patterns, and other forms of superordinate structure. In the famous dictum of the Gestalt psychologists, "The whole is greater than the sum of its parts." Thus some of the ideas described here are more than a century old, but it was not until the 1980s that social scientists began to acquire the analytic tools to exploit them fully.

When a whole become greater than the sum of its parts, it is exhibiting a bottom-up strategy, whereby the elementary parts are in place first, they interact by some means, and produce a whole that reflect the interactions that took place. Artificial life simulations capture some scenarios by which bottom-up emergence could occur. Whereas self-organization is a process, emergence is a result of the process. McKelvey and Lichtenstein (2007) outlined four different types of emergence. The simplest was the *avalanche* that produces power law distributions of objects of different sizes. The second was the phase shift. The internal organization that would be required for a species to differentiate into more specific types is more complex that breaking into little pieces; phase shifts also occur when the organisms hop from one niche to another suddenly. More generally, a phase shift is a reorganization of the internal structure of the system. Still it is not necessary for a hierarchical internal structure to occur.

The third level of emergence is the formation of a hierarchical relationship among the parts of the system. Driver–slave relationships are examples. A different type of example occurs when a person or group collects information about its environment and forms a mental model of the situation. The mental model does not exhibit the top-down supervenient force until people start acting on the model and the model persists after some of the original group members have been replaced. The presence of an active top-down component reflects the fourth level of complexity for emergent events. Arguably, the dynamics of bottom-up and top-down influences are matters of degree and relative balance.

Yet another type of emergence occurs when a variable that seems to play no role in the behavior of a system during one time period of the system's existence might indeed play an active role later on when some other aspect of the system underwent a change (Guastello, 2002). In time series analysis the problem is called *ergodicity*: Are the dynamics that are observed during one portion of the time series the same throughout the entire time series? Sometimes they are, but sometimes there are transient events that disturb the normal course of events. Because they are transient, the system reverts back to its earlier dynamics, but not exactly picking up where it left off (Gregson, 2013; Guastello & Gregson, 2011).

3.2.10 Synchronicity

Synchronicity is a special case of self-organization where the agents in a system act independently at first, but eventually start doing the same thing at the same time (Strogatz, 2003). One of the more dramatic illustrations was reported a few hundred years ago regarding a Southeast Asian firefly. In the early part of the evening, the fireflies would flash on and off randomly. After a few hours, however, they would all start flashing together so that the entire forest would flash on and off. *Synchronicity* can be produced even in nonliving systems with only minimum requirements—two coupled oscillators, a feedback channel between them, and a control parameter that speeds up the oscillations. The principle also has been demonstrated with electrical circuits and mechanical clocks.

The oscillators synchronize when they speed up fast enough. Once the speed of interaction passes a critical point, the system exhibits phase-lock, which is visibly tight mimicry of behavior. The two oscillators shown in Figure 3.10 are synchronized, but slightly out of phase with each other.

Agents producing chaotic data can also synchronize (Pikovsky, Rosenblum, & Kurths, 2001; Stefańnski, 2009; Whittle, 2010). When chaotic agents are substituted for oscillators, the sudden onset of phase lock is not as crisp. An example of a set of synchronized chaotic time series appears in Figure 3.11. The data were galvanic skin responses (a.k.a skin conductance or electrodermal responses) from a team of four people engaging in

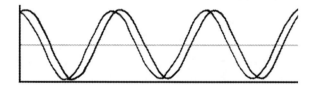

Figure 3.10 Two oscillators, slightly out of phase.

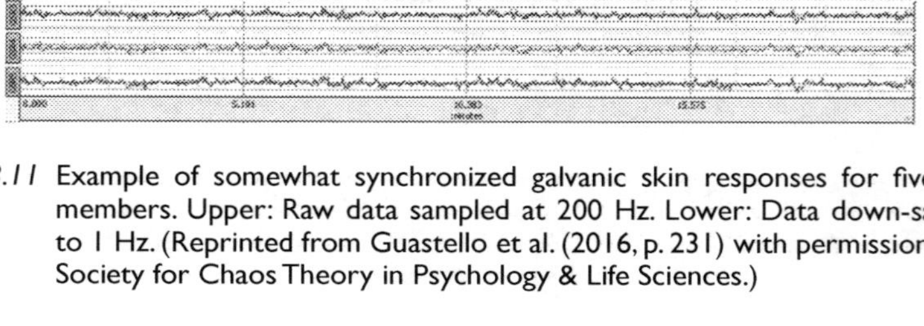

Figure 3.11 Example of somewhat synchronized galvanic skin responses for five team members. Upper: Raw data sampled at 200 Hz. Lower: Data down-sampled to 1 Hz. (Reprinted from Guastello et al. (2016, p. 231) with permission of the Society for Chaos Theory in Psychology & Life Sciences.)

an emergency response scenario against a single opponent. The opponent appears in the bottom row in each part of the diagram. More importantly synchronization plays an important role in human work performance and coordination and some of the social processes supporting collective work. The HFE applications appear in Chapter 14.

3.3 NETWORKS

Networks have become a favorite topic of study within the NDS and information technology communities, although the idea originated in mathematical social psychology. The concept of social networks was first introduced by Beveles (1948), and the study of networks, once known as *sociometric analysis*, was greatly augmented by the introduction of graph theory (Harary, Norman & Cartwright, 1965). Networks are represented as geometric configurations, or geodesics, that show the placement of each communicator relative to others in the network with connections drawn among those members of the network who communicate directly. The rapid growth of massively interconnected PMSs, data mining, autonomous agents, and new opportunities for system failure suggest that the basic principles (Barabási, 2002) would be relevant to HFE problems involving multiple PMSs.

3.3.1 Social networks

A social network is a set of individuals, or possibly groups or organizations, which communicate with each other in some fashion. Networks appear to be static structures, although they depict patterns of information flow. They can also grow and evolve. Different dynamics within networks often result from different network structures and internal properties (Boccaletti, Latora, Moreno, Chavez, & Hwang, 2006; Newman, Barabási, & Watts, 2006; Rubinov & Sporns, 2010). The limits of some networks may range in firmness from rigid boundaries to permeable and diffuse boundaries.

Two classic configurations appear in Figure 3.12a and 3.12b. The star configuration shows one central person with others radially distributed. The pentagram shows the same number of people who are capable of N-way communication. These and other configurations not shown represent the formal and informal communication patterns in organizations and any hierarchies that are involved. A formal communication network is typically whatever is drawn on an organizational chart of authority structures or a flowchart of a system operation. Informal communications are those that occur but are not drawn on the chart. The informal patterns of interaction are often influenced by idiosyncratic dyadic task interactions, friendship ties, or family ties. In many cases, the distinction between formal and informal

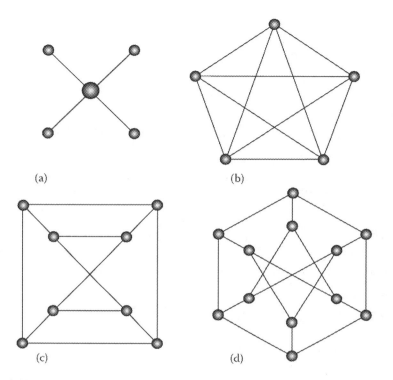

Figure 3.12 Example networks.

networks is irrelevant for understanding true communication patterns, and it is not unusual for networks formed by different interaction patterns to become superimposed on each other.

3.3.2 Nonhierarchical structures

There is a common tendency to equate hierarchical systems with the formal lines of authority associated with organizational charts. Authority is not a necessary assumption, however. Rather in the NDS context, a hierarchical relationship between two system components is unidirectional in its influence. When more than two components are involved, there is a one-to-many relationship as well. The latter feature is also known as *span of control* in organizational behavior and *fan-out* in human–robot interactions. Fan-out is explored in this Chapter 14. Nonhierarchical structures are considered next because they have a strong yet unexplored potential for making a multiple PMS more efficient. By the same token, these structures have been known for quite some time.

The wheel-shaped network in Figure 3.12a is hierarchical because of the asymmetry in communication flows among group members. The transition from the wheel to the pentagram is only the first of many possible nonhierarchical arrangements. O'Neill (1984) developed a set of nonhierarchical structures that contained some important properties: homogenous information flow throughout the network, a minimum number of meetings to convey the information, with a minimum number of people in each meeting. Figure 3.12c and 3.12d shows two examples from this set of [n, 3, 2] networks. In [n, 3, 2], an information packet flows across the network consisting of n people with a minimum of three meetings, each of which contains two people. This series can accommodate 4 to 20 people and allow for incremental growth. If more people are added to the network, another configuration can be formed that still maintains the [n, 3, 2] characteristic.

There is also a series of nonhierarchical models that require meetings of three people (not shown). The set of [n, m, 3] networks can accommodate 7 to 30 people. The number of meetings differs for different network members, but each meeting consists of three people. The structures are modular, meaning that two or more structures can be combined to form another nonhierarchical network. For instance, two [5, 3, 2] networks become a [10, 3, 2] network. Eventually there is an end to the series of efficient network structures. Nonetheless, the series is flexible with regard to the number of people involved, preferred meeting size, and modular organization. Thus hierarchical structures can be minimized.

3.3.3 Centrality

There are three primary concepts for measuring the centrality of a point within a network—degree, betweenness, and closeness (Freeman, 1979). *Degree* is the depth to which a point is interwoven with other points in the network. It is a function of the number of adjacent points to any particular point, or $a(p_i, p_k)$ that is relative to the total number of possible links that a network could have given the size of the network, n. Thus, centrality degree, CD, for a point (p_k) is (Freeman, 1979, p. 221):

$$CD = \left[\sum_{i=1} a(p_i, p_k)\right]\Big/ n - 1, \tag{3.7}$$

where n is the total number of nodes in the network. Note that in many discussions of degree, degree is characterized simply by the number of links; the denominator is ignored either because it is equal for all nodes by definition, or because the population size is not known.

Betweenness is the extent to which a point gets in between any two other communicating points. A betweenness indicator, *CB*, for a point p_k is (Freeman, 1979, p. 223):

$$CB = \sum_{i<j} \sum (g_{ij}(p_k)/g_{ij}), \tag{3.8}$$

where $g_{ij}(p_k)$ is the number of geodesics linking p_i and p_j that contain p_k, and g_{ij} is the total number of geodesics in the network.

Closeness is the extent to which a point enjoys the minimum number of points between itself and each other point in the network. It is actually an inverse function of *decentrality*, such that $d(p_i, p_k)$ is the number of edges between a point p_k and all other points. The closeness indicator, *CC*, for p_k is (Freeman, 1979, p. 225):

$$CC = (n-1)\Big/\sum_{i=1} d(p_i, p_k). \tag{3.9}$$

3.3.4 Small worlds

In a *random* graph, *n* nodes have at least one connection with another node to start. The connections occur on a random basis, meaning that there is no assumption that any node should have a preference for any other node. If the number of links between nodes is allowed to increase, again on a random basis, the network will eventually reach a phase shift in which there is suddenly a large cluster of interconnected nodes within the broader network (Newman et al., 2006). Clusters with higher betweenness centrality would be known as *hubs*, particularly if the overall network size is large, in which case there is greater differentiation between the cluster and the non-cluster nodes.

In a *regular* graph, all nodes have equal degree. Examples of regular networks might organize the nodes around a circle. Each node can connect to the node to its left and right and to the second node down in either direction. Thus in this example, each nodes would have a degree of 4, and the path between any node to any other node is equally long or short.

Next, take a regular graph and inject a couple of random links that do not exist already, and the result is a *small world* (Watts & Strogatz, 1998). Now the paths between some nodes are shorter than other paths. Interactions between nodes will follow the shortest path in the real world, thus some nodes will become part of more interactions than other nodes. Furthermore, the frequency distribution of degree for members of the network follows a power law distribution; a small number of nodes have many connections, while a large number have fewer connections. The presence of a power law distribution makes another connection to fractal structure, which can in principle be compared across networks or network configurations.

The presence of small world structure has been observed in a variety of contexts such as small organisms where the neural network is completely known, high-voltage power transmissions, and movie actors. In each case, one needs to settle on a definition of what it means to be connected. In one context, it might mean having made a movie with another actor. In scientific environments, it would mean whether a journal article cited another journal article, or whether people coauthored something together. In another context connectedness would mean whether someone knew someone else well enough to have started, a conversation or asked a small favor.

The small world phenomenon enchanted story writers and social scientists long before the formalities of small world networks were developed. The psychological studies that began with Milgram (1967; Travers & Milgram, 1969) in combination with later developments by network theorists led to the principle of *six degrees*: It is possible for anyone to reach anyone else by a median of six degrees. The trick is to find the six degrees that comprise the shortest path. The idea led to a parlor game of Six Degrees: How can I trace my connections to somebody famous?

Small world networks have some interesting properties. If there is a random attack on a network (e.g., a computer virus), small worlds are more resilient than random networks by virtue of their ability to find alternative pathways to maintaining all the necessary communications. However, diseases (or computer viruses) travel faster in small worlds if they hit a hub. It is also possible to "rewire" networks to balance the merits of small worlds and random configurations.

The power law distribution of links can be observed as chunkiness in the network layout. Figure 3.13 shows an example from Hazy (2008). Watts and Strogatz (1998) developed a clustering coefficient, which was later refined as (Newman et al., 2006, p. 288):

$$C = 6T / k, \tag{3.10}$$

where T is the number of triangles on a graph, and k is the number of paths of length 2. A triangle consists of any possible triple of nodes A, B, C. A path of length 2 exists if, for any triple A, B, C, the edges A $\rightarrow$ B and B $\rightarrow$ C are in the network.

Hazy (2008) observed, "Increased density [indicated by the dotted lines] carries risk of interaction catastrophe; leadership [can manage] interaction risk with agents through partitioning" (p. 294). Interaction catastrophe is similar to the complexity catastrophe and work overload related to all the communications. As a result, "The system is challenged in decision making, control, structuring, with respect to centralization versus decentralization,

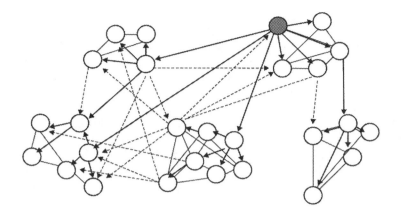

Figure 3.13 Small world network with internal clustering. (Reprinted from Hazy (2008, p. 294) with permission of the Society for Chaos Theory in Psychology & Life Sciences.)

[etc.]" (p. 294). Some of the interactions arise from the agents receiving different stimuli from the environment, many of which could be ambiguous in nature. The disambiguation process that occurs during communications should result in a unified and unambiguous response to the situation. The development of mental models and situation awareness is implied here, and these points are expanded further below in the context of collective intelligence and team coordination. Newman et al. (2006) also noted the potential for nodes to synchronize, lock up, or phase shift as a result of oscillating input through weak ties.

3.4 SIMULATION STRATEGIES FOR COMPLEX SYSTEMS

"Complexity theory" got its name from problems involving many, perhaps thousands, of interacting agents (Holland, 1995). Although there could be distinct rules of interaction between any two of them with specifiable outcomes, it was not possible to calculate the final outcomes by calculating all possible pair-wise interactions. Thus simulation tools were developed in the form of cellular automata, agent-based models, and genetic algorithms.

3.4.1 Classical system simulations

It is often difficult to envision the behavior of a system with many people–machine systems that operate both simultaneously and in concert with each other. The classical approach to system simulation begins with a flow chart of events. The flow charts are not dissimilar to the elementary information-flow diagrams that are used in Chapter 7 to describe serial, parallel, and hybrid organizations of cognitive processes. The difference this time is that many of the elements can be roped together, and the little boxes in Figure 7.7 can be complex processes involving people and machines.

The flow charts, fault trees, and Petri nets that is considered in Chapter 11 are the next closest approximations of a model that is viable for simulations. At the earliest stage in the thinking Forrester (1961) introduced a system simulation strategy called *system dynamics*, not to be confused with NDS, which featured the concepts of *stocks and flows*. His stocks are objects that accumulate at different points in a process, such as quantities of beer in a warehouse, which are released for use in other parts of the system at later times. Flows are movements of information or objects over time, and they are represented by the arrows that run between the boxes in a typical diagram of a system. It is sometimes debatable as to whether an entity, such as money or the accumulation of trash in one's house, is better represented as a stock or a flow. That decision may be relative to the questions that one is trying to answer.

The representations of flows and movements require some mathematical modeling. If the laws of physics are applicable to the problem being simulated, the equations are probably known, and they could (and probably should) be incorporated into the system. Other models would require some firsthand data collection and analysis to determine the optimal mathematical models and the statistical odds for various events. This task would not be substantially different from what one would build into an expert system. The challenges to the validity of the system are essentially the same as well.

3.4.2 Artificial life simulations

In classical system dynamics simulations, the overall structural representation of the system does not change over time. In a CAS, however, the network of boxes and arrows forms and regroups spontaneously when the system is in a high-enough state of entropy. This phenomenon of self-organization has been called *order for free*, meaning that the system takes on its own order without the intentional help of management, city planners, or similar others (Kauffman, 1993). From this perspective, each machine, machine operator, and other decision maker in the work unit is an *agent*. Work systems are the end result of a dynamical process of self-organization that was induced by the myriad bilateral interactions among the agents. With so many possible agents, and perhaps many rules of engagement, it becomes difficult to predict the outcome, or *fitness*, of a system by simply looking at it. Thus system simulations were developed to project potential outcomes of self-organizing processes.

There are three basic genres of artificial life simulation: cellular automata, agent-based models, and genetic algorithms. They can be constructed and evaluated as expert systems. The data consist of rules for agent behavior that the user can load into the program through an interface. The inference engine allows the user to tweak parameters on the model rules and then watch what happens. Displays may be numeric, but often they are pictorial

3.4.2.1 Cellular automata

The concept of *cellular automata* originated with von Neumann's vision of characterizing all life as the flow of information, which in turn can be represented as a logical program (Levy, 1992). The core applications emulated biological processes (e.g., Ulam & Schrandt, 1986). Cellular automata are iterative calculations that are defined by rules by which the information related to one cell, or pixel in a computer image, affects the behavior of each adjacent cell at the next discrete step in time. A simple example begins with the coloring of a single square, on a sheet of graph paper for instance (Figure 3.14). The second step is to color the adjacent squares in the 3 × 3 matrix that surrounds the initial square according to a rule of some sort. Patterns emerge as the process continues. Cellular automata are *spatially local*, meaning that the action of one cell on another is limited to the adjacent cells only. The system is also assumed to have no memory beyond the immediately preceding discrete step (Wolfram, 1986). Importantly, small differences in the initial values of rule parameters can produce some very different end results. Thus the concept of sensitivity to initial conditions, which is a hallmark of chaos, can be observed in cellular automata.

Cellular automata have acquired numerous applications that go beyond biological phenomena. A noteworthy example comes from studies in cryptography (Wolfram, 2002, p. 984). The goal of encryption is to obscure a message in such a way as to allow only an intended recipient to decode the

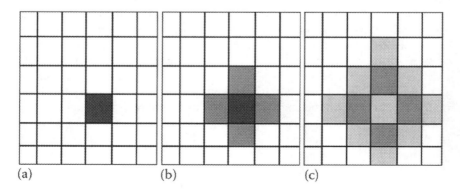

Figure 3.14 An elementary cellular automaton in three step of its evolution.

message. Encoding systems from the middle of the last century relied on mechanical devices that encoded and decoded the message. The encoding systems evolved to include sequences of encoding logic; for instance, the first time a message is passed the encoding would use a hypothetical System A, but the second iteration of the conversation would use a variation, call it System A1. It now appears, however, that cellular automata technology is, in principle, capable of cracking any encryption system.

3.4.2.2 Agent-based models

Autonomous agents are the primary ingredient for *agent-based models*, which simulate the activities of many such agents in a virtual environment. Although they are also based on the principle of adjacent pixels, agent-based models are not limited to rules that depict the literal location of objects (or conditions) in physical space, as do original cellular automata.

The agents can be programmed with rules for spatial mobility, economic interactions, and other forms of social connection such as genealogy and cultural similarity. One can also program *heterogeneous agents* that reflect different heuristics on the same basic decisions. This combination of rule systems, which are intended to emulate real-world behavior, gives rise to *artificial societies* (Epstein & Axtell, 1996) and studies of growing patterns of real urban development and industrial networks (White, Engelen, & Uljee, 1997). Among the more intriguing results from Epstein and Axtell are that if one puts many agents into a confined space, allows the agents to "trade" many commodities according to the same rules of trade, there will be a distribution of wealth that follows the inverse power law. If one were to compare terrains with high agent density, which resemble an urban area, with low-density terrains, the high density terrains support many more agents and many more products, but the distribution of wealth is more disparate. The disparities in wealth distribution arise simply from the large number of agents and commodities and are not based on any assumptions about capitalism or any other economic system; no such assumptions were programmed into the model.

The example in Figure 3.15 arose from a problem in economics (Bankes & Lampert, 2004). The end result consisted four qualitatively different outcomes that were distributed along the ranges of two other variables. Again the system user would be looking for color patterns that indicate self-organized behavior of some sort that is the end result of local interactions between agents.

Applications of agent-based models are widespread, although the research questions are usually framed from social-psychological, organizational, and sociological research questions (Ceschi & Fioretti, 2021; Ceschi, Sartori, & Guastello, 2018; Estévez-Mujica & García-Díaz, 2021; Frantz & Carley, 2009; Kashima, Kirley, Strivada, & Robins, 2016; Kiel, McCaskill,

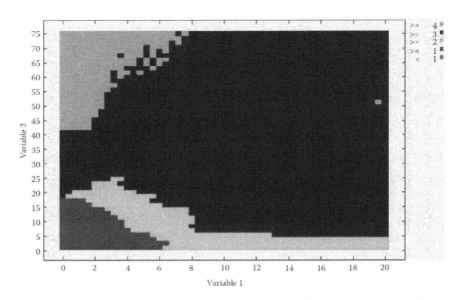

Figure 3.15 Example output from an agent-based model. (Adapted and reprinted from Bankes and Lempert (2004, p. 272) with permission of the Society for Chaos Theory in Psychology & Life Sciences.)

Harrington, & Elliott, 2021; MacCoun, 2016). A possible visualization of a network that evolved through agent-based modeling (Samuelson & Macal, 2006) appears in Figure 3.16. For network problems, it would be useful to think of an agent as a person–computer entity. Although the methodology for producing agent-based models is beyond the scope of this textbook, readers should be aware that the techniques are available for studying the evolutionary dynamics of complex PMSs.

If an agent-based model is designed effectively, the behavior of synthetic agents in the model will resemble autonomous agents operating live in the cyber world. Chapter 13 introduces an agent-based data mining program for investment decisions wherein the program incorporates the investor's trading strategy, collects appropriate trading data from the real world, and evaluates the efficacy of those trading decisions. When thousands of other investors using trading programs with similar or different trading strategies (heterogeneous agents), a high-frequency trading regimen is produced. The agents are reading what all other agents are buying and selling on a moment-to-moment basis, and making predictions and trades faster than any human can decipher the data and apply decision rules.

Unfortunately, all the trading agents can self-organize in ways that are not predictable from the analysis of a single trader's algorithms for data mining

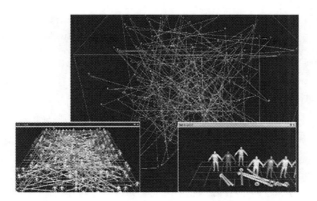

Figure 3.16 Agents interacting in a self-organizing social network inside a virtual simulation of social influence. (Image by Kurt Severance, NASA, and Kenneth Zick, U. Michigan (from Samuelson & Macal, 2006); in the public domain.)

and decisions, precisely because the myriad agents were not taken into consideration by all the trading programs. For instance a flash-crash was report on May 10, 2010, where the Dow Jones Industrial Average plummeted 900 points (almost the largest daily drop on record) within a few minutes, then regained itself 15 minutes later (Wilkins & Dragos, 2013). Prices of specific stocks went haywire in the interim. The financial damage incurred by the automated traders has not been ascertained, but estimated to have been enormous for some traders.

3.4.2.3 Genetic algorithms

Genetic algorithms are models of machine learning that are heavily inspired by biological processes. To stick close to the biological metaphor, characteristics of a population are ascribed to *genes*, which are recombined within the population according to a rule base that allows or constrains possible patterns of recombination. Mutation rates are introduced, and the fitness of resulting organisms can be assessed. Some algorithms involve compromises between genetic rules and problem-specific constraints. Hybridized models are known as *evolutionary programs*. Pure genetic algorithms have the advantage of theoretical support and wide applicability, but have been known to produce inappropriate results in some applications. Evolutionary programs, on the other hand, tend to be more viable for specific programs, but are narrower in generalizability (Michalewicz, 1993).

Nolfi and Floreano (2000) expressed a preference for genetic algorithms as models of robot learning. If the device were going to emulate a CAS, it would need adaptive capability. Neural networks are good for programming system learning with fixed (or known) end results, but the attractor strength of a learned behavior simultaneously introduces an adaptive cost, making the system unlikely to identify and select a novel response to a novel stimulus. The issue of adaptability brings us to the topic of a CAS. Further topics in system learning are considered in Chapter 14 in the sections on collective intelligence and human group coordination.

DISCUSSION QUESTIONS

1. What would be some events in the life of a PMS that would fit the description of chaos?
2. How many events can you think of that would be distributed (based on your first impressions), as an inverse power law distribution? What could that distribution tell us about those events?
3. A system might make a dramatic change when it becomes so brittle that it reaches the point of no return. What dynamical model would describe this type of change? What variables might be part of the process that could make change in the system easier or more difficult?

Chapter 4

Psychophysics

4.1 CLASSICAL PSYCHOPHYSICS

How bright is a bright light? How loud is the sound? Did anyone notice that the lights became dimmer suddenly? Did anyone notice that the sound has gradually become louder? Why was the bottlecap in the beach sand mistaken for a coin? Why was the blip on the radar screen mistaken for an enemy aircraft? It was discovered early on that the answers to the foregoing questions was not as simple as measuring the number of watts or candle-power in a light or decibels associated with a sound level. "Oops" may have been a good enough answer for the beach sand question, but not nearly good enough for the aircraft question.

The foundational contributions are credited to Weber (1846/1978) and Fechner (1964), although much credit should be given to those who participated in the debates that ensued (Murray, 1990). We consider next the basic concepts that they introduced and the general laws that they derived.

4.1.1 Threshold concepts

An *absolute threshold* is the minimum value of a stimulus that can be detected. A light does not get any dimmer before you cannot see it at all. A sound is never quieter, or weights on your hand ever lighter. In early times it was believed that the same absolute values existed for all people, and the goal of the research was to determine what those critical values happened to be.

A *difference threshold* is the minimum change in the value of a stimulus that a person can detect. This amount of psychological difference was further known as a just noticeable difference, or JND, and one may speak of stimuli differing by two, three, or more JNDs.

In the classical paradigm, the absolute threshold for detecting a sound (loudness) would be determined in a simple experiment, the general format of which is depicted in Figure 4.1. Human subjects would be presented with sounds of increasing loudness. When the subject first indicates that a sound

DOI: 10.1201/9781003359128-4

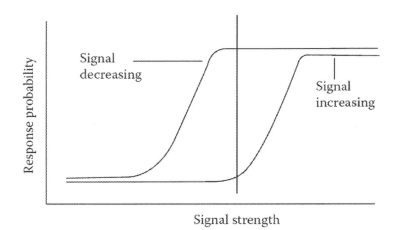

Figure 4.1 Charting absolute thresholds in a classical psychophysics experiment.

is present, an estimate of the absolute threshold is obtained. Next, the subject is presented with sounds of decreasing loudness. When the subject first indicates that the sound has disappeared, then the second estimate of the absolute threshold is obtained. Typically, the two estimates differ such that the first estimate is physically louder than the second. Thus the average of the two estimates was taken as the real value.

The *y* axis of Figure 4.1 is interchangeably known as the response criterion. Each person in the experiment has a criterion of signal strength that triggers the "yes" response. It is the experimenter's task to find out what that criterion happens to be. The notion of response probability was introduced somewhat later when it eventually became apparent that absolute thresholds were not so absolute.

Difference thresholds in the classical paradigm were obtained in a similar fashion, except that the human participants would be presented with a standard stimulus and a second stimulus. The minimum value of the second stimulus that would produce a response from the human saying that the two stimuli are different is taken as the JND.

4.1.2 Fundamental laws

Classical psychophysics produced two important laws. According to Weber's law, the ratio of the change in intensity to initial intensity is a constant:

$$\Delta I/I = C, \text{ or } \Delta I = CI$$

$$(4.1)$$

or, alternatively, that change in perceived intensity is proportional to the initial intensity, where ΔI = JND, and I = initial intensity. C is constant for all values of a particular type of stimulus, such as loudness of a tone, brightness of a conventional lightbulb, and so forth. Not surprisingly, Weber's work began with the perception of weights. It was not until electric power became available, along with equipment to plug into it, that it was possible to deliver controlled stimuli of many other varieties.

Fechner's law was derived from Weber's law. According to Fechner's law, psychological signal strength is a logarithmic function of physical signal strength:

$$\Psi_s = k \log(I), \tag{4.2}$$

where Ψ_s is the psychological magnitude, k: is a constant, and I is physical magnitude. Once again, k is specific to the type of stimulus (d'Amato, 1973).

A close variant of Equation 4.2 specifies that the sensation of a stimulus is relative to the absolute threshold, I' for the stimulus:

$$\Psi_s = k \log(I/I'). \tag{4.3}$$

Fechner devoted a good portion of his next 30 years conducting experiments that would define Weber's law, on which his own function was based (Murray, 1990). There were also some important criticisms to counter. First, if two different thresholds are obtained for ascending and descending series, the absolute threshold is not as absolute as the initial theory would indicate. Psychologists discovered later that the two-threshold effect was an artifact now known as *response set:* If the human participant has been habituated to responding "yes" for a while, the person would continue to respond "yes" even when the stimulus strength has fallen below the response criterion. Similarly, the "no" response persists after the stimulus exceeds the threshold. The response set problem goes away when the experiment is designed so that the stimuli are presented in random order with respect to their intensity.

The second problem was a critique offered by physicists, that it was not possible to measure psychological sensations with the same kind of meaning as one would associate with physical measurements. People do use measures such as "very bright," "not so bright," "very dim," and even "twice as bright as" in common parlance, but this form of subjective rating is not objective measurement. Psychologists with the requisite statistical skills later handled this problem, and the basic solution is presented in the next section of this chapter.

The third problem was that if a weak stimulus is presented with a strength below the absolute threshold (I' in Equation 4.3), the result is a negative sensation value. What is the meaning of a negative sensation? Although the Weber–Fechner laws lost accuracy for extreme values of I, they continued to be used with success for midrange stimulus strengths for simple 1-D stimuli. The classical model incurred greater difficulty (long after Fechner's productive era) when multidimensional stimuli were involved, such as in the perception of small color differences (Wandell, 1982).

The fourth problem was that a competing theory was already launched that posited that the relationship between physical stimulus strength and sensation was based on an exponential function. It took more than half a century to settle that particular issue in the form of signal detection theory.

4.2 SCALING PROCEDURES

This section elaborates a basic method for scaling the psychological values of physical stimuli that was refined by Thurstone (1927; Torgerson, 1958). The method is so basic that it can be applied to any stimulus, and the stimulus does not need to be limited to lights, sounds, weights, or other specific physical sensations. Indeed the method for scaling physical stimuli became the foundation of other psychological measurements such as attitudes. Although methods for attitude scaling are beyond the scope of this book, a brief foray is included nonetheless because there is a critical result that simplifies scaling procedures that can be used in a wide range of human factors applications.

4.2.1 Psychophysical stimuli

The *method of paired comparisons* begins with many stimulus values that are presented in pairs to a human observer. Each pair of stimuli must be presented many times. Alternatively, many observers may be involved to make the complete set of comparisons. In theory, these stimuli will have average (mean, or M) values given to them by the observers. These values will have standard deviations associated with them also, and the distribution of psychological values will be normally distributed. Furthermore, for any pair of stimuli, we will have two means (M_1, M_2), and the difference between the means (Md) will be normally distributed as well.

Figure 4.2 shows a possible distribution of Md. The shaded region indicates the proportion of times the observers responded that the strength of Stimulus 2 was less than the strength of Stimulus 1. The unshaded region represents the proportion of times the observers gave the opposite response. If the shaded area represents 80% of the area under the curve, then the observers agreed 80% of the time that Stimulus 2 was less than Stimulus 1.

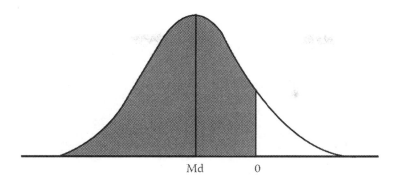

Figure 4.2 An underlying distribution of scaling comparisons.

Table 4.1 Frequency matrix for a scaling experiment

	A	B	C	D
A	—	30	60	90
B	70	—	70	80
C	40	30	—	60
D	10	20	40	—

Consider an example involving four stimuli, A, B, C, and D. The stimuli are presented in pairs, and each pair has been judged 100 times before the experiment is over. Count the number of times that each stimulus has been rated as stronger than each of the other stimuli, and put the frequencies into a frequency matrix as shown in Table 4.1. Each table entry represents the number of times the stimulus in the column was greater than the stimulus in the row. Note that it is not necessary to test each stimulus with itself.

The second step is to divide the frequencies through by the number of observations per cell (N). Because N is 100, this is an easy calculation to imagine. Insert .50, however, on the diagonals, before proceeding to the next step.

The third step is to convert the table of probabilities to Z score units, which can be done using a common normal curve table that is available in any basic statistics book. The result for this example is shown in Table 4.2. For the entry of column A, row B, the probability of .70 corresponds to a Z score of 0.52. Imagine that the shaded area in Figure 4.2 is 70% of the area under the curve. The corresponding unshaded area would be 30% of the area from the left of the normal curve, and the Z score is –0.52. The value

Table 4.2 Z matrix for a scaling experiment corresponding to Table 4.1

	A	B	C	D
A	0.00	−0.52	0.25	1.28
B	0.52	0.00	0.52	0.84
C	−0.25	−0.52	0.00	0.24
D	−1.28	−0.84	−0.24	0.00
ΣZ	**−0.98**	**−1.88**	**0.53**	**2.36**
ΣZ + 2	1.02	0.12	2.53	4.36

Note: The measurement scale is calibrated in arbitrary statistical (standard deviational) units. The origin of the scale is arbitrary also, and it does not need to be an integer.

of −0.52 is entered in the cell for column B, row A. Complete the table in the same fashion. Insert $Z = 0.00$ for the diagonals.

The fourth step is to add up the Z score entries for the columns, which are shown in bold in Table 4.2. For the final step, it is convenient to add a constant to the ΣZ values to eliminate negative numbers; a constant of 2 does that job nicely here. The results show that B was the weakest of the four stimuli, and D was the strongest.

4.2.2 Non-psychophysical stimuli

The earliest attempts to develop attitude scales, for example, a person's subjective impression of a social or political event, or liking for a job or a consumer product, were modeled on the method of paired comparisons. Research participants would be presented with pairs of statements such as, "The people who work the hardest are paid the least," and "The pay scales here are fair for what we do." Participants would indicate which of the two statements represented a more favorable attitude toward one's job. A few dozen statements would be compared in this fashion, and the calculation procedure for the method of paired comparisons would be deployed. The result would be a rating of favorableness for each survey item in the set, and a scoring system would be devised based on the favorability ratings. Then an independent group of people who were really evaluating their job would indicate which statements they thought were true, and their attitude levels would reflect the most favorable statements that they would endorse.

Likert (1932) made an important breakthrough that simplified the development of rating scales by proving that all the information that was necessary to measure an attitude could be obtained by use of the 5-point Likert scale as it is now known. In this procedure, the participants would be presented with a statement for which they would select one of five responses valued as 1 = strongly disagree, 2 = disagree, 3 = Neutral or don't know, 4 = agree, and 5 = strongly agree. The ratings would be summed over all the survey items to produce an attitude score. Although there are many important factors that contribute to the reliability (1 minus the error variance of all ratings on the scale) of such measurements, an important one for present purposes is that the target attitude statements cover a full range of favorability levels.

It is possible to adapt the 5-point rating system to other types of rating objectives by changing the anchors, so long as the relative intensities of the anchors remain approximately the same. For instance consider an ergonomic problem where the goal is to ask participants to rate the comfort or functionality of a leg prosthesis. One of the items might be: How stiff or flexible is your prosthesis? The rating options could be: 1 = Device is definitely too flexible, 2 = Device may be a little too flexible, 3 = It is comfortable the way it is now, 4 = Device may be a little too stiff, 5 = Device is definitely too stiff.

What is obviously missing in this example is any notion of what the physical features of the prosthesis could be that would generate any of the five levels of response. A study would need to be designed that would associate the design features with the ratings. Another difference between the Likert-type ratings and a conventional psychophysics study is that there are more than two response options, such as signal present versus signal not present. Instead we have gradations of experience. The issue of graduated rather than dichotomous ratings is resumed later with fuzzy signal detection theory (FSDT), which follows standard signal detection theory.

4.3 SIGNAL DETECTION THEORY

Signal detection theory was developed to address artifacts in the classical psychophysics experiments. We saw already that the random presentation of stimuli eliminated the response set problem, and that psychological ratings could be scaled into cogent measurements that could be used in other calculations and experiments. Other problems remained however. Participants in psychophysics experiments make judgments that are based not only on the nature of the stimuli themselves, but also on the basis of their sensory acuity. In other words, people with hearing problems might miss sounds (which we now call *signals* more generally) that other people might detect. There are also more subtle individual differences in sensory acuity that would not qualify as problems but that would have the same effect.

The human participants also work under a motivation to respond in the experiment in much the same way as they do in real life. There are gains and penalties associated with certain responses and errors. All this affects discriminative behavior, and thus subjects' threshold to respond to a stimulus. There still remained the problem of predicting responses to stimuli with strengths close to the threshold.

Signal detection theory separates the various influences on human response. Although it was an outgrowth of at least a half-century of work, its consolidation and its centerpiece, the power law, is usually credited to Stevens (1951, 1957).

4.3.1 Threshold concepts

Signal detection theory introduced several fundamental new concepts that distinguished it from the classical paradigm. The first major change from classical psychophysics was to redefine the threshold. Instead of absolute definitions of the absolute and difference thresholds, there are relativistic definitions. The absolute threshold in signal detection theory is defined as the point where 50% of the subjects perceive the stimulus. The difference threshold was similarly refined as the point where 50% of the subjects detect a difference. The frequency distribution of the number of people detecting a stimulus value as the threshold is thought to be normal (or Gaussian).

4.3.2 Discrimination index

The second major change was to redefine the task of perceiving a signal: We do not detect simply a signal, we detect a signal relative to noise. Figure 4.3 shows two normal-density functions on an axis defined by signal strength. The left distribution represents the absolute threshold for perceiving a standard noise, and the distribution on the right is the total strength of a target signal embedded in the noise. The difference in means is known as the discrimination index, or d'.

In a standardized signal detection experiment, the human participants are typically selected for having a qualifying level of sensory acuity, and they would thus be regarded as equal in ability (Robertson & Morgan,

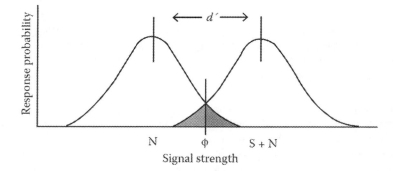

Figure 4.3 Distributions for thresholds to detect a noise or a signal plus noise.

1990). One may then proceed to assess the humans' response to the stimuli. Larger values of d' are obtained for signals of greater strength. For signals of greatest strength, all people will detect the signal; those are not the most interesting cases. Mid-range stimuli, which may be detected by only some of the people some of the time, are of much greater concern. Ultimately, a signal detection experiment would be performed to determine the error rates and to find a way to minimize them, especially if it is not possible to change the nature of the signal. On the other hand, it is a productive use of human factors knowledge to identify signals that need to be enhanced through some technological means and to determine how much enhancement they require to meet the needs of decision accuracy on the part of the humans.

4.3.3 Minimization of errors

Having defined signal detection as a statistical phenomenon, it was possible to express the results of a signal detection experiment in the form of a 2 × 2 contingency table. Figure 4.4 expresses the probabilities of a subject, or a group of subjects giving the right and wrong answers. The 2 × 2 contingency table forms the basis of several types of statistical tests that draw inferences about the overall percentage of right and wrong judgments. The table also forms the basis of decision theory where the decisions in question could be far more complex than determining presence or absence of a simple physical signal.

The *likelihood ratio* is defined as the ratio of the probability of a correct hit to the probability of a false alarm. The likelihood ratio is optimal at the

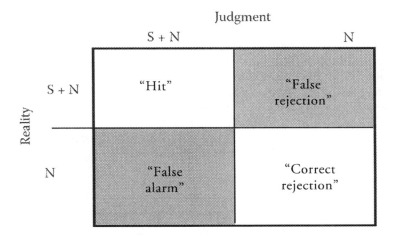

Figure 4.4 The 2 × 2 contingency table for signal detection responses.

point where the two normal curves cross (ϕ) in Figure 4.3. It is also the point where the two types of errors have equal probability of accuracy.

Probabilities, as defined in the chart, take into account subjects' response behavior under conditions where the noise and signal plus noise are equally likely. If the experiments were devised so that there was a 90% chance of a signal being present, a subject might be inclined to respond as if there were a signal on every trial, thereby obtaining a total accuracy of 90%; errors would be biased toward false alarms.

Subjects respond differently, however, when there are different payoffs attached to each type of response. For example, if a miner were panning for gold, a correct hit could be valuable, but a false alarm error would have a trivial impact; the coin and bottlecap example at the beginning of this chapter is another example of the same set of motivational conditions. In medical decision-making, however, a premature rejection of a disease diagnosis could be very costly, thus physicians might prefer to err on the side of overly sensitive testing, and then use follow-up tests to confirm the diagnosis (Swets, Dawes, & Monahan, 2000).

Jury decisions in criminal trials, on the other hand, are often subject to a reverse set of motivational conditions. In the U.S. legal system, the defendant is considered innocent until proven guilty. The evidence, or signals, favoring the guilty verdict should be sufficiently strong that jurors have little reasonable doubt about their decision. Jury decisions are typically made without the interaction with machine systems such as those considered in this book, although individual pieces of evidence may in fact have been the result of a forensic analysis using such equipment. A human factors expert should question the particular forensic decisions made in those contexts. It is a different, though forensically relevant question as to whether jury decisions would turn out the same if critical pieces of testimony were experienced directly in the courtroom or through videotape or telecommunications equipment.

In any case, the miner, the beachcomber, the physician, the forensic specialist, and the juror are making decisions with the same common elements. Something important is present in the real world or it is not present. A judgment is made as to whether the stimulus has occurred. The decision could be in error. There are two kinds of errors, and the goal of a human factors analysis is to minimize errors. Error is minimized when the motivational utilities associated with one or the other type of error are balanced to negate this bias (Robertson & Morgan, 1990).

4.3.4 ROC curves

A *receiver operating characteristic curve*, or ROC curve, is a plot of the proportion of correct hits, or $P(S \mid s)$ to false alarms, or $P(S \mid n)$, for all subjects in an experiment (Figure 4.5). The data points would configure as a set of

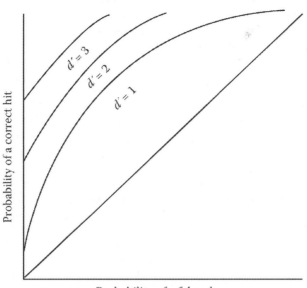

Probability of a correct hit

Probability of a false alarm

Figure 4.5 Receiver Operating Characteristic (ROC) curve.

arcs, where each arc represents a value of d'. Thus subjects of a given dis-crimination ability would respond anywhere along their arc. Position along the arc is the result of a response set. The points falling in the lower left of the arc would denote subjects who tend not to respond, and who accumu-late many false rejections and correct rejections. The points falling in the upper right would denote subjects who have a low threshold to response to any stimulus, and would accumulate a large number of correct hits and false alarms.

The ROC curves in Figure 4.5 were drawn with their peaks lined up more or less on a common axis. This arrangement is neither necessary nor suffi-cient. ROC curves can be shaped with their peaks skewed to the left or right of their center points. Analysts have found that the particular shapes of the curves result from the standard deviations of noise and signal-plus-noise curves, such as those appearing in Figure 4.3 (Grieg, 1990). As a result, d' should be calculated in such a way as to accommodate the standard deviations of the two distributions in addition to the two means. Thus Equation 4.4 emerged as an appropriate solution:

$$d_A = (\mu_s - \mu_n)/\sqrt{(\sigma_s^2 + \sigma_n^2)/2}.$$

(4.4)

In Equation 4.4, d_A is the adjusted discrimination index, μ_s and μ_n are the means of the signal and noise distributions, respectively, and σ^2_s and σ^2_n are the variances of the signal and noise distributions respectively. It is obvious that d' is a statistical value after all. Accurate values of d' will be obtained to the extent that the values of d' in the research sample approximate the theoretical distributions (populations values) for the stimuli. Population values are tenable when there are at least 50 observations for each particular signal strength used in the experiment (Robertson & Morgan, 1990). That means that each human participant will have to respond to every stimulus 50 times, or there must be at least 50 human participants who can respond to the set of stimuli at least once.

Next consider an application of the ROC curve to resolve, or at least shed some light on another forensic application, the best means of structuring a police lineup for identification of a perpetrator by an eyewitness. Faces and body shapes generate a lot of information that challenge an eyewitness's ability to identify the perpetrator from the lineup. This challenge often sits on top of the difficult circumstances in which the viewer saw the suspect. The standard procedure is for the witness to view a batch of perhaps five candidates simultaneously and pick out the perpetrator from the options presented. A new approach that has been gaining attention, however, is to view the candidates sequentially. Although the arguments for and against this method are complex, the relevant concern for present purposes is the contrast between probative value and the ROC method of evaluating the outcomes of the two procedures (Wixted & Mickes, 2012). Probative value is the odds of a correct conviction given the identification of a suspect.

To begin, it is possible that the instructions to the eyewitness could affect the correct hit and false alarm rates. Instructions could emphasize the goal of finding the perpetrator, exonerating the innocent, or both. The instructions, when combined with the witness's own proclivities, could result in liberal, neutral, conservative, very conservative, or extremely conservative responses by the witness, according to a hypothetical analysis by Wixted and Mickes. They compared ROC curves for the five response styles for the sequential and simultaneous lineup formats are as shown in Figure 4.6. "Extremely" and "very" conservative response styles produce hit-to-false-alarm ratios that were equivalent, but the ratios spread out for the other three styles. The best discrimination would be the ROC curve that was farthest away from the diagonal, which was obtained for the simultaneous approach.

4.3.5 Individual differences

The basic signal detection experiment, the 2×2 decision chart, and the ROC curve can be used to assess individual differences in acuity. For such an experiment the stimuli would have to be all alike with respect to signal strength. The chosen level of signal strength would need to be close to the

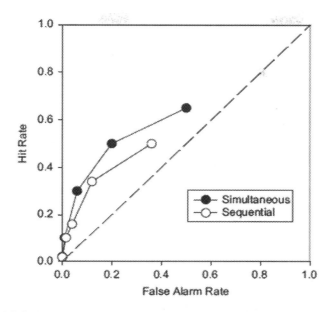

Figure 4.6 ROC curve comparing two methods for eyewitness identification. (From "The Field of Eyewitness Memory Should Abandon Probative Value and Embrace Receiver Operating Characteristic Analysis" by J.T.Wixted and L. Mickes, 2012, *Perspectives on Psychological Science, 7,* pp. 275–278. Copyright 2012 by the authors. Reprinted with permission.)

average person's absolute threshold or less. Given this setup, if the stimuli are all alike, then differences in d' would correspond to ability levels, with higher values of d' denoting people with greater sensory acuity.

Throughout this discussion it has been assumed that the participants in our experiments were paying attention to the experimental stimuli. That might not be an untoward assumption in a controlled experiment, but real life introduces distractions and forms of noise that are not part of the planned experiment. For the most part, we can assess the impact of extraexperimental noise by conducting a signal detection experiment in the environment where the humans are likely to work. The machine noise, verbal chatter among coworkers, and random yelling and screaming are uncontrolled; their contributions would be part of the noise curve in Figure 4.3.

Attention can be experimentally manipulated in a visual luminescence experiment, however, by varying the position of the stimulus on the display area. The experiment would begin with subjects being conditioned to respond to a stimulus or noise in the center of the display, thus focusing their attention on that location. Next the stimuli would be presented in either the target location or in the periphery of the display. A series of experiments by Hawkins et al. (1990) indicated that stimuli presented at the targeted location were more readily detected compared to stimuli that were presented in peripheral locations. Furthermore, the processing of luminance and location information was simultaneous rather than serial, wherein the brightness of the stimulus would be processed before the location or vice versa.

4.3.6 Power law

Finally the power law describes the relationship between the strength of a physical signal and the psychological perception thereof:

$$\Psi_s = k(S_s - S_0)^c, \tag{4.5}$$

where Ψ_s is the psychological rating of signal strength; k is a constant for the idiosyncrasies of the experiment, such as procedural errors that often appear minor or trivial or a particular payoff function; S_s is physical signal strength as before; S_0 is the absolute threshold value of S for each human subject; and c is the characteristic exponent for each type of stimulus. Examples of c are: loudness of a sound, 0.6; brightness of a light, 0.33; smell of coffee, 0.55; electric shock of 60 Hz through the fingers, 3.5 (Stevens, 1951).

The power law seems to hold for every type of signal detection except the detection of vibration, which seems to require a more complex function. Situations requiring multidimensional inputs and outputs also require more complex forms of analysis.

Although the power law is usually associated with Stevens' work, the first experiments that hypothesized such a law date back to Plateau in the 19th century, which were posed as serious contenders for the Weber and Fechner laws (Murray, 1990). According to Gregson (1998), Herbart published the first known theory of psychophysics in 1812, which contained the earliest version of the power law.

4.3.7 Decisions revisited

So far we considered the simple psychophysical judgment as a decision between two distinct outcomes—presence or absence of something important. We can now put these simple decisions into a broader context that includes not only simple psychophysical judgment, but also other forms of dichotomous decision such as the detection of a medical abnormality on the basis of a diagnostic test, or the forecasting of the weather conditions from meteorological data. Even though some of the information sources produced by machine systems produce results that are continuously valued, rather than categorical, the human operator is often charged with making a dichotomous action decision based on the information that is supplied. For instance, there may be a 40% chance of rain, but the behavioral decision is "Do I turn on the irrigation system today or not?"

Swets et al. (2000) presented an ROC curve for a real decision-making task showing the slope of the curve at two points to the left and right of the center. For combinations of low false positives (false alarms) and low hit rates, the respondents had set their response criteria so high that signals were never reported. For combinations of high false-positive rates and high hit rates, the respondents had set their response criteria so low that they hit all the signals, but the false-positive rate was very high. If the costs associated with errors are equal for false positives and false negatives, the point of optimal decision-making is located at the point where the slope of a d' curve is equal to 1.00; this point corresponds to ϕ in Figure 4.3.

If the errors are of unequal worth, then the decision makers would need to calculate the optimal slope on the ROC curve in order to maximize their utilities. There is a formula for making an exact calculation of the optimal slope for a given decision (Peterson, Birdsall, & Fox, 1954; Swets et al., 2000), but it would require specific values, perhaps actual monetary costs, associated with the decision errors. Often these numerical specifics are not available, and a decision maker may have to trade off a quantifiable monetary cost of one kind of error with the subjective cost of another.

4.3.8 Predictive power

Correct hits, misses, and false alarms can also be combined into metrics for positive and negative predictive power, which are useful for evaluating diagnostic decisions (Dillard et al., 2014; Swets et al., 2000; Szalma, Hancock, Warm, Dember, & Parsons, 2006; Warm, Finomore, Vidulich & Funke, 2015). *Positive predictive power* is the probability that a hit is correct; it is the number of correct hits divided by the sum of correct hits plus false alarms. For instance, we would like to know that if a diagnostic test or decision strategy finds the target or condition that it is meant to find, the odds are very high that the condition is really there and not a false alarm. Similarly, if the same test does not find the condition, then we would like to know if the condition is really not present, as opposed to saying the test was inconclusive and more tests are required. Thus *negative predictive power* is the probability that a "rejection" is correct; it is the number of correct rejections divided by the sum of correct rejections plus misses.

4.4 FUZZY SIGNAL DETECTION THEORY

Standard set theory and the use of Venn diagrams should be familiar to most readers. In standard set theory an object either belongs to a set or it does not belong. In fuzzy set theory (Zadeh, 1965), however, the boundaries of a set are indeed fuzzy; an object has a probability between 0 and 1 of belong to a set and can even have imperfect odds of belonging to more than one set. Fuzzy set theory has a history of applications in engineering, and its utilization bears a close resemblance to statistical methods for classification of objects into categories. It eventually did cross over into human factors applications in the form of FSDT.

In FSDT, the dichotomy of a signal being present or absent is replaced with a subjective rating of the probability (s) of a stimulus matching a target signal, with values closer to 1.0, or matching noise, with values closer to 0.0 (Masalonis & Parasuraman, 2003; Parasuraman, Masalonis, & Hancock, 2000). The response (r) can remain dichotomous with values [0,1] or be continuously valued, as when some signals are more urgent than others. In an illustrative example, an air traffic controller must keep planes 5 nautical miles (nm) apart horizontally, and 1 nm apart vertically (1 nm = 1.852 km). If two planes become too close together, a direction is issued to one of the pilots. If two planes are 4.9 nm apart, one plane is just barely within the window for a response, but if two planes suddenly become 3 nm apart, a response to that condition takes high priority. The evaluation of the severity of the situation could depend on close they are vertically as well as horizontally.

Hits, misses, false alarms, and correct rejections were valued [0,1] in conventional signal detection theory, but are now have partial values in FSDT:

$$\text{Hit (H)} = \min(s, r) \tag{4.6}$$

$$\text{Miss (M)} = \max(s - r, 0) \tag{4.7}$$

$$\text{False alarm (FA)} = \max(r - s, 0) \tag{4.8}$$

$$\text{Correct rejection (CR)} = \min(1 - s, 1 - r) \tag{4.9}$$

(Masalonis & Parasuraman, 2003, p. 1049). The rates for the four outcomes from a batch of stimuli are:

$$\text{Hit rate} = \Sigma(H_i)/\Sigma s_i \text{ for } i = 1 \text{ to } N \text{ stimuli} \tag{4.10}$$

$$\text{False alarm rate} = \Sigma(FA_i)/\Sigma(1 - s_i) \tag{4.11}$$

$$\text{Miss rate} = \Sigma(M_i)/\Sigma(s_i) \tag{4.12}$$

$$\text{Correct rejection rate} = \Sigma(CR_i)/\Sigma(1 - s_i) \tag{4.13}$$

(Masalonis & Parasuraman, 2003, p. 1050).

One motivation for reframing a signal detection task in FSDT form is to obtain a better understanding of the response to stimuli that are near the signal threshold. Near-misses would be classified as no-signal in the conventional approach, but would be recognized as such in FSDT. A trend that appears in the comparison of conventional and FSDT analysis is that FSDT produces lower hit rates and lower false alarm rates. Stimuli that are closer to the threshold can be ambiguous for multiple reasons. Luggage screening in airports is challenged by differences in luggage design, many possible shapes of several classes of threatening objects, and many forms of distracting objects. Signal detection ability is also influenced by time pressure and emotional context. Culley and Madhavan (2012) reported an experiment wherein 315 students were exposed to x-ray pictures of luggage in which they had to find one of five possible targets. Participants were exposed to the pictures for 2, 4, or 6 s. Prior to the main task, the participants were primed for an emotional condition of anger, fear, sadness, or a no-emotion control condition. The near-miss rate was significantly higher in the 2 s condition for all emotion groups compared to the longer exposures. Significantly different near-miss rates were obtained in descending order for fear, sadness, anger, and controls.

4.5 VIGILANCE

Watching radar screens and security camera feeds are prototypic vigilance tasks. Cybersecurity demands and new forms of automation with their ensuing reallocation of functions between people and machines have produced many new opportunities for more work in which vigilance is a dominant function. Vigilance tasks typically require operators to look at their visual displays for extended periods of time in search of threatening events that require remedial action. Performance on a vigilance task is usually evaluated in terms of correct hits, misses, and false alarms (Warm & Jerison, 1984), as previously discussed. Vigilance tasks are notorious for both boredom and stress; a good day on the job is when nothing important happens, and a challenging day is when a serious breach occurs. Vigilance tasks are often performed in conjunction with a second task, such as when a hospital nurse monitors videos from the patients' rooms while performing other tasks, or the watcher interacts with other people performing related security tasks. The soldier in Figure 4.7 is watching an array of security camera feeds looking for anything suspicious. He happens to be wearing an EEG helmet that is responding to the ups and downs of the workload he is experiencing; the role of EEG monitoring is described in later chapters.

An important, if not critical, concern in vigilance research is the origin, explanation, and control of a sharp decline in performance that often occurs

Figure 4.7 Operator performing a vigilance task while wearing an EEG headset.

within 30 minutes in laboratory experiments (Hancock, 2013; Mackworth, 1948). This effect might not occur for many hours in real-world situations, and is only rarely documented in industrial inspection studies (Drury, 2015; Hancock, 2013). The short-term decrement observed in laboratory studies can be attributed in part to high extrinsic motivation to maintain vigilance for a specified time with low intrinsic motivation; a life-threatening situation in the real world, however, would provoke different motivations. Other contributing factors include stimuli close to the threshold of discrimination, low frequency signals in a distracting background, and the overall lack of realism in laboratory display materials.

The performance decrement in laboratory vigilance is essentially a fatigue effect, defined as a decrement in work capability that results from extended time on task (Guastello et al., 2012; Guastello et al., 2014), not necessarily resulting from lack of sleep. Fatigue is addressed in a broader perspective in Chapter 10. In vigilance work, however, faster event rates (stimuli with and without a target signal), low signal probability, lack of knowledge of signal probabilities, and irregularity of the signals, produce larger performance decrements (Warm & Jerison, 1984). The performance decrement can be greatly reduced when vigilance tasks involve stimuli that reflect natural biological movements or 3-D stimuli (Greenlee et al., 2015; Szalma, Schmidt, Teo, & Hancock, 2014; Thompson & Parasuraman, 2012). Habituation, mindlessness (mind-wandering), and cognitive resource depletion are more possible explanations for the performance decrement (Dillard et al., 2014; See, Howe, Warm, & Dember, 1995; Parasuraman, Warm, & Dember, 1987; Warm et al., 2015), which can be reduced by rest periods (Ariga & Lleras, 2011; Thomson, Besnder, & Smilek, 2015).

4.6 MULTIDIMENSIONAL STIMULI

Conventional signal detection theory is rooted in the assumption that the signal has a 1-D characteristic for discrimination such as its brightness or loudness. Real-world stimuli, however, are seldom that simple. Brightness is only one characteristic of a light, which is its black-gray-white continuum. Lights also vary by hue (e.g., the chromatic colors that we know as yellow, red, green, blue, orange, and purple) and saturation (the relative amounts of chromic versus black-white content). We have also seen that lights are detected in a location, which thus adds one or more spatial components to the stimuli.

Sounds are not only loud, but they have pitch, timbre (set of harmonic frequencies that is associated with the fundamental frequency), and location as well. Loudness interacts with location such that our ability to locate the direction of a sound declines as the sound becomes louder (Soderquist & Shilling, 1990).

Taste and smell are also multidimensional. Although we seldom utilize these senses in PMSs, small differences in taste and smell are particularly important in food processing and perfume manufacturing. There are two notable exceptions, however: The gas utility companies put a chemical called mercaptan into the natural gas precisely so that we can smell it if gas is leaking in a system somewhere; natural gas is naturally odorless. The other exception is that if a machine system starts to produce an unusual odor, it is usually a signal that something undesirable is going to happen very soon.

In any case, it is possible to scale multidimensional stimuli. Multidimensional scaling techniques can determine how many dimensions are required to explain the apparent similarities and differences among various combinations of stimuli. They compute scale values on each of the resulting dimensions that look very similar to the unidimensional scales we produced earlier in this chapter.

4.6.1 Multidimensional scaling

One of the first multidimensional scaling experiments and statistical procedures used the spatial distance between U.S. cities as stimuli. Human participants were presented with pairs of major cities and asked to report how close they were in miles. All cities in the set were paired with all other cities in the set. The matrix of distances was then analyzed by principal component analysis. Principal component analysis is a close relative of factor analysis; it is an algorithm to determine the smallest number of underlying dimensions that account for the distances among objects (cities in this case). The final numerical result is an $N \times 2$ matrix of scaling coefficients, where N is the number of original stimuli, and 2 is the resulting number of underlying dimensions. Matrix entries are the scaling coordinates. In this example, the two dimensions corresponded to the geographic axes of North–South and East–West (Schifman, Reynolds, & Young, 1981).

If one were to do a similar experiment on chromatic colors, the stimuli would be the colors, which would be presented to the human participants in pairs. Participants would rate the pairs for how similar they appear to be. The results of the scaling analysis would produce a matrix of scaling coefficients in two dimensions. If we were to plot the pairs of coefficients for each stimulus we would obtain a circular arrangement, which is the familiar color wheel (Shepard, 1962), as shown in Figure 4.8. This is a favorite exercise for students to replicate. Multidimensional scaling programs can be obtained on popular statistical software packages.

After producing the set of scale values there may be further questions regarding how different the stimuli would appear to be. To determine the difference thresholds for, say, a pair of colored lights, a signal detection experiment would have to be set up in which the lights are presented in pairs of pairs. Subjects would then respond as to whether pair AB is as

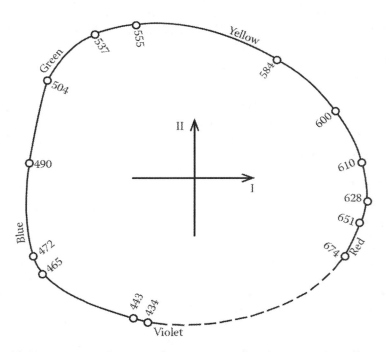

Figure 4.8 Multidimensional scaling of perceptual similarity among colors. (From "The Analysis of Proximities: Multidimensional Scaling With an Unknown Distance Functions II," by R. N. Shepard, 1962, *Psychometrika, 27*, p. 236. Copyright 1962 by the Psychonomic Society. Reprinted with permission.)

different as, or less different than, pair CD. The statistical analysis would be an extrapolation of the techniques already discussed; the interested reader should consult Wandell (1982).

4.6.2 Multidimensional nonlinear psychophysics

So far we have progressed from simple stimuli to multidimensional stimuli. Next we consider the dynamics of perception once the stimuli are perceived over time. The observer could be standing still while the stimuli move, or the observer could be moving through an environment where stimuli are occurring. It would be possible to invoke FSDT here where observers would rate the odds of an event based on their subjective notion of how often a stimulus appears in a series. The FSDT approach, however, renders the time series into static form, and thus does not capture the multistability of events over time or any temporal patterning. Furthermore, the ratings could be unduly influenced by illusory perceptions of connections between events over time (Botella, Barriopedro, & Suero, 2001). The illusory relationships are essentially autocorrelated errors. Both problems are handled particularly well with NDS models generally (Guastello & Gregson, 2011) and in nonlinear psychophysics specifically (Gregson, 1992, 2009).

Although the ogive curves drawn in Figure 4.1 have been assumed to result from a Gaussian probability function ever since they were introduced in the mid-1800s, they could just as readily result from a variety of NDS processes (Gregson, 2009; p. 114):

> It can be shown that the existence of a trajectory that begins close to an unstable fixed point ... is repelled from this point as [stimulus intensity] increases ... but then suddenly snaps back precisely to this unstable fixed point. This property of the dynamics is sufficient to imply chaos ... [I]t has [also] been shown that some neural networks can embody snap-back trajectories ... Bringing together these modeling features in one context is worth serious consideration in seeking dynamic representations of psychophysical stimuli.

It is noteworthy, furthermore, that the two ogives shown in Figure 4.1, when taken together represent hysteresis between the unstable (no signal) and stable (signal present) states, which further supports an NDS process taking place.

The theoretical development (Gregson, 1988) began with the limitation of preexisting theory to handle multidimensional inputs and outputs. He noted that olfactory stimuli, which are generally regarded as multidimensional, have a two-stage decay in effectiveness that is well known but not incorporated in standard psychophysical theory. The gamma recursive function was developed, therefore, to provide such an expandable model of psychophysical processes, which is based on the logistic map:

$$\Gamma : Y_{j+i} = -a(Y_j - 1)(Y_j - ie)(Y_j + ie) \tag{4.14}$$

(Gregson, 1992, p. 20). Equation 4.14 states that the strength of a response to a stimulus, Y, at time $j + 1$ is a function of the response at a previous point in time, j, a control parameter representing physical signal strength a, a situational control parameter e, and the imaginary number -1. Response strength might be measured as a subjective numerical rating, although the preferred method would be to measure response time. The shorter response times will be required for stronger stimuli. It is noteworthy that as the signal strength becomes sufficiently weak, the pattern of responses over time becomes chaotic.

Equation 4.14 can be expanded to two response parameters that may be useful for modeling cross-modality responses, such as interpreting a hue of a colored light from the perception of its brightness only, or studying the size-weight illusion. 2-D outcome expansion is accomplished by substituting a real number, x, for i:

$$X_{j+1} = a(e^2 - e^2 X_j + X_j^2 - Y_j^2 - 3X_j Y_j^2 - X_j^3) \tag{4.15}$$

$$Y_{j+1} = aY_j(-e^2 + 2X_j - 3X_j^2 + Y_j^2) \tag{4.16}$$

(Gregson, 1992, p. 27). Γ can be expanded further into multidimensional inputs and outputs that were applied to real situations such as perceiving the taste of wine. For superimposed stimuli Γ has the convenient property:

$$Y_{TOT} = \Gamma(a_1 + a_2) = \Gamma(a_1) + \Gamma(a_2) \tag{4.17}$$

in the parts of Γ involving real numbers (Gregson, 1992, p. 165). The presence of imaginary numbers in Gregson's equations may seem intractable for connecting psychological experience to real stimuli with real numbers. We saw already, however, that cross-modal perception can be tracked by replacing the imaginary number component with a second real component. In other words, the presence of imaginary numbers in the Γ function indicates that a second function *should* be present, even though we might be able to harness only one at a time. If the two stimulus parameters have different intensities, and the experimental effects e_i are also separate, one can expect further interactions between a_1 with e_2 and a_2 with e_1 (Gregson, 2009).

Every psychophysical process, according to Gregson (1992), is the result of at least two functions, one involving the proximal connection between stimulus and response, and one pertaining to the delivery of the stimuli over time. The vast majority of conventional experiments utilize fixed time intervals for stimulus delivery. No such regularity can be assumed in the real world. Rather, coupled oscillators should be prevalent. For instance, if a system consists of a chaotic slave, which is proximal to the perceiver's response, coupled to a periodic driver that delivers the stimuli, the overall pattern of response should appear as a periodic function that is peppered with bursts of chaotic behavior.

A plausible explanation for why stimulus perception (as opposed to stimulus occurrence) could be chaotic near the absolute threshold might be found in the neurological events that occur in the early stages of a perceptual process. According to Hess (1990), the neurons in the visual system are dedicated not only to specific tasks such as the encoding of brightness and color, but also of contours such as straight lines, curves, repetitive sequences, and so forth. If an incoming stimulus is just strong enough to activate neurons, it would follow that multiple recognition schemata could be activated at different moments in time, and the different cognition processes do not necessarily require the same amount of time to execute. The sensitivity to certain ranges of signal types becomes more acute if the human operator is charged with making distinctions among similar stimuli within that range. The many facets of a stimulus can be expected to interact in complex ways to produce a complex pattern of responses over time (Gregson, 2009).

4.6.3 Chaos and vigilance

Kern, Karwowski, Franco, and Murata (2018) investigated the temporal dynamics of performance on watch assignments on a U. S. Navy ship. The objective was to determine whether the performance decrement that is reported regularly in laboratory experiments bear any resemblance to the performance trends in a real vigilance situation. The study was done during peacetime conditions so that signals that suggested danger were not seen for days at a time. There were other types of events (e.g., weather) that needed to be watched and recorded every 30 minutes, however. The performance measure was whether the reports were made on time as expected. The watch spells ranged from two to six hours each.

Data were analyzed separately for high and low risk periods during the 24 h circadian cycle when sustained attention could be compromised by sleepiness, time on watch, and five ship locations in the Pacific Ocean where weather conditions and military demands could be different. Kern et al. found that the performance trends were chaotic, as detected by positive Lyapunov exponents, with less variability during the times of day when watchers would be most sleepy. This result is contrary to what would expect from conditions of fatigue (Chapter 10). Similarly, there was no performance decrement associated with time on the vigilance task, contrary to the laboratory environment once again.

Performance variability is an interesting phenomenon in its own right. It is explained in greater detail in Chapter 8.

DISCUSSION QUESTIONS

1. In 1986, a radar operator aboard the U.S.S. Vincennes noted a signal that he interpreted as an enemy missile. He responded immediately with the result that a civilian aircraft was blown out of the sky, killing hundreds of people on board. How could disasters like this one be prevented in the future? Hint 1: The representations of enemy missiles and civilian aircraft on a radar screen are very similar. The representations appear for only a brief period of time as the radar beam sweeps the sky. Hint 2: At least one of the possible remedies for this situation appears in a previous chapter.

2. The Air Force wants to know whether a new pilot–cockpit interface will help fighter pilots destroy enemy aircraft in air-to-air combat situations. How can signal detection concepts be used to assess pilot performance (in a simulator)? If training would help the situation, what indices should be compared before and after training? This was an actual problem that was reported in Eubanks and Killeen (1983).

3. A sophisticated printing company is in the business of printing wood grain designs on paper or vinyl that, after the print works are sold, are annealed to plywood substrates that are sold for decorating homes, offices, and furniture. The printing process involves four to eight separate colors. The original printing cylinder for each color is engraved separately and loaded onto the printing machine. After long-term use, the printing cylinders wear down and do not pick up enough ink; differences in print quality will eventually be detected. A problem situation arose in such a manufacturing operation when it was noticed that, for long print runs, the print quality had deteriorated so that the entire roll of printed work (anywhere from 5,000 to 20,000 feet) had to be destroyed. What is the signal detection problem here? How complex is it? What could be done to prevent this type of industrial waste? Note: The foregoing was an actual problem reported in Guastello (1982). Note that we are not considering any contribution of occupational stress at this time.

4. A medical school wants to train technicians to recognize the locations of cancers from magnetic resonance images. How can principles of signal detection be used to track the progress of the trainees?

5. Technicians use ultrasound equipment to detect possible cracks in air-craft wings. This method allows testing without dismantling the aircraft. Most of the results from ultrasound analysis indicate subcritical events or conditions in the metal. The technicians must make a judgment as to whether the conditions are severe enough to warrant taking a plane out of service to make a repair. What are the utilities associated with the judgment errors in the situation? How would the technicians set optimal thresholds or response criteria?

Chapter 5

Visual displays

5.1 THE SENSE OF VISION

Visible light is a relatively narrow band of electromagnetic energy. Its wavelength ranges from 300 nm for violet-colored light to 500 nm for red-orange light. The colors that people recognize as pure shades of blue, green, and yellow correspond to approximately 375, 525, and 625 nm, respectively. The color regarded as pure red is actually not on the visible light spectrum; it is produced as an additive mixture of red-orange and blue light. Additive and subtractive color mixtures are considered later in this section.

Figure 5.1 shows a diagram of the eye. Light, as reflected off objects in view, enters through the pupil, then is focused by the lens to project on the retina. The retina contains receptor cells. When light reaches the receptor cells, a chemical reaction occurs that is very similar to the chemical reaction on photographic film. As the receptor cells are activated by light, they stimulate adjacent neurons in turn. The stimulated neurons aggregate their signals in the optic nerve, and transfer the signal to the thalamus region of the brain. The thalamus is a relay center in the brain, and it is located in the central cerebrum below the cortex (part of the limbic system, located below the cortex). The visual signals are then transferred from the thalamus to the visual projection area in the occipital lobe (back of the head).

5.1.1 Visual acuity

The clarity of a visual image is the degree to which specific points on the retina are stimulated by the source image. The shape of the lens relative to the length of the eyeball affects visual clarity. Eyeglasses and contact lenses routinely correct deficits in clarity.

There are two different types of receptor cells, known as *rods* and *cones*. The rods have a lower threshold to respond to light, compared to the cones, and comprise the primary visual receptors in night vision or low light conditions. Rods are indifferent to color. Cones, on the other hand, do their best work in daylight conditions and are responsive to color.

DOI: 10.1201/9781003359128-5

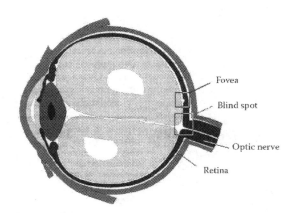

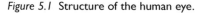

Figure 5.1 Structure of the human eye.

The *fovea* is a region of the retina that contains a high density of cones, relative to other sections of the retina. The *optic disk* on the retina is also known as the *blind spot*. It is the location where the optic nerve joins the retina, and there are no receptor cells in that particular region. Anyone who has learned to drive an automobile has learned the importance of the blind spot when attempting to change lanes. Once upon a time, automobiles did not have passenger-side mirrors. Drivers were required to turn their heads about 120 degrees in order to see if they were likely to hit another vehicle entering their blind spot. Passenger-side mirrors are helpful, although many drivers still rely on turning their heads if no danger is visible in the mirror. The mirrors often introduce a distortion in depth perception such that objects are somewhat closer than they might appear in the mirror. Turning one's head involves taking one's eyes off the roadway in front, which can be dangerous under some traffic circumstances, yet it is still the recommended response before changing lanes if the mirror looks as though the space is clearly open.

The quality of a visual image source can be defined in terms of the specific points within the image source that the viewer can discern. For instance, a common television screen is a matrix of 207 dots per cm (525 dots per inch, dpi). "High resolution" television is twice as dense. For the purposes of the following formula (Kantowitz & Sorkin, 1983) we only rely on the number of horizontal lines, 525 lines per inch. The quality of a visual image can be defined in terms of the number of discriminable lines, l:

$$l = 57.29 \, h \, N/R \, (fov), \tag{5.1}$$

where h is the projected image height, N is the number of lines inherent in the picture that is being displayed, R is the distance between the viewer and the target image, and *fov* is the system field of view, measured in degrees. The field of view is an angle created by the left edge of the image, the eye, and the right edge of the image. Its close relative, the field of vision, is considered later on in this chapter. It follows from Equation 5.1 that high-resolution television affords a clearer image at a longer viewing distance. For a period of time very large televisions (greater than 91 cm or 36 in diagonal) were popular enough to support their manufacture, but not universally so. Users placed them in rooms in homes that had shorter than optimal viewing distances. As a result, images projected from a 525 dpi source were too grainy for most tastes because *l* exceeded the line capability of the source and not necessarily worth the space in the room they would occupy. The combination of wall-mounted flat screens and high resolution source images seem to resolve the primary issues in image quality.

In an experiment with Air Force personnel viewing aerial photographs, Turpin (1981) systematically varied the relative amounts of blur and noise in the photographs. Noise affected image detection accuracy to a far greater extent than the blur.

5.1.2 Color vision

The cones are not all alike; they respond to different colors. Two lines of experimentation explain color vision: the trichromatic theory and the opponent process theory (Sternberg, 2004; Wandell, 1982). The trichromatic theory of color vision explains why primary colors are primary: Three wavelength thresholds have been isolated that correspond to red, blue, and green light. Thus the cones respond to wavelength at different thresholds.

The opponent process theory of color vision specifies that some cones are responsive to red and green, whereas others are responsive to blue or yellow. More precisely, one subgroup of cones is responsive to red or green interchangeably. If a particle of red light shines on that cone, it will respond as red, but if it responds as red, the same cone cannot respond as green. The adjacent red-green cone, however, can respond as green if green light hits it. The other subgroup of cones responds to blue or yellow. Unlike the red-green cone, however, it can respond as blue, but yellow is a default response from the result of "no red, green, or blue."

As a practical application, all chromatic colors can be decomposed as combinations of red, blue, and green light sources. The graphics industries have thus developed the four-color printing process, wherein any multicolor image, such as a photograph, can be reproduced as relative combinations of red, blue, green, and black dots of ink or light of different densities. The electron guns behind the television or computer screen produce a full-color image by projecting red, blue, green, and black onto the image shown to the viewer; this is the RGB (red, green, blue) color system.

Another form of support for the opponent process theory is the phenomenon of *negative afterimages*. The classic demonstration is to focus one's eyes for two minutes on an image of the U. S. flag that is presented in opposite colors, green, yellow, and black. Keep staring as the image is removed. One then sees an afterimage of the flag in the correct colors, red, white and blue. Prolonged activation of a color sensor in a cone triggers the activation of the opposite color once the stimulus is removed. The same process work spatially: An intense activation of red on one cone triggers a green response from the adjacent cone on top of whatever is supposed to be adjacent.

Graphic artists knew the secret of negative afterimages for years and could sharpen a multicolor painted image by applying a thin border of black between two adjacent color regions. As another example, the first color televisions were not universally accepted. Although price was a factor, many viewers preferred the sharpness of black-and-white to the fuzziness of color. The breakthrough came when Quasar (a television manufacturer) introduced an imaging technique that surrounded the projected dots of color with black, which negated the lateral afterimage. The result was a sharper color image.

There are two types of color-mixing processes, additive and subtractive. The additive case is exemplified by theater lighting, wherein white light is blasted through a colored gel and projected onto a target. If lights projected through red (actually reddish orange), blue, and green gels are projected onto the same spot on the stage, the viewer sees white. If, on the other hand, white light were projected through red, blue, and green gels (filters) simultaneously, the result on the stage would be black. An additive mixture of two opponent colors, such as blue and yellow, produces gray light. This is the basis of the RGB color system; a triple-screened light would be no light at all, hence it is not necessary to have a light generator specific to black.

The mixing of paints and dyes exemplifies the subtractive mixture. Each color in the mixture is actually a range of wavelengths. One range of wavelengths filters the other range of wavelengths. The only wavelength range that is allowed to pass is the color the viewer sees. Thus blue and yellow paint mixed together produce green. This is the basis of the CMYK (cyan, magenta, yellow, black) digital color system, which is intended for print, product packaging, or similar products for which the colors are produced by overlain inks. The tinted color elements produce more accurate renditions of the intended colors in a design or photograph.

Colors, more generally, are distinguishable in terms of three properties: hue, brightness, and saturation. Hue is the distinction among red, blue, green, yellow, and so on. Brightness is the relative black-white dimension, with all shades of gray located between the two extremes. Saturation is the relative of the hue plus the gray-scale dimension. Thus red plus white equals pink, and red plus black equals maroon. The K in the CMYK system is actually a grayscale axis.

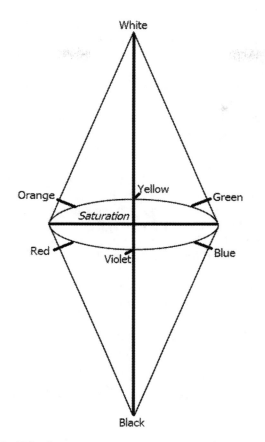

Figure 5.2 3-D Color Wheel.

The complete range of perceived colors can be represented as a 3-D display, shaped like two cones with the bases connected together, with the vertices at opposite poles (Figure 5.2). The cone bases in the middle correspond to the color wheel where the spectrum of hues is depicted along the outer edge. The brightness axis, which ends at the two points of the cones, is transverse to the color wheel. Highly saturated colors are positioned on the outer edge of the cone bases, and unsaturated colors are positioned closer to the central brightness axis. The color patch strips that are available in the paint departments of hardware stores are simply slices of this 3-D array.

5.1.3 Color vision abnormalities

Color vision abnormalities, commonly known as colorblindness, affect 8% of males and 0.3% of females. The most poignant deficit is where the individual cannot distinguish between red and green because of a missing pigment in the receptors. Blue-blindness is known, but it is relatively rare. More common, however, is the form of distorted color perception that is caused by deficits in the opponent process where color mixing takes place.

The Ishihara tests for colorblindness are often used to screen colorblind individuals for jobs where color vision is important. The stimuli are round stimuli that are composed of small round color patches of different sizes and tones. The colored patterns, which look a bit like mosaics overall, show a two-digit number as part of the pattern. The color-normal person sees one two-digit number, but a color-abnormal person sees a different two-digit number. There is also a version of the test that triggers either a recognition of the correct number or no number at all.

Ishihara screening has been used advantageously by the Air Force. People who fail the test fly as reconnaissance agents. Their task is to watch the ground for movements of enemy troops and equipment. Whereas the color-normal person would be fooled by camouflage, the Ishihara-blind person would see the disguised equipment or people clearly. As another example, colorblind people are often better than color-normals at detecting small woodland animals in a forest landscape.

Many industries that require tight controls over the color production of painted or printed products utilize Ishihara tests to screen out employees with the most blatant cases of color misperception. Industrial color matching, however, is subject to additional influences from the task itself and the environment that affect the accuracy of color matching by humans. These include habituation to a small color band, matching two or more color elements simultaneously in one display, time pressures imposed on the human to make the match, glare, and ambient lighting (Guastello, 1982). Additionally, computer-supported color-matching equipment can be foiled by the effect of the substrate into which the colored ink or paint pigment is suspended, and thus humans are required to make the final color adjustments to the paint or ink.

5.1.4 Photopic and scotopic functions

There are people who have abnormal color vision by the usual standards, but who can distinguish colors nonetheless. They do so by the relative brightness of the colors. In the case of traffic lights, the standard placement of red, amber, and green lights in the signal box also facilitates proper recognition of traffic lights.

People with normal color vision, however, also respond to the relative brightness of colors. In the case of lighted displays, it is also true that light bulbs of the same wattage will look brighter or darker depending on the color of the light. The *photopic function* is thus the effect of a color on subjective brightness of a light under normal daylight conditions. Its values, V_λ, can be found in tables, and range in value from 1×10^{-4} to 1.0, such that:

$$F = \sum P_\lambda \, V_\lambda. \tag{5.2}$$

In Equation 5.2, F is the total photometric intensity of a lighted display, P_λ is the physical intensity of the light (e.g., wattage), and V_λ is associated with a color (Kantowitz & Sorkin, 1983; Kaufman, 1974).

The *scotopic function* is similar to the photopic function, with the major exception that the viewer has been in the dark for 40 minutes before the color light stimuli were presented and judged for brightness. Its values, V_λ^i, range from 1.8×10^{-5} to 1.0, and they can be used in the same Equation 5.2 to calculate the photometric intensity of a lighted display when it is viewed at night.

5.2 PERCEPTION

Perception refers to how we make sense of a visual stimulus once the receptor cells and neurons have done their jobs. Psychologists have had some classic arguments as to how perceptual processes occur, but contemporary reasoning grants a lot of credibility to each of the major viewpoints. Helmholtz (1909/1962), the physicist who produced the first psychological theories of color vision and audition, argued that perception was a completely learned process.

Gibson (1950, 1979), on the other hand, argued that it would be a survival advantage for any organism to evolve with the ability to perceive the physical environment in as much detail as it actually contains. Furthermore, we have the ability to read the handwriting of someone whose handwriting we have never seen before. It must, therefore, be true that we have some innate capabilities and predisposition in our perceptual mechanism. This reasoning led to the discovery of *feature detectors*, wherein specific neurons or patterns of neurons in the optic nerve fire when exposed to distinct but elementary perceptual units such as vertical lines, horizontal lines, curves, diagonals, round things, lines that intersect, and so on. A sampling of elementary features appears in Figure 5.3. The correspondence between the elements and the capital letters of the Roman alphabet should be apparent.

Gibson (1950, 1979) also identified *invariant processes* in perception, which are basic processes of perception that are usually associated with depth and motion. The principle of size-distance constancy is an example. If an object occupies a small portion of our visual field, we regard it as small. If we know that it is truly a large object, such as an elephant, we interpret the elephant as located relatively far away. As the elephant gets closer to

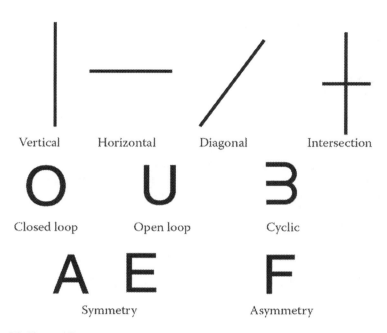

Figure 5.3 Feature Detection.

us, it occupies a greater portion of our visual field. Obviously the viewer must have a clue about the real size of the object in order to interpret its distance; fortunately, relative size is only one cue for distance perception. The presence of invariant processes, nonetheless, explains why optical illusions produce their illusory effects. In an illusion, some combination of invariant processes is invoked to induce a false conclusion about reality.

Kohler, Koffka, and Wertheimer, who were collectively known as the gestalt psychologists, also concluded that there are fundamental innate perceptual processes (Sternberg, 2004). The gestalt psychologists investigated the perception of form at a more holistic level than elementary feature detectors, which had not been discovered yet. In their reasoning, the brain (somehow) organizes the elements of a visual stimulus, and creates an interpretation of the whole image, hence the word *gestalt*. They argued that if we could understand how the brain organizes an image, we would be in a better position to understand how the brain works overall. In any case, their thinking led to the famous dictum, "The whole is greater than the sum of its parts," which actually had a longer history prior to cognitive psychology.

Hochberg, much later, introduced the constructivist perspective on perception that was, in a way, a throwback to some of the thinking of some 19th-century philosophers. In essence, our perceptions are not simply a view of what is presented, but a mixture of what is presented as it interacts with our past history. We store up elements that are experienced and learned in past history, which in turn assist in the perception of wholes even though we might be shown only parts of the complete image. From a human factors standpoint, Hochberg's thinking, at the very least, gives one more reason to be aware of the differences not only between novices and experts, but also among experts in different areas of expertise. When there is a wide variety of expertise within a population of system users, an image could readily produce different perceptions of the same stimuli (Bennett, Flach, McEwen, & Fox, 2015).

5.2.1 Gestalt principles of form

There are six basic gestalt principles to consider here: figure-ground distinction, similarity, proximity, good continuation, symmetry, and closure. *Figure-ground* distinction is the extent to which we can discern a visual target from a background containing it. Camouflaged objects are intentionally prepared to minimize figure-ground distinction. The vase-face perceptual illusion (Figure 5.4) that has become iconic in psychology works the way it does because the image equalizes the black and white portions of the image so that one can see a vase as the figure in one moment with the faces in the background portion, but the faces pointed nose-to-nose as the figure at another moment with the vase as the background.

Figure 5.4 Vase-Face Illusion.

The figure-ground distinction led to a concept of field independence versus field dependence that characterizes individual differences in the ability to separate figures from ground. The prototype measurement is the Group Embedded Figures Test (GEFT; Witkin, Oltman, Raskin, & Karp, 1971), which has a long history of use in research on "cognitive style." Field independence has gained some relevance for problems that are not literally perceptual in nature, but require the individual to identify relevant pieces of information for solving a problem from a mixture of relevant and irrelevant source material. Examples include solving chemistry problems (Stamovlasis, 2006, 2011; Stamovlasis & Tsarparlis, 2012), which require the chemist to isolate critical information from extraneous information and hold a number of pieces of information in mind when solving the problem. Field independent people are often better at solving problems in financial decision-making that require the investment analyst to isolate information that speaks to expected future value and the odds of future gains (Mykytyn, 1989; Guastello, Shircel et al., 2016). Witkin et al. (1971) noted, however, that the GEFT does not predict performance on all types of spatial perception tasks, only those that require the individual to isolate figure from a ground.

The remaining principles of form appear in Figure 5.5. The upper left display depicts the principle of *similarity*. The two groupings of dots are exactly the same in terms of their size, shape, and relative position. Their patterns of coloration differ, however, such that the grouping on the left induces the perception of rows, whereas the grouping on the right induces the perception of columns.

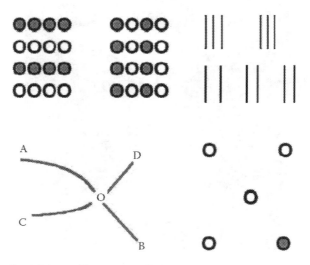

Figure 5.5 The Gestalt laws of form perception.

The upper right portion of Figure 5.5 depicts the principle of *proximity*. There are two rows, each of which contains six vertical lines. The relative proximity of the lines leads the viewer to perceive two groups of three lines in one case, and three groups of two lines in the other.

The principle of *good continuation* is analogous to a road map. If we were driving from Point A to Point O and wanted to continue on the same path, would we end up heading to Point B, C, or D? Most people would head toward B, as it makes the smoothest continuation from A.

The principle of *closure* is depicted in the lower right portion of Figure 5.5. Given the four unshaded dots, most people would place a fifth dot in the lower left of the image where the shaded dot now appears. There is a psychologically unfinished sense of the diagram without the shaded dot. Different graphic displays might produce different reasons why an additional graphic element would produce an experience of closure or not. In this case, the experience of closure relies in part on *symmetry*. People generally prefer designs that are symmetrically organized around a central axis rather than those that are not so organized. On the other hand, aesthetic judgments that are regarded as simple tend to be symmetric, whereas those that are complex embrace asymmetry. From a human factors standpoint, however, it remains to be shown empirically whether an asymmetric display or display layout are simultaneously more functional with regard to human performance criteria.

5.2.2 Depth

There are several means of perceiving depth. The first is *binocular disparity*, which involves the use of the two eyes. When a scene is viewed with both eyes, slightly different images appear on each retina. The two retinal images are similar to the extent that the portion of the scene is further away. Near-field portions of the image produce relatively different retinal images.

The first 3-D display technology appeared in 1847. The Brewster stereoscope facilitated the viewing of 3-D images by allowing the user to mount a pair of stereoptic photographs on one end of the device and look through the eyepiece at the opposite end. The photographs were taken with a double-lens camera that produced two images that corresponded to the binocular disparity that occurs when an image reaches the retinas of two eyes (Faris, 1997). A more contemporary version of the stereopticon (made of hard plastic) appears in Figure 5.6. The device was in use until after World War II. It became tedious to mount pairs of pictures that could be viewed by only one person at a time, when a simple photo album or projected slideshow were faster and more accommodating.

3-D photography resurged as a movie format in the 1950s. The viewer needed to wear disposable 3-D glasses in order to get the 3-D effects (e.g., vampire bats flying off the screen into the audience).

Figure 5.6 Stereoscope.

3-D photography resurged (slightly) again in the early 2000s as digital camera supported taking two pictures and holding the double image. Unfortunately the camera only had one lens, so the photographer needed to rely on personal judgment to position the separation of the two images. The viewer still needed 3-D glasses to get the effect. It does not appear that 3-D photographic skills become very widespread.

The remaining depth perception cues are *monocular cues*, meaning that it is possible to determine depth using only one eye. Thus it is possible for a person with only one eye to perceive depth. Depth can be perceived in terms of relative size. An object takes up a larger portion of the visual field if it is relatively close, and a smaller portion if it is further away. It is important to know the real size of the object in order to judge its distance correctly.

Linear perspective in drawings is the technique for slanting parallel lines so that the lines converge in the distance. Parallel lines, such as railroad tracks and straight roadways viewed in the near field look parallel, but they appear to converge in the distance at a *vanishing point* toward the horizon. Linear perspective combined with size-distance constancy explains the *moon illusion*: The moon appears to be its usual size when viewed overhead from earth, but looks comparatively huge when it seen rising over the horizon; the moon itself did not become any larger.

Occlusion is the result of one object blocking out part of another object in our view. The object that is blocked is regarded as farther away. The U. S. Department of Defense (2012) specifications warn against false occlusions in visual displays. Although they are not an issue with meters and numerical gauges, scenic graphics, which might be cartoonized in some virtual reality displays or synthetic vision systems, could provide opportunities for this type of design flaw.

Two objects may be positioned in a display to indicate different distances by placing one higher than the other relative to the bottom of the 2-D canvas. This technique of relative height is more convincing when the two objects are also drawn to different sizes in proportion to their alleged distances.

Texture gradients can be observed in visual images where there is a lot of repeating elements. For instance, if one were to view a field of corn, the corn closest to the viewer would be perceived as separate and relatively detailed stalks. The corn positioned toward the far field, however, would become progressively less distinct.

Lighting and shadows also provide depth information. Portions of the image that are meant to be closer to the light source appear brighter, whereas the portions that are receding from the light source appear darker.

Finally, *motion parallax* is a cue for determining depth while the viewer is in motion. Note that this is a depth cue, not a means of perceiving motion. If we were driving down the highway, the telephone poles at the edge of the road would whip by quickly, whereas the water tower in the distance would float by slowly. Thus the different speeds of movement of objects in our visual field convey signals of depth.

The U. S. Department of Defense (2012) specifications warn against mixed messages from conflicting depth cues. An example of a conflicting depth cue would occur if a dog was occluded by a house, but the dog was larger than the house. As the dog is behind the house, it implies farther distance; but because it is bigger, the dog would appear closer. Depth cues should represent natural depth; exaggerated depth cues should be avoided (p. 111).

5.2.3 Motion

The perception of motion requires a change in the perceived visual image. The first method is direct perception. When we are standing on the ski slope, the tree in our view becomes larger and larger as it moves toward us.

Stroboscopic motion involves the change in the position of a light source (Figure 5.7). The classic demonstration is to place a viewer in a darkened room with two lightbulbs positioned at a fixed distance apart. The experimenter then flips one light on leaving the other one off. Then the alternate light is turned on while the first light is turn off, and the process is repeated in alternation. The viewer perceives that the light is moving from one location to another.

The strobe lights seen in dance halls have the opposite effect because the light position does not move, but the people are moving relative to the light. As the high-intensity light flashes on and off at sufficient intervals, the viewer sees different views of the people in the room. The different views are sufficiently disjointed so that the viewer loses the continuity that is associated with motion. Movie film worked on this principle: The basic

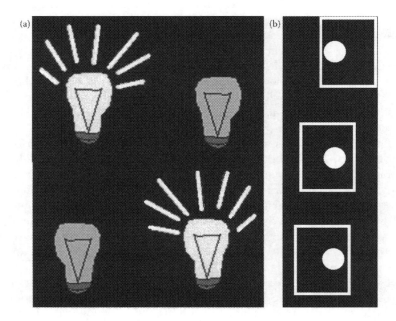

Figure 5.7 Motion perception: (a) Stroboscopic. (b) Apparent motion.

movie was a sequence of still frames that were rendered fast enough to provide the viewer with the same continuity of visual change as would be perceived from real-world events. Videotape turned the frame speed into a nonissue for a while. Digital video brought it back again when the transmission across the Internet becomes too slow due to high traffic volume and insufficient bandwidth.

Holding the light source steady and moving the contextual image around the light can induce *apparent motion*. The viewer perceives that the light is moving. A common example occurs when viewing the moon on a partially cloudy night. For a period of a few minutes, the moon is not moving its location in the sky, but the clouds are rolling by. It appears to the viewer, however, that the moon is rolling through the clouds—until the viewer notices that the moon is not really going anywhere.

5.3 PRINCIPLES OF DISPLAY DESIGN

Although it is convenient to describe classic display concepts one at a time, the real world at this time of writing is full of displays that are built on more than one principle. The designer's skill is required to pull the principles apart and detect which features of the display are desirable and which features could be improved. Similarly, it is tempting to discuss displays without any reference to controls, and that is the approach taken here so as not to confuse important issues. It should be remembered, however, that the control and displays often work together to produce a particular result.

5.3.1 Types of basic displays

Four classic display models are shown in Figure 5.8. The design in the upper left shows three system values in a piece of equipment located in an electric power plant. The needle on the left, which indicates a value of 65, displays a value that is currently in operation. The needle on the right, which indicates a value of 0, stays there until the operator turns the knob underneath. Turning the knob moves the needle upward to a desired value; the needle reverts to the 0 position when the knob is released. The third system parameter is the AUTO/MAN lights. The lights are mounted on the button itself. One would need to push the MAN button to allow the system

Figure 5.8 Common types of displays.

to accept a manual control operation. This display design was introduced in the early 1980s.

The display in the lower left is showing four values from related system areas in one display. Again the needle rides up and down over a convex drum. The relevant controls are not closely related to the display, which was designed in the mid-1960s. Other similar-looking designs have been built whereby the needle stays stationary and the numerical drum rolls under the needle. The convex drum in the two left displays predispose the operator to possible depth-perception errors while reading the displays if the indicated values are too extreme to the top or bottom of the display.

The display in the lower right is a circular meter wherein the needle crosses a flat calibrated background in a circular fashion. This design does not invoke the depth-perception problem, but it does require more physical space on the display panel than do some of the drum-type displays. It is susceptible to parallax effect (see later discussion) if the operator tries to read the display from an angle instead of head-on.

The display in the upper right uses a lighted bar to indicate the system value that is currently in operation. The electronic display method obviates the need to roll the needle over a convex drum, thus reducing the chances for errors of depth perception and parallax. The numerical value is also displayed digitally on the top of the display. This display design was placed into the power plant system in the early 1990s.

5.3.2 Display design criteria

Inevitably a designer would ask, "Which type of display is best?" The short answer is, "Pick the right tool for the right job." Testing is required to determine which method of display would work best in a given situation. Theoretical considerations include not only the properties of human vision, but also the context in which the device is used. Are there other controls and displays that go with it or surround it? How many other tasks are going on at the same time?

The top criteria for evaluating a display design are: (a) speed of interpretation or reading (b) accuracy of interpretation or reading; (c) speed of learning; (d) comfort; (e) absence of fatigue with long-term use; (f) small differences in human performance that can be traced to differences in the populations using the device; and (g) the stability of performance under poor environmental conditions or stress (Kantowitz & Sorkin, 1983). Fortunately, the displays that can be read more quickly are also those that can be read more accurately. Thus it is not always necessary to sacrifice one criterion in favor of another. Some of the more important aspects of display design that translate into performance outcomes are considered next.

5.3.3 Visibility

There are three aspects to visibility. The first is image clarity, which was considered earlier. Image clarity is perhaps most critical to photographic and computer-projected images.

The second aspect of visibility is the location of the display within the operator's field of vision. Figure 5.9 illustrates that the normal line of sight is not directly horizontal from the viewer, but is oriented at a 15° downward slope (Figure 5.9). The optimal orientation of a control panel would be a 45° angle to the angle of vision. The latter is the same as a 30° angle relative to the directly horizontal line of sight (from U.S. Department of Defense, 2012).

Next, consider a panel of displays facing the operator. The most important displays should be centrally located within the field of vision, and the less important displays located on the periphery. (The determination of "important" is another matter, which is considered in a later section of this chapter on display panel organization.) The amount of space within the central location depends on whether the operators are allowed to move their eyes only, turn their heads, or whether they are expected to turn their heads and roll their eyes. The horizontal field is ±15° to the left and right of direct horizontal viewing for eyes only, ±65° for turning heads, and ±95° for eye rolling and head turning (Figure 5.10).

The vertical latitude for the field of vision is also shown in Figure 5.10. If only eye movements are allowed, the accepted lower limit is 15° below the normal line of sight (30° below direct horizontal), and 40° above the normal line of sight. If head movements are allowed, the range is between 65° above

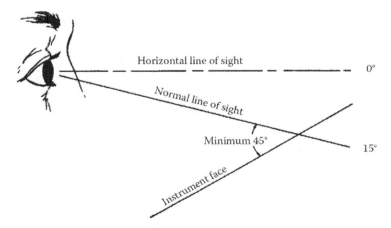

Figure 5.9 Line of sight. (From U.S. Department of Defense, 2012; in the public domain.)

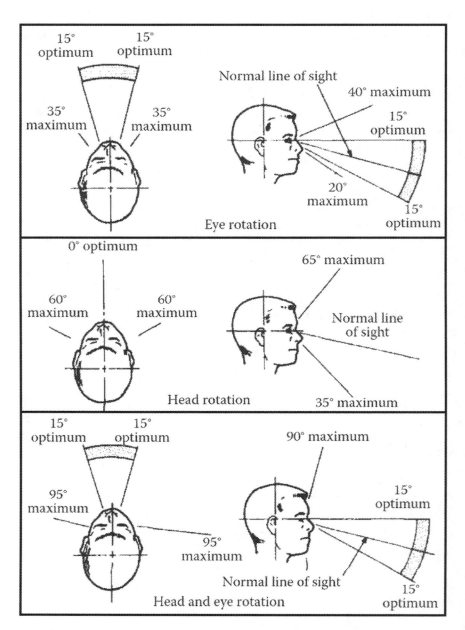

Figure 5.10 Field of vision. (From U.S. Department of Defense, 2012; in the public domain.)

the normal line of sight and 35° below. Note, however, that if head and eye movements are both allowed, the acceptable range is between 90° above the normal line of sight (75° above direct horizontal), and 15° below the normal line of sight. In other words, heads are allowed to turn upward and back to center, and eyes are allowed to roll downward and back to center.

Even in the case of a single display, the question of whether an operator could see the display could depend on the location of the display in the field of vision. Figure 5.11 shows two aberrant examples. On the left, a lonely display is mounted in a busy visual field in which the operator must walk up a ladder to see the meter reading. On the right, the meter is located within the vertical head-turning field, but it is upside down. This example of 1935 technology could not be installed right-side up because the flow of steam within the pipe flowed from ceiling to floor.

5.3.4 Distinguishability

Distinguishability is whether the important elements of information that the operator is supposed to capture from the display are sufficiently clear. Figure 5.12 illustrates the importance of making the tickmarks on an analog display distinguishable from each other. The U. S. Department of Defense (2012) standards specify heights, thicknesses, and spaces between tickmarks. Actual sizes in millimeters depend on the viewing distance.

Demand is the extent to which the display is indicating a serious condition in the system. Aircraft and environmental control systems often specify "alarm" conditions. Alarm conditions need to be easily distinguished from "normal conditions." Flashing red lights have good recognition value for conveying alarm.

Figure 5.13 is a reconstruction of a "starfield" display from a computer-based system for which this author was a part of the design team. The dots in the display indicate different sensor activity in five areas of the system. The central axis represents set points, and deviation from the center indicates the change in value for the sensor point over a period of time. Different colors were used to denote the separate parts of the system. An operator who desired more information about a sensor's activity could then click the dot and additional information would pop up. Alarms were initially denoted as white. The design recommendation was to make the alarms larger and to make them flash. Flashing white against a dark background was a good combination for that application. Military equipment standards, however, specify flashing red for situations where emergency action is required (U.S. Department of Defense, 2012).

Designers need to give some consideration to what value of a system function constitutes an alarm condition. Ideally, an alarm should mean that the operator must take action immediately or close to it. When alarm

(a)   (b)

Figure 5.11 Two displays with visibility problems.

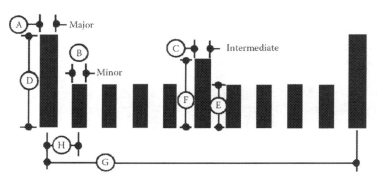

Figure 5.12 Analysis of tickmarks for display readability. (From U.S. Department of Defense, 2012; in the public domain.)

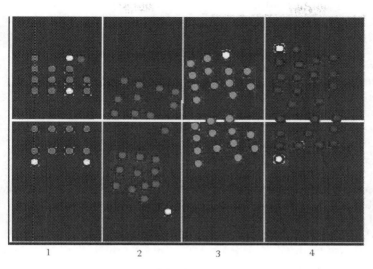

Subsystem

Figure 5.13 Prototype "starfield" display.

criteria are set too low, the operator's likely response is to reset the alarm to a higher value. If the response to reset the system comes too often or too easily, the operator runs the risk of missing an important condition in the system (Chapanis, 1975).

Military equipment design guidelines recommend that visual alarms be designed so that an alarm means only one condition. The nature of the condition should be apparent to the viewer so that the viewer is not required to read other displays in order to interpret the nature of the emergency (U.S. Department of Defense, 2012).

5.3.5 Interpretability

Interpretability is the extent to which the operator can view a display and get a sense of what the information means and what to do about it, if anything. This aspect of display design gave rise to *smart displays* in computer-based systems. Traditional displays indicate an aspect of the system from the system's point of view—its machinery, its levels of one thing or another. Those displays typically did not convey any direct information to the user regarding what to do with that information.

A smart display, however, will internalize some of the operator's decision-making strategy and produce information in a form that coincides with the decisions that the operator is trying to make. For instance, a military pilot needs to recognize another aircraft as friendly or unfriendly, its type of attack power, and whether it is in the firing envelope. A relatively dumb display might show that another aircraft exists within radar-detectable limits. A smart display will provide signals regarding the type and friendliness of the aircraft, and code its proximity within the effective firing envelope (Post, Geiselman, & Goodyear, 1999). This is a complex signal detection and perception problem that can now be assisted by the machine, rather than done by the human alone.

A smart display relies on a system of meanings that are shared by the population of operators (Bennett & Flach, 2013). If the situation is [X], what would the operators do in response, and why? Are multiple scenarios and outcomes possible?

Of course, knowing what to display in any form is a major step forward. The Model T Ford had no machine displays, thus no dashboard, and no fuel gauge! One can only imagine the joy of discovery when the system designers figured that out. Speedometers were yet to appear as well; perhaps they were not necessary in the days when traffic kept pace with the horses, and the horses were not equipped with speedometers either.

Similarly, there were airplanes before there were display panels. The phrase "flying by the seat of your pants" meant that pilots relied on the physical contact between themselves and the plane to determine changes in altitude. The altimeter was invented eventually.

5.3.6 Completeness

Completeness of a display denotes whether all the information that an operator requires and that is meant to be provided by the display is actually provided. A simple case would be to have a full range of numerical values shown if indeed all those values were both possible and meaningful (Kantowitz & Sorkin, 1983). The range of values shown on a display is the *space* of the display.

For many years, automobile speedometers were calibrated from 0 to 120 mph in the United States. A federal law was passed in 1975 making 55 mph the speed limit on interstate highways. The motivation of the law was to reduce gasoline consumption and highway deaths. Gasoline engines ran most efficiently at 55 mph at the time, which was a concern for reducing carbon emissions and improving air quality. The reduction in highway deaths due to speeding was a secondary benefit. Some automobile manufacturers responded by changing the speedometer so that "55" was placed about two thirds of the distance from one end of the meter to the other, with the upper limit shown as 85, instead of 120. When the car exceeded 85 mph, the needle continued to spin past the visible upper limit. The national speed limit was later rescinded because of the increased cost of traveling slower across large expanses of the country combined with improvements in engine efficiency.

5.3.7 Parallax effect

Parallax effect is the extent to which a reading of a display becomes distorted if the operator is viewing the display from an angle instead of head-on. Parallax effects are likely when the display involves the use of a needle that rides above the calibrated face panel, as in the example in Figure 5.8.

A common example of the parallax effect occurs in automobile speedometers. The driver sees one speed, but the front seat passenger sees another and asks, "How fast are you going? It sure looks faster than that from here." The design solution, which may have been inadvertent, was to shape the dashboard so that the passenger could not see the speedometer.

5.3.8 Color and contrast

Color is sometimes helpful, but there is a temptation for designers to apply it frivolously. For the most part, higher contrasts between the image components will produce better human performance, particularly among older users (Kantowitz & Sorkin, 1983). Black and white elements have the largest contrast. Color should be used only if it conveys actual meaning. The starfield display attempted to use color to represent components of a system.

Yet another speedometer story is applicable here. The speedometer in the 1960 Oldsmobile was laid out in a horizontal line from 0 to 120 mph. The indicator was a light that expanded from 0 to the indicated speed; this arrangement is not too different from the lighted display in Figure 5.8, except that it was horizontally oriented and changed colors. The indicator light, however, changed color from green to orange at 35 mph, and from orange to red at 65 mph. The contrast between the green and orange against the dark gray background were distinctive and perhaps pretty equivalently so, but the red might have blended too easily into the background. This was not a particularly good design choice for high-speed travel in a car that was equipped with an overdrive transmission (some models only). The contrast problem of red against a dark gray background resurfaced in other types of LED display technologies.

The tricolor speedometer design did not seem to have lasted long enough to support published usability studies. The population of manufacturers and drivers equilibrated on the semicircular meter and needle format that is still in common use. The meter and needle actually take up less space on the control panel, leaving more room for the diagnostic lights that were added gradually over time such as *check engine, low oil, change oil, seatbelt unfastened*, and *door ajar*.

5.3.9 Historical displays

Historic displays tell the operator something about the activity of a system over a period of time. Figure 5.14 shows two classic designs. In the upper left, a calibrated paper is placed over a drum. A pen marks the numerical value of the system as it changes over time. The paper in this example is changed every 24 hours, stored, and possibly examined later. A possible limitation to easy reading is that the single line of information is wrapped around a 3-D drum, thus requiring a bit of depth correction by the operator who is trying to read it.

In the upper right of Figure 5.14, a three-color pen assembly is recording values of a system on a rotating disk. The radius of the circle distorts the scale of the values; it is not possible to put all four indicators on the outer edge. On the other hand, the depth-perception conversion is not necessary, and the new values are centrally located on the display.

The disc design for a historic display is not appreciably different from the tachometer recordings on delivery trucks that were used in the mid-1970s. A tachometer would show the speed of the truck over an eight-hour period. Someone reading it could tell when the truck stopped and for how long while making deliveries. If a fleet of trucks was involved, which was usually

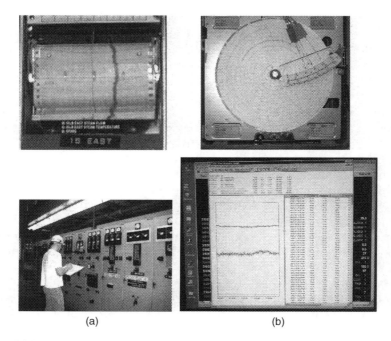

(a)   (b)

Figure 5.14 Historical displays.

the case, one would have to have the drivers' itinerary in hand to interpret the starts, stops, and movements. The truck dispatcher or operations personnel would have the itineraries handy anyway if the goal of collecting tachometer data was to match the truck's travel history against the expected routes.

In the lower left of Figure 5.14, a power plant operator is recording system values on a clipboard for other operators to examine later if necessary. This procedure, dated to the mid-1960s, was obsolesced by computer-based data-acquisition methods. An example of a computer-based display is shown in the lower right of Figure 5.14. The screen facing the viewer gives both a graphic and a numerical history of a system sensor. The historical information screen is a pop-up window that is accessed by pointing a mouse cursor to a line of information about a sensor on an alarm screen. The alarm screen, which is occluded by the historical display in this photograph, uses color coding to denote the levels of criticality associated with the sensors.

5.3.10 Predictive displays

A predictive display conveys the state of a system at a future point in time. Predictive displays are often necessary in slow-acting systems, such as the steering mechanisms of ships, in which there is a substantial delay between the time of the control action taken by the operator and the time the result is achieved. Slow-acting systems may induce the operator to make the control operation a second time and thus do too much of an otherwise good thing. The prototype predictive display in Figure 5.15 depicts a change in a ship's heading once a control action (change in direction) has been taken; it is a generic display design and not one in actual use.

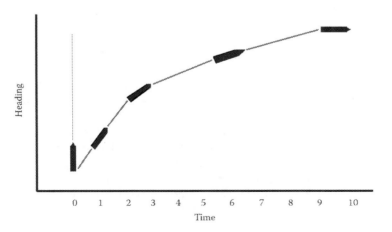

Figure 5.15 Basic format for a predictive display.

Predictive displays are also helpful in fast-acting systems, for example, aircraft, in which the available time for the human to make a forecast of the system's future behavior is much shorter than what would be required for a mental forecasting operation. Human computation of rates and acceleration are not particularly swift or sufficiently accurate (Yin, Wickens, Helander, & Laberge, 2015). Thus the temporal aspects of a predictive display design should be consistent with the time horizons of the events that are portrayed in the displays and the control decisions that the operator needs to make. The best-case scenarios are those in which the control action can be made precisely enough on the basis of the displayed information so that further tweaking of controls is kept to minimum, thus saving time for other actions.

The temporal resolution of the historic and predictive data also requires some fine-tuning for the variable the operator wants to control. High resolution displays collect and report data over very small time intervals. Although the high resolution gives the appearance of being very precise, it shows actual and forecasted values gyrating up and down very quickly. The operator needs to cut through the jitter mentally to arrive at a usable value, such as a moving average, for control purposes. The tech solution is to apply a filter to the numbers before presenting them in a display.

Resolution that is too low can be problematic if meaningful changes in events are occurring faster than the display. For instance, in a study of human forecasting and prediction that involved non-simple changes of events over time, the participants who used a 3-point moving average for their forecasting heuristic made more accurate forecasts than those who used an 8-point moving average (Guastello et al., 2021a). Yin et al. (2015) compared user performance on several variations of predictive display designs that were intended to control a manufacturing process. They found that the midrange resolution produced the best human performance; higher resolution helped to a lesser extent because of the trade-off between meaningful and irrelevant changes in the metrics.

5.3.11 3-D displays

Artists have been able to deliver high-fidelity renderings of 3-D information on a 2-D canvas for a very long time; the skill set expanded during the Renaissance. The advent of photography in the 19th century made another leap forward in this regard, as did movies and television nearly a century later. One must not forget the use of sculpture, which dates back millennia. An actual 3-D display, however is not a rendering of an image to a 2-D medium; it is a rendering of a 3-D image. For instance, the Brewster stereoscope and movies that allowed the viewers to experience bats flying off the screen into their faces serve as the prototype.

The technology for generating 3-D displays in line-graphic or photo-graphic form became available in the late 1970s. A review of the first stage of research indicated that 3-D displays were better than 2-D for displays involving terrain maps and perhaps other real-world situations where the original source is 3-D (Getty, 1983). 3-D imaging reappeared in the 1990s when computer display graphics technology improved and new needs were identified. For instance, projections of the human brain can be compiled from a functional magnetic resonance image (fMRI). They are 2-D displays in the sense that they convey depth through monocular depth cues rather than binocular disparity, although they are generated from true 3-D source information. Nonetheless, 3-D monocular images have considerable value in biomedical applications and elsewhere.

3-D displays also reemerged in the 1990s in conjunction with virtual reality applications where the user is actually immersed in the visual image, and not just looking at it (see Chapter 11). Other technologies have become available since that time that are intended as supplements to virtual reality construction. Although some require special glasses, one technological goal is to make the glasses unnecessary. Potential applications to toys and games proliferate. Work-oriented applications are often centered on terrain-based applications, such as aviation guidance, oil and gas exploration, telerobotic control, and visualization of room interiors.

There are some known assets and limitations to 3-D displays, according to Park and Woldstad (2000). On the one hand it is possible to perceive features of the image that would not be apparent in a 2-D rendering. On the other hand, perceptual ambiguity along the line of sight has been documented. The interpretation of one visual feature may be distorted by variations along a different visual parameter. Some of the perceptual ambiguity can be corrected, however, by adding graphic enhancements to the 3-D display such as calibrations and other symbolic markers.

On the frontier of 3-D visualization technologies, *volumetric displays* (will eventually) allow users to walk around the display and view all sides of the object (Figure 5.16; NEOS Technologies, 2004). The object is projected into the center with a system of laser beams (Nayar & Anand, 2007, 2013). A possible application could include air traffic control, according to NEOS Technologies, where controllers could visualize airplanes swarming within a contained simulated space. Some prototypes made by different manufacturers currently exist, with miniature buildings, toy characters, human heads, abstract geometric objects, or luminescent bunny-rabbits (not joking) projected into the display bulb as demonstrations.

Another format that is undergoing active experimentation would project the object into a free-standing space. In this format, a person could grasp an object that looks as though it is present but really is not, enjoy a hologram

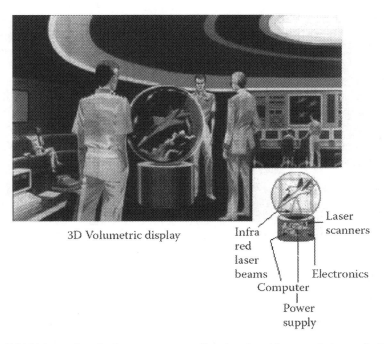

3D Volumetric display

Laser scanners

Infra red laser beams

Electronics

Computer

Power supply

Figure 5.16 Volumetric display prototype. (Reprinted with permission of NEOS Technologies.)

stage show, or encounter a salesperson that pops up out of nowhere in front of them and give a sales pitch.

The technology is still too raw to support meaningful usability studies. If the goal were to portray real-world events as they occur, such as air traffic, the sensor system would need to be very sophisticated, and the time that would be required to compute the holographic image might be too long to be useful. For the pop-up people paradigm, is the user the sender of the message, or the unsuspecting shoppers who could run into a pop-up salesperson everywhere they turn?

That said, the pragmatic advantage of a volumetric display would be to help users to discover or navigate through 3-D environments. For instance, fish find their way through volumes of water to reach food or other goals manipulated by the experimenter in a 3-D maze (Holbrook & Burt de Perera, 2013). Humans are usually surface-bound when navigating environments, although there are exceptions such as flying aircraft. Volumetric displays could have their biggest advantage for application where movement through 3-D environments is critical (Pastukhov, Lissner, Füllekrug, & Braun, 2014).

Other researchers are exploring the possible advantages of volumetric displays in medical imagery (Boer et al., 2018). The viewers are not immersed in the environment (unlike the characters in the movie *Fantastic Voyage*), but they could derive some advantage for targeting locations and neural tracts that require medical attention.

5.3.12 Digital versus analog displays

Perhaps the best-known example is the clock. There are the round ones with the hands, and the varieties that specify the time in digits. Which is better? If it is necessary to see how close one is getting to a critical time, such as the hour or half-hour, then the round clock is usually preferred because it automatically makes the subtraction. If the specific time is needed, however, perhaps in conjunction with programming an appliance, then the digital design is preferred. As a more general rule, analog displays convey a "big picture" of the situation that can be ascertained quickly; digital displays convey specific information that might require an extra second or two to decipher (Carswell & Seidelman, 2015). The example display in Figure 5.8 utilized both analog and digital representations of information; perhaps the operator relied on both forms of representation for different tasks. From another perspective the redundancy improves reliability in reading the display.

Developments in computer graphics in the late 1990s have made a significant impact on the design of airplane cockpits. In these applications, the digital construction is not confined to showing the digits; rather, it refers to the mechanism behind the production of the display. Figure 5.17 contrasts a strictly analog cockpit design with another that is equipped with digital displays (from NASA, 2004). According to NASA, the new digital display designs allow the pilot to zoom in on legends and details for better legibility. It is possible to produce graphic information as well as numeric information. Color coding can be utilized to distinguish different types of related information such as flight paths and traffic flows.

5.4 HEADS-UP DISPLAYS

Machine displays often compete with the real world for the operator's attention. There is a limit to how much attention a driver can divert from the road, or the pilot from the skyscape. Hence, the heads-up display (HUD) puts critical information in front of viewers' eyes at all times though use of a specially devised helmet and visual screen. Alternatively, the heads-up image can be projected onto the windshield of the automobile or airplane. The electronic trick, however, is to project the image in such a way that the operator does not lose the sense of depth and spatial information that

Figure 5.17 Analog and digital cockpit display panels. (From NASA, 2004; in the public domain.)

usually comes in from outside the windshield. In other words, HUDs can produce attentional tunneling, which must be assessed in any HUD design (Tufano, 1997). When properly projected, however, they can reduce the ocular readjustments that the pilot makes when viewing a close-up control panel and a full-depth view through a windshield (Fadden, Ververs, & Wickens, 2001). A windshield-mounted HUD for pilots is depicted in Figure 5.18 (NASA, 2004).

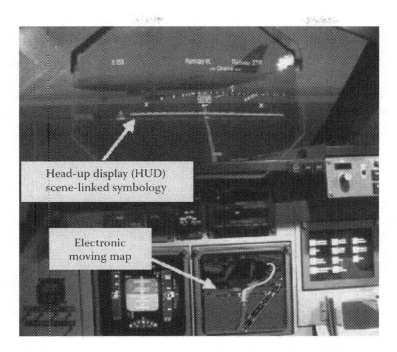

Figure 5.18 Heads-up display for aircraft windshield. (From NASA, 2004; in the public domain.)

5.4.1 Clutter

There is a trade-off between the clutter that occurs when too much information is placed in the heads-up display and the scanning-time cost associated with the heads-down (standard control panel) display. Although many HUDs that have been tested survive the trade-off, far-field vision becomes compromised when clutter becomes too great (Yeh, Merlo, Wickens, & Brandenburg, 2003). The difference between valuable information and clutter depends in part on what is task-relevant (Hammer, 1999).

Clutter, which can be a problem with both heads-up and heads-down displays (e.g., a standard display panel), can arise from several sources. One underlying theme is the *data-ink ratio*, which Tufte (1983) introduced in conjunction with print documents. Too much graphic embellishment or decoration detracts from the interpretation of the actual information that needs to be conveyed. A related problem is *numerosity*—having too many tickmarks labelled separately when long and short tickmarks with fewer anchors and labels would do the job relative to the decision that the operator needs to make (Carswell & Seidelman, 2015; Yeh et al., 2003). Do we really need to know or interpret all the gradations between a soon-to-die car battery, the good-enough range, and an overcharged (however that happened) range?

A second type of clutter comes from placing two displays with graphic similarities too close together so that the two displays produce conflictual meaning. This type of clutter arises from the gestalt law of proximity, which can be advantageous in some situations, such as the design of icons, but a distraction in heads-up displays if the wrong combination of items appear too close together. A good rule of thumb is that non-connected display items should be at least 1° of the visual field apart (Carswell & Seidelman, 2015). Icon design is described further in the next section of this chapter.

Other sources of clutter come from the general disorganization of the display, too many different functions and display formats crammed into a small visual space, poor distinguishability between target and background parts of visual images, and the task-relevance of the units of information within the display. These sources of clutter are unpacked in the next section on display panel organization. Overall, clutter from any of the foregoing sources would increase the cognitive workload required to process the displayed information, which could delay response time (Kim, Kaber, Alexander, Stelzer, Kaufman, & Veil, 2011; Moacdieh & Sarter, 2015). The amount of clutter could be detected by response time to a cue that involves searching a screen, the impact of display organization on performance indicators, subjective ratings, and eye-tracking techniques.

Multifunction displays can reduce clutter in a control panel (NASA, 2004). It remains an open question, however, whether stacking information from two or more individual displays into one display is always beneficial; see the section of Chapter 8 on "mode error."

5.4.2 Design interactions

Researchers have been experimenting with interactions between some of the newer display media and the heads-up environment. Lisle et al. (2019) found that volumetric displays in an automobile heads-up display improved distance perception compared to more traditional heads-up designs. Kim and Kaber (2014) compared synthetic vision displays terrain information based on geographical positioning systems with enhanced vision displays based on infrared sensors, and found that they had counteracting effects: GPS-type displays "increased overall [situation awareness] but degraded flight path control performance because of visual confusion with other display features. [Infrared-type displays] increased flight path control accuracy but decreased system (aircraft) awareness because of visual distractions" (p. 386).

5.5 ICONIC DISPLAYS

Visual icons are pictorial display elements, usually small relative to the total display panel or computer screen. They gained special notoriety in the early days of GUIs because the little pictograms represented a button that could be clicked with a computer mouse to enact an operation of some sort. Icons also get special attention because the pictures themselves are not conveying the type of data one gets from charts, graphs, maps, diagrams, or status lights that blink on and off. They have a symbolic or metaphoric value that conveys an additional layer of meaning regarding what the image represents. As such they have instant recognition value for system users who might be familiar with similar functions on different systems or trying to find a particular function in a menu of options. A menu of icons from a graphics program appears in Figure 5.19. Note that auditory icons have also come into use; their story is explained in the next chapter.

Figure 5.19 Icons from a graphics program.

5.5.1 Symbolic versus concrete pictograms

One of the first practical research questions was how to design a set of icons that do the intended job. Some pictorial representations of objects might be found in industry standard glyphs that are already in use in some domains, but their effectiveness requires some empirical validation. Guastello, Korienek, and Traut (1989) compared users' responses to icon designs for four industry vocabularies: building automation, finance, data processing, and engineering. There were five types of icons for each object: a short verbal abbreviation, a longer abbreviation or full word, the industry standard pictogram, a custom-made pictogram that was a more concrete representation of the object than the industry standard, and an icon design that included a concrete pictorial representation and a short verbal abbreviation. Human participants were presented with a real word for the object and asked which of the five icons was closest in meaning to the target. An example of a set of experimental icons appears in Figure 5.20.

Brain lateralization theory suggested that the mixed-modality icons would be more meaningful than the other types. The human brain contains two cerebral hemispheres. One side contains the verbal-processing center, and the other side contains the picture-processing center. The two sides of the brain work in parallel, although a particular person might be faster at processing pictures than processing words. With parallel processing for everyone, however, all system users would have access to their faster response times, assuming the verbal and pictorial content was compatible and not conflictual. The principle of redundancy was also applicable: Two brain mechanisms in operation are better than just one.

The results of the experiment showed that the mixed verbal and pictorial icons were rated as the most meaningful by the research participants. Although meaningfulness is not a terminal performance criterion, it is well known that meaningful stimuli are stored and retrieved from long-term memory more efficiently than nonsense stimuli (Chapter 7). The relative advantages of concrete pictures with the longer abbreviations were equivocal over the four content domains; this combination of elements soon became common practice (Galitz, 2003). The mouse-over technique

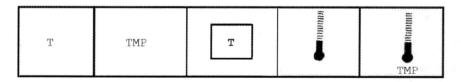

Figure 5.20 A sample visual icon for a temperature sensor evaluated by Guastello, Korienek, and Traut (1989).

quickly became available for computer interfaces, however. It allows a more comprehensible label to pop up, usually in a yellow box, to explain the meaning of the icon, thus permitting simultaneous verbal versus pictorial processing.

Some pictures are more concrete than others. In a later examination of performance on an interface with pictorial icons, Blankenburger and Hahn (1991) found that pictograms that were more concretely representative (their phrase was "small articulatory distance") of the target produced more efficient actions with regard to speed and accuracy. There is no rubric to tell us how concrete the icons need to be in a given context without running some tests with human operators.

Air traffic control systems are currently undergoing redesign to accommodate data communication links from the airplanes to the control systems. Air traffic controllers in different regions of the country do not always use the same system software, hence different sets of icons are in use. Thus Ngo, Vu, Thorpe, Battiste, and Strybel (2012) considered the question of how any industry standard pictograms should carry over to the new systems. The icons sets they tested tended to be more symbolic than concrete. Air traffic controllers were more likely to associate a double arrow pointing upward or downward as a symbol of increasing and decreasing airspeed, and there were a few other consistent associations with symbol shapes and commonly used pieces of information. Novel information units germane to the incoming systems, such as whether the airplane had particular data link equipment, did not have consistent icons. Color was good for distinguishing flight rules associated with an aircraft, but did not affect any other information parameter, nor did open versus filled shapes. Text icons were best for conveying heading and several other information parameters.

5.5.2 Emergent meaning

As mentioned earlier, too many pieces of information rendered in different formats can produce clutter instead of meaning. Additionally, clutter can also undermine the operator's ability to obtain a system-wide view of the system's functioning. To counteract this problem, a new class of iconic display has come under investigation that integrates information elements into a single composite display in such a way that the elements produce an emergent picture of the system's operation (Bennett & Flach, 2011, 2013; Bennett et al., 2015). A successful design will not only meet the traditional requirements of reading speed, reading accuracy, and interpretability, but also facilitate problem solving.

Two examples appear in Figure 5.21. Parts (a) and (b) are renditions of an iconic display for a power plant showing the flow of electricity in eight directions. If the display looks like (a), then the system is operating properly

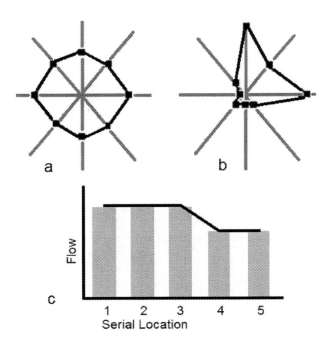

Figure 5.21 Iconic displays for power distribution and pipeline operations.

with power flowing in all directions at appropriate levels. If the display looks like (b), however, then a serious imbalance is occurring with load shifting from some directions to feed high demand from other directions; some functional areas could be on the verge of a blackout.

Figure 5.21c illustrates a rendition of an iconic display that was intended as a monitor for the flow of oil or gas through a pipeline that has several sections with sensors organized in a series. If the horizontal bar across the segments is flat, then flow is consistent throughout the pipeline. If the bar is bent as shown, the configuration would indicate an obstruction or a leak.

Figure 5.22 depicts a different example that did not work out as planned. The three bars are in a configuration that represented a well-operating system. The faint line connecting them is only present in the figure to illustrate the conjoint meaning of the bars that the operator perceives. The experimental icon was a reconfiguration of the three data points into a triangle that would become distorted if the system was operating suboptimally. Users' accuracies in reading the displays strongly favored the bars, which was the display already in use. The triangle required too

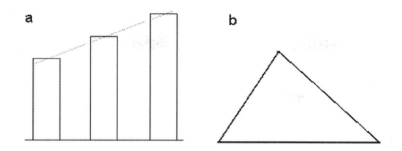

Figure 5.22 Triangular iconic display.

much additional effort to interpret and mentally compare against the optimal configuration.

Both icon designs in Figure 5.21 made use of the gestalt law of proximity to produce an emergent level of meaning. Elements that are positioned close together are seen as a whole. In addition (a) and (b) invoked the principle of symmetry, and (c) invoked the principle of good continuation.

5.5.3 Shared meaning

According to Bennett and Flach (2013; Bennett et al., 2015) the success of an iconic display relies in part on a concept of the PMS that is broader than the traditional one-person-plus-one-machine. The person is part of a larger system that includes industry standards and other sources of shared meaning that the icon designer can exploit. A good example of where the lack of shared meaning produced an unworkable icon design occurred in an attempt to configure several pieces of system information into a cartoon of a human face. The idea was that each informational element contained several discrete steps, and each element would map onto a shape for an eye, eyebrow, or mouth. If the system was happy the face would smile. But what would it mean if the system gave us a smile and a wink? Unfortunately, the participants in the usability experiment could not keep track of which condition or sub-condition contributed to which part of the facial configuration; the mapping of data elements to facial elements was too arbitrary, and the interpretation of facial expressions was not shared by the users to any sufficient level.

5.6 DISPLAY PANEL ORGANIZATION

It is now possible to state some broad rules for organizing a panel of displays (Kantowitz & Sorkin, 1983). The first is to organize the displays according to importance. How critical or central is the information from a display to a task that the operator is trying to perform? "Importance" is a vague term, however, although it is usually equivalent to frequency of use. The most frequently used, or otherwise important, displays should be positioned centrally in the field of vision. The constraints for the field of vision were defined earlier in this chapter.

Second, displays should be organized by order of use. If a task involves several sequential steps, the arrangement of displays should follow the same order.

Third, displays that are used together should appear together. For instance, in the case of the set of 2-D displays that convey a 3-D image as mentioned earlier, we would not want to separate the set of 2-D images.

Fourth, displays should be grouped by function. If a combination of displays pertain to one function of the system, they should not be mixed up visually with displays pertaining to another function of the system.

5.6.1 Linking analysis

Frequency of use, order of use, and togetherness can be determined empirically through linking analysis. One would need to set up a study, perhaps with eye-tracking equipment, whereby the movements of the operator are examined as the operator views one display after another while performing a standard task. How often does the use of one display follow the use of any other display? The most frequently appearing combinations of displays should be given the most central location on the control panel. The displays that are used less often should be located more toward the periphery, and in reasonable proximity to the displays to which they are sometimes linked.

Eye-tracking equipment requires head-mounted sensors that follow saccadic eye movements. Sample output appears in Figure 5.23. The operator is actually driving an automobile. If all the displays of concern appear on a computer screen, eye-tracking equipment is now available that can capture eye movements without a head-mounted device. A sensor is placed at the bottom of the computer screen. The operator stares at it until the sensor calibrates the eye positions. It then follows the eyes as they move around the displays. The movement information is stored on a second computer and does not interfere with the test display or related operations.

5.6.2 Ecological interface design

The smart displays that were introduced earlier become ecological interface designs when they are organized into a display panel, usually with some controls involved (Rasmussen & Vincente, 1989). They are "ecological" in the sense that they adopt principles from Gibson's (1979) ecological psychology. The design begins with a clear understanding of what the user's

Figure 5.23 Saccadic eye movements while driving a motor vehicle. (Reprinted with permission of BMW, Inc.)

environment contains and why any of it matters. The person is not a passive viewer, but is actively seeking answers to questions that matter for work-related objectives. Perception should be direct, meaning that the information portrayed is analogous in meaning to the real-world situation, and it should be actionable without any mental computation or interpretive analysis between the perception and action. Information should be formatted so that changes in a domain property correspond directly to what the viewer can observe. The overall organization of the display panel should correspond to the user's mental model of the actual work situation. The interface is still organized by function, but *whose* function? The operator's functionality is what matters (Bennett & Flach, 2019).

When all the parts of the interface meet the foregoing requirements and organized relative to the operator's intent, there should be an emergent picture of the system's properties taking form. The whole is greater than the sum of its parts in much the same way as the elements of an iconic display contribute to the emergence of the iconic quality of the display. (The emergence principle is actually a Gestalt idea.)

Bennett and Flach (2019) reviewed 37 studies in which user performance was evaluated for an ecological interface and compared to an alternative design format. Positive results favoring the ecological design were reported for 30 examples; six were neutral, and one was negative. The one negative case was later redesigned to enhance its ecological properties and became another strong positive.

5.7 SIGNS OF IMPORTANCE

Signs are static visual displays. The concern here is for their information-conveying quality, and not for the impact of decorative signs or advertising logos. There is a line of reasoning, nonetheless, that aesthetics affect usability because an aesthetically attractive display induces the user to pay more attention to the sign or device.

5.7.1 Design

Pictorial signs are particularly useful in multilingual environments such as a European railroad system. A good design will convey a concrete noun or verb rather than an abstraction. The design should capitalize on figure–ground distinction and the ease by which one sign image is distinguishable from another.

Verbal signs for instruction or emergency information should be prepared with clear and unembellished lettering. Sans serif type styles are preferable to styles with serifs. A comparison of four type styles is shown in Figure 5.24 that featuring the lowercase letter *r*. The example designated at 1 is Roman-style type such as commonly used in book print. The serif is the little flanges at the endpoints of the letters. The serif in 1 is circles. Example 2 is Helvetica type, which is a popular choice for figure captions and headings in book print as well as for informational signs; there is a

Figure 5.24 Serif on typographical characters.

family of related type styles that are equivalent with regard to their lack of serifs, hence sans serif.

Example 3 is a version of the famous Western (U.S.) style of type used for the word WANTED in a Wanted poster. It still attracts attention today, perhaps more because of its nostalgic reference than because of its clarity. Example 4 is a decorative sans serif type; because of its lack of simplicity, it is not recommended for informational or emergency signs. Further digressions into the dynamics of type styles are better reserved for the experts (e.g., Spencer, 2004).

In any case, the width-height ratio of lettering on informational signs should stick close to a 3:5 ratio. The signs should be visible from a realistic distance, which means that they should be large enough overall. The ambient lighting should be sufficient to read the sign.

5.7.2 Standards

Some standards exist in the United States for certain types of signs placed in public situations. The American National Standards Institute (ANSI) has devised a standard sign for *DANGER* advisement. The danger sign appears in Figure 5.25. The upper portion is a black field with a white oval encompassing a red oval. The red oval contains the word *DANGER*. The lower portion is a white field with the name of the danger in black lettering. If additional explanations are needed, they should be placed in black on a white field to the right of the main part of the sign.

An American National Standards Institute (ANSI) standard caution sign also appears in Figure 5.25. It is characterized by a bright yellow triangle containing a large exclamation point that appears to the left of the word *CAUTION*. It is followed by an explanation in black on a yellow or white field. The glyph with the triangle and exclamation point carries over to instruction manuals and related texts. An older style caution sign that might be found on some equipment still in existence is a yellow field with thick black spines around the perimeter. The text of the message is displayed in black beginning with the word *CAUTION*.

Occupational Safety and Health Administration (OSHA; 1996a) adopted standards for danger and caution signs in the workplace. The basic shape is square with a black border. The danger sign has a wide red stripe across the top with the word *DANGER* written in white. The description of the danger is written in black on a white field in the lower portion of the sign. The caution sign has a bright yellow stripe on top with the word *CAUTION* written in black. Again the nature of the warning is written in black on a white field in the lower portion of the sign. OSHA also has companion designs for safety (with a green stripe) and notice (with a blue stripe) signs.

The motivation behind the designation of standard sign formats is to condition an instant recognition of a danger situation on the part of the viewing

Figure 5.25 Danger and warning signs: danger sign in a public situation, currently used caution sign in a public situation, earlier design for a caution sign in a public situation, danger warning on a product label.

public. Other regulating entities have developed sets of standards as well. The U.S. Department of Transportation (2003) issued revised guidelines for the design and production of highway signs. Not only do they follow standard production formats for symbols, colors, numerals, and verbiage, but there are also specific sizes required for different roadways depending on the travel speed. The viewing distance must therefore be long enough for the driver to perceive the sign, register its meaning, and execute an appropriate action safely.

OSHA or ANSI designs are not always sufficient for warning labels that are placed on consumer products. Nonetheless, the standard designs are recommended for their recognition value. An example of an ANSI danger warning from a 6-foot stepladder also appears in Figure 5.21. Note the use of pictorial information that is placed in the box for additional explanations. Note also the angular drawing with heavy lines and the world-famous "do not" ring. Caution and danger warnings on consumer products play an important role in the threat of product liability litigation, as mentioned in Chapter 1.

5.7.3 Behavioral impact

Although voluminous research was performed long ago on the legibility and intelligibility of signs and labels, there is comparatively little information about their behavioral impact. One classic study that was conducted in a British steel factory (Laner & Sell, 1960) indicated that safety signs (with pictures) that stated the proper safe behavior were more effective at reducing a particular type of accident than signs that contained only generic safety awareness information. The rule of stating the proper safe behavior carries over to roadway safety signs also (Lonero et al., 1994).

In broader situations, people who might be prone to risk taking are making a trade-off between the benefits of breaking the rule and the likelihood of a dangerous outcome. The benefits of breaking a rule, or *counterdemand*, may be matters of convenience such as getting a job done faster, minimizing physical discomfort, or avoiding the task of figuring out an alternative method for performing the task.

5.8 ILLUMINATION AND GLARE

Proper lighting is required to view any visual display or sign or to find the correct controls. The light may be part of the display itself, which is a good idea if the display is used in an environment that is sometimes dark. Environmental lighting in the room should be sufficient for normal working conditions.

5.8.1 Illumination

Lumens are the measurement of the total amount of light given off by a source in all directions; lumens are closely related to the wattage of a new lightbulb. *Candelus* is the measurement of light in one direction of the light source. *Illuminance*, or light reaching a surface of an object, is measured in *footcandles*, which is the number of lumens per square foot of surface.

The *IES Lighting Handbook* (Illuminating Engineering Society, 1981) gives recommended lighting levels for a variety of work situations. In general, if there is relatively less contrast in the coloration of the work source or relatively small objects are involved, then relatively more ambient lighting is required (Kantowitz & Sorkin, 1983).

It is possible to conduct signal detection studies to determine the minimum lighting needed to read a display or do a piece of work. Sign legibility is optimal, however, at luminance contrasts of 10:1 between the target letters and background. Legibility actually tapers off by a small amount at greater contrast ratios (Sival, Olson, & Pastalan, 1981).

Lights that are not intentionally manufactured to produce colored light often produce almost-white light with hues of yellow, blue, or pink, depending on the materials involved. These trace amounts of color present little problem in most types of work, but an industrial color-matching task could be severely affected by the exact type of lighting used. If the manufacturer of a color-specific product and the customer do not agree to the reference lighting standard that they would be using, disagreements can ensue regarding the quality of the product.

Urban streetlights often give off a colored glow. Darnell (1997) reported street lighting in San Jose that was so yellow it could be confused with the amber light of a traffic signal. He wrote, "Imagine you are driving along at night and don't notice the yellow lights ahead are yellow traffic lights. You might be very surprised when they turn red."

5.8.2 Glare

Glare is the amount of light deflecting from a surface into the viewer's eyes. Discomfort glare is the amount of subjective annoyance experienced by the viewer. Disability glare is the reduction in visibility cause by the deflecting light source. Deflection patterns differ depending on the smoothness of the deflecting surface.

Glare can be reduced by placing the light source high over the viewer's head and behind the viewer. If the glare source is located too close to the viewer's line of sight and is located between the viewer and the surface, disability glare increases sharply (Sanders & McCormick, 1993). Typical solutions involve moving the work station relative to the ceiling lights, changing the location of the light sources, and working with glare-controlling surfaces.

Some standards from the DOD that would carry over to many other environments include: (a) "Surfaces adjacent to the display screen shall have a matte finish." (b) "The luminance range of surfaces immediately adjacent to display screens shall be between 10 percent and 100 percent of screen background luminance." (c) "In vehicles, ships, observation structures, and other spaces where users may need to be dark adapted (e.g., cockpit, ship bridge, vehicle cabs) the interior walls or bulkheads shall be a matte black. This will minimize specular reflection from interior light sources."

DISCUSSION QUESTIONS

1. Mirrors are often used to provide a display in locations where the actual events are not visible. When are mirrors valuable? What types of human perception problems can result from the use of mirrors?
2. Besides clocks, what other types of information can be represented with both digital and analog displays? When would one type of information be preferred over another in those situations?
3. Design a prototype experiment that would determine if there is too much clutter in a visual display. Describe some display design techniques that would reduce clutter if it were necessary to do so. Do the clutter-reducing designs suggest the potential for other types of display or control problems?
4. Suppose we had a group of displays like the one shown in Figure 5.26. Percentages can be assigned to the links in the diagram, which represent the proportions of times the operator views one display after viewing another. How should the display panel be reorganized if the operators spent 48% of their viewing time with C, 10% with F, and 26% with E? What if the operators spent more time viewing E after viewing A more often than they spent viewing B after viewing A?
5. Many late-model automobiles are equipped with an analog meter-type speedometer and a digital speedometer. Which one do you use more often?

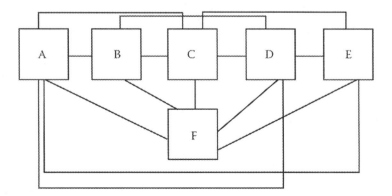

Figure 5.26 Hypothetical set of displays and links.

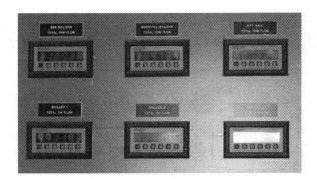

Figure 5.27 Display panel for a set of HVAC chiller units.

6. What would be some plausible new uses for volumetric displays once the technology has been perfected?
7. Figure 5.27 depicts a set of displays for chiller units in an HVAC system. What is wrong with this picture? (Hint: The photo was taken without a flash.)

Chapter 6

Auditory and tactile displays

6.1 SENSE OF HEARING

Sound is a vibration that creates disturbances in air pressure. Its three parameters are loudness, pitch, and timbre. A diagram of the interior of the human ear appears in Figure 6.1. The sound vibrations enter the side of the head where the cotton swabs usually go. It travels down the auditory canal until it reaches the eardrum. The eardrum vibrates, and in doing so transfers the vibrations to three little bones known as the hammer, anvil, and stirrup. The stirrup connects and transfers vibration to the cochlea. The cochlea is filled with fluid and contains the basilar membrane. The basilar membrane contains the receptor cells, which are also known as hair cells because of their antenna-like hair that protrudes and detects the exact nature of the incoming vibration.

The frequencies of the incoming sound are processed on the basilar membrane in terms of both the threshold of frequency that triggers a cell's activation and the location on the basilar membrane that is responding. High frequencies are registered in the area of the cochlea that is closest to the connection to the middle ear; low frequencies are registered further away. The basilar membrane does not contain enough cells to permit a one-for-one correspondence between all the shades of frequency that a person can hear and particular cells. Thus volley theory is the prevailing explanation for how we can hear so many shades of frequency: The particular frequencies we experience are a result of a pattern of cells firing; the specifics of the pattern and the pattern recognition process have not conclusively determined yet, however.

The responses of the hair cells are gathered together by ganglia in the cochlear nucleus and transferred to the auditory nerve, which connects to the auditory cortex in the temporal lobe. The auditory cortex is conveniently located close to the thalamus and speech interpretation area of the brain. It consists of three components: the core, the belt, and the parabelt (Snyder & Alain, 2007). The belt surrounds the core, and the parabelt is off to one side of the belt. The three components of the auditory cortex

DOI: 10.1201/9781003359128-6

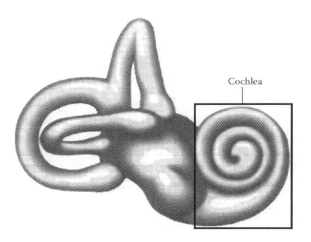

Figure 6.1 The inner ear.

are responsible for the early, middle, and late stages of auditory processing respectively. Sounds are encoded by frequency when they leave the cochlear nucleus, and frequency information is processed first. Later stages process harmonics (see below), temporal patterns, and location.

There are connections running in both directions between each two of the three components. There are further connections in both directions to the cortex and thalamus. The cortex contributes interpretations of the sound and directs attention to the incoming sound source. The thalamus coordinates sounds with visual cues and psychomotor responses.

6.1.1 Loudness

A sound is a wave, not unlike the one shown in Figure 6.2. A wave has an amplitude, a frequency, f, and a wavelength, λ, such that $f = c/\lambda$, where c = speed of sound, or 720 mph. The amplitude of the sound wave produces loudness.

Physical loudness of a sound is measured in decibels (dB). A whisper produces a sound of roughly 20 dB. A typical residence with the television off generates sound of about 30 dB. An office is usually producing sound in the range of 40 to 55 dB. A noisy city intersection is closer to 70 dB. A train produces sound close to 100 dB in the near field of the track. Aircraft typically produce sound in the 125 to 130 dB range.

If we were to ask people to increase the loudness of a standard tone until it is "twice as loud as" a reference tone, they would give us an increment of about 10 dB. The logarithmic relationship between physical loudness

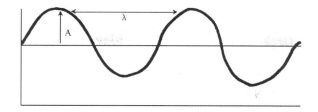

Figure 6.2 Basic sound wave showing amplitude (A) and wavelength (λ).

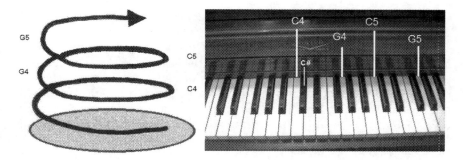

Figure 6.3 Pitch helix based on Deutsch (1992) and Shepard (1982).

and psychological loudness should be apparent. Psychological loudness is measured in sones, *S*, such that:

$$S = 10^{(0.03dB-40)}$$

(6.1)

(Kantowitz & Sorkin, 1983).

Our perception of loudness, however, is dependent on the frequency of the tone. Tones with high frequency have shorter wavelengths, and thus do not travel as far and fade out a bit before they reach our ears. An alternative measure of psychological loudness is the phone. It is measured the same way as the sone, except that it is figured relative to 1,000 Hz (cycles per second).

6.1.2 Pitch

Pitch is related to the frequency of the sound wave. The tone known as A above middle C on a piano (Figure 6.3) is tuned to 440 Hz. An A that is one octave above the reference A is generating a frequency of 2 × 440, that is, the reference frequency is multiplied by a factor of 2. An A that is one octave below the primary A is generating a frequency of $\frac{1}{2} \times 440$, that is, the reference frequency is divided by a factor of 2.

In Western music an octave is divided into 12 intervals of equal size, known as semitones. Thus the frequency of any note is the 12th root of 2 (approximately 1.06) times the frequency of the note immediately below it. (Note that this explanation is an oversimplification for some musical instruments. The note known as G-sharp is the same note on a piano as the note known as A-flat, but on a violin, the G-sharp is a little sharper and the A-flat is a little flatter than the piano rendition of the notes.) Eight of the twelve semitones comprise the common diatonic musical scale. Although the remaining four semitones may be used generously in a particular musical composition, they are interpreted by the listener as members of a different scale, or "accidentals."

Music of other cultures may divide the octave into more intervals or fewer intervals. A guitar allows us to bend the strings to reach the quarter-tones between the semitones. The point for our purposes, however, is that listeners have expectations of what tonal sequences are considered pleasant, unpleasant, familiar, or surprising. These expectations may carry over to the interpretation of auditory displays that are based on tonal sequences.

The perception of pitch has a well-known quirk. If we were to present listeners with two tones an octave apart, they would have little difficulty determining which tone was a higher or lower frequency. At the same time, they would recognize the two tones as the same note. If we were to present the listeners with two tones that were just one or two semitones apart, they would also detect which had the higher frequency without a great deal of difficulty. On the other hand, if we were to present two tones that were five semitones apart (or a half-octave) confusion would set in as to which was the higher note (Deutsch, 1992; Shepard, 1982).

The *pitch helix* (Figure 6.3) depicts these pitch–perception relationships. Although there is a psychological linear relationship between actual and perceived pitch from one octave to another, there is a nonlinear involution in the middle of an octave. Thus auditory display systems should not require the operator to make any important decisions based on some of the possible combinations of tone differences. Tonal patterns should contain more blatantly different pitch elements.

To make matters more interesting, if we generated two tones and captured the resulting sound wave on an oscilloscope, the resulting wave would take on a complex form that would be a merger of the two fundamental waves. The listener would hear not only the two fundamental tones, but also a third tone that is the sum of the two frequencies, and a fourth that is the difference of the two frequencies. The foregoing does not yet include the influence of harmonics.

6.1.3 Timbre

Why does the same musical note sound different on a violin, flute, or trumpet? One might immediately point to the materials used to make the instruments and the shapes of the instruments. Indeed shape and material are relevant because they produce different series of harmonic tones. *Harmonics* are integers multiplied by the fundamental frequency to produce additional frequencies. Different musical instruments produce different harmonics and emphasize them to varying extents. *Timbre* is the set of harmonics associated with a sound source.

The harmonics of various musical instruments are heard together in the form of different characteristic sound waves; some examples appear in Figure 6.4. Flutes produce sinusoidal sound waves. String instruments produce saw-toothed waves. Brass instruments produce square waves. Percussion instruments have noninteger harmonics.

The wave in the lower right of Figure 6.4 is a generic sound wave that forms the starting point for digital sound synthesis. The generic wave is divided into eight steps. The eight steps are allocated among three parts of the wave: the attack, the sustain, and the decay. Each step can be manipulated for relative amplitude and relative pitch. In this way any musical instrument can be mimicked in principle, and unique sounds can be generated as well.

Even with a well-designed interface, the demands on the user for sound wave knowledge and memory can be challenging. Thus the second gener-ation of sound synthesis equipment made much greater use of sampling techniques. A recording of the sound would be entered into the machine. The machine would take a statistical sample of the wave and reproduce it. Higher fidelity replications require higher rates of sampling.

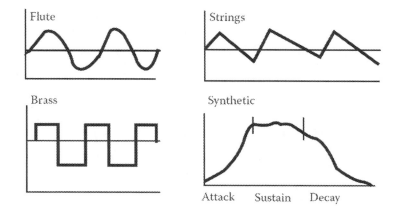

Figure 6.4 Typical waves for musical instruments.

6.1.4 Binaural hearing

The direction of a sound is detected by virtue of having two ears. The process is akin to binocular disparity in depth perception. The sound reaching each ear is not quite the same, and on the basis of that difference, the listener can locate the sound. A directional sound reaches one ear about 5 ms before it reaches the other ear. The sound waves reaching each ear thus become slightly out of phase, but only enough so as to allow the detection of direction. If the delay should become as long as 20 ms, the listener will perceive two separate sounds (Kramer, 1994).

Regrettably, humans' detection of sound sources is not entirely accurate. Vertical direction is more difficult to determine than horizontal detection. To make matters more challenging, sounds in the real world bounce off sides of buildings or mountains, or bounce off interior walls and wood floors. Thus the listener may be getting sound from reflected sources rather than original sources, or from both sources at the same time. Anyone who has ever driven an automobile through a city and wondered where the police siren was coming from can vouch for the complications that could exist in locating a sound source. Contemporary concert halls, on the other hand, are engineered so that almost every listener in the audience will receive the complete sound generated from the stage with a minimum of distracting reflected sounds and with minimum loss of sound information in any one seat (Kantowitz & Sorkin, 1983).

Recorded sound during the first half of the 20th century was produced monophonically, meaning that even high-fidelity recordings gave the spatial impression of coming from a single hole in the wall. High-fidelity recording allowed the reproduction of a wider range of frequencies and harmonics, which were intrinsic to the development of stereophonic sound. Stereophonic sound, which was developed in the 1950s, required at least two sound sources, for example, two microphones strategically positioned in an array of musical instruments. Different sound patterns reached each microphone. Two speakers were required for proper playback. The result was that the listener could detect where the sound elements were located in 3-D space. For any room equipped with a stereo system, there is an apex point in the room where the detection of full stereophonic sound is optimal.

6.2 NONVERBAL AUDITORY DISPLAYS

Auditory displays can be conveniently grouped into two types: those that contain speech and those that contain only nonverbal sounds. The nonverbal types will be considered first because they are more common in conventional PMSs and because the basic properties of the nonverbal displays carry over to the understanding of what makes a speech display work well.

According to the classic wisdom (Deatherage, 1972, p. 124), auditory displays are preferable to visual displays when the message is simple, short, or when the operator will not be required to refer to it later. They are particularly good for describing events in time or for calling an operator to immediate action. They are viable options when the visual system is degraded or overburdened. They can reach an operator who must move about continually.

In contrast, visual displays are preferable to auditory displays when the message is complex or long, or if the operator must refer to it later. Visual displays are preferable for conveying information about the location of objects in space and for when an immediate action is not required. They are preferable to auditory signals when the auditory system is degraded or noisy. Visual displays assume that the receiver remains in one position more or less.

There are two caveats to the foregoing recommendations. First, contemporary digital methods for generating auditory displays have triggered some experimentation for using complex sound patterns to depict complex visual information such as a chart of data from a research report (Kramer, 1994; Madhyastha & Reed, 1994). Such interfaces can be used in automatic reading devices for the blind, but they are also intended for sighted people who might be able to detect patterns in their data that might not be discernible visually.

Second, the mobility of the receiver can be managed with print documents in some cases. Laptop computers address the same issue posed by the desktop or mainframe computer. Anyone who has taken a romp through an airport will remember multiple panels of monitor screens displaying flight time arrivals and departures. People do need to refer to that information more than once if a flight has been delayed, and delayed, and delayed again. The auditory system in airports can easily become degraded with noise or overburdened with too many flight numbers, gates, and departure time updates spoken by multiple airlines workers.

6.2.1 Types of nonverbal auditory displays

Conventional auditory displays are usually produced as single tones, patterns of two or more tones, or tones that have a more complex wave envelope. A *beep* is a single tone displayed repeatedly with a clear definition between onsets; the beeping sound continues until an operator takes action, or the alert condition has otherwise ended. Examples of tones with complex wave forms are sometimes described as a *wails* or *yeows*.

Which type of display is better? Part of the answer lies in how many auditory displays are already in the system. Another part is connected to the intensity of the sound. Louder sounds prompt faster response times (Adams & Trucks, 1976), although particular sounds may be better than others at specific intensity levels. If in doubt, which may be often, a response-time experiment should be conducted where the operator is expected to produce a realistic response to the stimulus.

Masking experiments provide another means for determining the comparative effectiveness of auditory display designs (Teas, 1990). In a typical masking experiment, a display is presented for a brief period of time, followed by a noise source. The noise source may be white noise, which contains a balanced representation of the frequency bands within the audible range, or a conflicting sound that is more natural to the listeners' environment.

Three factors are involved in the detection of signals: the signal-to-noise ratio, the degrees of spatial separation, and the correlation between signal and noise sources (Doll, Hanna, & Russotti, 1992). As discussed in Chapter 4, louder signals relative to noise are more readily detected. Spatial separation is the extent to which sounds come from different sources. Greater separation usually allows the listener to attend to one source and ignore another. Alternatively, spatial separation allows the listener to monitor the secondary sound source in case anything interesting is happening.

Correlation in this context refers to the similarity of pitch and tone patterns in the signal and noise source. A pulsing beep signal against a background of speech babble would produce a low correlation, and greater detectability. A speech signal displayed against a background of speech babble would be highly correlated and thus less detectable. This concept of correlation is also known as *articulation* or *dissociability*; it is an auditory analogue to figure–ground distinction in visual perception (Sanders & McCormick, 1993; Williams, 1994).

According to the U.S. Department of Defense (2012) and U.S. Air Force (Doll, Folds, & Leiker, 1984), speech displays are superior to tones or complex sounds if auditory displays are to be used to convey quantitative, qualitative, or tracking information. Nonverbal displays are superior to speech when trying to convey status information. These trends were based on the assumption of relatively simple tonal displays rather than the complex sound patterns that can be produced through digital sound synthesis. Auditory displays that have been designed to convey trends in numerical data graphs can produce human interpretations that are equivalent in quality to interpretations of visual graphs (Flowers, Buhman, & Turnage, 1997). The applications to data displays often use pitch and loudness to represent independent facets of the data; sometimes interactions are detected whereby a particular pitch level will affect the perception and interpretation of loudness (Neuhoff, Wayand, & Kramer, 2002).

6.2.2 Gestalt laws of perception

Complex tone patterns can be characterized in terms of whether they conform to the Gestalt laws of perception of visual forms that were described in

Chapter 4 (Williams, 1994). In addition to similarity, proximity, good continuation, closure, and articulation, auditory patterns can be characterized with regard to their familiarity, common fate, and stability. Sound elements are similar or proximal with regard to their pitch similarity and closeness of presentation in time. Good continuation occurs when tones of different frequencies are connected by tonal glides or separated by silences. Closure appears to be evident when a tone pattern is started with a particular tone, which creates a bit of auditory fixation. When other tones are introduced, closure occurs when the sequence returns to the fixation tone. Articulation was mentioned earlier.

Common fate occurs when two or more tones have wave envelopes that produce increasing frequency, or decreasing frequency. Stability occurs when a pattern of two or three tones repeats in rapid succession with a pause between sequences. Introduction of unique tones within or between primary sequences would disturb the listener's sense of stability.

6.2.3 Streaming

Although the Gestalt laws of visual perception have not been applied to auditory perception in a systemic fashion (yet), they appear to be operating during auditory streaming. *Streaming* is the process of detecting a meaningful auditory pattern from a complex array of sounds. One of the earliest recorded examples in psychology was the cocktail party effect (Cherry, 1953). One walks into a room that is buzzing with voices and joins a conversation. Although the majority of one's attention is on the present conversation, a portion of one's attention is devoted to scanning the auditory scene for keywords and familiar voices that should be checked out next. The effect was initially reported as evidence that attention is *not* an all-or-none phenomenon; it can be divided among stimulus sources. The lingering debate is how much of the streaming process actually requires directed attention and how much carries on automatically (Snyder & Alain, 2007).

Streaming is a process of detecting similarities in frequency and loudness, proximity of frequencies (high versus low pitch), temporal continuity, and spatial source. At some point, the totality of similarities build up to produce a *fission* in which an auditory stream is distinguished from all background sound and recognized as a pattern. Fission is essentially a bifurcation in NDS theory. There are some unresolved theoretical issues, however, as to which parts of the streaming process, such as frequency patterns or timbre, are processed in which subareas of the auditory cortex. From a human factors standpoint, the current knowledge of the streaming process has not been systematically translated into principles of good display design, but streaming phenomena would appear to form the crux of pattern confusability and the effectiveness of 3-D auditory displays, as describe forthwith.

6.3 CLASSIC PROBLEMS AND SOLUTIONS

6.3.1 Localization

Auditory displays have some classic problems. First, they are often difficult to localize. Sometimes localization does not matter, but anyone who has driven down a city street and wondered whether the emergency vehicle siren was coming toward them, coming head-on, or crossing a nearby part of town on a completely different road knows the experience. Part of the problem with localizing sound in an urban environment is that one often does not perceive the sound coming directly from the source, but from bouncing off walls of buildings. The bouncing problem combined with the inherent limitations to sound location produces the common source of confusion.

6.3.2 Confusability

Auditory displays can be confused with other warnings. As a general rule, confusion sets in after five auditory displays in a system (Kantowitz & Sorkin, 1983). Hold that thought and consider that military fighter aircraft that were in use during the 1980s utilized 7 to 10 auditory displays (Doll et al., 1984), and up to 17 have been reported for other aircraft including some speech displays (Kantowitz & Sorkin, 1983). The problem compounds with multiple speech displays (Teas, 1990). Pattern confusion can be traced in part to the sheer number of tone patterns that a person must remember, similarities among the patterns, and similarities in the locations of the sounds. The Gestalt laws (Williams, 1994) may predict which tone patterns will be confused with others, although there is little empirical support yet for this suggestion.

Hospitals can be a cacophony of sound emanating from several pieces of equipment that monitor different patients. Nurses, who are likely engaged in a task requiring substantial mental concentration need to identify critical alarms, find the source patient, and respond. In an attempt to minimize the confusability of alarms, the International Electrotechnical Commission (IEC) (2005) issued standard IEC 60602-1-8 that consisted of specified tone patterns for use in hospitals to denote situations involving oxygen, ventilation, cardiovascular events, temperature, infusion, perfusion, and power failures. The advantage of creating industry standard alarms would be that staff who learned the system in one hospital environment could transfer their learning to a new environment without any proactive interference.

The disadvantage, however, is that the standards were issued without any prior human performance testing. Thus, in an experiment to determine the efficacy of the tonal patterns (Lacherez, Seah, & Sanderson, 2007) nurses were required to identify and IEC alarm that sounded while they were working on an arithmetic task. Nurses with prior musical training reached 80% accuracy on alarm identification, whereas the accuracy of untrained nurses was 50%. Nurses overall did not perform much better than non-nurses in spite of being aware of the medical implications of the alarms. Their accuracy dropped when the alarms were overlapping or partially overlapping. Although standard alarm formats could in principle have advantages over unstandardized formats, the IEC series did not induce the level of streaming necessary to make the alarms sufficiently effective.

Subsequent research that was also motivated by the confusability of IEC displays has ventured toward strategies that map physiological conditions, such as oxygen supply and heart rate to sound parameters such as frequency patterns (Janata & Edwards, 2013). Changes in the frequencies would then denote changes in the physiological condition. The usability study showed that there were substantial individual differences among medical personnel with regard to accuracy and interpretation of the meaning of the patterns. Some personnel were able to learn and respond to the displays with 100% accuracy, while others did not come close to the criterion level of 87.5%.

Figure 6.5 is a rendition of the industry standard medical alarms that have become so problematical, based on their text descriptions in Edworthy et al. (2018). The musical notation should be a reasonable means for visualizing the pitch and rhythmic patterns and comparing them quickly. The accent marks over some of the notes (power down, drug administration, oxygen) indicate a sharper attack and slightly louder note compared to the

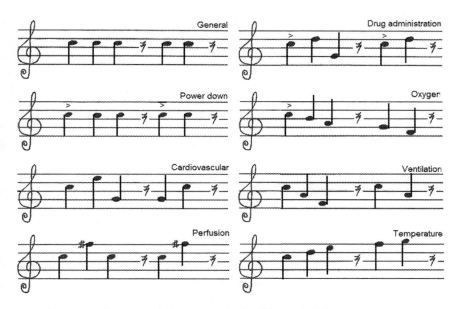

Figure 6.5 Tone, volume, and timing sequences for IEC hospital alarms.

surrounding notes. The symbols appearing after the third and fifth notes in each sequence are quarter-rests, which are pauses (silences) of durations equal to the quarter-notes used in the display. Note that the pitch patterns are very similar to each other. They all have the same rhythmic pattern. Edworthy et al. did not mention any timbre differences. Imagine a cacophony of alarms like these going off simultaneously. Location is yet another problem—where is the patient with the particular alarm?

New sets of alarms for hospital are currently under development. Edworthy et al. noted that some prototype alarms that were in use before the IEC standards were adopted were actually more distinguishable than what became the industry standards. Greer, Burdick, Chowdhury, and Schlesinger (2018) have been developing an alarm system that self-regulates loudness to the ambient noise level in the room, which should help with the signal-to-noise part of the problem. Further research on new sonification strategies is discussed later in this chapter under the heading of auditory icons.

6.3.3 Desensitization

Desensitization occurs when the listener hears so many whizzers and bells going off on a regular basis that the signals lose their attention-getting properties. Operators often respond by resetting the alarm to a higher critical threshold. The danger here is that a serious condition will be overlooked (Chapanis, 1975). A related phenomenon is the cry-wolf effect: Alarms that produce false-positive indicators of a real situation will eventually be ignored. The eventual problem is that the real danger condition will be ignored as well. The basic solution here is to set the alarm controls at a sufficiently high threshold to maximize response to true conditions and minimize false-positive indicators.

6.3.4 Recommendations

Sanders and McCormick (1993) identified some principles for good auditory display design in addition to keeping them few in number and distinguishable in content. The effectiveness of a display can be enhanced by use of a two-stage signal. The first signal should be a simple alert to the operator that a signal with important content is coming up next. For instance, an alerting tone could precede a speech announcement.

Invariance and compatibility are two more related considerations. Invariance means that the auditory signals always have the same meaning through the system or similar systems. Compatibility means that they have some relationship to the type of response that the operator is supposed to take. Again it should be recognized from one situation to another. An example is the national fire-warning standard depicted in Figure 6.6. The signal of about 400 Hz is presented in two bursts of 0.5 s, with 0.5 s in

Figure 6.6 Pulse sequence for standard fire alarm.

between, and followed by a 1 s burst and silence of about 1 more second (some variability from system to system is tolerated). The loudness should be about 10 to 15 dB greater than the ambient sound level. Some systems have the loudness cranked up to 100 to 115 dB just to make sure everyone's attention has been captured.

On the other hand, nonemergency auditory displays that are too loud can produce shock in the listener when such an emergency response is unwarranted. Although it is true that louder displays are more likely to be heard and processes correctly, there is an upper limit when discomfort sets in (Baldwin & Struckman-Johnson, 2002; Edworthy, 1994). A good rule of thumb, based on available experiments, is to produce the display at 10 dB greater than the ambient sound level of the room or system, and to conduct user tests if that amount exceeds 65 dB.

A final recommendation is to avoid extreme frequencies. In general, display tones of 200 to 3,000 Hz are good. Tones under 1,000 Hz are recommended if the listener is more than 1,000 feet from the origin of the sound. Tones under 500 Hz are recommended if the listener is located around a corner or otherwise not in a direct path of the traveling sound (Sanders & McCormick, 1993).

6.3.5 Adding redundancy

Auditory displays are helpful supplements to a primarily visual interface when a vigilance task is involved. An operator may be watching a radar screen for activities of the little blips, but may not notice a new one coming on to the screen. A simple tone would alert the operator to something new taking place (Colquhoun, 1975) and thus enhance performance.

A related application is the introduction of phone calls into the routine of a night watchman. On a good night, nothing happens, and there is little to watch. By 3:00 a.m. the watchman is likely to fall asleep. A series of phone calls through the night requesting a nominal report of activities keeps the watchman alert.

6.3.6 3-D auditory displays

The development of 3-D auditory displays was based on the principle of binaural hearing, and they are intended for the auditory component of a head-mounted display. The sound sources are processed to be slightly out of phase with each other as they would be in free field directional hearing. The signal processing is such that the soundtracks can be artificially located in space to the left or right, front or back, and horizontally.

Airline pilots are typically presented with three streams of speech displays coming from air traffic control, other aircraft, and communications within the airplane itself. The objective of the 3-D display is to project the sound sources in space so that the pilot can attend to one source or another without undue masking from the other sound sources (Wenzel, 1994; Wenzel, Wightman, & Foster, 1988).

Masking experiments in which operators' performance using 3-D head-mounted displays were compared against free-field directional perception show that localization perception was about as accurate with the 3-D audio device as with free-field perception overall, although there were more front-back reversals with the 3-D synthesized medium (Wightman & Kistler, 1989). Doll et al. (1992) determined, however, that greater degrees of sound separation produced lower thresholds of loudness that were required to detect a signal when a mask was generated. The effect was more pronounced when the signal and noise were less correlated.

3-D synthesized audio displays can enhance pilot performance in some types of tasks. Bronkhorst, Veltman, and van Breda (1996) prepared a 3-D audio track to accompany a primarily visual task on a flight simulator. The participating pilots were chasing another aircraft that disappeared at critical points in the flight. The participants were required to locate the target aircraft. The researchers found that the combination of visual and 3-D audio signals produced shorter search times than either visual or 3-D audio displays alone. Ratings of workload were not affected by the introduction of 3-D audio.

Wenzel et al. (1988) also found that 3-D audio displays can make serious alarms more salient without having to make them louder. The alarm preparation for wind shear involved tonal signals with a synthetic voice overlay saying, "Wind shear, wind shear." The signal was phased so that it would oscillate rapidly from the left ear to the right.

3-D audio also showed an advantage for detecting infrequent speech signals from a background stream of irrelevant speech (McAnally & Martin, 2007). The advantage of 3-D seems to be attenuated by a greater total density of speech, making the basic signal detection task more challenging.

6.4 AUDITORY ICONS

Auditory signals, such as beeps, squeals, bells, whistles, and sirens, are well known in conventional circumstances (Chapter 5). Eventually they appeared on computer interfaces. Gaver (1986, 1994) proposed a theory of auditory icon design that was made possible by digital sound synthesis. His taxonomy contained three types of auditory icon. *Symbolic icons* are those where the sound pattern has an arbitrary association with the machine process in progress. *Metaphorical icons*, however, bear some literal similarity. As with metaphors in literature, these icons capitalize on one feature of the process in question. For instance, a hissing sound might represent a snake. (Of course, the widespread utility of a snake to computer programs is another matter.)

Nomic icons, which Gaver (1986, 1994) preferred, are those that have a concrete similarity between the process or object of the icon. For instance, the sound of breaking glass might represent a glass object, the sound of wood hitting wood would represent wood, and so forth. Sounds such as glass and wood have a distinctive wave envelope, which would help in their fabrication. Although some investigators have expressed a preference for concrete rather than symbolic icon designs, it is sometimes a challenge to devise a concrete icon for every target idea or event that one wants to signal. Furthermore, no one really knows for sure how auditory or visual icons acquire their iconic quality (Jacko & Sears, 2003). If features other than concreteness are involved, however, iconic quality could be related to both the distinctiveness of the image or sound and its unique association with an object or event.

A close relative of the auditory icon is the earcon. *Earcons* are symbolic-type auditory displays that often accompany visual signs of a system operation (Blattner, Sumikowa, & Greenberg, 1989). They are typically found when an invisible operation has started or finished, or applications akin to where one might encounter a pop-up window. Some popular examples are drawn in Figure 6.7. Are earcons helpful to their users? The answer to this question is regrettably informal, but should suggest where formal research should head. Many system users prefer to work with the system sound off, unless they are listening to music in the CD drive. One design group that this writer had the pleasure of working with incorporated earcons into a software module that involved controlling a mechanical subsystem to fill up with water or empty it out. The earcons transpired over the physical operation time with an ascending or descending tone series to denote the flow of water. The users who would describe themselves as computer enthusiasts found the earcons amusing. The other users found them annoying and asked that they be turned off.

Earcons are popular elements in children's software. They often take the form of musical looney tunes that do not appear to serve an objective other than to hold interest, or perhaps add life to a potentially boring program. Some are mixed with speech displays, which might be necessary for giving children who are still learning to read some instruction about what to do in their program. Winters, Tomlinson, Walker, and Moore (2019) found auditory icons to be helpful learning aids in a software designed to teach some

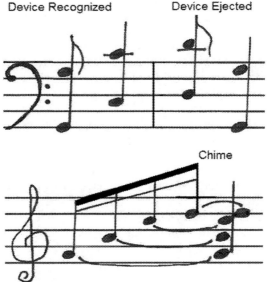

Figure 6.7 Some commonly used earcons.

principles of physics to middle school students. When possible, their earcon choices gravitated toward metaphoric, as opposed to strictly symbolic ones.

Edworthy et al. (2017) developed some new sets of alarms to improve on the IEC hospital standards. One set consisted of concrete auditory icons, such as drumming to signify cardiovascular events or a shaking pill bottle to signify drug administration. A second set consisted of *word rhythms*, which are nonverbal displays but with tones and accents that emulate spoken words. A sample of 194 research participants were assigned to learn one of the three sets or the IEC standards. After ten blocks of trials, auditory icons were learned and distinguished the best, followed by resilient, word rhythms, and IEC standards in last place. All other things being equal, harmonically dense sound patterns were more readily learned and distinguished than low density icons.

The one known example of word rhythms did not fare as well as some other icon sets in Edworthy et al., although they were still better than IEC standards. But what about icons built around actual words? Those would be *spearcons*, which are covered in Chapter 12 with multimedia computer interfaces.

Another property of sound that has been given some attention is the shape of the pitch envelope. Schutz and Stefanucci (2019) noted that many alarms in use have rectangular pitch envelopes (like horns or fire alarms), whereas realistic sounds more often have a sharp-enough attack and a more gradual decay. The jury is still out regarding where or when the nonrectangular "sound shapes" are the most helpful.

A new innovation of a more theoretical nature is the idea that alarm or icon design might benefit from adopting principles of music composition (Cave & Eyes, 2019). Perhaps principles of music composition could be redirected from expressive art to memorable icons. Specific examples of this idea have not yet taken form, although Cave and Eyes did mention that a set of icons that share a musical principle might not be expandable to accommodate a new icon target at a later time.

6.5 SPEECH DISPLAYS

The elementary connection between human speech and machine speech is the phoneme. The *phoneme* is the smallest unit of speech that contributes a change in the meaning of a word. Consider the set of words *pig*, *pit*, *pet*, *get*, *gem*. Each change in letter is a change in a phoneme. There are actually more phonemes in the English language than letters of the alphabet—long and short vowels, letters that sound alike, and diphthongs—but the elementary nature of the phoneme is about as elementary as a change in letter. Further up the linguistic scale we have *morphemes*, which are root words, prefixes, suffixes, compounded words, and similar attachments. Beyond the morpheme are the word, phrase, and sentence.

Speech displays occur in telephone systems, machine interfaces, and annotated computer programs. The criteria for system performance go beyond signal detectability to include intelligibility and acceptability (Voiers, Sharpley, & Panzer, 2002). Identification of the speaker is another important feature of telephone communication.

Intelligibility is the extent to which listeners can determine particular words from the source. Here it is possible to contrive experiments that require listeners to select the word they hear from a multiple-choice list. The real word and the distractors are typically only one phoneme apart. Factors that affect intelligibility include the signal-to-noise ratio, familiarity of the message, amount of vocal effort from the speaker, and electronic contributions such as distortion and peak clipping. Familiarity helps not only to recognize the particular words because they have been heard before, but also to provide some context for interpreting new information.

Acceptability is the extent to which noise or other electronic compromises are annoying. Acceptability is typically measured using a subjective rating. At the most favorable end of the scale, the speech is both intelligible and not annoying. In the middle ranges it becomes more annoying. At the opposite extreme the speech is both unintelligible and annoying.

Speaker identification can be assessed through an experiment wherein listeners are presented with clear examples of several target individuals' speech. They are then presented with speech samples from the display system that might or might not be examples of the targets' voices. The listeners should be able to associate test samples with the correct voices.

6.5.1 Speech spectrograms

A *speech spectrogram* (Figure 6.8) is a display of frequencies generated over time as a person pronounces a word or makes a longer utterance. The ordinate is time. The abscissa is the octave band of the frequency. The speaker in this example is saying, "Don't ask me to carry an oily rag like that."

In the early days of speech analysis and processing, scientists conjectured that it might be possible to link specific phonemes to specific frequency arrays. Unfortunately, the speech-to-signal correspondence is far more complex. People often emit different frequencies with the same word, depending on mood and context of the sentence. For instance, we can make a statement sound like a question by simply raising our voices at the end of the sentence without changing a word.

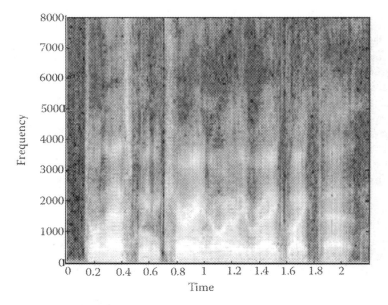

Figure 6.8 Speech spectrogram for "Don't ask me to carry an oily rag like that." A phonetically balanced sentence used as normalization, from TIMIT database. (Copyright 2004 Michael Johnson, Speech Laboratory, Marquette University School of Engineering. Reprinted with permission.)

The creation of synthetic speech displays begins with the access of the spoken information. The technology is similar to the digital sampling method used to assimilate synthetic musical instruments. The complexity of frequencies emitted by voice, however, had made digital sampling a challenge. The common solution, however, is to record spoken keywords, phrases, and statements into little files. The computer program will then execute the right combination of files to produce the message. Telephone directory assistance in the United States is an example. The recorded files include, "The number is," words for each of the 10 digits, "hundred," "thousand," "repeat," and the closing message about automatic dialing. The recorded files preserve the original speaker's vocal inflections and regional accents. The listener can detect a little choppiness between the elementary word files, but not enough to present a problem. The difficulty, however, is to extend that technique to situations where a more extensive vocabulary is needed to compose many possible individualized messages.

The literature on speech communication and processing is extensive, although much of it pertains to the engineering aspects of the system rather than the human factors involved. Of course, human factors limitations have undoubtedly spurred the electrical engineers to greater heights. Speech-related controls are considered in Chapter 8.

6.6 NOISE

Noise, for the purposes of understanding auditory displays, is any unwanted sound. Chapter 4 already explained the role of noise with respect to the detection of signals. This section elucidates a few more important aspects of auditory noise.

6.6.1 Colors of noise

Acoustic specialists characterize noise as having four basic colors: white, pink, brown, and black (Ward, 2002). White noise is an equally weighted composite of sound from the full range of frequency octaves. If we were to plot a frequency distribution of the number of times a frequency was emitted from a white noise source by the frequency of the particular tone, the distribution would be square. If we were to take repeated random samples of white noise from a square distribution and add the samples together, a Gaussian or normal distribution would be obtained.

The colored noises occur when the distribution of frequencies from a noise source are not as random as "random" should be. Instead there is some internal deterministic structure to the noise source. Distributions of colored noise can be expressed as:

$$F(H) = k \, H^{-a}, \tag{6.2}$$

where H is the frequency of the sound in cycles per second, $F(H)$ is the number of times the tonal frequency occurs in the sample, a is a constant of the distribution that determines the color of the noise, and k is a constant of no particular consequence.

The noise is designated as pink if $|a| = 1$ The noise is brown if $|a| = 2$. The noise is black if $|a| > 2$. As the noise changes from pink to black, there is greater determinism in the source. At the black extreme, we have an unwavering, sustained, and probably annoying single tone. Colored noise with $|a| < 2$ thus indicates that a self-organized process is taking place; it often characterizes the statistical deviation from a deterministic linear pattern in nonlinear time series analysis. Parameter a is, incidentally, a fractal dimension for the distribution of observations. In any case, white noise is typically the noise of choice for the preparation of experimental stimuli, although colored noise is closer to what occurs in reality.

6.6.2 More signal detection

Although the loudness of a noise can obviate the perception of a stimulus, the frequency of the noise tones can also interfere with the perception of the true stimulus. As one would expect, the interference is greater when the noise and signal are close in frequency. For reasons not yet known, noise tones that are slightly lower in frequency than the signal induce more interference than noise tones that are slightly higher in frequency than the signal (Kantowitz & Sorkin, 1983).

Just as it seems safe to conclude that noise is always bad for signal detection, we must introduce the concept of *stochastic resonance*. This is a signal detection phenomenon whereby a very weak signal can actually be detected more often if a modicum of noise is added to it. Ward (2002) characterized stochastic resonance as a double-well phenomenon, which is essentially a cusp catastrophe function as depicted in Figure 6.9. The curve in Figure 6.9 represents an underlying potential energy function. As we travel across the ordinate we move from one low-entropy well to another. The well on the left corresponds to the perceptual conclusion that no signal exists. The well on the right corresponds to the conclusion that the signal plus noise is present. The addition of the noise pulls the perceiver out of the fixed conclusion that the signal does not exist and introduces enough entropy to suggest that perhaps it could exist. The addition of noise to the weak signal, however, does not guarantee that the perceivers will always detect the signal. It only enhances the chances of them doing so.

6.6.3 Hearing loss and noise exposure

Extremely loud blast noises can damage the eardrums, but the majority of occupationally related hearing loss is gradual and chronic. A good deal of

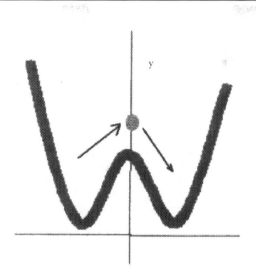

Figure 6.9 Two-well phase shift for stochastic resonance.

what has been learned about hearing impairment comes from studies in which a temporary threshold shift has been induced. If a person is exposed to a loud noise source, their threshold to detect a sound becomes elevated after the noise has been removed. The threshold drops backs to normal soon. With prolonged and repeated occupational exposures of 10 years or more, however, the threshold eventually does not return to normal but stays elevated. The perception of frequencies greater than 4,000 Hz are impaired first. With additional exposures the impairment spreads to the lower frequencies as well (Kantowitz & Sorkin, 1983).

Environmental noise levels can be measured with sound meters. There are two settings that pertain to OSHA standards (1996b). The FAST or SLOW setting allows the device to change its reading as the operator moves through an environment; the OSHA standards are set to SLOW. The contour control places different weights on the various frequency bands when the bands are sampled and the overall sound level is calculated. The standards require the use of the "A" contour, which places greater weight on frequencies greater than 1,000 Hz compared to the other contours, to measure overall sound level. OSHA's exposure limit is 90 dB for the composite total noise for an 8 h daily exposure. The allowable total exposure decays rapidly thereafter, so that only 15 minutes per day are allowed at 115 dB of ambient noise.

Figure 6.10 depicts a noise criterion curve for the OSHA standards. The various curves indicate a distribution of frequency intensities associated with a particular dB level that is going to be used for allowable exposure.

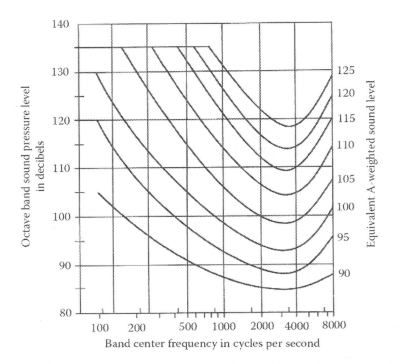

Figure 6.10 Noise criterion curves for OSHA standard 1910.75. (In the public domain.)

The goal of system engineering is to control noise within each frequency band so that all noises stay below the curve. The basic concept of a noise criterion curve can also be applied to any work environment that is subcritical for total noise as defined in the OSHA (1996a) standard.

Noise can be controlled with structural engineering or personal protection equipment. In the case of the environment, the architects can build sound-resistant rooms for machine operators if their work patterns do not allow them some chances to escape from the noise of the machines. Such rooms are built with sound-absorbing walls, double-pane glass, and seals around the doors. Employers who are building a factory from scratch might consider designing the factory floor as a series of platforms that are supported separately from below. Heavy machinery that creates vibration should be placed on separate pieces of flooring, and the pieces of flooring can then be connected with vibration-absorbing seals. In this fashion the vibration from one machine will not transfer to all parts of the factory environment.

Roof fans produce another source of noise. In short, fans with more blades are better than fans with fewer blades. Fans with a greater number of blades still produce noise, but they do so at a higher frequency (pitch) that does not travel as far as the low-frequency noises (Kantowitz & Sorkin, 1983).

If structural redesign is not possible, operators need to use ear protection. Earplugs are cheap and disposable. Their advantage is that they are inexpensive and can be handed out to visitors readily. Their disadvantage is that the users often do not install them properly, thus they do no good and perhaps give people a false sense of protection. A more certain alternative are earmuffs that are designed like a set of headphones, except that they do not convey sound; they insulate against it. They are sturdy and effective. They cost more to purchase, but they are reasonably permanent devices.

6.7 TACTILE DISPLAYS

6.7.1 Haptic perception

Touch is primarily an active process in human factors applications. This principle extends to *haptic perception*, which is the perception of objects' shapes based on tactile stimulation alone. Subcutaneous sensors respond to stimulation from mechanical, thermal, chemical, or electrical sources. Free nerve endings are stimulated, usually mechanically, and connect to the Pacinian corpuscles located deep in the skin. There does not appear to be a direct one-to-one correspondence between the stimulation of particular nerve endings and particular sensations, however, and the means by which a person can assemble the sensations into a perception of an object's form is mostly uncharted territory. In spite of the thousands of articles that are somewhat related to haptic perception and haptic feedback or displays (Song, Lim, & Yun, 2016), generalizable principles are sparse. The majority of associations between tactile stimulations produced by tactile displays and some form of meaning is learned.

Applications of tactile displays over the years have included the shapes of knobs and handles, keyboard designs, navigation aids for visually impaired and unfamiliar environments, virtual braille, prostheses to assist the restoration of body balance, implants in shoes to assist balance control and postural stability, and interaction with virtual environments. Sensations on separate areas of the body, however, can be used as information elements that are conveyed by vibrotactile displays. The topic has also attracted a lot of attention since the designs of virtual reality programs often require engaging a synthetic sense of touch to give the individual the experience of handling real objects within the computer program. It is usually not the primary source of information that a person utilized while performing a task, but a form of feedback from control actions that have been taken.

Roughness, or texture, is a basic perceptual element in haptic perception, along with volume, surface area, and weight. Perceptions of weight are often distorted by object size, similar to the size–distance constancy or size–distance illusion (Kahrimanovic, Teist, & Kappers, 2011). Larger objects are perceived as weighing more, which would be true if the objects were made of the same material.

6.7.2 Knobs

Tactile displays are far less common than visual or auditory displays, although there are a few classic applications (from Kantowitz & Sorkin, 1983). Airplanes of the World War II era were equipped with several joysticks that were equipped with handles of similar if not identical shapes. Too often the pilot who was trying to keep an eye on the sky reached for a stick and operated the wrong one with disastrous results. The solution to the problem, credited to Alphonse Chapanis was to mount handles of different shapes on each of the joysticks so that the operator could feel the correct stick (Roscoe, 2006). It worked, and aircraft manufacturer quickly responded with new knobs for all occasions.

A similar situation occurred in an electric power plant where two control knobs that were shaped exactly alike and located side by side. The knobs often foiled the operators when they were monitoring something and reached for a control. The operators' solution was to attach beer tap handles, two different brands and shapes to controls (Norman, 1988). The problem was solved once again.

Navy underwater divers often have to rely on the feel of knobs to know which control they are operating. Although the shape of knobs is not a display in the usual sense, the tactile sense is used to substitute for the labels that cannot be seen in that environment. A set of knob shapes or joystick handles that the U. S. Department of Defense (2012) considers maximally distinguishable is shown in Figure 6.11.

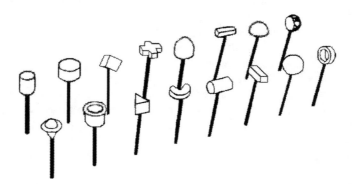

Figure 6.11 Maximally distinguishable knobs. (Reprinted from U.S. Department of Defense (2012); in the public domain.)

6.7.3 Vibratotactile displays

Vibrotactile displays can be useful additions to a PMS if the auditory and visual channels are overloaded. They can also add redundancy to visual and auditory signals. Some of the more common applications convey spatial information to the operator when the driver is moving and potentially disoriented, signaling the direction of an incoming event or object, and signaling the need to shift attention.

A classic aircraft application involved the use of vibrator seats and shaker sticks. The old expression, "fly by the seat of your pants," originated at a point in aviation history where the airplanes were not equipped with altimeters. Inasmuch as they were flimsy vehicles compared to contemporary designs, the pilot could literally feel the elevation and descent in his seat. As aircraft became more comfortable and insulated the human from the environment to a greater extent, it became necessary to put vibrators in the pilots' seats so they could feel the elevation. Of course, the same epoch of aircraft designs also included altimeters, but redundancy improved reliability, especially for pilots who had learned from the old school.

Cell phones that can be switched to vibrate in the operator's pocket instead of ringing audibly are now fairly common. They can give the operator a modicum of privacy when the message arrives and not annoy other people in the operator's immediate environment.

Diving suits and other environmental suits insulate the operator from direct control of the environment, and there is often an advantage to building in synthetic sensory information so that the operator can react with the external environment in a manner more similar to direct contact. Similarly, virtual reality gloves (Chapter 12) replace the tactile sensations that are missing in a virtual environment.

Whereas knobs and vibrating phones convey simple pieces of information that does not change over time, vibrotactile displays can convey more complex information that does change. Systems of tactile stimulators can be worn as objects strapped to the operator's thigh (Salzer, Oron-Gilad, Ronen, & Parmet, 2011), or in the form of belts and vests (Jones & Sarter, 2008). Tactors, which are the objects generating the vibration signals, should be separated by at least 25 mm on the forearm, and 60 mm elsewhere on the body. Tactors located on the central front torso are better recognized and distinguished than those in more peripheral locations of the body.

A firm set of rules for making optimal connections between elements of information and aspects of vibration is still a work in progress, but a few useful principles have come to the foreground. Information can be mapped to any of five aspects of a vibration—frequency, duration, amplitude or intensity, location on the body, and rhythm pattern. Sets of seven or eight signals that are readily distinguished and learned are not uncommon (Jones & Sarter, 2008). It appears that people can distinguish 15 levels of intensity,

9 levels of frequency, and various patterns. There are possible interaction effects, however, between the design of tactile display sets and the tasks that the operator is engaging while wearing the system.

6.7.4 Tactons

If we can have visual and auditory icons, earcons, and spearcons, we can also have—you guessed it—tactons, which are tactile icons. Nomic or concrete tactons have not been forthcoming, as vibrational signals are too far removed from the object or condition that they refer to. Some designs in use, however, bear some similarity to manual signals, although they are not exact matches (Barber, Reinerman-Jones, & Matthews, 2015).

Some tactons are more complex than others. Brewster and Brown (2004) identified three levels of cognitive complexity associated with tacton design. There is the *simple tacton* that has a particular meaning that is dissociated from any other tactons that might be present. *Compound tactons* are combinations of two or more simple tactons, such that different signal frequencies map to different system conditions that need to be represented. *Hierarchical tactons* are larger sets of signals with general and specific properties—different subsets with different signals within each subset. *Transformational* icons are those that can display several pieces of information about an object or condition using different types of contact such as location on the body, frequency, and rhythm (Jones & Sarter, 2008, pp. 104–105).

6.7.5 Gloves

The sense of touch is often compromised by the use of gloves. Gloves can be compared on the amount of sheer force that they exert on the operator. Sheer forces block incoming information and the more obvious aspects of the glove often inhibit manual and finger dexterity.

DISCUSSION QUESTIONS

1. A classic problem in human factors is to determine what information needs to be displayed. Of course this question arises earlier in the design process than the choice of display type. How would you go about determining what information needs to be displayed, either aurally or visually?
2. Redundancy improves reliability. How would two modes of display help, and in what situations would an auditory display assist a visual display?
3. How many different types of noises does your desktop or laptop computer make? What is the system signaling? Are any of these sounds actually necessary? If so, which ones?

4. About 90% of the auditory alarms that go off in a hospital patient-care facility are false alarms. What is the impact of the false alarm rate? How might the false alarm rate be curbed?
5. A recent innovation in auditory displays is to present them in the form of a few bars of a popular song. Comment on the assets and limitations of this technique.
6. How many auditory signals come from your automobile? How many of them are necessary? Should others be added?

Chapter 7

Cognition

7.1 ORGANIZATION OF HUMAN MEMORY

Figure 7.1 shows a schematic for the human memory process. At the first stage, an incoming stimulus is retained on the sensory register for a fraction of a second. The sensory register would be the retina of the eye or the basilar membrane of the ear. This is just enough time for the remainder of the process to move into action if the human is paying attention to the stimulus.

A neural representation of the stimulus, or *memory trace*, starts to form when the initial stimulus is received. Once the memory trace is formed, the information item enters a two-stage process by which it could be committed to permanent memory or lost. It should be mentioned here that no direct correspondence has been made yet between the configuration of neural impulses and the thoughts that are produced by them, as we experience the thought. This unanswered question is known as the *binding problem*. In recent years, there has been some attempt to solve the binding problem by drawing on principles of quantum physics that are then combined with some clumsy notions of "consciousness." This approach has not yet paid off to any great extent (Koehler, 2011).

7.1.1 Short-term memory

The first stage after forming the memory trace is the short-term memory. The short-term memory can hold information for a range of 30 seconds to 2 minutes. During that time the information is being processed for meaning. Although we do remember the original sights and sounds of events, the majority of our memories consist of the sense or meaning of what happened more so than a recording of the exact event in real time.

The further transition of an information item to long-term memory depends on rehearsal. Here the operator is repeating the information, probably silently, enough times to either retain the information long

DOI: 10.1201/9781003359128-7

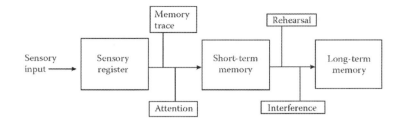

Figure 7.1 Memory schematic.

enough to use it, or for more in-depth processing for long-term memory. Anyone who has ever repeated a phone number to themselves often enough to actually get to the phone without writing the number down has experienced the first type of rehearsal. Anyone who has ever studied for an exam and gone over the material several times has experienced the second type of rehearsal.

The transition from short-term to long-term memory can be interrupted, however, by interfering stimuli. There are two sources of interference: more incoming stimuli through the sensory register and short-term memory, and old information coming from the long-term memory. When a flow of incoming information occurs, the information items that occur at the beginning of the sequence are remembered best; this is known as a *primacy effect*. The information items occurring at the end of the series are remembered second best, so long as the opportunity for recall is relatively immediate; this is known as the *recency effect*. The information items in the middle of the sequence are at the greatest risk for being lost.

The primacy and recency effects were first discovered in experiments where human participants were asked to memorize lists of meaningless items. There were opportunities for timed rehearsals in those experiments. In real-world applications, there is little time for planned rehearsals, and the information is flowing in over time. The operators often lock onto the first items coming in for proper processing. At this point attention is diverted from newer incoming stimuli, which are likely to be lost if the machine system does not contain a memory storage facility of some sort. This special case of the primacy effect is known as an *anchoring effect*.

The short-term memory has a limited storage capacity in the amount of 7 ± 2 chunks (Miller, 1956). A psychological chunk is not easily translated into information units such as bits and bytes. Rather, a chunk is better conceptualized as an intact information unit of some sort. It is also possible to cram more than seven (or nine) units into short-term memory; people do remember longer lists of items. A process of chunking (verb) as depicted in Figure 7.2 accomplishes the short-term memory enhancement.

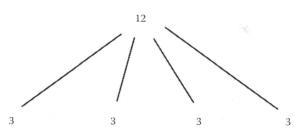

Figure 7.2 Chunking in short-term memory.

7.1.2 Long-term memory

The long-term memory has an indefinitely large storage capacity, and it can hold information for an indefinitely long period of time. Information that resides there has been processed for meaning during the rehearsal process of short-term memory. The meaning of "meaning" is subjective, as not every idea has equal meaning to everyone. Nonetheless, meaning can be interpreted as the number of connections that an idea has to other ideas already in long-term memory (Kohonen, 1989).

Information can be retrieved from long-term memory on demand by means of a *retrieval cue*. A retrieval cue is a stimulus–response association between one idea element and other ideas with which it may be connected. Figure 7.3 describes a plausible configuration of idea elements in the long-term memory. Imagine that the dark circle in the shaded area denotes the target memory. A proper retrieval cue, perhaps a clear question, will activate the idea. Once the target idea is activated, adjacent ideas become somewhat activated as well, as depicted by the other circles in the shaded area. The ideas with secondary activation are linked to more ideas in the unshaded area, but the latter ideas do not become activated until the secondary ideas are more substantially activated themselves.

An important side effect of the retrieval cue process is that change in retrieval cues can produce a memory failure. It commonly happens in technical language, for instance, that a concept may be known by more than one name or word. If the speaker uses a synonym with which the listener is unfamiliar, the listener will not remember the idea when it is needed.

Retrieval from long-term memory can be inhibited by interference. With anterograde interference, old learning may be so entrenched that it inhibits some element of new learning. With retrograde interference, new learning may inhibit the retrieval of the old learning. Again, relative meaning of the material that is being learned could account some anterograde interference effects. If one person already has learned one set of meanings, incoming new information that conflicts with the original in some way could produce

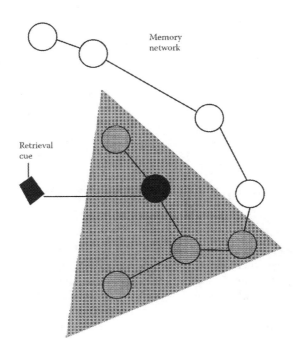

Figure 7.3 Retrieval in long-term memory.

interference, or might have very reduced incremental meaning, whereas someone who has not had the prior learning experience would not have the same experiences with the new material.

7.1.3 Types of memory

There was some debate in earlier decades as to whether memory functions were localized to particular areas of the brain or whether memory was distributed throughout the brain. Both interpretations are now considered partially true. The hippocampus in the limbic system is active when any memory process is at work. Specific types of memory, however, such as verbal or pictorial memory, engage different specific regions of the cortex in addition to the hippocampus; mapping out the brain circuitry that supports this variety of memory functions is an ongoing challenge (Guastello, Nielson, & Ross, 2002).

There are several specific types of memory. There is *semantic memory*, where meanings of words are encoded and processed. The semantic memory makes use of categories, and the formation of categories. In a classic experiment (Bousfield, 1953), experimental participants were given a list of animals, plants, professions, and names of people in random order. After they were given a period of time to memorize the list, they were asked to repeat the list items in any order they preferred. The head dump revealed that animals, plants, names, and professions were retrieved in clusters that corresponded to semantic categories and not in random order. Categorical thinking is revisited in Chapter 12.

Propositional memory pertains to relationships among concepts. *Spatial memory* pertains to pictures and relationships. Propositional and spatial memory appear to be closely connected; perhaps for this reason it is not uncommon for people to draw diagrams of ideas.

Finally, there is *episodic memory*, which is the memory for events, and *procedural memory*, which is memory for sequences of one's actions. Procedural memory will be considered further in this chapter under the section on cognitive analysis of a PMS, where one of the objectives is to extract the content of an operator's procedural memory in order to understand how a task is performed and what resources are used to do so. Episodic and procedural memory are often aggregated as *implicit memory* to refer to the relative involuntary nature of memories we accumulate from experiences, as opposed to *explicit memory* that is engaged when we deliberately try to remember something. Semantic, propositional, and spatial memory would be explicit forms of memory.

7.1.4 Working memory

Working memory is "assumed to be a hierarchical system involving short-term memory, representational components, plus a general executive attentional component" (Kane & Engle, 2002, p. 638). It is analogous to random access model (RAM) in a computer, which is the amount of workspace available to manipulate programs and data input. It is considerably less than the total memory storage space, which is generally a static entity that does comparably little additional processing by itself; yet anything in it can be called into working memory (if it hasn't been forgotten) when given a viable retrieval cue.

The representational components are workspace modules that are associated with specific types of memory and cognitive functioning (Logie, 2011). The executive function recruits workspaces and the functions needed to perform a task. The executive function also maintains the focus of attention. When we tell someone, "Watch what you're doing!" we are recommending a reengagement of the executive function, which might have lapsed due to fatigue or distraction by non-task events. Whereas human factors engineers examine properties of tasks that result in performance stability or decline under different types of load conditions, cognitive psychologists have sought to unravel the structure of working memory itself (Baddeley, 2003). Channel capacity, according to Logie (2011) is likely to be an emergent result of the working memory components that are required by a particular task. For more demanding tasks, the domain-general function of attentional control supplements the domain-specific resources. Furthermore, memory and processing are two different functions with different capacities, although they normally operate together to varying degrees.

Working memory capacity is usually measured by span tasks; in a common example, a test administrator would read sequences of numbers of varying length, and the respondent would have to remember the numbers that were spoken and repeat them back in order, or in reverse order. Span tasks can also involve verbal, pictorial, or episodic material (Conway, Kane, Bunting, Hambrick, & Engle, 2005; Guastello, Boeh, Shumaker, & Schimmels, 2012).

The relationship between working memory capacity and performance is somewhat enigmatic. Kane and Engle (2002) reported that "high spans engage attentional processing to achieve results while low spans rely on automatic processing. High spans are expected to be more taxed by a secondary task" (p. 641). McVay and Kane (2009) reported that working memory capacity was correlated with performance on a 45-minute version of a task in that it kept the participants' minds from wandering, but capacity had little or no impact on performance on a 2-hour version of the task. This finding led Redick, Calvo, Gay, and Engle (2011) to conclude that time on task "does not appear critical to the working memory capacity relationship with task performance" (p. 322). In other words, working memory capacity was interpreted as relevant to the management of cognitive workload. Working memory capacity and prolonged time on task produce fatigue, however, which is covered in greater detail in Chapter 10.

Next, to follow up on an idea that was introduced earlier in this section, working memory is hierarchically organized such that the upper level contains executive functions, and the lower level is associated with any of several content domains (Oberauer & Kleigel, 2006). The primary executive functions are inhibition of responses, focus of attention, task switching, updating, and allocation of workspace to the lower levels (Miyake, Friedman, Emerson, Witzki, & Howerter, 2000; Ilkowska & Engle, 2012). The executive function also calls up routines for use by the lower-level domains. The lower level domains include verbal, acoustic (also known as the *phonological loop*), spatial computational, and procedural activity (Baddeley, 2003). To make matters more interesting, workspace needs to be allocated to the executive functions themselves; the resources for executive and other functions appear to be drawn from a common pool (Barrouillet, Portrat, & Camos, 2011). When the executive function detects that a lower level function does not have enough workspace, it allocates space from lower levels that are not being used for the task (Drag, & Bieliauskas, 2010; Schneider-Garces et al., 2009). Reallocation of space occurs more

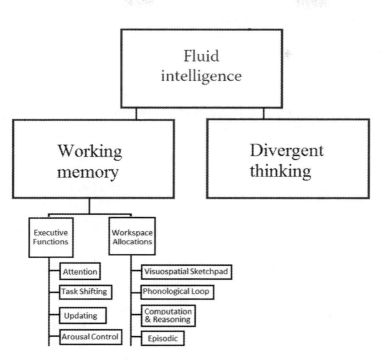

Figure 7.4 Hierarchical organization of working memory.

often with aging. The apparent organization of working memory is shown in Figure 7.4; the role and positioning of fluid intelligence is discussed later in this section.

7.1.5 Attention and perception

The cocktail party experiments (Cherry, 1953) provided a clear illustration of what became known as auditory streaming. They also raised the question as to whether attention and perception were two different processes and whether there was a filtering process involved. Two decades later, Neisser (1976) drew the connection that attention and perception were part of the same process, and there was no need to assume that a filtering process was taking place. Divided attention and distracted attention were sufficient explanations for incomplete or degraded perception–action outcomes.

Perception is not a passive process; people (and animals) are looking for something that could meet their goals and objectives or might affect their survival or induce injury. This search for affordances is consistent with Gibson's (1979) ecological perspective on perception that was introduced in Chapter 5. The perceptual cycle is a sequence of anticipation, explorations, and information pickup. If the perceived target connects to something meaningful to the active search, the response ensues. The ability to recognize and respond to affordances requires *schemata*, which are mental structures of perceptual elements that are essentially template recognition processes. The perception–action combination is eventually learned and consolidated as the perception cycle is repeated and completed. The schemata can be reused whenever the environmental situation and demands warrant.

Schemata facilitate the anticipation of what to expect and do next. A possible negative consequence of the cycle is that incorrect schemata can be activated, at which time the incorrect action sequence proceeds automatically without any further control (Plant & Stanton, 2012; Stanton, Chamber, & Piggott, 2001). The amount of time it takes to find and execute a schema can be substantial (Spivey & Dale, 2006). The time requirement would depend on its rule complexity, and thus its demand on working memory, and the ambiguity of the perceptual target. The availability of an existing schema could preempt the development of a new schema when needed, although new schemata can eventually differentiate given enough experience.

In an illustrative experiment, Ye, Cardwell, and Mark (2009) showed their experimental participants a variety of objects, each of which had one or two affordances that were not part of the original design intention of the object. An example would be the use of a drinking glass that could be used to dig sand or pound a cracker or other object. The participants were able to find one affordance, but often failed to find two when the two were present. Once one schema was activated it was not readily changed. Dual affordances were more difficult to detect if the affordances required grasping the object in different ways; in other words, the action portion of the schemata was different. Ye et al.'s experiments illustrated that perceptual cycles are relatively intact, and one schema needs to be executed before a different one (for another affordance) can be engaged.

7.1.6 Attention tunneling

Attention tunneling occurs when a person fixates on one channel of information and ignores the content of other channels (Wickens & Alexander, 2009; Kortschot & Jamieson, 2020). When a person repeatedly searches for a nugget of information that always appears in the same portion of a visual display (or computer screen), there is a tendency to zero-in on that portion of the display and ignore the rest of it. We saw the potential for it in the design of heads-up displays: The operator becomes fixated on the synthetic images in the near visual field that is superimposed on the view of the real world, and fails to notice something important in the real field.

Attention tunneling can result from efficient schemata becoming overly automatic or overused, or as an adaptive response to cluttered visual (or other) stimulus environment. The downside, however, is that the operator becomes likely to overlook meaningful nuances elsewhere in the environment.

7.1.7 Task switching

Numerous studies, summarized in Andreadis and Quinlan (2010) and Rubinstein, Meyer, and Evans (2001), have shown that switching tasks incurs a *switching cost* in the form of lower response time. The time loss was traced to the activation of rule groups when anticipating the next task. Rule complexity adds to the switching cost, such that a switch from Task A to Task B might not be the same as vice versa. Some of the cost is attentional, and some of it is computational (Lorist et al., 2000).

Andreadis and Quinlan (2010) noted that experimental paradigms for task switching typically relied on predictable sequences of tasks, for example, blocks of trials that are all repeated tasks or alternating tasks. They found that predictable patterns incurred lower switching costs as did cuing the impending task and lengthening the lag time between the cue and the task. Ambiguous task cues increased switching costs, because both rule sets remained activated when the cue was presented. Such "crosstalk" between tasks was equivalent for both predictable and unpredictable sequences. When cues were ambiguous, the switching costs were actually less during the unpredictable conditions than under predictable conditions (p. 1788).

In the foregoing studies, task switching regimes were planned by the experimenter. They might mimic conditions where task selection is controlled by automation reasonably well. Voluntary task switching could be a different matter entirely. Often people have several tasks to choose from and many of each to do in their work queues. Although involuntary task switching can induce fatigue (Hancock, 2007; Lorist et al., 2000), voluntary task switching can be a means for alleviating fatigue or boredom (Alves & Kelsey, 2010; Lorist & Faber, 2011).

Some logical questions at this stage are (a) "What are the rules and patterns that people naturally engage for voluntary task switching?" (b) "How do task switching schemata shift in response to externally induced constraints on the task goals, such as task performance quotas?" In a study of 54 undergraduates, the research participants performed 7 different computer-based cognitive tasks 7 times to produce sets of 49 responses. They worked under two instructional conditions; one required a quota of 7 examples of each task, and the other did not impose a task quota beyond doing 49 examples of the tasks in any combination (Guastello, Gorin et al., 2012). The sequences of task choices were analyzed using symbolic dynamics to extract types of patterns and pattern lengths. Patterns were then classified and compared with regard to Shannon entropy, topological entropy (related in principle to the type of turbulence measured by a Lyapunov exponent), number of task switches involved, and overall performance. The results indicated that similar but different patterns were generated under the two instructional conditions. There were four patterns in the task quota condition: "task-first," which was to do all examples of one task before moving onto the next task; "set-first," which was to do set of seven different tasks then repeat the set until finished; random selection of tasks, and a mixed strategy that showed some indications of task-first and set-first strategies. In the nonquota condition, the task-first strategy was replaced by the "favorite task" strategy, where the participant stuck with a task that seemed likeable for a while then switched to something else; sometimes there were two favorite tasks represented in the series. Set-first, random, and mixed strategies were also used in the nonquota condition, although the set-first strategy was adopted less often than in the quota condition.

Better performance was associated with task sequence choices that exhibited lower topological entropy, but only about half the participants adopted a low-entropy strategy (Guastello, Gorin et al., 2013). Both entropy metrics were correlated with the amount of voluntary task switching. There appeared to be a trade-off between switching costs and minimizing entropy. The set-first strategy actually produced the lowest amount of entropy because it involved the repetition of a long unchanging sequence, but it did involve the greatest number of switches. Task-first and favorite task strategies produced the lowest number of switches, but not the lowest entropy in task patterns.

A possible explanation for the dominance of task-first strategies, other than being a favorite for some reason, is that some people have more difficulty than others switching tasks when demands change. Gopher (1993) reported that pilots who were previously trained on a computer game that involved task switching due to situational demands eventually switched more fluidly during real aircraft operations.

7.1.8 Fluid intelligence

Individual differences in working memory capacity have been traced to individual differences in fluid intelligence. Some condensed history of intelligence would be helpful here. When the concept of the intelligence quotient (IQ) was first introduced by Binet in the 1890s, it was treated as a single construct. By 1930 there was a debate as to whether intelligence consisted of one general factor, g, or multiple independent factors. The arguments in favor of one conclusion or another was based on differences in the methods of factor analysis that researchers, primarily Spearman and Thurstone, were developing and using, the notion that different brain areas had different dedicated functions, and the validity of intelligence measurements for predicting behaviors of different types.

Guilford (1967) introduced a different approach to defining intelligence, and a factor analytic strategy to go with it, that was based on different types of inputs, different types of operations on those inputs, and different types of outputs. The net result was a taxonomy of 120 factors of intelligence, each of which consisted of a unique combination of input, process, and output. One might parenthetically see a parallel here with the notions of input, process, and output in basic computer science. In any case, one important product of Guilford's work was the discovery of the difference between divergent thinking and convergent thinking. The former being more germane to creative thinking, whereas the latter was more germane to optimizing problems; this distinction is expanded further later in this chapter.

The next major development was introduced by Cattell (1971) who resolved the debate between *g* and multiple factors, and the omission of divergent thinking prior to Guilford in the form of a theory of hierarchical intelligence (Figure 7.5). At the top of the hierarchy was *g*, for general intelligence, which separated into two broad components, *crystallized* and *fluid* intelligence. Crystallized intelligence describes the mental abilities and knowledge bases that have been accumulated over time. Fluid intelligence involves the adaptation of mental schemata and knowledge to new situations; it includes the divergent thinking processes that are involved in creative thinking (Hakstian & Cattell, 1978; Nusbaum & Silvia, 2011). The other more specific abilities that had already been identified in earlier

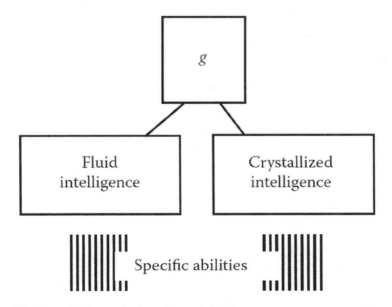

Figure 7.5 Hierarchical organization of mental abilities.

intelligence research were then thought to form a third tier in the hierarchy (Hakstian & Cattell, 1978). That said, contemporary research now shows a robust relationship between working memory capacity and measures of fluid intelligence (Kane, Hambrick, & Conway, 2005).

7.2 TYPES OF DECISIONS

There are several fundamental types of decisions that are relevant to human factors applications. Here they are grouped as simple binary decisions, optimizing decisions, and nonoptimizing decisions. Cognitive biases fall somewhere between optimizing and nonoptimizing decisions. Afterward we consider the production paradox and troubleshooting strategies where combinations of optimizing and nonoptimizing decisions are involved.

7.2.1 Simple binary decisions

Perhaps the most basic decision is the simple detection of a stimulus, such as a light or a sound. For the most part the dynamics of signal detection were covered in Chapter 3. It was also noted in Chapter 3, however, that some very complex decisions culminate in a binary decision. Is event X a member of category A, or isn't it? In both the simple and complex versions of signal detection, the accuracy of a set of decisions can be represented by the fourfold table for correct hits, correct rejections, misses, and false alarms. Inspection tasks are common examples of decisions that intensively involve signal detection.

7.2.2 Optimizing decisions

Given a set of possible choices, which one will produce the best results? Any student who has taken a multiple-choice test has great experience with this class of decisions. In the simpler examples of optimizing decisions, the thinker has all the information and reasoning processes available to determine which outcome will be optimal. In many real-world decisions, however, the thinker does not have all the information, but may be able to get the information if it were available along with the time to find it. In less fortunate situations, the thinker does not know what information is unknown but necessary. In flimsier situations, the thinker does not know what all the options are, let alone how to evaluate them.

7.2.2.1 Expectancy theory

The foregoing compromises to optimality lead to a broad class of optimizing situations involving decisions under risk. Given the situation where there are defined options and rational behavior, optimal outcomes (Max

[O]), and odds of an outcome coming to fruition (Pr[O]), a strictly rational decision takes the form:

$$E[O] = Max[O] \times Pr[O], \tag{7.1}$$

where $E[O]$ is the expected outcome.

The concept represented in Equation 7.1 dates back to 17th century economics (Vroom, 1964). It surfaced in psychology with cognitive learning theory (Tolman, 1932). The principle was that the rat *knows where* the cheese is. The notion that a rat could know anything ran counter to the basic tenets of the strict behaviorism of the time, which assumed nothing whatsoever about mental processes or structures. Strict behaviorism only considered the movement of people or animals through time and space, which, in principle, was observable to anyone who cared to observe the event or repeat the experiment.

The notion that a rat knew where anything was also ran counter to strict behaviorism because it assumed cognitive structures that Tolman named cognitive maps. *Cognitive maps* are mental representations of a physical environment that are embodied in the individual and which the individual uses to solve problems concerning the location of objects or one's personal location in space and how to move to another location. In the most pointed experiments, Tolman would first allow his rats to explore a radial maze without any reward. A radial maze has several walkways organized in a circle that all converge on a central area in the middle. He would then place cheese of different amounts in different fingers of the radial maze, and the fingers of the maze would be used at varying probabilities. The rat would be allowed to enter the maze from any of the openings, would run toward the middle, then take whatever left or right turns were necessary to go to the maze finger that had the greatest likelihood of containing the most cheese, as expressed by Equation 7.1. If there was no cheese in that target space, the rat would return and go to the second most likely location, and so on. If the rat entered the maze from different points, and the location of maximum cheese was the same in each case, the rat would need to make a different set of left or right turns to get the cheese. Because the rat was able to do so, the conclusion was that the rat had a cognitive map of the maze. Cognitive maps arise again in Chapter 14 on the topic of navigation through spaces.

Equation 7.1 would be (mentally) constructed for each option. The option that produces the best $E[O]$ wins. In situations with less than perfect information, however, one does not know the $Pr[O]$ and must guess it somehow. *Overconfidence bias* occurs when the individual overestimates the odds of success.

7.2.2.2 Prospect theory

Equation 7.1 defines a strictly rational approach to optimization. Prospect theory (Kahneman & Tversky, 1979), however, found three common compromises to strict optimization in human decision-making. One form of suboptimality occurs when the decision maker is willing to *pay for certainty*: The decision maker is inclined to accept lower $E[O]$ if there is a higher $Pr[O]$ associated with the option.

On the other hand, there are also conditions where risky behavior can be expected. If, in a series of investment decisions, the decision maker is accumulating a lot of losses, the tendency is to accept more risk in hopes that the large payoff will actually occur.

The second example of suboptimality is that losses weigh heavier than gains. Suppose Equation 7.1 were extended to include a cost for participation in an option:

$$E[O] = Max[O] \times Pr[O] - C[O] \tag{7.2}$$

and $C[O]$ was different for each option. Whereas strict rationality would simply make the subtraction in each case and look for the best $E[O]$, the suboptimal thinker would place greater weight on options that involved lower participation costs at the expense of expected outcomes.

Another version of the same bias is related to expected losses. Whereas the strictly rational decision maker would simply calculate Equation 7.1 for all the options, and make a subtraction for expected losses, and select the option with the highest net $E[O]$, the conventional thinker would regard $500 lost as larger than $500 gained. Thus, decisions would be biased toward loss aversion.

The third type of suboptimality is seen in insurance-buying behavior. Insurance is a protection against risks. Insurance costs are structured to reflect higher premiums in cases where the odds of having to cover a loss are high, and lower premiums in cases where the odds of a payoff are lower for the insurer. Hence there is a deductible amount on many types of policies; the deductible amount represents an almost guaranteed loss. Smaller losses are much more likely than larger ones; the distribution of loss sizes follows an exponential distribution. The suboptimal insurance buyer, however, is inclined to protect against as much loss as possible, and thus chooses policies that cost more in the long run than the expected losses from out of pocket. The advisement would be, therefore, to buy a policy with as large a deductible as the buyer can afford, and protect against the losses of more devastating amounts.

One might rightly ask how investment decisions became entwined with human factors. One reason is that there are other types of decisions that follow the same basic framework of expected odds and maximum possible outcomes. Another reason is that the sheer volume of information that person needs to process to make some of these decisions has inspired the development of decision tools (software) that are meant to assist the decision maker.

7.2.2.3 Incomplete information

When the decision makers do not know either or both $Max[O]$ or $Pr[O]$, they have to make something up for the missing amounts or reframe the problem in some way. In some cases, the missing information takes the form of a statistical distribution which, in most cases, can provide information about $Pr[O]$. In other cases, probability distributions can support forecasts of what to expect over time. Consider the following problem suggested by Robert Gregson (personal communication, 2008):

> You are a medic who for the first time is doing volunteer work in Africa, and you get posted to a lonely hospital. When you arrive there are ten dead bodies of Ebola virus victims in the hospital courtyard waiting to be buried. You have to order drug kits for possible N other patients who have not yet arrived. Kits are scarce and very expensive. How many kits do you order? How did you arrive at your estimate of N?

If the medic were to answer this question with the hopes of any degree of accuracy, it would be helpful to know whether the medic entered the situation at the beginning, middle, or end of an epidemic cycle. In many cases epidemic cycles start at low frequency, then the frequency builds to a peak and declines. The next question would be what sort of $Pr[O]$ produced the ten dead bodies in the courtyard, and how quickly did the ten casualties accumulate.

7.2.3 Nonoptimization decisions

Nonoptimizing decisions are not predicated on choosing a best option out of several options. In these contexts it matters only that a decision will fulfil an objective. Arguably nonoptimizing decisions might involve subdecisions that have an optimization component to them.

7.2.3.1 Planning

Here the thinkers are defining a sequence of actions that are necessary to fulfill an objective. There may be more than one viable solution, and if there is sufficient time, alternatives can be explored. We often think of plans as cognition sequences where contingencies and options have been identified and selected in advance of any actions taken. There are times, however, where unexpected events occur, and either a Plan B goes into effect or the operator must make instant revisions on the fly.

In many cases, it is sufficient to say, "This will work." In emergency situations, where time is not a luxury, "minutely wrong decisions are better than no decisions at all" (Flin, Slaven, & Stewart, 1996, p. 272).

7.2.3.2 Predicting a future state

One form of nonoptimizing decision is to determine what a system will be doing in the future based on knowledge of its situation at the present (Eberts & Salvendy, 1986). The actions that one might take in response to this prediction may involve an optimization component, but the prediction of the future state is, by itself, not an optimizing task. Evaluating a situation and anticipating needs is thus one large group of nonoptimizing decisions that appear in both benign and emergency conditions.

Professional forecasts, such as weather, political, or economic events, are formed in conjunction with decision aids, data sources, and data visualization resources that are typically more elaborate than what is usually possible in laboratory experiments (Hoffman, LaDue, Mogil, Roebber, & Trafton, 2017). Professional forecasts also require substantial domain knowledge, some of which impacts the forecaster's optimal use of decision support tools (Endsley, Bolté, & Jones, 2003; Hoffman et al., 2017). Importantly, "…there is really no evidence that subjects … can discriminate a probabilistic task from a deterministic task" (Brehmer, 1974, p. 3). This discrimination would appear to require some relevant domain knowledge. Professional forecasters can also vary in their strategies for interpreting data (Hoffman et al., 2017; Rosser, 1997). Thus, multiple heuristics and mental models could be engaged when producing forecasts, particularly when teams of forecasters work together on a complex problem.

Some investigators have explored the types of heuristics a forecaster might use to predict future states. Ritz, Nassar, Frank, and Shenhav (2018) utilized stochastic (rather than deterministic) stimuli and a procedure that produced learning with feedback. Stimuli consisted of a series of numbers generated from a mean plus Gaussian noise. Participants forecasted the next step ahead. At random times, the series shifted to a new series with a discontinuously different mean. The shifts were intended to emulate a real-world time series that is nonstationary. In a *stationary* time series, the variable X under study has the same relationship to time, autocorrelation, and connection to other variables in all parts of the time series. In a nonstationary time series, those relationships change; for instance, X could show greater or less variability over time, or show sudden changes in mean values, or new variables could start to change and have an influence of X.

Ritz et al. (2018) examined how participants made adjustments to their forecasts based on a rule of proportionality of their recent errors (delta rule), an integrative cognitive operation, or a derivative operation. They found that the participants actually adopted a combination of these strategies. Importantly, differences in reward structure could induce a bias toward maintaining the stability of the time series or exploiting recent change. The former type of forecast incorporated more historical data than the latter and was less accurate overall.

Schulz, Tenenbaum, Duvenaud, Speekenbrink, and Greshman (2017) examined biases in the learning and forecasting of deterministic trends in temporal and spatial data. It had been previously established (Bott & Heit, 2004; Delosh, Busemeyer, & McDaniel, 1997) that forecasters could exhibit biases toward smoothness, linearity, periodicity, and symmetry. The central objective of Schulz et al.'s experiments was to assess the role of compositional structure and modularity as heuristics wherein the person viewing the experimental stimuli parsed the function into linear, periodic, or other curvilinear trends as both a means of developing a mental model of the function and for making forecasts. An alternative explanation was that people viewing the data would mentally parse the trends into oscillators and make forecasts based on the dominant oscillator. The results showed that assuming a deterministic function and then parsing it somehow provided a more accurate description of the forecasts made during experiments involving forecasting numeric trends. The opposite occurred, however, when the participants were required to position and draw a data cloud (a static representation of data points) that was a continuation of a given data cloud.

There are many types of work in which chaotic patterns of events is the norm, however. They are not readily decomposable into linear, periodic, or symmetrical components, and they do not stay smooth for very long. Examples would include managing a supply chain, a manufacturing process, medical surgery, ecological systems, a natural disaster, a swarm of robots, and numerous situations in finance and macroeconomic policy. Forecasts might entail identifying possible negative side effects (e.g., of a medicine), revenge effects (Tenner, 1996), in which a plausible solution to a problem actually makes the problem worse, or the disruptive effect of an event on the status quo and the implication thereof (West & Scafetta, 2010). The ability to forecast chaotic events could be an important factor in maintaining the functionality of such systems.

The unpredictable nature of chaos in a chain of observations is paradoxical. On one hand, the mathematical-statistical techniques for identifying and predicting it in real-world time series data is possible, even though the techniques are continually under development. On the other hand, people live and work in systems that exhibit chaotic behavior, and they need to predict and control what will happen next using a heuristic rather than an a calculation on data that might not yet exist. They need to recognize patterns extemporaneously, and stopping the system to build a database and conduct numerous statistical analyses is often infeasible. So how do they manage it?

Some experiments that involved tasks requiring a participant to forecast chaotic numbers from a short sequence of stimulus values showed that people can learn chaotic functions and make accurate forecasts accordingly (Neuringer & Voss, 1993; Ward & West, 1998). Others were able to establish that the ability exists when the experiment was designed to preempt learning through feedback (Heath, 2002; Metzger & Theisz, 1994; Smithson, 1997), a small minority of participants were able to do so, but they did so nonetheless.

It remained an open question for many years as to whether the most successful participants employed a simplification strategy or had some other unknown heuristic for forecasting chaotic numbers. It was eventually shown that the first heuristic was to orient toward whether a function was persistent or anti-persistent. A persistent function would have an autocorrelation approaching 1.0; the autocorrelation of a random function would be closer to .00, and the autocorrelation of an anti-persistent function would be closer to –1.0, approximating an oscillator (Hurst, 1951). Anti-persistent chaotic functions, such as the logistic map (see Figure 3.4), were more difficult to forecast than the more persistent attractors (Figure 7.6; Guastello et al., 2021a).

The second heuristic distinguished low, medium, and high-performing participants across four different types of chaotic attractor. The participants were shown a short trend of eight number, and were asked to predict the next four numbers. The low-performing participants made their forecasts based on a moving average of all eight stimulus values. The medium-performing participants made their forecasts based on moving average of three values. The high-performing participants used three-point moving averages to some extent, but they appeared to have intuited the actual chaotic pattern more clearly than what three-point moving averages would have provided (Guastello et al., 2021a).

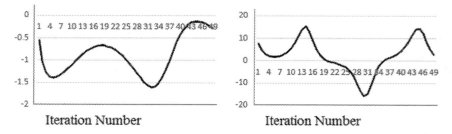

Figure 7.6 Persistent chaotic attractors. Left: Sprott-B attractor. Right: Lorenz attractor.

7.2.3.3 Divergent thinking

Whereas optimization problems require the selection of the best option, divergent thinking, which characterizes the creative thinking process, involves the identification of many possible solutions or courses of action (Guilford, 1967). Divergent thinking is eminently valuable in the system design process (Mavrommati, 2001), creative thinking more generally (Guilford, 1967), forecasting (Guastello et al., 2021b), and in unique situations where well-worn decision paths are not available.

Three decades ago, however, divergent thinking resided outside the realm of decisions relevant to human factors. The advent of new types of computer programs that are intended to assist engineering design and creative exploration of options has given creative thinking a new relevance to human–machine interaction.

Other types of nonoptimizing decisions involve maintaining communications, delegating authority, and dealing with stress issues (Flin et al., 1996). The role of stress in human performance is addressed in Chapter 10. Dynamic decisions can be both optimizing and nonoptimizing, and they are explored in their own section of this chapter.

7.2.4 Production paradox

Somewhere between optimizing, nonoptimizing, and suboptimizing is the *production paradox* in technology choice (Carroll & Rosson, 1987). In this class of situations, humans have already become accustomed to their machine systems and learned how to obtain maximum production and reliability out of them. A new technology choice is introduced as a possible productivity enhancement. Although the new production system may indeed work out well, the users of the current system are reluctant to disrupt their well-worn and successful production strategies to learn the new system and make the usual learning mistakes along the way. The production paradox thus results switching costs on a larger scale.

Thus, optimality may be different in the short run than in the long run. A PMS that is making a cold start-up may be better off with the new system. A system in operation, however, would have to reach a utility crossover point at which the expected gains from changing the system sufficiently outweigh the disruption and inconvenience involved.

7.2.5 Troubleshooting

Things break or stop working, and someone has to figure out why. Buck and Kantowitz (1983) identified some basic cognitive strategies for system troubleshooting, which may be assisted to greater or lesser extents by diagnostic tools and equipment.

7.2.5.1 Fault isolation

A system consists of parts, and parts have functions. System parts may have many subparts with more specific functions. Diagnosticians who have a good working knowledge of a system's components can isolate a problem. The problem within the problem, however, is to determine how broadly or specifically the problem can be defined. A broad-level diagnosis may be quick to perform, but a detailed diagnosis may become extremely labor intensive and thus costly to repair before the repairs actually begin. Systems that are designed in a modular fashion can allow the replacement of chunks of the equipment without having to resort to a repair that is targeted to the smallest possible failing components.

"The car won't start" is not a very helpful problem statement. The noises that the car makes (or refuses to make) will help, however, to isolate the problem to a dead battery, a faulty ignition, faulty fuel-injection system. In this case the diagnostician is following a sequence of automotive system performances to isolate the failure. On the other hand, if the car starts and the "check engine" light goes on, another automotive subsystem is indicated, although the possibilities are still vast.

7.2.5.2 Template matching

Some user manuals for consumer products contain a chart of common faults with common fixes. They take the form "If fault X is present, try correction Y." Although troubleshooting pages that appear in the operating instruction for consumer products only address simple or peripheral repairs, a system diagnostician of greater skill may rely on the same kind of thinking but can reach a greater level of specificity in the diagnosis. Sometimes the diagnostic windows are known to the repair folks but not to the system user.

7.2.5.3 Statistical template matching

Sometimes an observed fault indicates more than one problem. The diagnostician then plays the odds. The most likely cause of the fault will be checked first, followed by the less likely possibilities. In such situations, it helps to have an odds table handy, mentally if nowhere else, which would be based on previous experience with the same or similar systems.

Sometimes the statement of one system fault does not provide enough information. It may be helpful to know what other parts of the system are still working. Based on that additional knowledge, the odds of one cause or another being relevant may shift.

7.3 COGNITIVE WORKLOAD

Widespread interest in cognitive workload problems started with Simon's (1957) concept of bounded rationality and Broadbent's (1958) research on cognitive capacity. In the case of *bounded rationality*, the observation was that many types of decisions require the individual to find pertinent and process large quantities of information in too short a period of time. As a result, suboptimal decisions are usually adopted in a process called *satisficing*: The solution will solve the problem, but it is probably not the best solution possible.

The earliest research on cognitive capacity attempted to measure workload in bits and bytes of information, with the goal being to draw conclusions such as, "The human information-processing capacity is 10 bits per second." This type of statement would be tidy and convenient, but it turned out to be incorrect. Rather, the number of bits per second that a person can process is highly dependent on the type of information that is involved as well as with the person's knowledge and skill for working with it. Thus the relevant issues of human mental capacity must be understood in relative terms. Relativism, however, produces workload measurements that are not comparable across types of tasks (Lin & Cai, 2009; Morineau, Frénaud, Blanche, & Tobin, 2009; Neerincx & Griffioen, 1996). For instance, cognitive workload for air traffic controllers is directly related to the number of planes taking off or landing in a given time period (Chatterji & Sridhar, 2001; Loft, Sanderson, Neal, & Mooij, 2007). Bad weather adds another level of workload because of the way takeoffs and landing need to be managed and because of rescheduling flights in the more demanding weather conditions. Such conclusions are not particularly helpful in situations that do not involve airplanes.

On the other hand, professional services are also involved. The goal of a workload investigation, a "diagnosis" so to speak, is to identify specific tasks, conditions, and decisions that produce the greatest workload, and figure out why uncomfortable delays and errors are occurring. The solution step might be directed toward simplifying tasks, providing decision support through automation, improving the human–machine interaction where automation already exists to some extent, or enhancing the level of coordination within a team of operators. Measurements of workload can be performance measures under different types of conditions, subjective ratings, or physiological measures; they are discussed in a later section of this chapter. Work team issues are discussed in Chapter 14.

As all the relativity came to the foreground, the theory question morphed into a broader question: Is the human channel capacity fixed or variable? The implications of limited capacity and variable capacity theories of mental workload (Kantowitz, 1985) are considered next.

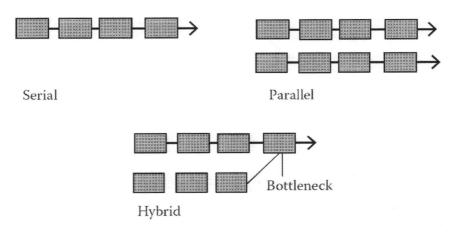

Figure 7.7 Configurations of channels and stages.

7.3.1 Channels and stages

Figure 7.7 depicts three basic configurations of cognitive channels and stages. The *serial process* is the simplest. There is a sequence of mental operations that usually begins with an input of some sort and ends with a product of some sort. The *parallel process* indicates two mental operations going on simultaneously. The two processes could be completely unrelated tasks, or they could be related such as processing the audio and visual cues of a situation while writing something down or clicking a mouse at an auspicious moment.

The *hybrid process* contains aspects of serial and parallel processes. Of importance, there is a bottleneck in the process whereby two mental streams must converge with their products at the same place in time. Bottlenecks are likely places where a process can become seriously slowed or derailed.

Although Figure 7.7 is intended to represent varieties of cognitive processes within one particular human, it can be used to depict production systems that involve multiple PMSs just as readily. The latter constitute complex systems, which are considered in Chapter 14.

7.3.2 Limited capacity theory

The limited and variable capacity theories of cognitive capacity are considered next. According to Kantowitz's (1985) review, both views have support from cognitive experiments. According to limited capacity theory, there is a rigid fixed upper limit to the human mental channel capacity. The total capacity may be divided among primary, secondary, and even tertiary tasks or channels, where mental efforts are allocated among the tasks. The allocation function is thought to function like a time-sharing computer; this is a metaphor that was held over from the days when mainframe computers were king. The mainframe would work on jobs that were submitted by many users at the same time. A certain amount of time would be allocated to processing each person's job. Run times for a job on a mainframe were substantially faster after midnight when most system users were not working.

In any case, if a person's total channel capacity was really 10 bits per second, and the primary task only required 6 bits per second, then 4 bits per second were left over for the secondary task. If the secondary task did not require the full 4 bits, then some small amount was left over for a tertiary task.

Task difficulty is defined in terms of the number of bits per second of a particular type of information that a person can process. Note here that difficulty and complexity are different concepts. Complexity only refers to the number of channels and stages involved in a task. Difficulty pertains to transmission speed. If two easy tasks required a total processing bandwidth that was less than total human capacity, there would be no slowing of performance on either task. If one task was more difficult, however, slowing would be observed.

The limited capacity theory would identify two kinds of bottleneck conditions. In one case, if the human tried to process more jobs simultaneously than the total channel capacity would allow, any job in progress would be slowed. Another type of bottleneck point occurs where two parts of a parallel processes converge in a hybrid system. An added increment of processing time is related to the integration of the two flows of information in addition to what is required to process each one separately.

An implicit assumption running through the research on limited channel capacity theory was that the limits to channel capacity are fixed by the nature of the channels. Naturally occurring channels in human cognitive processes may be related to the use of different brain functions, such as pictures versus words, or visual tracks versus auditory tracks. If the goal is to process pictures and words in two channels, the process might flow freely. A good picture could help the words to make more sense, hence we illustrate textbooks and design visual icons; the redundancy can improve the reliability of the human decision process or trigger two neurocognitive pathways, one of which could be faster than the other. If the two channels of communication involve processing two conversations at once, however, the total information is crammed into a narrower channel and might not be processed well at all. Hence we hear the familiar phrase, "One at a time, please! I can't listen to you both at the same time!" The colloquialism *information overload* applies to both types of situation.

7.3.3 Variable capacity theory

In variable capacity theory, processing space is allocated on an intentional basis. The operator sets priorities and thus allocates processing space to the incoming tasks in the desired order. The operator is then assumed to define tasks as those that must be done first, those that get some attention while the first task is really getting done, and those that benefit from a little downtime on the first two tasks.

Importantly, maximum channel capacity increases with demand. There are no firm rules as to when a bottleneck will occur, but an upper limit to the human's capacity can be reached eventually. Ralph, Gray, and Schoelles (2010) likened the phenomenon to squeezing a balloon: Constraints in one place produce stretching in another place, but one cannot predict exactly where in advance; eventually too much constriction pops the balloon. Emergency response (ER) is such an example. Search and rescue teams and their supporting associates could work very long hours in potentially dangerous environments to find survivors of a building collapse or volcano eruption. They push themselves far beyond the norms of daily life until the job is under control, and they are exhausted and not quite ready to move so intensely for some time afterwards.

According to Kantowitz (1985), empirical evidence supported both the fixed and variable capacity models of channel capacity. The next research questions would investigate where the variability in the capacity comes from, and what part of the cognitive system is fixed overall. Several sources of variability came to the surface.

One partial explanation favoring variable upper limits involves coping or resilience (Guastello, Marra, Correro, Michels, & Schimmel, 2017; Hancock & Warm, 1989, Harris, Hancock, & Harris, 2005; Hockey, 1997; Matthews & Campbell, 2009; Sheridan, 2008); people find ways to adapt to uncomfortably high or low workload levels. There could be *many* possible adaptive responses that someone could make to control fatigue or excessive workload demands, however, some of which might be afforded in some situations but not in others. For example, people working under fatigue conditions often respond by slowing down their output, that is, reducing speed stress (Lorist & Faber, 2011), in an effort to maintain their work quality standards. Attention tunneling, which could be voluntary or inadvertent, could be another type of response. There may be other ways to regroup a task process. Task switching is another source of possible strategies. The options for adapting lie somewhere in the interaction between the person and machine system, and we are looking for the degrees of freedom that allow us to do something differently, and hopefully more efficiently.

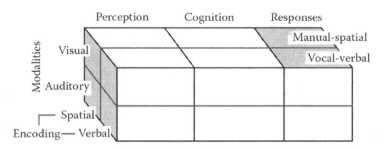

Figure 7.8 Resource competition model based on Wickens (2002, p. 163).

7.3.4 Resource competition model

According to Wickens (2002, 2008a), the decrement in performance that occurs when two or more tasks are performed simultaneously depends on whether the tasks draw on the same perceptual, cognitive, or psychomotor resources. For instance, an audio-based task and a visually based task could be processed simultaneously rather well, consistent with Baddeley's theory (2003) of working memory, but two visual tasks of the same complexity and information density could gag each other if the operator tries to do both simultaneously. It is possible to carry on a telephone call while watching things on a screen, but two different phone calls at one are not going to work unless one of the callers is willing to spend a lot of time on hold.

Another important feature is that the tasks are performed in stages. If the stages are cascaded so that they do not absorb the same resources simultaneously, so much the better for performance overall. The relationship between task resources and stages is shown in Figure 7.8. The two phone calls that were just mentioned are much more likely to move forward as planned if they were conducted in series rather than in parallel. The resource competition model explains how and when dual task or multitask performances can be effective. It does not explain by itself, however, the individual differences in total load capacity are affected when a single, large intact task is performed and capacity limits could be reached.

7.3.5 Attentional blink

The relative independence of auditory and visual channels and the relative benefits of serial processing have their limits. Experimental conditions have come to the foreground whereby auditory and visual signals arriving simultaneously can have inhibitory effects on each other. The phenomenon, known as *attentional blink*, has been observed at the levels of responses to single-mode stimuli presented in rapid succession (~300 ms), responses to dual-mode stimuli presented in rapid succession, and neural activation (Dux & Marois, 2009; Haroush, Deouell, & Hochstein, 2011; Kelly & Dux, 2011; Marti, Sigman, & Dehaene, 2012; Raymond, Shapiro, & Arnell, 1992; Theeuwes, van der Burg, Olivers, & Bronkhorst, 2007).

The attentional blink is one of the two similar phenomena that have been observed during the first half-second of the presentation of the signals or stimuli; the companion phenomenon is the interstimulus refractory period (IRP). Like the single neuron that cannot fire twice in response to two stimuli that arrive too close together in time, neural circuits in the prefrontal gyrus appear to have the same limitation in reorienting to a stimulus appearing at an interval less than one half-second. IRPs are more evident when either auditory or visual stimuli appear in rapid succession. When interstimulus intervals exceed the IRPs, which are less than 300 ms, however, an attentional blink with either the auditory, visual, or cross-modal audiovisual stimuli can occur (Marti et al., 2012).

Not all experimental participants exhibit attentional blink under the same experimental conditions, leading some researchers to study individual differences (Kelly & Dux, 2011; MacLean & Arnell, 2010; van Dam, Earleywine, & Alterriba, 2012). If the participant does not miss the stimulus altogether, there is still a slightly extended response time to one or the other stimulus. When cross-modal blink does occur, however, auditory stimuli interrupt visuals more often than vice versa (Haroush et al., 2011; Marti et al., 2012). The alleged source of conflict between stimuli is that, although the auditory and visual circuits start in separate places, they both bottleneck through a common circuit that is associated with the executive functions of working memory (Figure 7.9). The circuits that are responsible for IRPs are thought to be the same as those responsible for attentional blink. The circuits separate again as later processing continues.

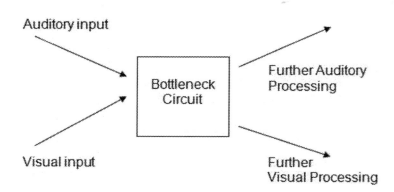

Figure 7.9 Bottleneck producing cross-modal attentional blink.

That interpretation is consistent with experimental results. One such bottleneck was identified in the earliest processing stage by manipulating the asynchrony in stimulus onset. When the presentation times of two stimuli were too close together, the response time to the second stimulus was impaired, but as the asynchrony became greater, the impairment reduced to an asymptotic minimum (Pashler & Johnston, 1998).

The typical blink experiment utilizes a stimulus recognition task that does not appear to place a high demand on working memory. One research team investigated what could happen when the task, such as an N-back task, is more demanding (Guastello, Reiter, Malon, & Shircel, 2015). In an N-back task, the research participants attend to a series of auditory or visual stimuli and need to remember whether they saw a particular stimulus just previously (1-back), two events previously (2-back), or three events ago (3-back). N-back tasks place a strong demand on the executive function of working memory (Kane & Engle, 2002), particularly the updating function.

The stimuli in the N-back experiment (Guastello, Reiter et al., 2015) contained an auditory, which was a spoken letter of the alphabet, and a visual element, which was one of 24 cartoon faces. Participants completed 15 blocks of 20 trials in the 2-back mode, which was followed by 15 blocks of 20 trials in the 3-back mode on a desktop computer. The visual exposure on screen was 500 ms, and the allowed response time was 3 s; thus the duration of a 3-back trial was 9 s, well beyond the usual range of blink. Results showed that some participants showed 0% accuracy on the auditory channel, whereas about four times as many performed with 0% accuracy on the visual channel. There were fewer errors on the auditory channel overall, which was consistent with previous stimulus recognition experiments. An important new conclusion, however, was that blink in the first 500 ms can have a cascading effect on subsequent levels of processing that could extend as long as 9 s/stimulus.

7.3.6 Multitasking

A good quantity of theoretical and applied research on cognitive workload utilizes the *dual-task methodology*. The usual objective here is to determine the workload impact of a particular task or to compare two alternative tasks. The first step is to define a task for the participants that all participants will perform. This task is intended to soak up a large amount of channel capacity so that the critical second task will push against the cognitive boundaries. The dependent measure is usually the response time or accuracy associated with the second task. As an example of dual-task methodology, Thackray and Touchstone (1991) gave their research participants a simulated air traffic control task as a primary task. The secondary task was to respond to another signal for of intrusions into the airspace. The intrusions were presented in two experimental conditions, flashing lights and colored lights. Flashing lights produced better detection over a prolonged period of time.

Multitasking is simply a case of doing multiple tasks at once, and has become a popular buzzword in recent years. One difference between multitasking and the dual-task methodology is that multitasking in the way most people do it, does not clearly designate a primary or a secondary task. Another is that it is not so methodological. Any two tasks that need to be done can be combined into a multitasking event, and when it is completed another task can be engaged to the extent possible. The keywords, however, are "to the extent possible." There are limits to channel capacity, and the simultaneity of the two tasks can be problematic as well. If one task involves a flow of inputs over time that is slow and irregular, there is more opportunity to fit in a second task. If the input flow from one task is high and steady, however, there is less opportunity to work on the second task.

Success at multitasking often depends on the timing demands for the two tasks. De Pontbriand, Allender, and Doyle (2008) distinguish between explicit and implicit timing demands. *Explicit timing demands* are system-driven such that the pace of the inflow is controlled by the system and the person needs to respond to events quickly. This is again a scenario that does not favor performance on the second task. *Implicit timing demands* are set by the individual who finds a way to coordinate two tasks by managing the wait times on the two tasks. Sometimes it is possible to let work on one task pile up to a critical mass before switching tasks.

Task switching is thus a part of multitasking, and it can become more critical in cases where the operator must tend to the affairs of several machine simultaneously, especially when the need arises to shift focus from one type of machine to another. When the requirements of the two operations are very different, there is a greater demand on working memory as a result.

One of the more controversial scenarios for multitasking is found in the contemporary automobile where resource competition is high. According to Angell and Lee (2011, p. 3):

> In contrast with 1977 – when the first vehicle to implement a microprocessor was introduced (controlling for a single function) – today's high-end vehicles contain 70 to 100 distributed computer processors that run more than 100 million lines of code, 20 million of which may be required for the navigation system alone ... And ... drivers bring their own technology into the vehicle to use while driving: cell phones, smartphones, music players, and portable navigation systems, just to name a few.

Numerous studies published have assessed the impact of distracting devices on driving. The key points that arise from the research are that (a) the non-driving tasks are visually demanding and sometimes cognitively demanding as well. (b) The distractions slow response times to suddenly appearing road hazards. (c) The drivers' attention becomes restricted such that they are most likely to miss events in their peripheral vision rather than in their central visual field (Horrey, 2011). The new technologies incur substantially different demands compared to conventional distractions such as talking to a passenger, playing the radio, or lighting a cigarette with the in-vehicle lighter. Rather, the focus of attention is often shifted away from the user–road interaction to the outside world that is not physically present (Strayer, Drews, & Johnston, 2003). The driving error rate associated with cell phone use while driving was found to be comparable to the error rates associated with driving under the influence of alcohol at the legal-critical rate of 0.8% blood alcohol content (BAC) (Strayer, Drews, & Croutch, 2006). Furthermore, hands-free telephone devices, while required (if the driver uses a phone at all) in 24 out of 50 U.S. states, Washington DC, and the U.S. territories, do not mitigate the driving errors (Governors' Highway Safety Association, 2021; Strayer et al., 2006). The use of handheld phones is banned for novice drivers (age 16 or 17, or holding a license for less than 6 months), which is expected to curb some of the potential disasters.

Text messaging while driving is perhaps the most controversial distraction at present; 48 U.S. states, Washington DC, and the U.S. territories have laws prohibiting it, and 1 other has a ban for drivers under the age of 21 (Governors' Highway Safety Association, 2021). Of the other distractions beyond the cell phone, text messaging tops the list with a relative risk of a crash or near-crash in excess of 2300% while driving a truck compared to driving without using the device (Dingus, Hanowski, & Klauer, 2011). Heavy trucks require more attentional demand than standard automobiles; a comparable figure for texting while driving an automobile is not available. The use of cryptic, highly abbreviated and often misspelled words, or "text-speak" puts an added demand on the person who is receiving and trying to understand the text (Head, Helton, Russell, & Neumann, 2012). By the same token, people do other questionable things behind the wheel. Applying makeup and other personal grooming, or reading books, work papers, and newspapers while driving are good for a relative crash risk of 300–450% (Dingus et al., 2011).

7.4 MEASUREMENT OF COGNITIVE WORKLOAD

As mentioned previously, the early attempts to quantify cognitive workload in terms of bits and bytes were not successful; it had become clear that metrics that were framed from a machine's perspective were mostly irrelevant to the human experience and processing capabilities. A more helpful approach to cognitive load assessment is to rely on overly aggregated indices of workload such as the number of airplanes hovering over an airport or the number of customers at a restaurant. Those measurements are situationally specific, usually at a particular point in time, and do not translate well for assessing workload in other contexts or for comparing PMSs for their relative impact on load. Another point to consider is that there are often several tasks transpiring simultaneously, each of which is on its own time horizon regarding when it starts, how long the parts require, when it ends, and whether it helps or conflicts with anything else going on (Hancock, 2017).

A good first step in the translation process is to distinguish complicated systems from complex system. *Complicated* would refer to many channels and stages of cognitive processing, but those components could operate in a predictable mechanistic fashion. A person working within a *complex* system, however, is susceptible to influences from other subsystems and makes an impact on other agents within the system and its environment, particularly as one task or subtask phases into another. Having at least drafted an outline of the cognitive demands, one can design studies around the different demands of the task and the particular demands on the executive functions channel capacities of working memory. There are three classes of measurement solutions available, nonetheless—behavioral, subjective, and physiological—which are discussed next.

7.4.1 Behavioral indicators

The objectivity of behavioral indicators has always been attractive to psychological researchers. In a workload experiment, the participants would perform tasks under different conditions, and the dependent measures would be performance quantity or quality, error rate, or response time. Gawron (2019b) gives numerous examples of performance measures that have been useful in a variety of applications. A workload assessment would then look for conditions that produced significant decreases in performance or response time or increases in error rates.

The downside of behavioral indicators, however, is that they might not always indicate the unseen part of the process. There are many work ecologies where the task in question is not likely to consume total workload capacity, and thus differences in workload between different PMSs would not be reflected in performance outcomes. People can juggle their available degrees of freedom in task performance to compensate for changes in load and keep their performance consistent as a result. Learning and automatization can play a role in keeping performance buoyant while workload is increasing (Hockey, 1997; Szalma & Teo, 2012) as do other types of coping strategies (Guastello et al., 2017; Hancock & Warm, 1989).

The dual task methodology was developed to address the hidden processes in workload–performance relationship. The basic strategy is to assign one task that will absorb much of the workload capacity, and present the test task as a secondary task. Then the performance differences on the second task would be reflective of differences in the underlying workload.

In cases where the target task absorbs the majority of the workload, drops in performance on the secondary task would reflect differences in workload demand from the primary task. Gopher (1993) noted that it is important to clarify to the research participants that they should maintain performance on one or the other task during the experiment.

The dual task methodology adapts well to multitasking environments by shifting the emphasis on how much of one task is done well to how well two or more particular tasks are performed together. In some of those examples, it might be preferable *not* to specify which task is more important, and watch what happens on a voluntary or ad lib basis.

7.4.2 Subjective indicators

The most widely used subjective indicator of workload is the NASA Task Load Index (TLX) form (Hart & Staveland, 1988). Although the NASA-TLX is not always used in the same manner in all applications, a commonly used version consists of six 20-point scales. After completing a task or task segment, the research participants mark their ratings of mental demand, physical demand, temporal demand, performance (How successful were you in accomplishing what you were asked to do?), effort (How hard did you have to work to accomplish your level of performance?), and frustration (How insecure, discourage, irritated, stressed and annoyed were you?). The extreme ends of the scale are simply labeled "very low" and "very high" without any other anchors in the middle. Sometimes the NASA-TLX is used as a total scale giving a single number for the event, but many researchers like using the scales separately for diagnostic purposes (Matthews & Reinerman-Jones, 2017).

The NASA-TLX has good reliability, which reports Cronbach's alpha values of .81–.83 for internal consistency across the six-item survey (Braarud, 2001; Hart & Staveland, 1988). In another report, alpha reliability was .72–.77 without the performance subscale, and split-half reliability of .70 when all six scales were used to rate two related but different task situations (Guastello, Correro, & Marra, 2018); the performance subscale does not always follow the same trends as the other scales (Sellers, Helton, Näswall, Funke, & Knott, 2014). TLX has good sensitivity to experimentally induced changes in workload (Dey & Mann, 2010; Guastello & Marra, 2018), and it correlates as expected with physiological measurements of workload (Funke, Knott, Salas, Pavlas, & Strang, 2012).

The NASA-TLX has also been used in a computer-delivered format rather than paper-and-pencil format, which seemed to make sense for evaluating experiences with a computer program. It appeared, however, that the computer-delivered format produced higher workload ratings than the paper-and-pencil version (Noyes & Bruneau, 2007). Some variations on the scale format involved the use of 10-point or 100-point scales and a system of weighted scales. Neither the weighting system nor the numerical ranges on the scale seem to have made any difference in experimental results, and are not considered further here.

There are other subjective rating systems for workload, which are catalogued in Matthews and Reinerman-Jones (2017) and Gowran (2019b). Some are close variations of themes found in the NASA-TLX, while others examine other aspects of workload in formats more similar to a survey or questionnaire. There is also a group-level workload survey with scales that parallel the NASA-TLX in rating style and content: coordination, communication, time sharing demand, team efficacy, team support, and team dissatisfaction (Helton, Funke, & Knott, 2014). These group-level constructs are better appreciated in later chapters on workload stress and group dynamics.

In a different type of rating concept, the Multiple Resources Questionnaire (Boles, Bursk, Phillips, & Perdelwitz., 2007) measures the demand on 17 different physiological indicators of cognitive workload in auditory and visuospatial channels, such as: responding to spoken words, numbers, and symbols; focusing attention on the location of an object; judging time intervals; and recognizing faces (p. 33). Each of the items is rating on a 0–4 scales, ranging from no usage to low, medium, heavy, and extreme usage. The underlying idea follows Wickens' (2002, 2008a) resource competition model. When implemented with a dual task research design it should be possible to determine the extent to which two tasks involve independent resources, and which pairs have greater overlap. There was some debate over how well those objectives were accomplished by this type of rating, however, and how well it captures the underlying cognitive processes (Boles & Phillips, 2007; Vidulich & Tsang, 2007; Wickens, 2007).

The Subjective Workload Assessment Technique (SWAT; Reid & Nygren, 1988) contains three measurements: time load, mental effort load, and psychological stress load. Each measurement is a rating on a 0–100 scales with three anchor levels. Low workload corresponds to often having spare time and few interruptions, mental efforts that are almost automatic, and little confusion, risk, or frustration. High workload corresponds to having almost no spare time and many interruptions, extensive mental effort for a complex activity, and intense stress. The technique has seen a lot of use in aircraft environments, but is in principle more widely applicable.

7.4.3 Physiological indicators

Several types of physiological indicators of workload have been examined for their sensitivity to differences in response to workload. They include EEGs, measurements of autonomic arousal, visual tracking, and near-infrared scanning. They were introduced as part of a broader vision for *neuroergonomics* (Parasuraman, 2003, 2013), one goal of which was to build three-level theories that integrate behavior, cognition, and neural activity. The intended applications included studies of workload (Gowran, 2019b; Matthews & Reinerman-Jones, 2017), building adaptive interfaces that respond to the operator's level of attention (Schmorrow & Stanney, 2008), and exploring linkages between coworkers during a task (Kazi et al., 2021).

EEGs are measurements in electrical conductance of the brain that are taken from electrodes attached to the skull. Up to 64 electrode channels have been usually used, but modern high-density equipment can accommodate 128 site readings. EEGs have been useful for decades for clinical diagnostic purposes or general study of the brain that can be done unobtrusively with humans (compared to surgeries and chemical ablations of brain tissue that had been performed on brains of animals). The chief limitations of the EEG medium are that (a) they are topical readings and are not likely to register functions from deeper within the brain, and (b) the research participants (or real-world employees) have limited mobility with all the wires attaches to their heads connected to a machine that reads the data. Current wireless technology, however, has solved the second problem so that research participants only need to wear a light skull cap containing the sensors while performing tasks that might involve moving above or driving a vehicle. Military personnel can wear the sensor caps under their helmets while on maneuvers. EEG signals are apparently sensitive enough to make the question of deeper brain activity less of a concern to researchers studying workload.

The P300 wave is the signal that is most sensitive to changes in workload within a lag of 300 ms between the onset of a stimulus and the electrical response. The most sensitive readings are taken from the parietal lobe, although readings from the parietal-to-frontal and parietal-to-temporal areas have also been shown to activate in response to mental processes (McEvoy, Smith, & Gervins, 1998; Schmorrow & Stanney, 2008; Verbaten, Huyben & Kemner, 1997). The P300 signal increases in amplitude when a stimulus produces an attention demand, and decreases in amplitude when working memory load is higher (DiNocera, Ferlazzo, & Gentilomo, 1996; Fowler, 1994; Gontier et al., 2007; McEvoy et al., 1998; Morgan, Klein, Boehm, Shapiro, & Linden, 2008; Pratt, Willoughby, & Swick, 2011).

Alpha and beta waves, which are actually wave shapes, are also under investigation for usable responses to workload demands. Beta waves are prominent in a waking and alert state. They are more "tightly wound" than the alpha waves (Figure 7.10), which are prominent in a relaxed state and often just prior to falling asleep. Deeper stages of sleep produce yet other wave forms. The latest approaches to workload analysis involve observing the interplay between alpha and beta waves, where the alpha waves signal cognitive fatigue (Desari, Crowe, Ling, Zhu, & Ding, 2010; Schmorrow & Stanney, 2008). The P300 also shows variability over time on sustained attention tasks; Smallwood, Beach, Schooler, and Handy (2008) showed that

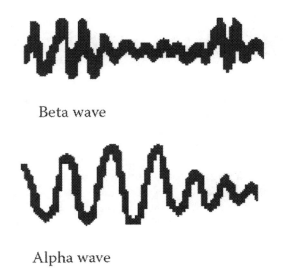

Beta wave

Alpha wave

Figure 7.10 Examples of beta and alpha waves.

subjective indicators, error incidence, and P300 amplitudes varied together over time leading to the conclusion that operators' level of awareness, attention or focus naturally varies over time.

7.4.4 Autonomic arousal

The primary measures of autonomic arousal are the electrodermal response (also known as galvanic skin response), heart rate, and respiration rate. The simple idea is that workload demands produce arousal, and the arousal would occur just prior to the cognitive activity and would provide more information about workload responses that would be dampened out by the time the operator makes subjective workload ratings after the task is completed.

Electrodermal response is a change in electrical conductance of the skin caused by microscopic activity of sweat glands, and is the most responsive of the autonomic measures to cognitive workload manipulations. Some successes in experimental studies have been recorded also for heart rate and heart rate *variability*, but not so often for respiration rate (Gowran, 2019b; Johannsdottir, Magnusdottir, Sigurjonsdottir, & Gudnason, 2018; Lindholm & Cheatham, 1983; Matthews & Reinerman-Jones, 2017).

The conventional analytic strategy for EEGs and electrodermal responses is an event-based method that comparing a person's reaction to a stimulus against a baseline level of activation, noting the difference, and comparing

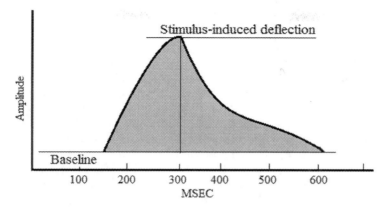

Figure 7.11 Deflection in EEG or autonomic response.

differences across stimuli as shown in Figure 7.11 (Bach, Flandin, Friston, & Dolan, 2010). Bach et al. (2010) presented some evidence that measuring the area under the arousal curve produces a better indicator of responses to stimuli; the extent to which the arousal deflection sustains over time apparently matters. For continuous tasks, particularly those in which team members can have substantial effects on each other's physiological processes during the activity, a continuously streaming EEG or electrodermal response is collected and then subjected to time series analysis (Guastello, Correro, Marra, & Peressini, 2019; Stevens & Galloway, 2016).

The interest in heart rate variability is an idea that crossed over from NDS research, notably Goldberger (1996). The general principle is that human physiological and performance systems contain natural amounts of variability that are deterministic in nature, for example, detectable with a fractal or chaotic analysis of time series. Healthy systems have a modicum of variability that exists to give the person some degrees of freedom to make adaptations to nuances in the environment. Unhealthy system either become rigid and lose this variability or become excessively variable to the point of losing almost all sense of order (Schuldberg, 2015; West, 2006). Although workload experiments do not manipulate their research participants into unhealthy conditions, a loss of variability is expected when workload becomes too high.

Conventional analyses of heart rate (and other) variability data use the standard deviation of the time series as their primary metric of variability. NDS researchers prefer metrics such as sample entropy or the fractal dimension because they separate noise (statistical error) from deterministic variability, quantify the level of complexity in the time series, and preserve the dependency between observations over time, making the NDS metrics more sensitive to changing workload conditions. Standard deviations are indifferent to the order in which the data points are observed.

Eye-tracking data were mentioned in Chapter 5 as a means for assessing the patterns by which operators view a complex display or a collection of displays. For workload assessment, the primary indicators are *dwell time*, *eye blink rate*, and *pupil dilation* (Matthews & Reinerman-Jones, 2017). Viewers will spend more time on displays or parts of displays that require greater workload to process. Similarly, greater visual workload produces slower rates of eye blink and larger pupil dilation.

The measurement and interpretation of dwell time appears to work well enough for carefully controlled stimuli delivered at regular intervals. The interpretation becomes more challenging in more naturalistic settings, however, because saccadic eye movements have natural breaks in their focus time, and perceptual cycling could interrupt the fixation time. In a recent study that used eye-tracking metrics to measure workload in tasks performed with robotic surgery equipment, greater *entropy* in gaze fixation was associated with greater workload (Wu et al., 2020). The interpretation was that novice users in the study did not know where to focus their attention in the visual field, and experts could focus on the best locations easily.

Near-infrared spectroscopy is one of the newer physiological measures in the toolbox. It measures brain activity in the form of oxygen metabolism in much the same way as fMRI, but with greater flexibility of movement and much less expense. Instead of immersing the research participant in a large tube, probably with movement restraints, it only requires a sensor mounted on the person's forehead. It can produce continually streaming data instead of snapshots taken at 2 s intervals, which is the usual speed of fMRI data collection.

Near-infrared spectroscopy is approximately as sensitive to workload manipulation in event-based tasks as EEG, heart rate variability, and eye-tracking data. Yet all four metrics are not completely effective and each tend to be more responsive to different features of the stimuli (Matthews, Reinerman-Jones, Barber, & Abich, 2015). This situation led to the conclusions that multiple physiological measures should be used in workload studies.

7.4.5 Augmented cognition

The success with the response of the P300 wave and the alpha-beta distinction led to a broader plan of augmented cognition and adaptive interfaces. Augmented cognition takes the form of an adaptive paradigm and an integrative paradigm (Schmorrow & Stanney, 2008). In the adaptive paradigm, the system senses elevated workload and goes into automated mode. The automated mode turns off when workload returns to normal. The switching between automated and normal modes makes sense if the human performs better than machine under normal circumstances. Alternatively, if one designed a better interface, the need would go away.

The integrative paradigm would program the machine to detect stimuli that are in high demand, for example, when the operator is analyzing or scanning a set of photographs. The machine would return high-demand stimuli to operator for a second inspection. Importantly, the demand is detected from physiological data rather than from analyses of the stimuli themselves.

There are at least three unanswered questions at this juncture: (a) Given that attention and the P300 indications thereof wax and wane over time, how would the machine discern between a stimulus that might have received insufficient attention and normal fluctuation? (b) Is the technological vision here to test all tasks for P300 fluctuations and then use the knowledge some other way, or is it to have every operator everywhere wear a skullcap all day long so a machine can prod them into increased levels of attention? (c) EEGs contain a lot of unexplained variance, which is often characterized as "noise." Data analysis techniques for nonlinear dynamics address the problem by separating deterministic chaos from noise, thereby reducing the amount of noise or unexplained variance (Guastello & Gregson, 2011; Shelhamer, 2007).

7.5 AUTOMATIZATION OF COGNITIVE PROCESSES

7.5.1 Telegraph operation

Two of the first empirical studies that could be considered human factors studies pertained to the effect of training on the proficiency of telegraph operations (Bryan & Harter, 1897, 1899). Telegraph operation required the use of Morse code to encode, transmit, and decode messages. Morse code required the operator to encode each letter of the alphabet into a system of long and short strikes on a telegraph key. The same messages would eventually be sent by telephone or telegrams that used the standard alphabet. In the days of the telegraph, however, channel bandwidth was very narrow and the message had to transmit a long distance.

Bryan and Harter's (1897, 1899) results are redrawn in Figure 7.12. Their first finding was that the speed by which a message could be sent was generally faster than the speed with which it could be decoded. Decoding required writing the message down by hand, and a skilled operator was often required to keep an entire sentence in short-term memory while writing it down. Their second basic finding pertained to the bumps shown in the learning curves in Figure 7.12, which are not consistent and gradual. At first the operators would send and receive messages on a letter-by-letter basis. After a period of time they could mentally manipulate entire words,

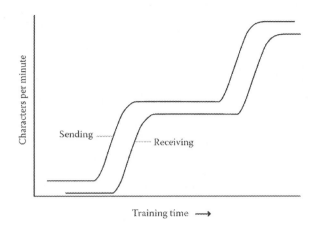

Figure 7.12 Automatization of telegraph operation, a cognitive task.

beginning with the familiar words such as *train, arriving, late,* and so on. Additional training did not seem to improve for a while until suddenly the operators could manipulate entire phrases and short sentences.

Automatization was the general principle that emerged from Bryan and Harter's (1897, 1899) studies. Cognitive processes that may be separate in early phases of training merge together into a flow of mental operation where the separate phases are not distinct to either the performer or the observer. Kantowitz (1985), in his review of research on channels and stages, also noted that stages of learning progress might not always be as distinct as Figure 7.7 would indicate. Nonetheless, we encounter at least one difference between novice and expert system users: Experts can work in larger chunks.

7.5.2 Controlled processes

Ackerman's (1987) review of the literature connecting abilities to training results showed that some basic mental abilities and job-relevant knowledge bases did predict performance in the early stages of training, but some of them became less relevant as greater levels of training took over. Importantly, as learning progressed, task performance separated into automatic processes and controlled processes that still required problem solving that is specific to the situation. Later research showed that the executive function is more engaged in the early stages of learning; it then phases out to varying extents as learning progresses (Chein & Schneider, 2012).

Bryan and Harter's (1897, 1899) skill-acquisition curves indicated clear phase shifts in the underlying mental activity. Phase shifts in turn indicate that a self-organization process is taking place. According to Juarrero (1999; see also van Orden, Holden, & Turvey, 2003), self-organization of mental activity is propelled by the intentionality of the mental activities. The thinker has a goal. A sequence of actions is taken, and feedback is received for each action that informs a "go, no-go" decision as to whether the next action in the sequence should be undertaken.

The foregoing analysis of automatization assumes that a visible control action (e.g., tapping the telegraph key) is taking place. Otherwise it would be very difficult to observe the mental process directly. Control actions are covered in Chapter 7, but for now it is only necessary to anticipate that control actions are behavioral results of the mental operations that are implicated in a particular work process.

7.5.3 Recognition-primed decision-making

In military operations, many courses of action require a quick assessment of a situation and the selection of a single course of action that is known to work under the given circumstances. The decision maker must think through the intended operation first, however, to ascertain whether some plausible configuration of events would prevent the intended result (Klein, 1989). This thinking strategy is known as recognition primed decision making (RPD), and could account for more than half of the decisions made in NASA flight simulators (Flin et al., 1996). In principle RPDs can eliminate time lost to discussion and debates, particularly in emergencies and promote the use of learned optimal responses (Bond & Cooper, 2006; Smith & Dowell, 2000).

The use of RPDs constitutes another distinction between experts and novices. The experts would have a larger inventory of experiences to draw upon, and can quickly evaluate a situation and pull out a response. The effectiveness of RPD is limited by the *coherence* of the problem situation. Put simply, clear problems more often lead to clear answers. Another limitation is that RPD is a mental shortcut where a solution is identified without actually analyzing the problem. The decision maker might in fact be trained to jump to incorrect conclusions.

The RPD is implemented in training programs using a three-step process. The first step is a CTA of the specific situations that require training. The goal of the analysis is to isolate the knowledge base and decision processes of experts who have the greatest proficiency at the tasks. Styles and methods for CTA are explained later in this chapter. The second step is to organize the learning objectives into modules so that elementary modules are presented first and followed by modules requiring a greater level of integration. The third step is to train the trainees and evaluate their performance (Staszewski, 2004).

One salient application of the RPD technique resulted in a military training program for landmine detection equipment (Staszewski, 2004, 2006). The earlier generation of equipment worked well for detecting the older generation of metallic mines, but it performed poorly against the newer generation of landmines that replaced most of the metal with plastic and were often smaller in size. A design for detection equipment was developed with a new sensor system at a cost of $38M over nine years, but first-run tests with operators who were already trained on the older equipment showed that the new system was not appreciably better than the old one for detecting low-metallic mines. The solution was to redevelop the training program for the new equipment based on RPD principles. Results for the training treatment group showed an average accuracy of 85–95% correct detection of low-metallic mines depending on the mine design compared to 15–25% accuracy for the control group. Significant gains were recorded for the detection of metallic mines as well. Of course, one should bear in mind that a 5% miss rate could be deadly several times over in the course of a day's minesweeping operation.

7.5.4 Degrees of freedom principle

Self-organizing dynamics typically take the form of information flows among the subsystems. The concept of degrees of freedom was first introduced in conjunction with physical movements (Bernstein, 1967; Marken, 1991; Rosenbaum, Slotta, Vaughn, & Plandon, 1991; Turvey, 1990), and it has cognitive implications as well. In any particular complex movement, each limb of the body is capable of moving in a limited number of ways, and the movements made by one limb restrict or facilitate movement by other limbs. For this reason, we do not walk by stepping both feet forward simultaneously, for instance. More generically, degrees of freedom are the number of component parts, such as muscles or neural networks that could function differently to produce the final performance result. The notion of internally connected nodes of movement is substantially simpler and more efficient than assuming that all elements of movement are controlled by a central executive function (Turvey, 1990); see Figure 7.13.

When a movement is in its earliest stages of learning, the individual explores several optional combinations; but once learning consolidates, the movement combinations gravitate toward conserving degrees of freedom, which is essentially a path of least resistance (Hong, 2010). The gravitation is actually a self-organization dynamic involving muscles and neural pathways. Residual variability in the movement persists, however, to facilitate new adaptive responses. Substantial changes in goals or demands produce a phase shift in the motor movements, which are observed as discontinuous changes in the sense of catastrophe models (Mayer-Kress, Newell, & Liu, 2009). Even simple tasks such as reaching and grasping an object could involve testing possible neurological pathways before the individual locks on to a pathway that is most functional; this phase transition is

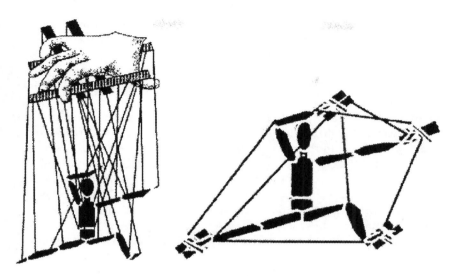

Figure 7.13 Degrees of freedom in (left) an executive-controlled and (right) a self-organized system. (Reprinted from Turvey (1990, p. 939) with permission of the American Psychological Association.)

important in rehabilitation contexts especially when connecting intention to actual movement (Nathan, Guastello, Prost, & Jeutter, 2012).

Cognitive behaviors are thought to operate on similar principles with regard to stages of schematic development, the role of executive functions, and the principle of conserving degrees of freedom (Hollis, Kloos, & van Orden, 2009). Although the full complement of possible degrees of freedom is not known at present, the current thinking about the structure of working memory and the allocation of resources for the various mental abilities should provide some indication regarding what could be involved in any particular task. Cognitive components would be involved in combination with psychomotor skills.

Cognition is often tied to action. Cognitive psychologists would introduce the constructs of embedded and embodied at this juncture. A cognitive process is *embedded* in the sense that the individual is interacting with the environment when the process is occurring, either through manipulation of objects or navigation through space in search of some goal object or condition. The process is *embodied* in two respects: One is that natural cognitive processes, thought to be the outgrowth of evolution and which have the limits of various types, act automatically in what Gibson (1979) called invariant processes. The other aspect of embodiment is that the mental operations are in fact connected to the body that move through the physical world, and the body's available movements play a role in the formation of a behavioral scheme that one can observe. Thus the possible degrees of freedom in a cognitive process span the whole perception–action sequence. Here one might look to resource theory (Wickens, 2002) for the best ways to arrange and rearrange the channels and stages.

When overload results from reaching a fixed upper limit of cognitive channel capacity, the sudden decline in performance would be the simple result of hitting a barrier. As such there would be little room for the elasticity associated with variable upper limits. If variable upper limits were operating, however, the principle of conserving degrees of freedom would have a few implications: In the case of adding tasks or new demands to existing tasks, a change in one cognitive-behavioral motion would impact on the other motions in the system or sequence. If it were possible to conserve degrees of freedom further, a phase shift in the cognition–action sequence would result in a catastrophic effect on performance. An example of an adaptation might occur when an increased demand for visual search results in a shift from an exhaustive search schema to an optimized first-terminating strategy (Townsend & Wegner, 2004) in which, the search terminates when it produces something good enough for the purpose.

A similar process of juggling degrees of freedom occurs when the individual plans sequences of discrete tasks in the context of a longer work period. Changing priorities could be involved. Task sequencing is essentially rule-governed, although not completely so. Again if there are N tasks there are N-1 opportunities for a task selection with possibly multiple options at each switch point. The use of rules for task selection, nonetheless, reduces the number of degrees of freedom as a task sequence become more automatic. Walker and Dooley (1999) showed that a system is more stable in the sense of consistently high performance to the extent that the system does not have to respond to error corrections, interruptions, or disruptions of input flows. A CAS would be successful at responding to such events, but it would need to have a few other rules (schemata) in place to do so.

7.6 DYNAMIC DECISIONS AND SITUATION AWARENESS

7.6.1 Dynamic decisions

Dynamic decisions involve a series of decisions that are not independent of each other, a problem situation that changes either autonomously or by virtue of the person–system interaction, and decisions that are made in real time (Brehmer, 2005, p. 77). The time-phased inflow of information induces dynamics that increase the complexity of the decision situation. Currently we know that time pressure, feedback delays, and reliability of incoming information place demands on the human operator that affect their performance (Brehmer, 1987; Jobidon, Rousseau, & Breton, 2005; Omodei, McLennan, & Wearing, 2005).

The computer programs that are typically used to generate scenarios for the study of dynamic decisions are alternatively known as "scaled worlds" or "low-fidelity simulations" (Schifflett, Elliott, Salas, & Coovert, 2004). The situations and scenarios usually involve a complex system of some sort, such as the operation of a sugar factory or a beer distribution center (Osman, 2010). There is a reduced concern for the realism of the peripheral features of the scenarios, for example, the layout of a graphic user interface (GUI) and a strong emphasis on the psychological constructs that the experimenter wants to assess; cognitive constructs of interest would involve learning the rules of operating a given situation. Realism is thus regarded as relative to the research objectives (Cooke & Shope, 2004). The systems often allow flexible reprogramming for desired experimental conditions.

There has been some expressed concern, however, whether the unreliability of the performance measures that have been used in research on dynamic decisions, which typically consist of a single number at the end of the simulation, is undermining attempts to test conventional hypotheses such as the relationship between general intelligence and performance (Brehmer, 2005; Elg, 2005). NDS theory would suggest here that the apparent unreliability of simulator performance measures could be related to the time-phased nature of the task and might not be a psychometric problem at all. As with other forms of individual and group learning, chaotic behavior occurs before the self-organization and stabilization at the levels of neural networks, individual behavior, and group work performance (Guastello, Bock, Caldwell, & Bond, 2005). Unlike the typical learning experiments, however, the specific decisions within dynamic decision sets are not independent of each other. Choices made in an early stage can affect options and utilities of options later on. The latter point is particularly true when the actions of one operator affect the actions of others in what could, or should, be done in a coordinated fashion.

Small-world simulations are usually designed to be *opaque*, meaning that the individual working the simulation is not given explicit instructions regarding the rule structure or mental model embodied in the computer program. Rather, the individuals must figure all those things out for themselves. This form of learning is known in other contexts as *implicit learning*, which is a form of learning that takes place while the individual is explicitly learning to do something else (Frensch & Runger, 2003; Seger, 1994). Although the trainees might be less attentive to the implicit learning process than they are to explicit learning goals, they are acquiring knowledge about processes that are instrumental to executing the more obvious goals.

7.6.2 Control of dynamical systems

One of the more classic simulations that involves opaque rules, implicit learning, and dynamical decisions is the beer distribution simulation by

Sterman (1988). The participants' tasks were to place orders, receive deliveries from breweries, receive orders from customers, and make deliveries, all without overflowing the warehouse or running out of stock. Unpredictable events occurred meanwhile, such as "strikes," "transportation problems," and "demand shifts." Respondents' beer inventories were chaotic over time, and only 12% of the players were capable of maintaining a beer inventory between the two limits for the duration of the game.

The low rates of effectiveness in the beer simulation and later studies were interpreted as indicating that the cognitive skills needed to control chaos in real-world applications was in short supply (Guastello et al., 2021a). Gonzalez, Vanyukov, and Martin (2005) later extracted four principles of small-world simulations—complexity, dynamics, dynamic complexity, and opaqueness—which is tantamount to saying "chaos that no one explained to you yet." The elusive skill apparently does not arise from greater working memory capacity. Lerch and Harter (2001) found that people with high working memory capacity tended to focus on a small range of the decision space very carefully, and not consider the full range of possibilities; their attention was overly focused and not sufficiently broad, which is the classic issue with attention tunneling.

Sterman (1989) noted, however, that there is ample opportunity for misinterpreting feedback in a small-world situation. Atkins, Wood, and Rutgers (2002) isolated three different types of feedback, that might not be readily distinguished by the operator: *process feedback*, which could take the form of a list of mistakes the operator made during a shift; *system feedback*, which is the function of the whole system including collateral events; and *outcome feedback*, which more clearly reflects the results of actions made by the operator to the extent that clarity is actually possible. An independent line of research showed that when feedback is controlling behavior in a continuous ongoing process, delays in feedback can produce a great deal of volatility in performance that would otherwise be relatively steady (Matsumoto & Szidarovszky, 2012). Delays in feedback in complex system simulations have become an issue by itself that is addressed in Chapter 12.

Another systematic source of error stems from a problem in domain knowledge. Small-world simulations that have been developed for training purposes or entertainment do not require prior knowledge of specific real-world systems. In real-world systems, a lapse in the user's knowledge base could be critical, particularly when it is important to know whether the dynamical decisions in play are stochastic or deterministic in nature (Hommes & Rosser, 2001; Sorger, 1998). The *perfect forecasting equilibrium* occurs when agents are aware of the actual dynamics, and their decisions produce smooth dynamics and a steady state. The *consistent expectations equilibrium* occurs when agents assume a stochastic process, and their decisions produce chaotic results as they try to compensate and adjust their decisions when using an autoregressive forecast strategy. Sorger

(1998) examined how the two assumptions would produce steady or chaotic results using a household finance model in which households adjust their spending in the face of income, taxes, expected interest rates, and need for savings. He used a mathematical analysis to compare the results of the two decision strategies. Hommes and Rosser (2001) performed a similar analysis of management strategies for fishing harvests, which have particularly complicated chaotic dynamics, in the face of fluctuating market prices and competition with other agents for open access supplies of fish. They found that the stochastic thinkers actually produce chaos in the economic environment, leading to a self-fulfilling prophesy that the process is chaotic.

7.6.3 Situation awareness

Effective control of dynamical systems appears to require an accurate and sufficiently complex assessment of the situation. Thus *situation awareness* evolved as a system design concept in its own right (Endsley et al., 2003) and is defined as, "the perception of the elements in the environment ... the comprehension of their meaning, and the projection of their status in the near future" (Endsley, 1995, p. 36). It is a good example of the whole being greater than the sum of its parts. Although it is an important matter to design information displays well, but the right combination of displays and controls gives the operator a sufficiently complete sense of everything meaningful that is going on in a form that makes immediate sense. As a very elementary example, the earliest automobiles did not have fuel gauges, but someone got the idea that a fuel gauge would be very helpful to the operator. Similarly, the earliest airplane pilots flew "by the seat of their pants" because they did not have an altimeter. How did these displays, even if crudely designed, help the operators gain awareness of their vehicle's situation and their own needed actions? Recent thinking on the nature of situation awareness recognizes it as both an individual and a group or team process. For present purposes the rendition of situation awareness in this chapter is confined to individual-level experience. Group or team processes are explored in Chapter 13.

Endsley et al. (2003) elucidated eight principles of display design that contribute to situation awareness. The first principle is to organize displays around "*the operator's major goals*, rather than presenting it in a way that is centered on the sensors and systems that generated the information" (p. 83, emphasis added). This is the principle behind smart displays covered in Chapter 5.

The second principle is to *present information directly*, rather than require the operator to make calculations to determine the needed nugget of information. The third principle is to provide *assistance for system projections*. This recommendation is essentially the principle behind predictive displays, which are thought to be especially helpful to the less experienced operators.

The next three principles involve the global and local aspects of system operation. The fourth principle is to *display the global system status*. Operators could become too focused on information pertaining to limited portions of the system's functionality and could miss the big picture. The fifth principle, however, is to *support global–local trade-offs* where the emphasis is not letting the big picture overwhelm the recognition of more specific information that either requires action or the interpretation of system status. The sixth principle is to *support perception–action schemata*. Displayed information should, whenever possible, translate directly into control actions.

The last two principles recognize the limits of operators' cognitive capacity. The seventh principle, *taking advantage of parallel processing capability* recommends not overwhelming the visual system but instead making use of tactile or auditory displays. The last recommendation is to filter information judiciously. On the one hand, reducing information to the essentials lightens the demand on working memory capacity. On the other hand, dropping out the wrong details could have a negative impact on situational awareness.

Ultimately situation awareness is a psychological state rather than a property of the contributing displays (Stanton, Salmon, Walker, & Jenkins, 2010). Salmon et al. (2008) reported that different research teams focusing on different aspects in the topic area do not quite agree on the definition of situational awareness, with some emphasizing the tracking of information as it changes over time and combining it into working memory, while others (notably Smith & Hancock, 1995) emphasizing the embedded nature of the interaction between the individual and the environment. Furthermore, situation awareness, no matter how accurate, is necessary but not sufficient for effect performance. Performance also requires actions that go beyond the state of mind. Sometimes there is a conflict between designing for situation awareness and performance. The use of smart displays and other elements of automation could serve to reduce workload, but there is, in principle a critical point where the benefits of workload reduction counteract actual situation awareness (Patrick & Morgan, 2010; Wickens, 2008b).

The psychological components involve three levels of abstraction: the elementary information appearing in displays, the perception–action schemata at the middle range, and the global picture of the system's status and the means of controlling it at the broadest level of abstraction (Endsley et al., 2003; Patrick & Morgan, 2010). The three levels of abstraction boil down to three questions: (a) What is the system doing now? What will the system do if no action is taken to change it? (c) What will the system do if a particular action is taken? As the operator's awareness transits the three levels, the cognitive operation shifts from perceiving info, to sensemaking, to an accurate mental model of the system and how best to control it.

7.6.4 Measurement of situation awareness

There are three basic types of situation awareness measures. One is the Situational Awareness Global Assessment Technique (SAGAT), which was first introduced by Endsley (1995, 2015). The second is the Situation Present Assessment Method (SPAM), which was first introduced by Durso and Dattel (2004). The third is distributed situation awareness, which was introduced by Chiappe, Strybel, and Vu (2012). SAGAT and SPAM have been used by other research teams beyond their original proponents, and there are situationally specific variations to them (Endsley, 2021; Gawron, 2019a; Vu & Chiappe, 2015). SAGAT and SPAM methods are discussed next. Distributed situation awareness is better appreciated in Chapter 14 in conjunction with team dynamics and complex systems.

The SAGAT method measures situational awareness in a form that will inform the design of the contributing displays and controls and the possible need for design modifications. It is executed in laboratory environments where it is possible to construct simulations around structured situations of interest and freeze the action at different points in time so that the participants can answer some knowledge questions. The questions would be drawn from all three levels of abstraction, and they would be specific to the system undergoing evaluation, and they would be scored as correct or incorrect. Endsley (1995, 2015) made a particular point of separating knowledge of a situation from the process by which the operator puts information together, and again from the product, which would be the operator's performance. A limitation of the SAGAT method, at least in some researchers' point of view, is that freezing the scenario produces a form of task switching that would interrupt the engagement of the operator with the scenario and equipment.

The SPAM method is based on some different assumptions about the situation awareness process. Like SAGAT, it samples factual knowledge about the situation at the same three levels of abstraction. It differs, however, in that it assumes that process and product are inextricably related in real-world high-stakes situations, such as military operations, air traffic control, and operation of drone aircraft (Vu & Chiappe, 2015). The knowledge that comprises situational awareness is embedded in the ongoing person–system interaction, rather than embodied in the operators themselves. The research procedure does not freeze the simulation to ask questions. The questions are sometimes posed in such a way that research participants can look up the information on the system rather than know the answers outright. Answering questions during the scenario places additional cognitive workload on the operator, which could have a negative impact on performance. Endsley's (2021) review, however, showed that SAGAT and SPAM questionnaires have relatively equal correlation with performance, which are relatively high in both cases.

The literature on situational awareness appears to assume that the functional mental model of the situation remains fixed while the particular situations within the model are changing. Such an assumption might be reasonable in a relatively closed system, in which there is little or no information exchange with the world outside the system. A more open system would have more opportunity for perturbance from environmental events and might be more challenging to comprehend. Sharma and Ivancevic (2010) considered the possibility that too much information could distort the global evaluation of situational awareness by channeling attention toward two or more possible global states. Their analysis was drawn from a computer simulation based on the logistic map as a model for information inflow intensity (Figure 3.3 and Equation 3.2), in which the bifurcation of awareness into multiple states is a distinct possibility of having too much information to process. The uncertainty associated with fast-moving and changing information could seriously undermine awareness.

7.6.5 Sensemaking

Once the incoming information has produced a state of situational awareness, the next cognitive step is to make sense out of it. Experts within a given situation (e.g., military, weather, medical emergency, industrial process) have a good stock of mental models that would explain why events happened up until the present moment, and what antics of the system could have produced those events. Experts then match the situational evidence against their stock of mental models, and pick the one that applies best. This process of sensemaking is not fundamentally different from the template matching that might be used in troubleshooting a system malfunction. The nuance, however, is that troubleshooters already know they have trouble and what it the malfunction looks like. In other environments, the trouble has not come to the foreground yet, and the necessary skill is to notice a pattern of events shaping up and invoking a model that accounts for what is known and generates questions regarding what is still unknown. Once a viable model has come into consideration, the decision maker then implicitly asks questions and looks for more information that would be expected if certain missing pieces were no longer missing. In other words the hypothetical model, once it is tentatively adopted, dictates what new information is useful and what is irrelevant (Snowden, 2011; Stewart, Dominguez, & Way, 2011).

Two cautionary notes are in order, however. The first is that decision makers can be prone to *confirmation bias*, meaning that they only accept the information that fits the model as relevant. The consequence is that the inconsistent or ignored information that is probably piling up by degrees could be consistent with a different model of the situation. A different model could have different implications for controlling and correcting the system.

Thus, critical questioning and exploration of alternative models are necessary parts of the process.

The second cautionary note is to guard against mindlessness. The automaticity that facilitates RPD and timely response can backfire if the decision maker does not question the applicability of a particular model and whether the available information adds up in the right way (Klein, Snowden, & Pin, 2011).

There is no law of physics (or anything else) that requires that all the relevant information be present at the same time. It is more likely to trickle in at unexpected times (Baber & McMaster, 2016). With the right amount of attention to incoming data and critical thinking, a switch in mental models can occur, and the switch might be more sudden than gradual. Once a hypothetical model has self-organized, there is little entropy of motion within the idea, and it takes a substantial amount of entropy to hop to a new one. More information does not necessarily support the existing model or offer a better one until there is an insight by the human. The insight involves the construction of *meaning* or new meaning. The construction of meaning and the switch to a new model is not algorithmic in nature, and thus does not lend itself to automation; it is a uniquely human thought process (Arecchi, 2011).

7.6.6 Anticipatory thinking

The third aspect of situation awareness is correctly anticipating not only what the system will be doing in the near future, but also the effects of possible control actions. The foregoing authors all recognize that naturally occurring situations tend to be complex and dynamic with uncertainties looming large. Snowden (2011) for instance found it useful to triage problem situations that are ordered and mechanical (some are because they were built that way), chaotic (which some people would confuse with "random"), and self-organizing. Self-organized systems are more ordered than chaotic ones, but less ordered than strictly mechanical systems. An important principle for making any sense out of a self-organized system is to understand the interconnectedness of the various aspects of the environment and the agents within it. In essence, complex systems need to be understood as such (McDaniel & Driebe, 2005; Weick, 2005).

Anticipatory thinking has a lot in common with forecasting as a mental operation, but it differs in some ways (Klein et al., 2011). Anticipatory thinking is framed around what could be done by the decision makers and the affordances in the task situation. The forecasting part of the mental operation involves following a trajectory of events as they change over time, anticipating where the system is going to go, and identifying a solution that negates or encourages the trend. The trepidations of assuming a linear versus nonlinear model, or a stochastic versus deterministic mental model, were discussed earlier. The entire process of situation awareness, sensemaking, forecasting, and action would need to transpire quickly enough for the decision makers to implement the action effectively. If XYZ is the trend, what is actionable now that would have the desired effect?

The last question might be naïve. Situations evolve, mental models coevolve with the situation, and the actions taken affect the system which coevolves in response. A self-organized system survives because it has interconnected subsystems that respond to assaults to one or more parts of the systems—hence the notion of CASs. The door swings both ways, however: Attempts to operate on a self-organized system—even when a legitimate improvement is intended—can be countered by the system adapting to maintain itself in spite of the attempts to correct and improve. A better approach, according to Klein et al. (2011), would be to take actions that steer the (deterministic) system in a desired direction and be ready to implement further nudges as the interactions between the target system and change agents evolve. Regarding the action and the effect themselves, one of the frequent errors in anticipatory thinking is overestimating one's ability to make a response and how quickly the situation is changing.

7.7 COGNITIVE ANALYSIS OF A PERSON–MACHINE SYSTEM

Neerincx and Griffioen (1996) recommended that systems engineers take the totality of cognitive tasks performed by the operator into consideration, and not just focus on single tasks out of context. Tasks must be understood in a hierarchical context. The broadest level of analysis would take the form of a job description, which frames the analysis around the entirety of a job. The middle level of analysis would focus on particular tasks within the job description with the intent of describing the qualitative substance of the workload for each task. The third level would take the form of a time chart showing the probable times and circumstances where the various tasks and subtasks are performed and exposures to mental overload occur. Corrective actions that would follow from these analyses could involve revising the single task, but they may involve improving training or reorganizing the tasks to prevent the transient times of mental overload; the interventions could have a critical impact on system performance or safety.

This section of the chapter starts with techniques for broad-level job descriptions, and then narrows the focus to the cognitive demands and strategies for ascertaining the cognitive work analysis (CWA). They are followed by two interesting examples that illustrate the value of these techniques for identifying and solving problems. All of the techniques require input from subject matter experts, the most prominent of whom are the incumbent employees themselves. The final section explains a procedure for extracting the decision processes used by these experts.

7.7.1 Job descriptions

Job descriptions are building blocks of many personnel functions, some of which fall outside the realm of this book. The list would include explaining in normal and accurate language what activities are part of what jobs; deducing relevant measures of performance; identifying the knowledge, skills, and abilities and other qualifications required; identifying training needs, and forming a basis for comparing jobs, for example, with compensation in mind, on a fair and rational basis. For present purposes, however, we consider briefly two commonly used types of job description and the type of information they provide. The roles of cognitive processes and physical actions become apparent in both cases.

7.7.1.1 Functional job analysis

Functional job analyses consist of clearly written paragraphs that detail the worker's actions. The *Dictionary of Occupational Titles* (U.S. Department of Labor, 1977, with updated versions currently available online) utilizes this type of job analysis in an inventory of all known jobs that exist in the United States. The textual definitions are tagged with a hexadecimal numerical code. The first three numbers to the left of decimal indicate an industry category. The three numbers to the right of the decimal are values on scales that indicate the job's demand level regarding interaction with Data, Things, and People.

The body of the definition usually consists of two main parts: a lead statement and a number of task element statements. The lead statement summarizes the entire occupation. It offers essential information such as: worker actions; the objective or purpose of the worker actions; machines, tools, equipment, or work aids used by the worker; materials used; products made; subject matter dealt with or services rendered; and instructions followed or judgments made.

Task element statements indicate the specific tasks the worker performs to accomplish the overall job purpose described in the lead statements. The sentences in the example beginning with "Turns handwheel ... ," "Turns screws ... ," "Sharpens doctor ... ," "Aligns doctor ... ," "Dips color ... ," and so on are all task element statements. They indicate how workers actually carry out their duties.

The scales for Data, People, and Things (Table 7.1) denote the level of sophistication associated with each of the three categories. Fine and Getkate (1995) introduced four more scales to accompany functional job descriptions. The four scales pertain to Worker Instructions, Reasoning, Math, and Language.

A job that rates a low level on Worker Instructions is one where all the goals and details are presented to the worker for execution. A job that rates

Table 7.1 Dictionary of Occupational Titles scales

Use of Data	
0	Synthesizing (integrating analysis of data to discover facts and/or trends)
1	Coordinating (determining time, place, and sequence of operations or actions)
2	Analyzing (examining and evaluating data)
3	Compiling (gathering, collating, or classifying information about data, people, things)
4	Computing (performing arithmetic operations and reporting on and/or carrying out a prescribed action in relation to them; does not include counting)
5	Copying (transcribing, entering, or posting data)

Interacting with people	
0	Mentoring (advising, counseling, and/or guiding others)
1	Negotiating (working with others to arrive jointly at a decision, conclusion, or solutions)
2	Instructing (teaching subject matter to others or training others)
3	Supervising (determining work procedures, maintaining harmony, promoting efficiency)
4	Diverting (amusing others)
5	Persuading (influencing others in favor of a product, service, or point of view)
6	Speaking-signaling (including giving assignments and/or directions to helpers)
7	Serving (immediately responding to the needs of requests of people or animals)
8	Taking instructions or helping (no responsibility involved; includes nonlearning helpers)

Use of things	
0	Setting up (adjusting machines or equipment to prepare them to perform their functions)
1	Precision working (involves use of body members or tools and considerable judgment)
2	Operating or controlling (setting up and adjusting the machines as work progresses)
3	Driving or operating (involves controlling a machine that must be steered or guided)
4	Manipulating (using body members, tools, or devices and involves some judgment)
5	Tending (starting, stopping, and observing machines and equipment; little judgment)
6	Feeding-off bearing (inserting, throwing, dumping, or placing materials in or removing from machines or equipment that are automatic or tended by other workers)
7	Handling (using body members, hand tools, and/or special devices to work, move, or carry objects or materials); involves little or no latitude for judgment

a high level on Worker Instructions is one where the needs are presented to the worker, and the worker must figure out what information, tools, equipment, and communications are necessary to fill the need and to get the job done. Reasoning ranges from simple algorithms, in which the worker executes the same steps in a specified order for each example of the task, to mental operations that require technical reasoning and a large proportion of controlled process thinking. Math ranges from simple counting to advanced calculus. Language ranges from pointing to signs and checking badges to the writing and interpretation of cutting edge scientific material.

7.7.1.2 Task-based job analysis

Task-based job analysis starts with an inventory of tasks that a group of workers might be performing, and then asking workers how often they perform the particular task. The responses are factor analyzed to produce groupings of tasks that are likely to go together. Each worker is then likely to be performing one group of tasks more often than they would perform tasks from other groupings. In fact, it is the clumpy association between people and tasks that give rise to the final factors (or components) in the statistical analysis. If desired, it would be possible to draw a profile for each employee or job title summarizing their activities on each of the task factors.

The task-based analysis does assume that a master list of tasks can be compiled, and that all the tasks are defined to the same level of specificity. It may be necessary to follow up with a cognitive analysis to decompose each task into parts and examine the mental challenges in each part of the job.

7.7.1.3 Benchmark jobs

A benchmark job is one that exists in many organizations, such that the management from the different organizations can compare notes as to how the job is done, and what combinations of human input and technologies work best under what conditions. In such circumstances, management may be comparing PMSs as whole units, and it may be up to the human factors engineer to figure out why certain combinations of people and machines work as well as they do (or do not). It would be unusual to find benchmarks for more than a fraction of jobs within an organization, but benchmarks can be helpful when they are available.

7.7.1.4 O*NET online

The *Dictionary of Occupational Titles* can be interpreted as a big thick catalogue of benchmark jobs, although sticks close to the goal of describing the content of all possible jobs and steers clear of recommending best practices for any of them. Administrators can use the structured paragraphs unchanged in many cases, or they can make the necessary alterations to fit their own idiosyncratic way of organizing work.

As one can imagine, the catalogue has morphed substantially in response to technological changes in the work environment over the past few decades. The current version is now a website that contains the traditional style job descriptions, but also additional information and searching facilities that could assist with career planning. According to the U. S. Department of Labor (2021).

> The O*NET Program is the nation's primary source of occupational information. Valid data are essential to understanding the rapidly changing nature of work and how it impacts the workforce and U.S. economy. From this information, applications are developed to facilitate the development and maintenance of a skilled workforce. Central to the project is the O*NET database, containing hundreds of standardized and occupation-specific descriptors on almost 1,000 occupations covering the entire U.S. economy. The database, which is available to the public at no cost, is continually updated from input by a broad range of workers in each occupation. O*NET information is used by millions of individuals every year ... The data have proven vital in helping people find the training and jobs they need, and employers the skilled workers necessary to be competitive in the marketplace.

7.7.2 Cognitive task analysis

CTA shifts the focus from the global definition of the job to the contents of the more specific tasks that comprise a job. Different tasks make different demands, and there was a useful insight years ago that the more enriched jobs—those that were more motivating overall—actually required a variety of skills and competencies (Hackman & Oldham, 1976). Other aspects of job enrichment are doing the whole task instead of only part of it, having sufficient autonomy to make job-related decisions instead of having to ask permission to respond to every nuance, doing a job that has some impact on the work of other people, and deriving feedback about the results of one's efforts from the job itself. Inasmuch as the old wisdom still holds true, the focus of attention here on the cognitive demands could play an additional role for making jobs more motivating than they might be at present.

To complicate matters further, the unique association between an individual and a cluster of tasks is a lot less prevalent in today's workforce than it was in the manufacturing-centered economy of the not-too-distant past, as some job and cognitive task analysts have noted (Grant & Parker, 2009; Lesgold, 2000). Rather it is more often the case that tasks are performed by teams, even though individuals often play unique roles within teams and could be members of multiple project teams. Thus once again, because the focus of this chapter is on individual cognitive processes, it is necessary to filter the various CTA strategies that have been advanced over the years to those that speak to the individual experience, with emphasis on possible designs for person–system interaction and to represent some of the more distinctive varieties of approaches. A review of 20 *review* studies on CTA by Schraagen, Chipman, and Shute (2000) was very useful in this endeavor.

7.7.2.1 Cognitive inventory

Olson (1987) composed a list of cognitive functions that should be addressed in a cognitive analysis of a person–machine system. The basic functions include perceptual and motor skills, rule-based decision-making, and the opportunities for analytic thinking and problem solving.

The first step in the analysis is to define the goals of the system and the tasks that are nested within it. Cognitive operations are identified in the second stage of the analysis. At the simplest level, the worker is simply moving information around. At the second level, information is transformed from one form to another. At the third level, information is processed through changes format, but not in content. At the fourth level, information is interpreted; here the information is placed in a new format that adds some value to the data to facilitation the extraction of meaning. At the fifth level, data are analyzed so that new information is extracted. At the sixth level, a new information base is created.

Human assets and limitations are considered in the third stage of cognitive analysis. What are the important visual and auditory perceptions in the work situation, and are they taxing the extremes of human capability? What are the demands on motor skills, response time, and overall action? Response time and motor skills are considered further in Chapter 8. What are the demands on short-term memory or long-term memory?

7.7.2.2 Hierarchy of rules, skills, and knowledge

The contents of Table 7.1 illustrate a range of cognitive involvements ranging from simply to complex, which forms the basis of an approach advocated by Rasmussen, Pejtersen, and Goldstein (1994). At the simplest level, the operator follows simple rules—essentially perception–action schemata—repeatedly and not much else. Skills fall in the middle of the spectrum. The word "skills" denotes individual differences in proficiency that comes with experience and practice. Skills combine with a variety of potentially applicable rules to execute tasks of a much greater level of complexity overall.

Knowledge sits on top of the hierarchy and informs the individual of what to do when novel circumstances arise, particular when planning and design issues are involved. Although knowledge without some supporting skills does not usually go very far, the level of abstraction that is associated with a sound knowledge base can make an individual very effective at a variety of complex and demanding tasks that draw upon a particular knowledge base. Insight in problem solving requires more than an accumulation of information. It requires a system of meanings that can reconfigure to accommodate a new context.

7.7.2.3 Hierarchy of goals

The first instances of techniques for specifying systems of goals date back to the early 20th century (Adams, Rogers, & Fisk, 2012). The objective was to specify goals from the top-down, finally arriving at the simplest goals that it was possible to define. Although this style of goal setting has seen many incarnations in management practice, it tends to be light on specifying actions that will attain the goals, and even lighter on cognitive content.

A more actionable technique for human factors purposes is the method of goals, operators, methods, and selection rules (GOMS; Kieras, 1988, 2004). This method would take a task and decompose it into the most elementary actions with the idea of programming a computer to do part of the job. This goal of automating a process, unfortunately, requires an analysis that is as microscopic in perspective as the code, and does not communicate a big picture to humans that might require one, nor does it directly address cognitive processes in the action sequences. Adams et al. (2012) gave a pointed example for a short-order cook: The task goal hierarchy consisted of reading the order, preparing each ingredient to be used as part of another ingredient base, planning cooking time so the whole order will be ready at the same time, and so on. (Fortunately for some programmers, cooking is relatively modular if one looks at it the right way.) GOMS models were, nonetheless, friendly to the symbolic computer languages that were fashionable at the time GOMES was first introduced, which was essential for automating a wide range of tasks.

7.7.2.4 Ecological task approach

The ecological task approach (ETA; Kirlik, 2006) follows Gibson's (1979) basic premises of ecological psychology fairly closely. In this perspective, the task undergoing analysis is embedded in the environment more explicitly. ETA frames the analysis and tells the story from the point of view of the people doing the task and making the decisions, rather than the perspective of what can be done to automate them or make comparisons for administrative purposes.

Key questions for an ETA task analysis would identify areas of expertise, cues and strategies used, and some consideration for why the task could be difficult sometimes or for some people. Expertise would involve all the forms of knowledge, skill, and rules discussed so far, although additional areas of knowledge or skill that involve noticing changes in one's working environment would count also. For cues and strategies, Adams et al.'s (2012) example of the short-order cook makes a good comparison against GOMS: The cook would anticipate larger volumes of popular orders, plan the ingredients accordingly, juggle cooking times, and respond to the overall rowdiness of the hungry crowd to minimize their wait time. The difficulties often revolve around managing multiple orders simultaneously, staying calm, and keeping track of all the customers' special requests.

7.7.2.5 Illustrative examples

7.7.2.5.1 Helicopter landing

Landing a helicopter on a ship is a cognitively challenging task and potentially very dangerous for well-trained pilots. Tušl et al. (2020) started the CTA with in-depth interviews with ten helicopter pilots from the Italian Navy, all of whom were very familiar with the task. The researchers' first objective was to understand the sequence of actions that were part of the landing process and identify the crucial elements of each phase of the landing. The second objective was to delineate the cognitive operations in each phase along with the information displays that were most relevant for each phase. The third objective was to identify potential areas of improvement in the PMS.

The first phase is to locate the ship on which they are supposed to land. In the better cases, the ship is equipped with a tactical air navigation system that sends a homing beacon to the helicopter. In the more challenging cases the ship is not equipped with such a system, the helicopter is not equipped with radar during a covert operation, and the flight team has to work with a GPS at night during bad weather.

In the second phase, the flight team locks onto a descent path and establishes communication with the ship. Communication with the ship can be problematical during the more difficult or emergency scenarios because the pilot, copilot, and third crew member are also in close verbal communication with each other, and verbal communications coming from the ship can be distracting.

The third phase is to follow the path to approximately the touchdown location. The critical demand here is for the pilot to switch focus from the helicopter instrumentation to lighting and other cues from the ship, which is still in motion in often-turbulent waters, and the surrounding physical environment. This phase carries with it a high potential for vertigo and spatial disorientation if the pilot vacillates between internal and external displays.

The challenge continues to build in the fourth phase, in which the helicopter needs to align with the ship and landing position. A critical role is played by the third crew member who is in a position to see what is happening at the back end of the helicopter and advises the pilot and copilot. The pilot's field of vision is restricted to the front end.

In the fifth phase the helicopter hovers over the landing location while the pilot stabilizes the helicopter to land on the moving ship. According to one pilot:

> A common error that an inexperienced pilot can do is to move the helicopter following the roll of the ship when he should stay in line with the real horizon. Because you have to keep scanning outside and you cannot follow the artificial horizon inside the cockpit, my reference point becomes the ship, but it should be the real horizon.
>
> *p. 248*

The real horizon is visible in the daytime, but pilots in the dark need to rely on the visual display (horizontal reference bar) on the ship that corrects for the motion of the ship.

The final phase is the actual landing, which must be done very slowly to prevent collision damage. The pilot's vision is restricted by the helicopter naturally tilting backward by a few degrees because of its weight distribution. The last two phases of the landing operation presented the greatest workload demands on the pilots.

Some recommendations for system improvement were uncovered through this analysis. The flight training program should be revised and expanded to give pilots and crew a better understanding of the cognitive demands of each phase and how to make best use of the available visual displays and audio communication channels. Another recommendation was to develop better visual landing aids. The display panels in some helicopters required pilots to gather information from multiple displays. The better designs, according to the pilots, produced the information they needed in one aggregated display.

7.7.2.5.2 Medical emergencies

Hospital care starts with standard operations, which work well until an unforeseen emergency condition develops. The team is then required to take adaptive action as a group, which is a self-organizing process by itself that sometimes does not work out very well (Gorman et al., 2020). Workload changes substantially during the phases of critical tasks, such as establishing a cardiac bypass, even when the medical status of the patient does not reach emergency proportions (Kennedy-Metz, 2021).

Against this background, Morineau and Flach (2019) synthesized a CWA of hospital care. They targeted episodes where the patient's condition would be heading toward an emergency condition. They used a six-point CTA scheme that they adapted from Vincente (1999): (a) a work domain analysis, which would be essentially the same as the functional job analysis that was mentioned earlier; (b) goals and boundary conditions for what is considered an acceptable range of biometrics; (c) methods of controlling the task, which would include use of displays, controls, and feedback loops that connect from action back to display; (d) strategies used to select procedures and equipment and heuristics related to speed-accuracy trade-offs; (e) division of labor and information channels among medical team members; (f) worker competency requirements.

Morineau and Flach (2019) noted that an implementation of the CTA is essentially a top-down endeavor, especially where standard operating procedures are involved. Emergencies are bottom-up processes that disrupt the standard procedures and any other specific plans that are in motion. They use NDS constructs to characterize the stabilities and disruptions in the work system:

> Any event in the work system can change the configuration of forces shaping trajectories. For instance saddle plays the role of both repellor and attractor; an oscillator alternatively repels and attracts agents. These forces can potentially lead to bifurcation in the agents' trajectory ... Adaptive control is only marginally stable and can produce trajectories that result in risky margin or boundary crossings (e.g., leading to overload or to violation of safety margins.
>
> *p. 100*

From their analysis of hospital emergency room events, they compiled a matrix of emergency activities. One axis consisted of three boundary conditions, which were requirements for patient care, information processing, and task management. The other axis were five phases of activity during an emergency: (a) handoff, in which the available relevant medical information is prepared and transferred from the intake desk to the emergency medical team; (b) preparation for examinations; (c) clinical monitoring; (d) resuscitation, and (e) post-emergency assistance. Each of the five phases required working with only one of the three boundary conditions, except for resuscitation which involved all three. The resuscitation phase thus required more attention because of the high workload involved. Morineau and Flach also found that raising and lowering the hospital bed when needed during an emergency did not always turn out the right way because of equipment control issues or conflicting human judgments regarding optimal bed height for the occasion.

At a more abstract level of analysis, Morineau and Flach concluded that a CWA of a work routine should be regarded a continual work in progress, precisely because of the nonroutine elements that occur. Sensemaking and action often occurs from the bottom-up, based on the immediate conditions of the patient and the available resources, and not necessarily from top-down planning and standardization. Thus, the cognitive work in demanding environments should be reevaluated periodically for any meaningful changes that might have evolved.

7.7.3 Think-aloud technique

It would be tempting to forgo the job descriptions, rating scales, and laundry lists of cognitive functions and simply ask the operators what they are doing. Unfortunately, the verbal reports are likely to be incomplete and not necessarily in the actual order of the real functions. The experts in the group, who might be best equipped to tell the story, are also those whose cognitive processes are the most automatic. Thus Card, Moran, and Newell (1983) developed the think-aloud technique for analyzing the cognitive processes in a human–computer interaction. The basic concept can apply to other types of systems as well.

The think-aloud technique asks the operators to explain what they are doing three times: before doing it, while doing it, and after doing it. Surprisingly, the three reports may come out a bit differently. Operators may initially chunk a couple of steps together and take them apart on a later account. Some parts of the process may have been left out of the first report, but the actions themselves can bring some of the missing aspects of the work to mind.

One should bear in mind that the process of verbalization itself might introduce some distortions (Olson, 1987). If the task places a high demand on channel capacity, talking will slow down the execution of the task. Talking also distorts the perception of time such that people perceive longer elapsed time for a task compared to when they perform the task without narrating (Hertzum & Holmegaard, 2015). Similarly, talking may interrupt the automatic processes and distort both the thinking process and the report. The distortion is particularly notable when the operator is trying to narrate actions associated with a spatial task (Gilhooly, Fioratou, & Henretty, 2010).

Distortion is generally less when the talk is confined to verbal and nonverbal information that is already in attention. Operators should confine their narrative to what they are doing in the moment and not digress into the rationale for why they are taking an action. The investigators, meanwhile, should not interrupt with questions; the questions should be saved for the retrospective part of the think-aloud session (Hertzum & Holmegaard, 2015). Video recording of the action phase can be helpful for posing questions such as, "What were you trying to do when ... ?" or "What did you have in mind when ... ?" The recordings would be helpful to the operators also for composing their answers.

Talking can facilitate the think-aloud process if the task places a low demand on channel capacity. It can function as a form of rehearsal, and thus enhance the transition of information from short-term memory to long-term memory compared to normal working conditions. If the task is performed in an environment where there is a great deal of noise, especially verbal noise from other people, verbalization of the task aids performance. This form of verbalization is called *shadowing*. In other words, talking to yourself helps.

DISCUSSION QUESTIONS

1. Your car will not start. What do you observe that tells you how to localize the problem?
2. How would you build a computer program to assist with a classic (pick your favorite) optimization problem?
3. *Multitasking* is a contemporary management buzzword. What is it supposed to mean? How might you use multitasking efficiently in cases where: (a) you are sitting at a desk or computer workstation as your primary physical setting, and (b) you are required to move around and to do things other than press computer keys.
4. Is it possible to say, "In every non-optimizing problem there is an optimizing problem"? Consider a non-optimizing problem that you know something about. Does optimization play a role in any part of the thinking?
5. Consider texting while driving and your answer to the previous question. What are the implications for product liability suits? How about wrongful death suits?
6. The National Traffic and Highway Safety Association (NTHSA) currents sets an advisory limit of 30 characters for incoming text messages to a driver for non-driving information. Non-driving information would include messages from friends (who might be unaware that the recipient is driving) or coupons from advertisers. Is the 30-character limit reasonable? If not, what limit would you recommend instead? How did you come to your decision?
7. An individual who might be sympathetic to the information technology industries states that the question above regarding the 30-character limit is the wrong question to ask. Rather, we need to consider the trade-off between increased productivity and the risks of accidents. What do you think of this idea? Whose productivity are we talking about?
8. Consider the earlier question about the virus kits: When you arrive at the rural hospital there are ten dead bodies of Ebola virus victims in the hospital courtyard waiting to be buried. You have to order drug kits for possible N other patients who have not yet arrived. Kits are scarce and very expensive. How many kits do you order? How did you arrive at your estimate of N?

9. The TLX appears to produce higher ratings of workload when it is delivered through a computer application rather than by paper and pencil. Why do you suppose that might be the case?

10. Wireless EEG skullcaps are promising tools for research on cognitive workload, but how would you design a study that would produce results that could be implemented to improve normal operations? Could it be done without the continued use of EEG hats during normal working hours?

11. Do you think the day will come when everyone will be required to wear an EEG skullcap at work all day every day? How would the nature of the work change as a result?

12. How would noise affect the effectiveness of augmented cognition systems? What are some implications of false positive and false negative decisions made by the machine?

13. How would situation awareness play an important role in landing a helicopter on a ship? Give specific examples.

14. How would situation awareness play an important role in hospital emergency room activities? Give specific examples.

Chapter 8

Psychomotor skill and controls

8.1 REACTION AND RESPONSE TIME

The reaction time between a stimulus and a response was a critical dependent measure in the earliest days of experimental psychology (Cattell, 1886). The thinking at the time was that the length of the delay between stimulus and response signified the complexity or extensiveness of the mental process that was taking place. Since then we learned how some tasks are more complex or difficult than other tasks, and the amount of a specific type of information that a person can process in a fixed amount of time greatly affects the reaction time. Indeed, we now use *response time* instead of *reaction time* to denote the totality of mental processes that are occurring. *Reaction time* is now reserved for situations involving predominately psychomotor responses with minimal cognitive input; the acronym RT is used here for either purpose. Response time is a critical performance outcome in HFE; it is sometimes known as *movement time* (MT) to denote human motions that are psychomotor in principle, but could involve more actions that are more complex than simply pushing a button, such as walking over to the control panel where the button is located and then pushing it.

8.1.1 Donders' RT

The first concept of RT that is still useful today, traces back to Donders in the late 19th century (Kantowitz & Sorkin, 1983). Donders developed a prototype RT experiment that allows the experimenter to separate RT into two components: identification time and selection time. The experiment requires three separate measurements. In the first, the human is presented with one stimulus and must make one possible response when the stimulus occurs; the time elapsed is $RT(A)$. In the second, the human is presented with two or more stimuli and the human must choose among two or more possible responses to the stimuli as they occur; the time elapsed is $RT(B)$. In the third, the human is presented with two or more stimuli, but only one

DOI: 10.1201/9781003359128-8

stimulus maps onto a response; the time elapsed is $RT(C)$. Stimulus identification time is $RT(C) - RT(A)$. Response selection time is $RT(B) - RT(C)$.

In practical application, an experiment would involve a representative sample of people who are presented with many examples of stimuli in the A, B, and C conditions. The data analysis would compute averages for each person in the A, B, and C conditions and then compute identification and response selection times. Alternatively, one could examine whether different stimulus and response combinations are producing markedly different RTs in the A, B, and C conditions. One might then aggregate the RT measures according to the type of stimuli or responses, make the statistical comparisons, and thus determine what aspects of the display–control (stimulus–response) set overall need improvement.

8.1.2 Type of stimuli

There are several groups of determinants of RT. One is the type of stimuli that are involved. For instance, RT increases linearly as the amount of information increases if the information is composed of digits. RT increases logarithmically as the amount of information increases if the information is composed of other types of information (Kantowitz & Sorkin, 1983).

Considering the broad scope of possibilities for display content, the logarithmic relationship between RT and quantity of information is the general rule, and the linear function for digits is the special case. The experiments that produced these relationships (summarized in Kantowitz & Sorkin, 1983) characterized information as the number of choices facing the operator. This operationalization of "information" was consistent with the Shannon (1948) definition. Cluttered displays induce larger RTs because of the time required to search for information (Yeh, Merlo, Wickens, &Brandenburg, 2003).

8.1.3 Stimulus–response compatibility

Consider the knob and dial combination in Figure 8.1. If the operator wanted to move the pointer from its present position to the location to the right, in which direction should the knob be turned? The knob would be turned to the right in a compatible situation. The knob would be turned to the left in an inversely compatible situation. Outboard motors on small boats are common examples of inverse compatibility. If the driver wants the boat to turn to the right, the tiller arm must be turned to the left.

In a completely incompatible situation, the knob in Figure 8.1 might not be a knob at all. It might be a pushbutton that could be pushed one or more times to move the pointer a small number of increments to the right. To move it to the left, one would need another button, or a toggle button with an up arrow and a down arrow drawn on the button. All other aspects of

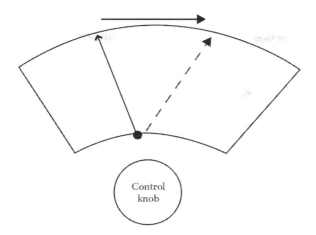

Figure 8.1 Stimulus–response compatibility: Which way should the knob be turned to move the pointer to the right?

the situation being equal, RT is fastest when the stimulus and the response are compatible (Fitts & Seeger, 1953). RT would be second-fastest, but by a significant margin, when the stimulus and response are inversely compatible. Incompatible stimulus–response combinations produce the slowest RTs.

8.1.4 Population stereotypes

A *population stereotype* is an expectation within a population of users that a system should be controlled in a particular way. The expectation would come from experience with different but similar situations. For instance, in a kitchen or bathroom or anywhere else, is the faucet handle for the hot water on the left or on the right? Imagine your surprise if you stopped at a cheap motel one night, tried to take a shower, and found that the opposite was true.

Similarly, if you wanted to screw a screw into a wall or into a device of some sort, which way would you turn it? Left is loose and right is tight, right? Imagine my surprise when I encountered a toy with a screw that was lathed in the opposite direction. If someone had not suggested, "Try turning it the other way," I would not have found the solution for a very long time.

In a more serious example, suppose two aircraft were approaching each other head-on. Which way should they turn? The correct answer here is to turn to the right; the convention is an outgrowth of maritime law for two ships approaching head-on, dating back a couple of centuries. Unfortunately, Berringer (1978) discovered that only 85% of professional pilots in a simulator turned right. The others turned left, and the mistake was unrelated to any of the pilots being left-handed. Consider the implications here when two pilots are involved: there is a 28% chance of a collision because one of them did not turn to the right, and a 2% chance of no collision occurring because they both made the wrong move.

There were also comparison groups of non-pilot students in the experiment, of whom 75% turned to the right when faced with a head-on collision. The conclusions were that a population stereotype existed among pilots and non-pilots and that safety in the skies would improve with training on this particular maneuver.

8.2 LEARNING AND SKILL ACQUISITION

8.2.1 Skill acquisition

RT improves with practice. The curve shown in Figure 8.2 is an example of a skill acquisition curve in which the RT declines as the number of trials progresses. It is actually an inverse learning curve. Instead of showing an increased amount of learning over time, up to an asymptotic level, learning here is characterized as a decrease in performance time.

Not evident from the curve is another long-standing relationship in learning theory: spaced trials produce better learning than massed trials. In other words, if one were to deliver 100 trials in a learning set, better results would be obtained if the 100 trials were spaced out over several days than if they were delivered all in one session.

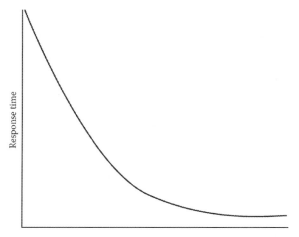

Figure 8.2 Skill learning curve.

The equation of the skill acquisition curve of the type shown in Figure 8.2 is:

$$RT = k \log(P) + a, \tag{8.1}$$

where k and a are constants, and P = number of trials or units of practice. It can also be characterized as a power law relationship:

$$RT = aP^{-b}, \tag{8.2}$$

where b is the shape parameter for the power law function and signifies the rate by which training speeds up response time; a is a constant once again.

In the earlier times of human–computer interaction, control functions were commands that were typed into the system with a keyboard. Commands were kept brief under the thinking that shorter was better. McKendree and Anderson (1987) modeled the control action as a combination of mental and physical time as:

$$RT = \text{physical time} + \text{mental time} = nkP^{-a} + cmP^{-b}. \tag{8.3}$$

In this case, n = number of keystrokes, k = time per keystroke, P = units of practice once again, m = mental time, c is a constant; a and b are separate speed-up exponents associated with physical time and mental time respectively.

Models such as Equations 8.1, 8.2, and 8.3 can be used to compare possible stimulus–control designs. One of the more memorable findings from the keystroke research, however, was that fewer keystrokes are not always better. Rather, if the commands are too brief, the commands become ambiguous and errors in control operation or longer think times are introduced (Landauer, Galotti, & Hartwell, 1983).

8.2.2 Dynamics of learning processes

Another point that is not evident from Figure 8.2 is how much variability exists within the learning curve; it is not as smooth as it is traditionally shown in textbooks. The nonlinear dynamics of learning can follow one of two basic patterns depending on how one frames the problem. One principle is that learning processes are chaotic in their early stages until they reach the asymptote where self-organization occurs (Hoyert, 1992). The neurological explanation is that neural firing patterns are themselves chaotic in the early phases of learning while the brain is testing out possible synaptic pathways. Once learning has progressed sufficiently, the brain locks onto a particularly pathway to use consistently (Minelli, 2009; Skarda & Freeman, 1987). The variability and reduction thereof can be explained as the result of the number of degrees of freedom inherent in the neural pathways or neuromotor connections that could comprise a motion (Friston, 2010; Hong, 2010). The system gravitates toward minimizing the degrees of freedom during the learning process.

The second dynamic principle involves the cusp catastrophe model. If we extend the baseline of the learning curve (Figure 8.3, left) prior to the onset of the learning trials, two stable states become apparent: the baseline and the asymptote. According to Frey and Sears (1978), hysteresis that occurs between learning and extinction curves cannot be explained otherwise. Different inflections in learning curves can be explained as a cusp bifurcation manifold (Guastello, Bock, Caldwell, & Bond 2005) as shown in Figure 8.3 (right). The cusp model for the learning process applies Equation 3.6 and Figure 3.9, in which control parameter a (asymmetry, governing proximity to the sudden jump) is the ability of a person or the number of learning trials, and control parameter b (bifurcation, size of the sudden jump) would be the difference between treatment and control groups, motivation, or differences in schedules of reinforcement, or any other variable that would contribute to making some learning curves stronger or steeper than others.

The cusp model is particularly good for training and program evaluation. If a statistical cusp effect turns out to be better (larger R^2 = percentage of behavior variance accounted for) than the next best alternative model (which is usually linear) it would denote all the features associated with a cusp model. Here the idea of *stable* end states adds a desirable feature to program evaluation: We want stable improvements to behavior targets, not simply statistically significant differences. "Stable" does not mean "without variability," however. A bit of variability is necessary if it will ever be possible for the person, group, or organization to attain greater levels of performance (Abbott, Button, Pepping, & Collins, 2005; Mayer-Kress, Newell,

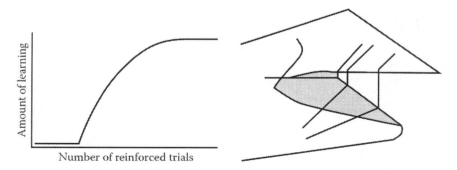

Figure 8.3 Cusp catastrophe model for learning events.

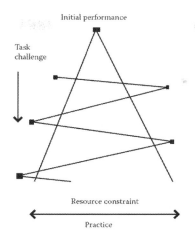

Figure 8.4 Hysteresis during performance improvement.

& Liu, 2009). Figure 8.4 illustrates the dynamics of performance improvement. The person (or team of people) addressed by the model encounters a new task that cannot be readily assimilated into old or crystallized learning. The new learning is attained with practice, and the level of hysteresis across the cusp manifold increases with repeated new challenges.

8.2.3 Speed–accuracy trade-off

"Haste makes waste," as the old saying goes. People can work faster than they usually do without making more mistakes, but only up to a critical point. Beyond the critical point the error rates increase dramatically, as shown in Figure 8.5. The engineering strategy would be to find a critical point that is located just before the sharp increase in errors and set the work pace to that point. For instance, the whole work team for a manufacturing process would coordinate to that production pace. Sometimes the errors take the form of accidents and injuries, and the trade-off then becomes production versus safety.

Sometimes the trade-off shows up in single decisions. In military operations, speed in firing weapons matters to get the jump on the enemy or to keep pace with the enemy's movements. The accuracy problem takes the form of casualties from friendly fire that has not had the chance to sort out the good guys from the naughty ones (Head, Tenan, Tweedell, Wilson, & Helton, 2020). We return to this dilemma later in this chapter.

The S shape of the curve in Figure 8.5 denotes a cusp catastrophe function once again. Work speed would function as the asymmetry parameter, which

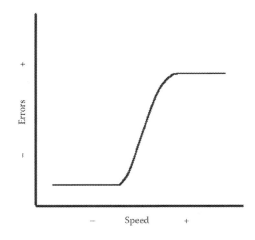

Figure 8.5 Speed–accuracy trade-off.

brings the system closer to the critical point. The bifurcation factor would be a situational variable that promotes larger or smaller increases in errors. One possible bifurcation factor would be the number of opportunities for error that are inherent in the work situation, which could be tied to a complex task with several steps in the process. This is the principle of degrees of freedom again. Another possible bifurcation factor would be the standard of accuracy that constitutes a meaningful error. In the process of manufacturing television sets, for instance, the system can tolerate missing 40 tiny welds out of 10,000. That same standard of accuracy would be completely unacceptable for the assembly of guided missile systems. The situation with the more rigid standard would be a high bifurcation condition.

There is evidence favoring the existence of multiple performance processes that explain work performance under conditions of different speeds. Signal detection tasks offer the opportunity for two types of errors, misses and false alarms. As explained in Chapter 4 the errors can be biased in favor of one error or another by the rewards or penalties associated with them. Balci et al. (2011) designed an experiment that induced participants to make a trade-off between accuracy and rewards. They found that bias toward accuracy dominated the early stages of learning and performance, and a bias toward higher overall rewards dominated later on.

In a visual search example, Zenger and Fahle (1997) manipulated load by increasing the search target area to produce a pressure on response time. Missed targets were more common than false alarms, and they were traceable to a speed–accuracy trade-off.

Szalma and Teo (2012) constructed a vigilance task that varied the number of screens containing possible targets and the speed of screen presentation.

For that set of conditions, there was a linear decline in accuracy as a function of speed for low-demand content, but the classic breakpoint was observed as the number of screens increased from one to eight. False alarms declined with increasing speed for the eight-screen condition only.

According to Koelega, Brinkman, Hendriks, and Verbaten (1989), research prior to their study had shown that the density of targets and the memory load inherent in the task affected whether error rates set in during slow or fast working conditions. They assessed performance on four vigilance tasks that varied in memory demand and by type of stimuli for speed, accuracy, electrodermal response, and subjective ratings. A principal components analysis of the measurements produced two components that indicated that speed and accuracy could result from two different underlying cognitive processes. The outcomes of working too slowly are revisited in Chapter 10.

8.2.4 Taxonomy of psychomotor skills

Fleishman (1954) contributed a taxonomy of psychomotor skills that was based on a wide range of tasks performed by military personnel. A large sample of personnel performed the full range of tasks, and the taxonomy of 11 basic skills was produced from the factor analysis. The common factors, or themes, are listed next, following by a quick explanation of factor analysis and what it produces.

Control precision refers to the precise setting of a control, such as the calibration controller shown later in this chapter or the control–display combination in Figure 5.8 (upper left). *Multilimb coordination* involves the use of arms and legs simultaneously, such as one might do when playing the drums.

Response orientation is a person's first response to an alarm signal, such as the stimulus identification time in the basic Donders experiment. *Reaction time* is a separate psychomotor measurement that was discussed here already, except that the reactions were localized to the use of hands. *Speed of arm movement* is another distinct psychomotor skill.

Rate control involves following a moving object at the same speed as the object's, and changing direction when the object changes direction. Several types of tracking tasks fall into this category; the pursuit rotor task is an example. The operator must maintain contact, usually with a stick or other handheld device, with a sensor that is mounted on a wheel. The wheel spins at different speeds and changes direction. The criterion would be the length of time the operator can maintain pursuit without breaking contact.

Manual dexterity involves the production of complex motions with one's hands, such as the use of some of the more challenging hand tools. Similarly, *finger dexterity* involves the production of complex motions with one's fingers; tying a knot would be an example. *Arm-hand steadiness* occurs in tasks requiring the combined use of arms and hands. *Wrist-finger speed* involves the combined use of the wrist and finger in repeated motions, such as tapping keys on a keyboard. The last psychomotor skill on the list is *aiming*; this skill might be expressed in the use of firearms, threading a needle, or putting a screwdriver into a screw that is already on the wall.

8.2.5 Factor analysis and principal components analysis

Factor analysis is a statistical analysis that takes a large number of variables or measurements and reduces them to a small number of common factors. The variables could be measurements of performance on dozens of psychomotor tasks, questions on a personality inventory, or a cognitive ability test. The procedure starts with a correlation matrix of all the variables (or measurements) with each other (Figure 8.6a). An iterative computational

(a)

	v1	v2	v3	v4	v5	v6	v7	...	vN
v1	1.0								
v2		1.0							
v3			1.0						
v4				1.0					
v5					1.0				
v6						1.0			
v7							1.0		
...								1.0	
vN									1.0

(b)

Variable	Factor 1	Factor 2	Factor 3
v1	.93	-.07	.26
v2	.93	.22	.04
v3	.91	-.21	-.03
v4	-.07	.90	.26
v5	-.09	.89	.04
v6	-.20	.88	-.03
v7	.26	.26	.79
v8	.04	.04	.65
etc	-.03	-.03	.54

Figure 8.6 First and final steps of a factor analysis.

process occurs that produces a matrix of correlations between the original variables and the underlying common factors isolated in the analysis (Figure 8.6b).

The matrix entries in Figure 8.6b is a rather ideal example in which each variable has a high association with one of the factors, but a negligible relationship to the others. The factors themselves are statistically independent, but they could be correlated at the same time. Statisticians apply further computations (factor rotation) to tweak the factors in the correlated or uncorrelated condition to produce the most interpretable set of final factors. If a set of variables share a hierarchical relationship, correlated factors are more likely.

A principal components analysis is essentially the same as a factor analysis computationally, but there is one key difference that can lead to a different configuration of final factors, which are now called "components." Principal components analysis starts with a correlation matrix as configured in Figure 8.6a with 1.0 on all the diagonal cells, applies the algorithm, and produces a set of factors that parse all the variance in the original matrix of variables. Factor analysis starts with the same correlation matrix, but replaces the 1.0 entries on the diagonal with a multiple regression coefficients that reflect the predictability of each variable from all the other variables. The resulting factors reflect all *common* variance among the original variables. Both strategies have a long history of use in intelligence theory, personality trait theory, and the development of psychological measurements of all sorts. For more about these techniques the interested reader should consult textbooks on multivariate statistics, such as of Tabachnick and Fidell (2007), or texts devoted to factor and component analysis.

8.3 TYPES OF MANUAL CONTROLS

Kantowitz and Sorkin (1983) characterized controls as discrete versus continuous and linear versus rotary. A discrete control is comparable to an on-off switch, or a similar switch with a fixed and small number of options. Examples would include pushbuttons and toggle switches. Some prime examples appear in Figures 8.7 through 8.9. The example in Figure 8.7 is one of the first pushbutton control designs that were introduced in the early 20th century. It required two buttons to do its job, one for "on" and one for "off." When "on" was depressed, the "off" button popped out ready for the operator's next move.

The earliest light switches in homes worked like the example in Figure 8.7. The home unit often required less physical force than the controllers for some industrial systems did at this time. This design for light switches was eventually replaced by the two-pole toggle switch that has been in common use for at least the last 50 years. A different style of toggle switch appears in Figure 8.8, which requires the operator to turn the handle from one setting

Figure 8.7 Early pushbutton design (left) and its modern counterpart (right).

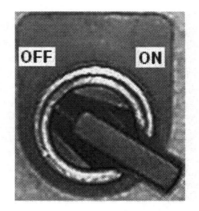

Figure 8.8 Discrete control with two states.

to another. This design is less likely to be swiped up or down, left or right, into the wrong position accidentally because of the arc of motion required to change the setting.

Figure 8.9 depicts two other forms of pushbutton that were encountered on an emergency override panel in an electrical power plant. The top-level control system was operated from a computer terminal. The button on the left in Figure 8.9 is an example of a button that has been protected against accidental operation with a collar around the button; the finger pushing the button has to push below the outer surface. The button on the right is an

Figure 8.9 Pushbutton with protective collar and pushbutton designed for emergency access.

emergency stop button. Note that the shape of the emergency button allows the palm of a hand to slam against the surface of the button at a moment's notice.

Figure 8.10 depicts a pushbutton panel in which the controls are hierarchically organized. The operator selects a mode. The name of the selected mode appears in the LED display at the top along with the numeric setting associated with it. The operator may then adjust the setting using the up and down buttons. The up and down buttons are located on the last two rows of buttons, second button from the left. The designers saw fit to put the up button above the down button.

Figure 8.10 looks like a common keyboard with additional function buttons. A keyboard is a familiar example of a system of discrete controls. This particular example comes from an HVAC system in a municipal building complex. The top-level controls appear on computer terminals, which the humans monitor on a regular basis; control operations can be made from this point. The button panel shown in Figure 8.10 is mounted on one of the chiller devices, and changes to the behavior of that particular unit can be made on the unit itself. The keyboard in Figure 8.11 is an intermediate level of control between the device-specific controls and the global computer controller. At this point the operator can intervene into many aspects of the system's operation from the keyboard panel. This control panel was once the top-level controller, but it was eventually superseded by the computer system with a GUI.

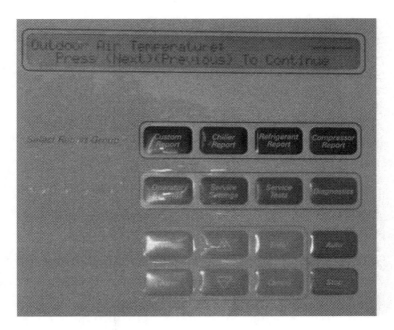

Figure 8.10 Discrete controls for a set of hierarchically organized functions with numerical settings.

Figure 8.11 Wall-mounted keyboard with a generous supply of function buttons, numeric keypad, and position controller in the lower right portion of the panel.

Figure 8.12 Virtual slideswitch emulates a real-world control.

Continuous controls typically take the form of knobs and slideswitches. A knob would be a rotary control because it turns. A slideswitch would be a linear control. A slideswitch would do the same job as a knob, but it would require movement along a vertical or horizontal path. Figure 8.12 depicts a virtual slideswitch on a computer interface. It gives the appearance of an actual slideswitch although it is actually controlled less directly by a mouse. Continuous control motions in many cases can be entered through a keyboard: position the cursor, hit the desired keys, and then press "ENTER." The example in Figure 8.10 allows for operation through either the virtual slideswitch or the keyboard using the white panel directly below the slide.

The device shown in Figure 8.13 is also a rotary controller of the continuous type. It calibrates the flow of steam through a complex pipe system in a building complex. It was installed in the mid-1930s and was still in use at the time the photograph was taken for this book.

The development of computer-based technologies introduced some new types of control mechanisms. The buttons discussed thus far have a real-time function, in which a flow of electricity starts or stops. In computer-based technologies, the buttons shown on computer screens are only virtual representations of familiar controls. Touchscreens and virtual slide switches are other examples. Other novelties such as gestural interfaces, have become prominent in virtual reality systems and human–robot interaction and are discussed in later chapters.

Meanwhile a good technique for making effective virtual buttons is to make the space on the screen that is responsive to a mouse-click somewhat larger than the space that is visually occupied by the button or icon itself. That way an operator moving a mouse cursor to the button will have a larger target to land on, which would in turn reduce response time and errors produced by buttons that were not pushed. An expansion of this principle appears later in this chapter in conjunction with Fitts' law.

Figure 8.13 Continuous calibration control.

Figure 8.14 The joystick is a 2-D control for positioning a security camera. It is shown here with a set of discrete function buttons associated with the camera system.

8.3.1 Multidimensional controls

The most common example of a multidimensional control is one that moves an object around a 2-D space. The joystick in Figure 8.14 belongs to a security camera system that allows the operator to adjust the camera remotely. A panel of additional buttons is used in conjunction with the joystick, several of which pertain to zooming or camera selection. 2-D buttons can do the same job in a continuous fashion. Up-, down-, left- and right-arrow keys can often be substituted for a joystick or 2-D button; in this case the movement of the object has been discretized.

The HFE wisdom in this case is that a 2-D control should be employed if it corresponds to a 2-D display. If the two intended parameters of control are depicted as 1-D displays, two 1-D controls should be used (Kantowitz & Sorkin, 1983). Figure 8.15 shows three renditions of a common 2-D control. The example on the left only navigates a menu with the *enter* (or *select*) button in the middle; the four points on the circle are analogous to the up, down, left, and right arrow keys on a common computer keyboard. The example in the middle uses the four-directional configuration to manipulate the two controls that most often used on a television— channel (station) and volume; this control configuration runs counter to the standing wisdom because volume and channel are really independent functions. In the example on the right, which also operates a television, volume and channel are two separate 1-D controls. Fortunately, these are low-consequence household entertainment devices, and the user acclimates to the device quickly.

Figure 8.15 2-D controls for remote control of televisions.

Another way to manage multidimensionality in a control system is by creating hierarchical arrangements of control functions using modes. If one controller has set the system into Mode A, the remainder of the buttons will perform functions, and generate displays, pertaining to Mode A. If the mode controller is set to Mode B, the remaining buttons will perform their usual functions, but with respect to Mode B. The danger here is that the operator must be fully aware of the mode in operation. If the operator imagines that the system is in Mode A, but it is really in Mode B, the correct information will not be introduced to Mode A and new wrong information will be introduced to Mode B.

Display design is critical to helping the operators keep track of the mode that is in operation. In one surprising and fatal airline crash, the cockpit crew wanted to insert information to the automatic control system that signified an angle of descent, measured in degrees. They did not notice that they were in the wrong mode and entered the numerals that they wanted, which were read by the system as speed of descent, measured in km per second. The only difference in the display for the two modes was the placement of an automatic decimal point, which the operators did not notice (Straub, 2001). The aircraft collided with the side of a mountain instead of flying over the mountain and gently descending to the airport that was located on the other side of it.

The cargo arm on the space shuttle requires a 3-D controller. The arm is manipulated from within the life-support pod while the action is taking place outside. The arm is manipulated through 3-D space until it reaches the grapple connection on the cargo object (perhaps a satellite). A second controller opens and shuts the grapple mechanism, and the operation of the arm in 3-D space continues to relocate the object in the desired position.

8.3.2 Size

Historically, large-size controls were large because they moved large physical forces. Their size was dictated by the physical configuration of the equipment. A sense of importance was thus attached to size. Figure 8.16 depicts an early 20th-century wheel that is now located in the basement of an electrical power plant. It is still available for use if the layers of automatic control fail to function.

The device shown in Figure 8.17 is a piece of equipment that converts physical activity to signals that the computer upstairs can process and display for the operator. Note that it needs to be precisely set and calibrated, and it has controls and a display of its own. Errors at this stage in the process of information flow can upset what is recorded and interpreted later on.

Size continues to have symbolic value with current technologies, where larger is more visible and probably more important to the system in some fashion. Size also matters if the operator is required to push a button or

Figure 8.16 Large controls like this one control large mechanical forces.

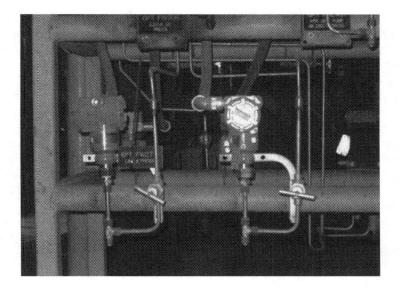

Figure 8.17 Equipment that translates a physical activity of a system into something that a computer can process.

flip a switch when approaching from a distance, especially during an emergency. For a panel of pushbuttons or switches, each one should be large enough and separated far enough apart from the other buttons and switches to allow for accurate operation. The width of the human finger is a limiting size requirement. The relationship between size and distance is expanded further in this chapter in the context of Fitts' law.

Touchscreens have been replacing button panels on a variety of devices for work-related and leisure technology products. Touchscreens are commonly used on banking kiosks, cell phones, tablets, navigation and entertainment modules in automobiles; they have been replacing most uses of buttons. Some agencies such as ANSI have issued recommended minimum button sizes ranging from 9 to 22 mm, based on performance studies or anthropometric tables for the size of human fingers (Sesto, Irwin, Chen, Chourasia, & Wiegman, 2012). Sesto et al. compared forces applied to button of sizes ranging from 10 to 30 mm to determine if size or spacing between the buttons affected impulse or dwell time; they also investigated how people with motor deficits would respond differently to the touchscreen panels. The experimental touchscreen was equipped with force detection sensors under the surface.

Impulse was the cumulative amount of pressure applied over time for the duration of the button-push. *Dwell time* was the amount of time the finger spends on the button before releasing it. It should be noted here that some contemporary button designs utilize dwell time to move the control into another control mode. Another variation is to use dwell time to adjust the amount of the control motion, such as the buttons in Figure 8.15 for adjusting the volume of a television set.

The results of the study (Sesto et al., 2012) showed that the force measured were about 6% harder or longer for the smaller size buttons. Button spacing did not show an effect for the range they considered. People with motor impairments tended to apply more force than unimpaired people. Overall, the participants applied about four times as much force on the buttons than was actually necessary to activate the control. The added force was thought to be related in part to participants' experiences with a broader range of buttons. Real buttons have a stroke length that is a lot longer than what is needed for a virtual button on a touch screen. Some button controls have built-in levels of resistance that vary. Thus the forces that might be reasonably applied to some types of buttons might be overkill for touchscreen buttons.

8.3.3 Shape

Designers might exploit the shape of controls to induce greater tactile feedback to the operator that indicates that the operator used the correct control. Control operations that are performed in low-visibility conditions, such as underwater, would benefit from shape cues on the controllers (Carter, 1978). If both the auditory and visual systems are overloaded, shape cues on the joystick handles can make marked improvements in operator accuracy.

In another classic example, numerous pilot errors on World War II military aircraft were resolved by tactile displays. Importantly not all types of error occurred in all types of aircraft in use at the time, which was a clue that cockpit design was contributing to the error rates. For instance, a common error in some aircraft was raising the landing gear after the plane had landed. The problem was solved by Alphonse Chapanis. Roscoe (2006) reported retrospectively:

> Chapanis realized ... that the problem could be resolved by coding the shapes and modes of operation of controls. As an immediate wartime fix, a small, rubber-tired wheel was attached to the end of the wheel control and a small wedge-shaped end to the flap control on several types of airplanes; the pilots and copilots of the modified planes stopped retracting their wheels after landing. When the war was over, these mnemonically shape-coded wheel and flap controls were standardized worldwide, as were the tactually discriminable heads of the power control levers found in conventional airplanes today.
>
> *p. 12*

Seminara, Gonzales, and Parsons (1977; Norman, 1988) reported a do-it-yourself retrofit of controllers in a nuclear power plant. The operators were required to reach for two important controls and wanted to avoid confusing them. They replaced the indistinguishable control handles with beer tap handles, which had different shapes and were much larger than the standard issue equipment.

8.3.4 Space of controls

The total amount of action potentially resulting from control movements is known as the *space of the control*. The system is capable of a minimum and maximum output. A good control design will allow for a range of control settings (i.e., with a continuous control) in which the full range of action is evenly distributed across the full movement of a knob or slideswitch (Kantowitz & Sorkin, 1983).

A common violation of this principle occurred years ago in the design of volume (loudness) controls on guitar amplifiers and stereo equipment. The knob would show calibrations from 1 to 10, but the effective action of the control was constrained to a shorter range, for example, 1 to 4. Values 5 through 10 did not accomplish much. The prospective purchaser would be told by the salesperson, "This is how it sounds on 4. We can't play it any louder in the store." True, but they could not play it any louder anywhere else either. Imagine the surprise and delight of this writer who, after becoming accustomed to this design flaw persisting for a couple of decades of audio equipment, encountered a knob control that was actually properly designed.

8.3.5 Labels

Kantowitz and Sorkin (1983) identified four rules for the use of labels on controls, besides the basic idea of labeling the controls. First the labels should be located in a consistent position relative to the controls: always above the control, or always below or to the side of the control. Shifts in the positioning of labels can mislead the operator. It is sometimes possible to engrave the label on the control itself. Here one runs the risk of the lettering wearing off with extended use.

The second rule is to avoid abstract symbols. Clear verbal labels are better. A true tale of two motorists illustrates the problem. Motorist A had owned a German-made automobile for decades when the event occurred around 1970. His next-door neighbor, Motorist B, had owned a U.S.-made automobile for as long a time. Motorist A was planning an extended trip to Europe and asked B if he would start A's car once a week or so and take it for a ride just to keep it running. B agreed, but when B eventually got behind the wheel of A's car he found he could not interpret any of the glyphs on the controls and thus could not get it out of the garage. Eventually A returned from the trip and rented a car to travel from the airport to his home. The car rental desk gave him a Buick convertible to drive, which he managed to start and get onto the expressway. A could not interpret most of the glyphs he saw either, and ended up driving home with the top down on a cold and windy day in second gear, substantially below the highway speed limit.

The third rule is to use color coding if and only if the colors have meaning. Decorative use of color only introduces irrelevant information. One strategy for colorizing a control panel might use different colors for controls that pertain to different groups of functions. Another strategy might use an unremarkable black button for most of the controls, but a special green control for "start" and red for "emergency stop" functions.

The fourth rule is to be sure there is sufficient lighting around the control panel to see the labels. Otherwise the labels are pointless. On the other hand, shape cues could mitigate poor visibility, as mentioned earlier.

8.3.6 Resistance

Controls are often built with a small amount of resistance to the operator's motion. The resistance is sometimes a result of the physical forces that are involved. At other times resistance is deliberately introduced to protect the device from accidental operation or from damage during operation (Kantowitz & Sorkin, 1983).

There are four types of resistance. Static resistance gives the operator a bit of a counterforce in the early milliseconds of operation, but operates freely afterward. Elastic resistance feels like a spring-loaded knob; once the control has turned to a desired setting, it reverts to its initial position as soon as the operator releases it.

Viscous resistance is proportional to the speed of motion. The dashpot on a utility door allows the door to close at a preset speed once the user has set it in motion. The objective is to prevent the noise or damage that could arise from slamming the door or the operator's failure to give a manual door enough force to close properly.

Inertial resistance is proportional to the acceleration of motion. This utility prevents unwanted consequences from sudden starts. The accelerator pedals of many makes of automobile were redesigned in the mid-1970s to prevent "jackrabbit starts," which wasted fuel. In the older design, the metal bar behind the pedal that was ultimately connected to the engine was suspended from the floor. In the later designs the same metal bar was suspended from above. As a result, if the operator "put the pedal to the metal" the physical forces would translate into fewer forces to the carburetor.

8.3.7 Control panels

The same rules for good organization that apply to a panel of displays also apply to a panel of controls. Control panels should be organized by their frequency of use and their order of use. Controls should be grouped by function. Controls that are used in combination should appear together. Linking analysis may be useful for determining the usage patterns.

The controls in the panel shown in Figure 8.18 are primarily grouped by function. In this particular case, several buttons in a function group are used together. The buttons within a function are colored differently to enhance the segmentation of the layout. The visual displays for this digital sound synthesizer are modest. Little red status lights appear above the control

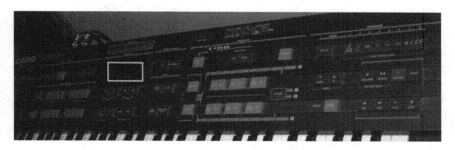

Figure 8.18 Control panel for a sound synthesizer, late-1980s vintage. The box with the white frame indicates the location of a central LED display.

Figure 8.19 Power plant control room, mid-1970s vintage.

button. Alphanumeric information appears in a multimodal LED display. The display is centrally located and very small compared to the space occupied by the full range of controls.

A control panel for a power plant dating back to the mid-1970s appears in Figure 8.19. It is not possible to tell from the picture to what extent the currently known rules for control panel organization have been observed. The sheer morass of controls and displays, however, strongly suggests that a better organization of the panel design would greatly reduce operator error and response time. The connection between the displays and control is not readily apparent. Operators would need to retain a mental map for which displays on the vertical back connect to displays on the horizontal panel.

In newer systems, functions such as those shown in Figure 8.19 are controlled from a computer terminal. The program's screens are organized hierarchically by function. In the typical hierarchical organization of screens, the top-level screen shows broadly defined situations and content, and the screens further down the sequence become progressively more specific in control. With a windowing interface it is possible in principle to use different combinations of functions together as desired. Figure 8.20 depicts a control station for a liquid and gas pipeline operation. The operator is apparently controlling 11 screens' worth of material simultaneously. Industry standards are now being developed to assess the cognitive load on the operators during specific phases of operation and to design systems to facilitate their decisions regarding how to mitigate an impending emergency

Figure 8.20 Control room for pipeline operation, late-2000s vintage. (Reprinted from U. S. Department of Transportation, 2012; in the public domain.)

(McCallum & Richard, 2008; U. S. Department of Transportation, Pipeline and Hazardous Materials Safety Administration, 2012).

Figure 8.21 is a different type of control panel layout that operates several pieces of equipment used in open-heart surgery. The machine segments are strategically distributed around the operating room and require multiple humans to monitor them. The researchers (Kennedy-Metz et al., 2021) were particularly concerned with the surgery team's cognitive workload, which they found to vary substantially with the patient's heart rate variability (HRV) and actions that needed to be performed simultaneously. The interface was not implicated as a major contributor to peak workload issues; future research directions would be directed toward designing robotic surgery equipment instead, which would imply a substantially different interface.

8.3.8 Voice control

Voice-controlled systems are often sought for disabled persons (Noyes & Frankish, 1992) and for situations where the operator's hands and eyes are overtaxed such as piloting a military aircraft (Corker, Cramer, Henry, & Smoot, 1992; Reising & Curry, 1987) or driving an automobile with a navigation system or cell phone (Zaidel & Noy, 1997). Voice controllers also have some potential for distinguishing users' voices when the voice is intended as a locking or security system. Although advances in signal processing have made it possible for systems to recognize single words with sufficient accuracy, there is substantial within-person variation in vocal production, perhaps resulting from stress, mood, or ambient distraction that could compromise the effectiveness of voice-controlled locking systems.

Figure 8.21 Distributed control panel for equipment used in open heart surgery: (1) arterial pump module, (2) secondary arterial pump module, (3) pump and vent controls, (4) data management system, (5) heater/cooler, (6) cardioplegia delivery system, (7) blood gas monitor, (8) isoflurane vaporizer, (9) oxygen monitor, (10) pump timers, (11) pump pressure sensors, (12) gas blender, (13) blood reservoir. (Reprinted from Kennedy-Metz et al. (2021), p. 762; in the public domain.)

The typical system uses the word as the unit of analysis rather than the phoneme. Systems tend to discriminate best among words when the system vocabulary is small, and when the speaker is given an instruction to something specific: "To help us direct your call ... you can say, 'place an order,' or 'pay a bill,' or 'something else.' " In this telephone application, the voice control is substituting for pushing a button on the touch-tone phone (Baber & Noyes, 2001). The limited vocabulary compensates for the wide range of voice registers that the system is anticipating. Fortunately, voice-control users automatically make their speech commands more concise than they might do when talking to another human (Amalberti, Carbonell, & Falzon, 1993). More versatile systems allow users to use either touch-tone buttons or talk.

Noise on the channel and noise in the operator's environment induce two sources of error that limit the use of voice controllers to noise-controlled environments. If the system requires individual users to enter template commands in their own voice, the templates should be entered while the user is in the environment in which the voice-controlled device is going to be used (Baber & Noyes, 2001).

Usability studies should compare voice-controlled systems against other control modalities. In one application involving the disclosure of potentially sensitive medical information in a survey, respondents disclosed more information using an interactive voice response system than they did using a web-based survey (Evans & Kortum, 2010). Here one could say, however, that the user was controlled *by* the system rather than controlling the system. In a different type of system in which the users were accessing weather information in an air traffic control environment, voice control compared unfavorably against stylus and touchscreen versions of the system with regard to exploration, gathering, and recall of information (Dang, Tavanti, Rankin, & Cooper, 2009); the users found the slow rapport time for the voice controlled system frustrating. In a military environment, there are advantages to having multiple control modes available as some communication objectives are more compatible with some controls systems than others (Walker, Stanton, Jenkins, Salmon, & Rafferty, 2010); digital voice input worked well enough for transmitting *data*, but information that is supposed to convey *meaning* was communicated better through conventional person-to-person analog systems such as radios and telephones.

Levulis, DeLucia, and Kim (2018) compared the effectiveness of voice, touchscreen, and multimodal (voice plus touch) controls for three types of tasks conducted with a cluster of two helicopters and three unmanned aerial vehicles (UAVs). The tasks were photograph processing, a power supply problem, and a communication task. Touchscreens and multimodal controls outperformed voice control on all three tasks with regard to accuracy and response time. The gap between voice and other controls was greater for versions of the tasks that involved higher subjective workload. Participants' spatial ability was also a substantial predictor of performance on the tasks.

8.4 FEEDBACK AND CONTROL

The operation of machine controls should result in feedback to the operator concerning the results of the control action. The feedback can be visual, auditory, or tactile. Visual feedback can be constructed in three ways: a display indicating that something has happened as a result of the control action, the observation of one's finger pushing the button on the control panel, and the effect of the system as one watches it do what it is supposed to do. A common form of auditory feedback is the status tone that signals to the operator that a control has been received. Auditory feedback can also take the form of a change in the sound of the machine itself, for example, an engine turning on or off, or the sound of some other mechanism starting to move.

Tactile displays are often produced as feedback from control operation. Diving suits and other environmental suits insulate the operator from direct control of the environment, and there is often an advantage to building in synthetic sensory information so that the operator can react with the external environment in a manner more similar to direct contact. A mechanical form of tactile feedback delivery occurs when an operator types on a keyboard. Note that the surface of the keys is slightly beveled to accommodate the curvature of the fingertips. The keys are spaced so that an operator can feel the edges. As a result, some errors can be corrected when the operator notices that a keystroke did not *feel* correct. This is an important source of feedback because expert typists do not need to looks at the keyboard when typing information from a source sitting on their desks.

8.4.1 Open and closed loops

Feedback can be positive or negative in character. Negative feedback tells the operator to reduce or stop the control action. Its purpose is to hold the status of a system to a fixed point; a thermostat is a prototypical example. Positive feedback tells the operator to increase the quantity of a control motion. If there is no negative loop kicking in, the control action and the result escalate to an indeterminate point. Other sorts of dynamics could occur under strictly positive feedback conditions such as the destabilization of the system (Jagacinski & Flach, 2003). A diagram of an open loop system is not fundamentally different from the serial processing schematic shown in Figure 7.7.

The feedback patterns, or *loops*, can be characterized as open or closed. In an open loop system, the operator exerts a control motion, and the effect is carried out without the effect of feedback changing the operator's actions in any way. In a closed system, information from visual, auditory, or tactile displays affects the operator's motions, plan of action, or hypothesis concerning the nature of the problem that needs attention. A diagram of a closed loop system is not fundamentally different from the diagram of a self-organized system in Figure 3.7, except that loops can be inserted at several different points in the control–action system to possibly affect changes in psychomotor motions, plans, or hypotheses.

8.4.2 Fitts' law

Fitts' law is a fundamental description of control response time. An operator's response is conceptualized in a physical space where the finger, hand, or tool is applied to a button, control, or target location (Fitts, 1954). It is a precise movement toward a specific target in space, where movement time (MT) is a function of distance to the target (D) times the width of the target (W):

$$MT = a + b \log(2D/W), \tag{8.4}$$

where a and b are constants. The constants are calculated through linear regression analysis. In practical application, researchers modify Equation 8.4 to include separate regression weights for distance and target width (Liao & Johnson, 2004):

$$MT = a + b \log(D) + c \log(W). \tag{8.5}$$

The additional regression weight accounts for some individual differences, either from different people, for example, experts versus novices, with regard to how the two facets go together for different types of tasks.

Kantowitz and Sorkin (1983) noted that a flow of information occurs between the operator and target as the finger approaches. The operator gauges the distance between the target and the current location of the finger, and adjusts the direction of the moving finger to hit the target. This simple-sounding task can be challenging for cluttered control designs and tiny buttons. In those cases, the operator is required to process greater amounts of information.

The speed of finger movement changes from the early phase of control motion to the later phase. Movement is relatively fast in the early stage, but as the finger gets closer to the target area, more information needs to be processed and the speed of movement slows down. Woodworth (1899) defined the two velocities in movement as *impulse amplitude* and *impulse duration*. The phenomenon was later reframed as a simple nonlinear model in which the square of movement time was proportional to the distance between the finger and the target:

$$MT^2 = a + bD. \tag{8.6}$$

The nonlinear model was also known as *bang-bang theory* (Jagacinski & Flach, 2003; Kantowitz & Sorkin, 1983).

If one wanted to fix a movement time to a desired level and do the same for the distance between the operator and the target, the target would have to have a critical width (W_c) to meet the distance and movement time objective. The relationship is:

$$W_c = k \ D/MT \tag{8.7}$$

(Jagacinski & Flach, 2003; Schmidt, Zelaznik, Hawkins, Frank, and Quinn, 1979).

Fitts' law and the bang-bang theory assume that a very simple movement is occurring. If the movement involves multiple submovements, Fitts' law generalizes to:

$$MT = k_1 + k_2 (D/W)^{1/n}, \tag{8.8}$$

where n is the number of submovements (Jagacinski & Flach, 2003; Meyer, Smith, Kornblum, Abrams, & Wright, 1990).

The foregoing models do not consider the weight of the operator's arm while making a movement. It is not unlikely that most of us take the weight of our body parts for granted and simply know how to move them. On the other hand, the arm might not be just an arm, but an arm plus a tool, such as a pair of pliers, welding device, chainsaw, or rocket launcher. According to Flach, Guisinger, and Robinson (1996), the amount of force (F) required to make a control motion is proportional to a function of the distance and movement time required:

$$F = kAD/MT^2,\tag{8.9}$$

where k is once again a proportionality constant, and A is the weight of the arm.

Jagacinski and Flach (2003) likened the effect of F to the effect of water pressure in a dripping faucet. At the lowest levels of flow, the drip is periodic and irritating. With greater force, the drips are aperiodic or chaotic over time. With still greater force we no longer have a dripping faucet, but running water instead. The analogy here is that insufficient force will produce an ineffective control motion or at least a poorly executed one.

Fitts' law can also be applied to the operation of foot pedals. In the 1990s, many drivers of the Chrysler Jeep Cherokee reported cases of unintended acceleration. In automobiles more generally, the brake pedal is located to the right of the centerline of the steering wheel, as it is meant to be operated by the right foot. In the Cherokee, the accelerator and brake pedals were located too far to the left, with the brake pedal left of the centerline, so drivers hit the accelerator instead of the brake. Here the situation was combination of too long a distance to the brake pedal and the population stereotype for where the brake is supposed to be located (Lidwell, Holden, & Butler, 2011).

The computation of distance across three dimensions is actually an extrapolation of the calculation of distance of objects across a 2-D plane. The computation for a 2-D orientation is the application of the Pythagorean theorem:

$$D_{a,b} = [(a_1 - a_0)^2 + (b_1 - b_0)^2]^{1/2},\tag{8.10}$$

where a_0 and b_0 are the coordinates of the starting position, and a_1 and b_1 are the coordinates of the landing position. For three dimensions,

$$D_{a,b,c} = [D_{a,b}^2 + (c_1 - c_0)^2]^{1/2}.\tag{8.11}$$

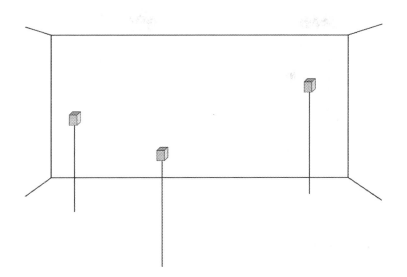

Figure 8.22 Droplines on 3-D display. (Reprinted from Liao and Johnson (2004, p. 479) with permission from the Human Factors and Ergonomics Society.)

The transition in the computation of distance from Equation 8.10 to Equation 8.11 can be extrapolated to four or more dimensions, but the need has not arisen in concrete applications to control systems. It has been used, however, with multivariate statistics when the goal is to detect an outlier on the totality of measurements dimensions that were involved.

Liao and Johnson (2004) assessed the accuracy of Fitts' law for 3-D arrays of objects that are portrayed on a 2-D screen rather than through a stereoptic viewing system. One of the difficulties of the 3-D application of Fitts' law is that target size and distance are confounded visually by virtue of the size–distance constancy in perception. Thus, objects that vary along the depth axis could be more difficult to reach accurately. Liao and Johnson also considered the possible advantage of placing a dropline on the display for orienting an object on the horizontal axis. Their example appears in Figure 8.22. Their results showed that Fitts' law accurately represented the movement time data for different experimental subjects and for movements along the X, Y, and Z axes which were assessed separately. The R^2 coefficients for degree of fit between Equation 8.5 and the data were all equal to or greater than .90. The droplines produced a small but consistent effect on reducing movement time, but did not affect the degree of fit between the equation and the data.

The latest frontier on this topic involves positioning objects in a 3-D virtual space. Deng, Geng, Hu, and Kuai (2019) found that they needed to decompose movement into three parts—acceleration, deceleration, and correction. The correction is needed to accommodate target tolerance, which is analogous to target width in the familiar 2-D case. Deng et al. were able to develop regression models that were similar to those introduced by Fitts and later researchers that were based on three variables: distance to target, target tolerance, and object size. Object size played a critical role in the acceleration phase and correction phase, but not in the deceleration phase. Smaller objects required longer movement times to position. The use of a checkerboard background in the object placement area did not affect the accuracy of the statistical models that were generated without the checkerboard. The accuracies of the models were not affected by whether the operator used a manual control or a head-tracking control.

8.4.3 Motor control

It is evident from the discussion of Fitts' law that if the analysis involves anything but the simplest movements, the mathematical description of those movements becomes complex in a hurry. Complex movements involve coordination between body movements. The timing and phasing of each movement is at least as important as the movements themselves. The acquisition of motor coordination for a given task involves a process of self-organization among the motor movements. The exact pattern of movements varies with the specific task, the range of environmental constraints under which the task is performed, and the demand for stability in the behavioral outcome that is desired (Turvey, 1990).

Motor coordination is an example of the broader class of synchronization phenomena that were described in Chapter 3. In the simple task of bipeds walking, our left and right legs are making reciprocal movements at precisely the correct speed; rehabilitation is probably required if not. In the less simple task of juggling (e.g., one person, two hands, three rubber balls), the two hands are tossing balls in the air to a consistent height and pattern; the balls are caught and tossed back into the air when gravity returns them to the vicinity of the hand.

Motor control was an underresearched topic in human factors and psychology more generally (Guastello, 2006; Rosenbaum, 2005), although there have been some new developments in rehabilitation science, particularly where concepts from NDS have been involved. The information technology industries have not been shy about simplifying motor skill requirements in the interest of improving human performance speed, quantity, and accuracy. Machines and automation software often replace old skills. High-skill tasks, such as those found in the graphic arts, have been reduced often to two motor skills—point and click. The commercial interests that might be served by enhancing motor skills have been served instead by minimizing the demand for those skills.

It is possible, nonetheless, that the industrial trends are now reversing. One of the latest frontiers in the computer industry is haptic perception and motor control. The perception questions revolve around the means by which touch sensations are translated into mental representations of objects. The motor control aspect attempts to describe what the person does as a result of those mental representations, bearing in mind that there is a continuous two-way flow of information between the person and the external object while the event is taking place. Virtual reality programming requires modeling information about simple-looking tasks such as how one bounces a rubber ball off a wall and catches it (Burdea & Coiffet, 2003). Motor control is, furthermore, strongly influenced by the distribution of fluid in the inner ear (*otolithic system*), which is in turn strongly influenced gravity, which tells us which way is up. Simple physical tasks that might be taken for granted on earth become scrambled in outer space, as the astronauts acclimate to new relationships among visual cues, motor movements, and disrupted touch sensations (Bock, Fowler, & Comfort, 2001).

8.4.3.1 Walking

Walking is more complex than it looks to the untrained eye, and the tools of nonlinear dynamics have been useful for gaining insight into the process. Most notably, there is a lot of variability within and between people as measured by gait camera systems in biomedical engineering laboratories. Gait cameras capture horizontal hip swing, vertical hip swing, floor clearance, leg muscle contraction, and stride length. Abnormalities along any of these parameters are often detected by the camera more readily than by the person walking (Postman, Hermens, De Vries, Koopman, & Eisma, 1997).

An excessive amount of variability in a gait sample taken from a single individual would reflect a level of abnormality, but variability in the healthy or normal case is not random. Healthy variability is chaotic in much the same way as the variability found in skilled performance (Mayer-Kress et al., 2009; Stergiou, Harbourne, & Cavanaugh, 2006). The loss of variability should be considered as a cause for alarm. Loss of variability means loss of adaptive capability, for example, to various irregularities in walking surface conditions or other environmental conditions. Thus the loss of variability should not be equated with "stability" (Stergiou & Decker, 2011). Excessive variability that meets the analytic definition of random, however, strongly suggests that neural pathways have not stabilized or have been severely disrupted. The more common rehabilitation challenge, however, is to overcome the loss of variability. The principle operating here is the optimum variability construct that was explained in the previous chapter.

The rehabilitation objective of restoring a healthy level of variability is made possible by increasing the level of variability, additional degrees

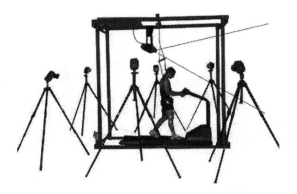

Figure 8.23 Virtual walking environment and gait camera. (Reprinted from Katsavelis et al. (2010a) with permission from the Society for Chaos Theory in Psychology & Life Sciences.)

of freedom in the person's movement are also induced (Vaillancourt & Newell, 2000). In one promising system (Katsavelis, Mukherjee, Decker, & Stergiou, 2010a, 2010b) a virtual reality display was devised so that a person walking on a tread mill could watch the pathway in front of them change gradually as if they were walking in real space and time (Figure 8.23). The idea was that by controlling the optic flow on the visual display, the visual–motor feedback channels would induce improved variability in gait. The technique produced the desired result in the laboratory. The long-term efficacy of the technique for rehabilitation has yet to be determined, as many factors related to a patients' underlying disorders are involved.

8.4.3.2 Reaching and grasping

Stroke survivors can have any of a number of cognitive or psychomotor deficits. For many, the simple task of reaching and grasping a common object is a challenge. Thus, robotic therapies are currently being developed to help retrain the cognitive–motor system. The essential idea of the therapy is that the patient engages in reaching exercises while attached to an electro-mechanical device (Figure 8.24). The robot matches the human movement as it transpires against a mathematical model of normal movement and guides the arm when it gets off-course. The theoretical challenges are currently: the development of appropriate mathematical models for the reaching movement, the fact that most prototypical PMSs capture the reach portion of the movement but not the grasp, and the definition of the mathematical model that coordinates the opening and closing of the hand with the reach

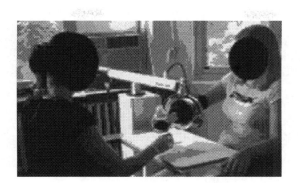

Figure 8.24 Robotic therapy device. (Reprinted from Wisneski and Johnson (2007) under the terms of the Creative Commons Attribution License.)

portion of the movement. Part of the problem was resolved in principle by the development of a sensor glove that is fitted at the hand's end of the arm controller (Nathan, Johnson, & McGuire, 2009).

Mathematical models for movement are developed in the laboratory by attaching sensors to the arm and hand that transmit information to a recording device that produces digitized data for the movement along X, Y, and Z axes of motion. One of the candidate mathematical models for the reach segment of the motion is the *minimum jerk*, which is a set of three equations for X, Y, and Z axes, all of which are fifth-order polynomial functions of time to complete the motion (Flash & Hogan, 1985). Another candidate is a seventh-order polynomial function of time that operates along one axis of motion (Amirabdollahian, Loureiro, & Harwin, 2002).

The latest candidate model reframed the movement as an NDS process where position at time 2 is a function of position at time 1, in contrast to earlier models that used time as an external variable (Guastello, Nathan, & Johnson, 2009). A movement cycle that involved reaching for a cup of water, putting it to one's mouth, and replacing it to the table was divided into four "events": reach, lift, pull to mouth, replace. According to their operating theory (p. 106):

> The perception of the goal object triggers the movement. The visual information stream provides a continuous feedback flow that directs the hand to the object. The grasp response is coordinated to both the visual stream and arm movement to open the hand when it is in the near vicinity of the object and close it again upon reaching the object. Tactile and visual feedback at the last step signals the human or robot that the goal has been reached, and triggers the next movement ...

Attractors can be conceptualized in two forms. In the situated form, the object [of the grasping motion] constitutes the attractor. The perception-action sequences are organized around the capture of the object. In the embodied form, the attractor is the stabilized pattern of behavior that is, for the most part, insensitive to variations in irrelevant environmental cues. (Ijspeert, Nakanishi, & Shaal, 2003; Nolfi & Floreano, 2000)

The models themselves were Lyapunov structures:

$$x_2 = a \exp(bx_1) + c \exp(dy_1)$$
$$y_2 = a \exp(by_1) + c \exp(dx_1) \qquad\qquad (8.12)$$
$$z_2 = a \exp(bz_1) + c \exp(dx_1),$$

where x, y, z are positions the Cartesian space coordinates, and a, b, c, and d are nonlinear regression weights (Guastello et al., 2009, p. 107). As the results turned out, motion along any one axis required input components along two spatial axes. The two exponents captured two different rates of movement simultaneously, which is consistent with the two-velocity feature of Fitts' law. The Lyapunov model provided a more accurate fit to the data than the alternatives, and with fewer parameters (Figure 8.25). The R^2 coefficients for goodness of fit ranged from .940 to .999 for the various tasks and events in the study.

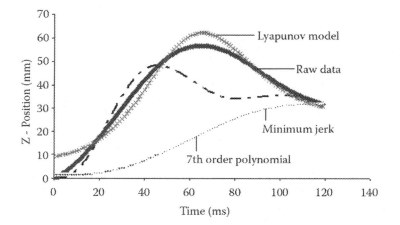

Figure 8.25 Relative accuracy of Lyapunov, minimum jerk, and 7th order polynomial models for arm movement. (Reprinted from Guastello et al. (2009, p. 114) with permission of the Society for Chaos Theory in Psychology & Life Sciences.)

The Lyapunov structure was also useful for developing the first mathematical model for the grasp phase of movement. When the arm extends toward an object, there is a critical point located at approximately two-thirds of the path toward the object when the hand opens up. The goal of the model is to predict the critical point in order to send a signal to the patient's nervous system if the hand is not opening properly:

$$g_2 = a \exp(bg_1) + c \exp(dx_1), \tag{8.13}$$

where g is the aperture of the hand, x is the position along the x axis, and a, b, c, and d are nonlinear regression weights (Guastello et al., 2009, p. 108).

The best procedures for signaling the nervous system is a work in progress. The backwards pathways between the hand leading to the brain are still a black box, but the goal of robotic therapy is to activate the plasticity of the nervous system to find new degrees of freedom in the form of circuits that be used to replace the ones that were damaged (Nathan, Prost, Guastello, & Jeutter, 2014).

8.4.3.3 Aiming

Fitts' law provides an accurate prediction of movement time if the operator's actions are also accurate at a level of 96% or greater. Wijnants, Bosman, Hasselman, Cox, and Van Orden (2009) considered the case of much lower levels of accuracy, which one could expect during the early stages of a learning process. They found that movement time had considerable variability under conditions of low accuracy, but the variability is deterministic in nature such that it can be quantified using fractal metrics. They found that the distribution of movement times fit a $1/f$ distribution, and the degree of fit improved over blocks of trials of an aiming task. Attractor strength increased with practice. Sample entropy, which is an NDS metric based on the fractal dimension and entropy principles, decreased over blocks of trials. The underlying dynamics that produce the self-organization of the perception and action patterns is thought to be similar to those described in conjunction with arm movements.

8.5 COMPLEXITY OF CONTROL SYSTEMS

8.5.1 Order of controls

Controls have different effects over time, and part of the learning process is to learn the behavior of the system once the command has been given to it. Figure 8.26 compares some common control functions as they transpire over time.

A *zero-order control* is comparable to the effect of adding a constant to a column of numbers. The control motion in the upper left of Figure 8.16 is probably turning a system, or a function of a system, on or off. A *first-order control* exhibits an effect that is proportional to time; hence the relationship between the control output and time is linear. First-order controls display velocity; if we drive a car at 50 kph for a half-hour, we will have traveled 25 km.

A *second-order control* exhibits acceleration. The middle panels of Figure 8.25 show two types of acceleration. The growth curve could be the result of a positive feedback effect, and could signify a system that is heading out of control in some fashion. The decay curve could be the result of a negative feedback effect that could be induced to hold the system at a constant level of output.

The accelerator pedal on an automobile certainly does accelerate the vehicle, but it is also used to maintain a constant speed. This type of control is actually an acceleration-assisted control. It has a first-order component built into it as well as the second-order component.

Third-order controls involve a change in acceleration, which we would experience as a sudden jerk motion. As mentioned previously, accelerator pedals on automobiles were redesigned years ago to dampen jackrabbit starts, or sudden jerk motions. Jerks are unfriendly to many systems, but they might be needed in others when an extra initial force is needed to start the system, but not for long afterwards.

Fourth-order controls are rare. One example occurs in ships, however, in the mechanism that changes the heading of the ship while the ship is in motion. The position of the ship during the control sequence can be imagined as a 4-D function of time. One dimension is a change in heading in degrees off the bow. The second is the rate of change in the rudder position, which actually steers the ship. The ship is already moving, but if the ship is maintaining a constant speed going around in a (partial) circle, it is actually accelerating. The movement associated with this type of control is complex, and the mechanism is comparably slow. A predictive display would thus be needed to inform the operator if the ship will be headed in the correct direction once the mechanism has completed its work.

The sinusoidal function in the lower portion of Figure 8.25 has the potential for third- or fourth-order control functions. The operator may wish to adjust the signal that is displayed in the oscilloscope by its frequency or amplitude. Frequency is rate over time, or velocity. It could be made to accelerate over time if the operator wanted to change the frequency over time, for example, change the pitch of a sound. The amplitude may be held constant, or it can be modulated also at a rate of acceleration.

8.5.2 Ashby's law

Some control functions are obviously simple, whereas others are just as obviously complex. Which is better? According to Ashby's (1956) law of requisite variety, a controller must be at least as complex as the system it is meant to control. *Complex* in this case refers to the number of system states or combinations of states that the system could produce or process in some way, or the system's movement pattern over time.

Remote-controlled military weapons can be deployed faster than weapons requiring standard human operation, all other things being equal. A significant problem with all military weapons, however, is the potential for casualties due to friendly fire. *Inhibiting* trained automatic responses is not a good idea. A better solution is to *complexify* the response, for example, changing a one-button system to a three-button system, to allow more cognitive processing time of the ambiguous visual signals that often lead to friendly fire (Helton & Kemp, 2011). A further investigation showed that a more complex three-button sequence produced more accurate responses with fewer commission errors than a simple sequence (Head et al., 2020). A possible downside to complexifying the response tasks would be that slower response could give a faster-acting enemy an advantage.

There are limits to the amount of complexity that ensures survival. Excess complexity could result in wasted energy or loss of time when responding to important events, which would be antagonistic to survival in severe circumstances. In less severe circumstances, having more complexity available than what is needed would do no harm so long as it did not interfere with the work the system was trying to do. Some strategies for decomplexifying systems are discussed in Chapter 14.

Figure 8.26 decomposes the optimum variability principle and illustrates that the optimum is a saddle point between two forces. One force is the increase in complexity of the controller that is needed to match up with the complexity of the system being controlled. The other force is the minimum entropy principle where many of the neuromotor pathways needed to get the job done reduces with learning and automaticity.

8.5.3 Chaotic controllers

Ashby's law and managing complexity brings us to the topic of chaotic controllers. There are four basic approaches to controlling chaos: anticipation and contrary motion, adding more instability, forcing periodicity, and manipulating control variables that are known to induce dynamical effects in the system.

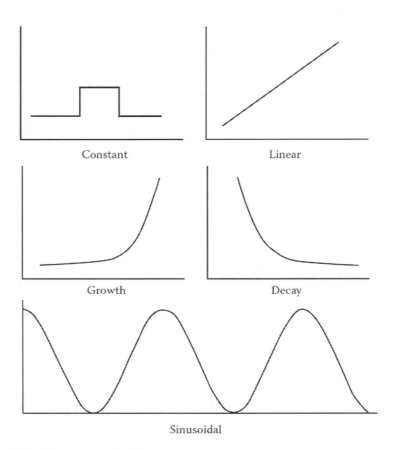

Figure 8.26 Effects of some basic controllers as they occur over time.

8.5.3.1 Anticipation

In the most advanced stages of situation awareness, the operator can antici-pate what the system is likely to do and forecast the impact of a control action that could be used to redirect the system. Chaotic systems are wide-spread and difficult to forecast, but it is humanly possible to do so without computational aids. The heuristics that work and some that do not were described in the previous chapter.

Another use of the anticipation principle is to offload the chaotic con-troller to the machine, leaving a simpler task for the operator. For instance, one of the leading causes of death in agriculture is tractors tipping over on their drivers. The problem is related to both the slope of the hill and irregularity of ground; the latter has a fractal or chaotic character to it. Roll bars can be factory installed on the tractors or as retrofits. Roll bars reduce the problem of the slope, but not the instabilities coming from the land under the wheel, which have a greater impact on light-weight excavating equipment. Thus mechanisms in the wheels have been developed that can interpret the incoming pattern in the land and induce a correction to the wheel–axle assembly (Sakei, 2001). The human still needs to avoid pushing the improved system beyond the extent that it can reasonably go.

8.5.3.2 Adding instability

The idea of controlling chaos with chaos seems counterintuitive, but it works in theoretical systems (Breeden, Dinkelacker, & Hubler, 1990; Jackson, 1991a, 1991b). This type of chaotic controller operates by first adding a small amount of low-dimensional noise into the system. The sensitivity to initial conditions is not uniform throughout the attractor space. Sensitivity is greatest in the outer edge of the attractor basin and least in the center. The level of sensitive dependence may be affected by the attractor's proximity to another attractor. Adding noise to the system allows the attractor to expand to its fullest range thereby giving the observer a more complete picture of the system's dynamics. The technique could be regarded as a variation of stochastic resonance, except that what was noise in one context is now deterministic variability here. The controller can then respond in one of two ways. One option is to apply a filter that allows a special range of values to pass through. The second option is to mimic the chaotic motion of a point, predict where it is going, and respond in a strategic manner.

One application that was devised involved the development of a special shoe for people who had difficulty walking due to motor impairment. As mentioned previously, their gait typically suffers from insufficient variability. The prototype shoe would apply vibration to the foot with the intent of encouraging greater variability and adaptation of movements by the wearer (Priplata, 2006). The efficacy of the design is still under study (Collins, 2008).

8.5.3.3 Periodic entrainment

The method of controlling chaos through periodic entrainment (Ott, Grebogi, & Yorke, 1990; Ott, Sauer, & Yorke, 1994), or OGY, is based on the relationship between oscillators and chaos: Three coupled oscillations are minimally sufficient to produce chaotic time series. Many economic and biological events exhibit periodic behavior, or result from coupled periodic oscillators. Although not all of the possible combinations of parameters for these oscillators produce chaos, it is still a viable analytic strategy, nonetheless, to decompose chaos into its contributing oscillating functions when it needs to be controlled. The controlling agent would then identify the strongest oscillator and amplify it by organizing activities around its timing sequence. Because of the connections between the dominant oscillator and the others, the other oscillators will entrain more strongly to the dominant one bringing the entire behavior pattern into a simpler and more predictable regime.

One common application of periodic entrainment is batching tasks. If an operator receives task of a particular type on a random or chaotic basis, and the task is mixed up with other tasks of different types, the operator would let tasks of a given type pile up to a critical mass before going to work on them. Then once the batch is completed, the operator can move on to a batch of a different type of task. Another variation is to wait until a specified time in the work period to do whatever is in a batch. Batching has the effect of minimizing task switching costs and all the loss of momentum that goes with it.

The downside of the OGY method is that it is a top-down form of control. As such it is susceptible to disruption from noise coming from the bottom up. An external shock will be carried through subsequent iterations of a chaotic system and become magnified if it is not sufficiently controlled.

8.5.3.4 Use of control parameters

The logistic map (Equation 3.2 and Figures 3.3 and 3.4) receives a lot of attention in complexity studies because its one control parameter governs changes in system behavior from fixed to periodic, aperiodic, and chaotic states. Although there is no guarantee that a real chaotic system can be controlled through all its dynamical states by just one parameter, it is intriguing, nonetheless, that a wide range of behavior patterns could be controlled by a very few well-chosen parameters.

Cusp catastrophes are controlled by two parameters. One brings the system closer to or away from the critical point where behavior changes suddenly. The other governs the size of the change. Although the most frequently discussed catastrophe models only assume fixed points or oscillators as their attractors or stable states, the attractors can be chaotic also. The same control principles would apply nonetheless: move the system away from the critical points or turn off the bifurcating effect.

DISCUSSION QUESTIONS

1. Take another look at the control–display combination shown in the upper left of Figure 5.8. The positions on the pointers in the display are set with the use of the knobs appearing below it. How would you assess the compatibility of the control with the display?
2. Consider some tasks where error rates increase dramatically with work speed. What are the consequences of the errors? What characteristics of the situations promote large or small opportunities for error?

3. What types of psychomotor skill are involved in driving an automobile?
4. What are some good methods for preventing controls from accidental operation? A few methods have been mentioned in this chapter. Can you think of some others?
5. Several types of discrete controls were introduced in this chapter. Describe the assets and limitations of each. How would you determine which is the best one to use? (Hint: Pick the right tool for the right job.)
6. The notion of redundancy improving reliability was introduced in Chapter 2. How many applications of this concept could pertain to the design of control systems?
7. Some researchers have considered the possibility that people could become fatigued by pushing too many buttons in the course of a day. Is this really a meaningful concern? Why?
8. How do astronauts determine which direction is "up"?
9. One science fiction fantasy that could become a reality soon is the "smart house." Imagine a house or apartment that is voice-controlled so that lights and security systems would turn on and off, appliances would operate, and heat and air conditioning would adjust if you spoke the magic words. Would such a system really be effective?
10. We usually think of users controlling the system, but are there cases where the users are controlled *by* the system? What would be the advantages and disadvantages of controlling the user?

Chapter 9

Anthropometry and workspace design

9.1 BODY MEASUREMENTS

The study of bodily measurements began with some of the earliest uses of descriptive statistics by Adolph Quetelet (Howell, 1992). Quetelet discovered that the measurements were normally distributed, an idea that could not be taken for granted during that time period. The discovery gave statisticians direction for handling and interpreting large populations of measurements and samples taken from interesting subgroups.

Figure 9.1 shows several aspects of bodily measurements that are used for common workspace design problems: operator standing, operator's reach envelope on a flat workbench, and operator's seated area. The arrows mark commonly needed measurements. For the most part, workspace engineers can rely on tables compiled in known documents such as U.S. Department of Defense (1991) and Roebuck (1995) for U.S. military and civilian populations, Pheasant (1996; Pheasant & Haslegrave, 2005) for British populations of varying ages, and PeopleSize (Open Ergonomics, 2000) for a compendium of international samples. In addition to the linear measurements denoted in Figure 9.1, circumference measurements such as the wrist and forearm, are also included in some tabular sources.

Some occupational groups could differ from the general population as a result of the people who are attracted to the job and how they are selected. Electric utility workers, for example, tend to run larger than the average population (Marklin, Saginus, Seeley, & Freier, 2010). The U.S. military has its own norms tables that need to be utilized when supplying any equipment to the military (U.S. Department of Defense, 1991). A web search of *DOD-HDBK-743* would turn up anthropometric standards for military of some other countries.

Note that most tables provide measurements for unclothed people. People will be wearing clothes when the system is put into use, however. Pheasant (1996) recommended adding 25 mm to each linear dimension for males to compensate for shoes, and 50 mm for females. Other adjustments should be

DOI: 10.1201/9781003359128-9

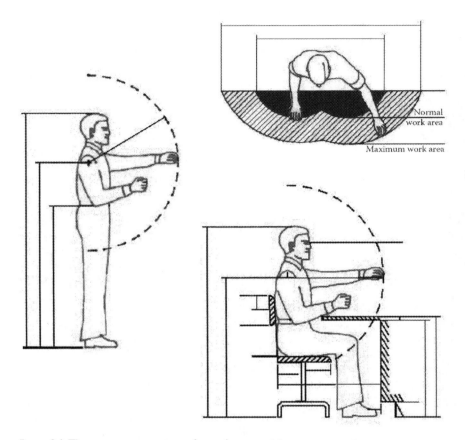

Figure 9.1 Three common postures for anthropometric measurement.

made for the thickness of clothing, as the situation requires. Swimwear and winter clothing will have different standards, obviously.

If the measurement goals can be accomplished from a single measurement taken from a single population source, the engineering goals would only require needs some statistics: the mean, the standard deviation, and a chart of percentiles associated with each measurement. Systems are typically engineered for the central 90% of the distribution, that is, for the range between the 5th and 95th percentiles. If the goal of the design is to make sure that operators have at least enough space, the design process ends here. If too much space is a potential problem, such as objects being placed out of a person's reach or a driver not being able to reach the pedals of a vehicle, some flexibility for adjustments in seating or object positioning need to be built into the design.

If a composite of populations is involved, the engineer needs to introduce some statistical sampling techniques that draw from the measurements of specific populations. For instance, in the early days of the space program (late 1950s), astronauts' capsules were designed for a highly selected population of 45 men who were candidates for space flight (Roebuck, 1995). Space flight crews of the late 1990s and beyond include women, civilians, and people of other national origins. The range of body movements has also changed from constricted seating to spaces that support work requiring movement around the spacecraft and exercise.

The goal of an anthropometric design often requires involving multiple measurements on a human, however. On the one hand, sizes of body parts are correlated. If one part is larger, the other measurements are larger in proportion. On the other hand, the correlations are imperfect and sometimes low. Thus the correlations among the measurements are important information. If the set of required measurements were completely unrelated, one could determine the bounds of a workspace by simply looking at the 5th and 95th percentiles of each desired measurement. If the relevant population contains males and females, which is almost always the case, the design range is for the 5th percentile female and the 95th percentile male, which is wider than a population of males and females mixed together (Marklin et al., 2010). Inasmuch as statistical independence among bodily measurements is seldom realistic, reliance on simple percentile data could produce a design that is too large or too small for the real central 90% of the population. One way to get around the problem of correlated measurements is to collect new data from a representative sample of the user population; local population data would capture all the ethnic variability represented in the sample.

9.1.1 Designing for bodies in motion

The discussion of body dimensions thus far has been predicated on the assumption that little physical movement is actually involved. In the real world, however, tools and work objects must be manipulated in many ways. Figure 9.2 depicts a system for determining a work envelope in 3-D space (Roebuck, Kroemer, & Thomson, 1975). The worker is positioned in a space that is fitted with movable poles in all directions. The worker then acts out the likely work tasks, moving the poles sufficiently out of the way to accomplish the actions. The amount of protrusion of each pole into the workspace can be measured. The data for each pole can be compiled and tabled as for other anthropometric measures.

Once the space requirements have been determined, the next phase of the design is to build the prototype workspace. The iterative design concept that is used for controls and displays is also applicable here. The engineer would build a mock-up of the workspace with inexpensive materials. Any

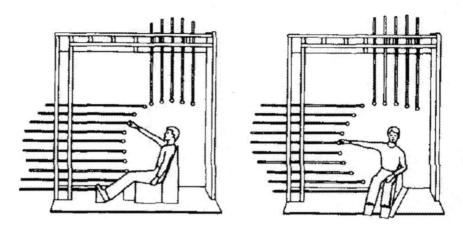

Figure 9.2 Equipment for determining a 3-D work envelope. (From Roebuck et al., 1975, p. 72. Reprinted with permission of the authors.)

controls and displays that would be part of the workstation do not need to be functional at this point. Small, medium, and large people should stand or sit in the mock-up and go through the motions of work activities. Designers should then determine whether activities could be executed comfortably, whether the users can reach the controls or tools, and so forth. If the workspace is an environmental pod such as the driving compartment of a motor vehicle, the designers should determine whether users can enter and exit quickly and comfortably and whether there is enough head clearance.

Various components of the system tend to interact. Once the suitable space for the operator has been determined, the next questions are: Can control and display panels can be fit in comfortably? Does the operator need to reach over one control in order to grab another? Is accidental activation of one of the controls likely? Can the displays be made large enough relative to the viewing distance?

If the system were expected to move through space and time, it would be helpful to use mannequins instead of people for testing some aspects of the system's integrity. Hence a few crash test dummies would be worthy investments. Mannequins can be built with internal sensors that detect impact on various parts of the body. Small annoyances such as rubbing or chafing can be detected with simulations that assess the results of extended use of the system as well as traumatic impact.

As each layer of questions is answered, the next layer needs to be addressed and tested before the design can be considered finished. Although human factors research and development may seem laborious and expensive in some respects, it is a far better investment to place the resources into a proper design than to manufacture a product that needs to be redesigned to compensate for user dissatisfaction.

9.1.2 Facial anthropometry

Developments in artificial intelligence for face recognition have expanded into a new area of recognizing emotions, resulting in new and evolving anthropometric databases specific to the human face (Boesch, 2022; Fulmer, 2022). The photo of a face is decomposed into a wireframe consisting of 38 points. The software detects what is happening to the mouth, eyes, and the rest of the face when an emotion is expressed. The ratios of movements at the various measurement points have been associated with the six basic emotions that are known to be universal—happiness, sadness, anger, fear, surprise, and disgust (Ekman, 1993)—using a library of representative pictures of faces. Emotional recognition is thought to have advantages in health care applications, marketing, and better service delivery when service providers and clients are connected through cell phones or computers equipped with video channels.

The current state of the science is not sufficiently accurate, however. Different cultures have different norms for how and when to express an emotion. Within a single culture, a person could laugh because something is funny, but people also laugh just to be polite or to express discomfort with what was just seen or heard. Emotional recognition is estimated to be 65% effective at best at the present time.

9.2 SAFETY AND RELATED CONCERNS

The previous section of the chapter considered the engineering of a single workspace. In this section we consider the juxtaposition of one workspace with another, or with hazardous objects positioned nearby. The soldier in Figure 9.3 has a workspace problem. Can you figure out what it is?

9.2.1 Machine guards

Blades, other sharp edges, protruding structures, and pinchpoints on machines are all examples where a person–machine contact should be avoided. Figure 9.4 illustrates two hazardous configurations where two machine rollers have turned inward (based on Smith, 1977). If a hand or piece of clothing becomes caught it could drag the operator into the machine, and the odds of a significant injury would be very high. Another form of pinchpoint occurs where a drive chain connects to a sprocket; an example appears on almost any bicycle. Machines with pinchpoints should be equipped with guards to prevent the operator from inadvertently entering the mechanism. For the equipment shown that contains a number

Figure 9.3 A workspace design problem. This picture circulated the Internet without any identifiable source.

of pinchpoints, a good solution might be to construct a guard that surrounds a large portion of the mechanism (Haigh, 1977).

Figure 9.4 also illustrates what would happen if the operator should accidentally slide under the machine or try to do so while performing maintenance on it. Safety, comfort, and practical amounts of workspace are just as important for maintenance processes as they are in regular machine operation.

Emergency shutoff switches are as important as machine guards to workspace design. Emergency switches should be reachable from any operator position relative to the machine. In the case of maintenance functions, it is standard practice to turn off motors before servicing the machine. There must also be sufficient notice given to operators of machines in the immediate area that someone is performing maintenance and that the system should not be turned on. Some factories install lockout devices on a cluster of nearby machines to prevent any one of them from operating during a maintenance task.

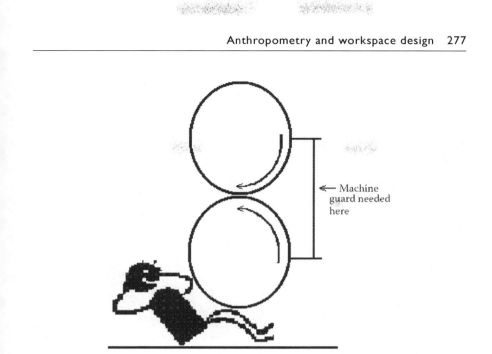

Figure 9.4 Machine configurations illustrating pinchpoints.

9.2.2 Overcrowding and confined spaces

The previous sections of this chapter outlined the necessary steps for giving the workers just enough space. Repeat the steps too many times, however, and something dismal happens, as shown in Figure 9.5. Crowding is stressful. The impact of stressors of different types on occupational health and performance is explained in the next chapter.

Confined spaces is a formal term given by NIOSH to denote a type of space that workers are only expected to enter on a very occasional basis. Their options for entry and exit are very limited, and they are usually bedeviled by insecure standing platforms, structural instability, insufficient oxygen, and highly toxic air pollutants. Examples include grain silos, trenches, water and chemical tanks, wells, manure pits, pipelines, and sewers. The CDC (2016) reported that confined spaces are responsible for about 92 fatalities per year in the United States.

NIOSH (2011) has a system of guidelines for working in different types of confined spaces. Some of the more generalizable ones, which appear to have been overlooked in fatal situations include: (a) have a written plan for entering, exiting, and working in the confined spaces; (b) equip the workspace with lifelines and harnesses; (c) test air quality before entering; (d) ventilate the space from contaminants; (g) supply oxygen masks (self-rescuers) on site; (h) have an emergency rescue plan developed that should be known to local fire and emergency medical teams.

Figure 9.5 Tight workspaces contribute to crowded and stressful conditions. This picture circulated the Internet without any identifiable source.

9.3 PHYSICAL ABILITIES

This section describes some other human measurements that are particularly relevant to physically demanding work. Fleishman (1964) identified nine physical abilities factors. The objective of the study was to establish physical fitness norms for military personnel and to define training requirements for them. Some of the concepts transfer well to civilian industrial settings.

9.3.1 Strength

Static or *isotonic strength* is the amount of force in pounds per square inch (or metric equivalent) that a person can exert on a target. The measurement does not actually involve moving the target. Static strength is measured by a dynamometer; some simple examples for measuring arm, back, and leg strength appear in Figure 9.6 (from Chaffin, Herrin, Keyserling, & Foulke, 1977). For the measurement of arm strength, the person stands upright, holding a metal bar. The bar is attached by a chain to a load cell that is mounted on the floor. The posture and chain length allow only arm strength to register on the load cell when the person pulls upward. The load cell is attached to a digital meter that displays the strength measurement. The

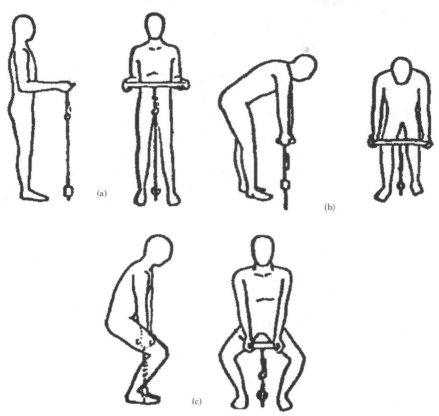

Figure 9.6 Dynamometer positions for (a) arm, (b) back, and (c) leg strength measurement. (From Chaffin et al., 1977; in the public domain.)

body position and chain length for back strength and leg strength are also shown in the diagram.

Dynamometers are also available for other strength sources, such as vertical pull and handgrip. The importance of handgrip strength is well-known to anyone who has ever tried to open a jar of pickles. From an engineering point of view, norms for handgrip strength from a relevant population are also important when designing mechanisms such as motorcycle clutches. Popular models can vary by a factor of 2 in their requirements for user strength. These differences in strength requirements are large enough to produce a schism in the markets for motorcycles, such that different subpopulations gravitate toward different models.

Dynamic or *isometric strength* describes the ability to move objects. It is calculated in terms of weights and distances and is the human equivalent of horsepower as it is applied to engines and motors.

Contemporary norms in factories restrict the weight of objects moved by humans to 50 pounds (or 25 kg) without the use of mechanical lifting devices. This is a guideline rather than a rule, and it is an impossible objective to maintain in some circumstances, particularly in cases where the physical environment is not conducive to positioning the lifting devices. Roofers in the construction industry are often able to load packages of roofing materials to desired locations on a roof with a crane, but the 100-pound (50-kg) packages must be moved short distances manually often enough. Firemen are typically trained or screened for their ability to carry a 150-pound (75-kg) body out of a burning building. People seated in the rows of airplanes adjacent to the emergency exits are informed that the door that they must remove weighs 65 pounds (30 kg).

Explosive strength is the ability to deliver exceptional force over a brief period of time. It is analogous to sprinting versus running, but often with a sledgehammer or other tool. In this writer's experience in the steel industry, emergency conditions arose occasionally where the operators had to break off a machine fitting with a sledgehammer before an explosion occurred. Two people working at explosive proportions for 2 to 5 minutes could get the job done, but they reported being exhausted for the next 20 minutes.

Trunk strength is the limited ability associated with the stomach muscle. The best indicator of this ability is the number of sit-ups a person can perform.

9.3.2 Flexibility

Extent flexibility is the ability to extend one's limbs as far as they can reach. *Dynamic flexibility* is a similar ability associated with repeated flexing.

9.3.3 Body coordination and equilibrium

Gross body coordination involves tasks requiring two hands or hands and feet. Some example tasks would involve packaging and manipulating boxes. Drummers in musical ensembles typically require extensive coordination of hands and feet, and some have become famous for their exceptional ability.

The ability to maintain *gross body equilibrium*, or *balance*, is typically measured in the rail walk test, the apparatus for which is shown in Figure 9.7. A 2" × 4" × 8' piece of wood (standard construction material) is firmly lodged to the floor with the narrow side up. The person being tested walks the plank three times forward, three times backward, and three times sideways. The score is the number of times the person walks the length of the whole beam without stepping off. The rail walk test tends to produce

Figure 9.7 Apparatus for a rail walk test.

higher scores for men than for women on the average because the body center of gravity is lower to the ground for women than for men.

9.3.4 Stamina

The last physical ability in Fleishman's (1964) taxonomy is stamina. Stamina is the ability to withstand physical effort over a long period of time. It is the opposite of fatigue, which is considered as a separate issue in the next chapter.

9.3.5 Lean body mass

Lean body mass did not appear as a variable relevant to physical fitness and work capacity physical ability until after Fleishman's (1964) taxonomy was published. Pound for pound, muscle is more useful than fat. Given equal strength, people who are higher in body fat are more likely to fatigue to greater extent (Guastello & McGee, 1987).

The original method for measuring percentage of body fat was to weigh the individuals then submerge them in a task of water, record the amount of water displacement, and then calculate the person's density. Fat is less dense than lean body mass. The obvious inconvenience of this method led to the development of a method using skin fold calipers. The person taking the measurements pinches a fold of flesh at critical body locations and uses the calipers to measure the thickness of the fold. The critical points on the male body are just above the knee and on the back just below the arm socket. The critical points on the female body are folds on a triceps and the midriff. The conversion table that comes with the device translates the skin fold measurements into percent body fat.

9.3.6 Personnel selection and physical abilities simulation

Several of the large steel manufacturers conducted studies to determine which job applicants would be best suited to the physically demanding work. It was not possible to determine just by looking at individuals whether they would be capable of doing the work. Medical examinations can determine if a person had a prior back injury, thus making the person unsuited for certain types of assignments, but they cannot determine suitable strength without the use of dynamometers for persons who have never been injured.

The physical abilities studies in occupational settings are formatted like standard personnel-selection studies (Guion, 1998): The independent variables consisted of measurements that could be taken during a preemployment interview, and the dependent measures were work performance measurements taken under controlled conditions. Campion (1983) reviewed personnel selection studies with physical abilities from industrial settings as well as for police and fire personnel. The following is a synopsis of a similar study reported from steel manufacturing (Guastello & McGee, 1987).

At the time the study was conducted, several of the major steel producers were working under a legal judgment from a suit brought by the Equal Employment Opportunity Commission requiring the employers to increase the proportion of women in traditionally male jobs to 20%. They needed a fair system for selecting applicants. The independent variables were thought to predict later work performance were height, weight, percentage of body fat, the rail walk test, and dynamometer measurements of arm, back, and leg strength. Other indicators were the number of hours of physical exercise the individual engaged in during the course of a week (self-reported) and whether the individual was currently a member of the labor pool.

The sample consisted of 129 volunteer employees, both male and female. The full complement of two production mills were included plus others who volunteered from the management staff. The objective of the sample composition was to produce the widest range of measurements of both the independent and dependent measures. Unrestricted score ranges will maximize correlations between independent and dependent measures and allow for the generalization of the final results to the widest range of job applicants.

The dependent measure was a composite of measurements taken during a structured two-hour period of simulated work that started immediately after the dynamometer readings were taken. The first task was to shovel

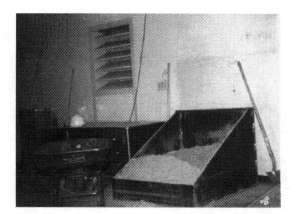

Figure 9.8 Shoveling station that was used in the physical abilities study reported by Guastello and McGee (1987).

enough furnace slag into a wheelbarrow at level height, then empty the wheelbarrow; the measurement was the time required to perform the task. The shoveling station is shown in the photograph in Figure 9.8.

The second task was to push a wheelbarrow through an obstacle course three times with loads of sandbags weighing 100, 200, and 300 pounds (Figure 9.9). The obstacles were designed as segments of railroad track that naturally occurred in the real work environment. The dependent measure was time elapsed. The other tasks were sledgehammering, lifting and carrying a drum of material used in the plant that weight a little over 100 pounds, pushing and pulling a heavy bar of steel across a greased track, hauling sandbags, and dumping gravel.

The results of the study indicated that the composite of measures of work performance could be predicted by six variables (multiple $R = .87$). The two most effective variables were arm and leg strength. Arm strength was not surprising given the amount of upper body strength generally required in this type of work. Leg strength often acts to compensate for arm and back strength. Back strength was also correlated with work performance, but it accounted for no unique variance beyond that accounted for by arm and leg strength; thus it did not enter the multiple regression model. This was a fortunate result because it was desirable to avoid the task of measuring back strength among job applicants to prevent injuries arising from unknown conditions.

The remaining four predictor variables that showed unique associations with work performance were height, balance, age, and whether the person was presently a member of the labor pool. Height was important for two reasons: (a) Taller people have more skeleton on which more muscle is

Figure 9.9 Wheelbarrow obstacle course used in the physical abilities study reported by Guastello and McGee (1987).

attached, and (b) the added height gave people greater leverage over some of the objects that they had to manipulate.

Older participants performed more work than younger ones; this was an unexpected finding. In normal circumstances, muscle strength declines about 5% starting at age 45. The explanation in this situation was that the older workers, who were either currently in the labor pool or who once were part of the labor pool but had moved on to other jobs, used their body motions and tools more efficiently than the younger workers who had been employed in the industry for less than six months.

Body fat did not have a direct impact on the work performance measure. It was related to greater fatigue, however. Fatigue was measured as the difference in arm strength between the time of the initial measurement and when the measurements were taken again after the two-hour work sequence. The topic of fatigue resumes in Chapter 10.

9.3.7 Exoskeletons and exosuits

Exoskeletons and exosuits are wearable technologies intended to support posture and muscle strength in work environments requiring substantial physical demands for extended periods of time. Exoskeletons are made of hard materials and can be powered or unpowered using levers and springs. Exosuits are made of soft materials that are essentially smart textiles. Both types of devices are also being explored actively for rehabilitation purposes, for which they would be used with carefully crafted therapies and supervision (David, Reid, Rempel & Treaster, 2020).

Bequette, Norton, Jones and Stirling (2020) examined the comparative effects of powered, unpowered exoskeletons with a control condition of neither types of device on physical and cognitive performance of 12 military personnel. The research participants carried loads of equipment at two different weights specified by military standards while following a research confederate at a standard military pace over a path containing step-over obstacles. During the walk they performed auditory and visual tasks. The focus of the investigation was whether the exoskeletons aided physical performance, and thus aiding cognitive workload capacity, or whether the technology distracted attention from the cognitive task. Results were mixed overall with substantial individual differences associated with both types of exoskeletons; physical agility was impaired by the devices for some of the participants, while others were distracted by the operation of the devices. Bequette et al. concluded that exoskeletons and training to use them needed improvement before widespread adoption would be warranted. Other research teams emphasized the importance of gaining buy-in from intended user populations with specific environmental and task-related demands (Alemi, Madinei, Kim, Srinivasan, & Nussbaum, 2020; Ármannsdótir et al., 2020; Elprama et al., 2020; Stirling et al., 2020).

9.4 SOME COMMON BIOMECHANICAL ISSUES

This section considers some topics in biomechanics that are frequently encountered during workspace design—lifting, design of seating, hand tools, and work conditions that are often associated with carpal tunnel syndrome. The biomechanical issues are often sources of musculoskeletal disorders (Chapter 11). NIOSH (1997) identified potential sources of musculoskeletal disorders in several industries. In meatpacking, for instance, the potential is associated with cleaning metal tubs, shank trimming, and removing lard and internal organs. In warehousing, the potential risks are associated with lifting and carrying containers of assorted weights. The point here is that each industry has its own array of potentially hazardous situations.

9.4.1 Lifting

In spite of the importance of back strength, there is a widespread advisement in labor environments to lift with one's legs, not with one's back. The weight per cm^2 on the vertebrae when the person lifting the object stands upright from a bent-over position is twice as great as the load on the back when the person stands upright with the load while unbending the knees.

Figure 9.10 illustrates an instance of work design that lessened the load on the worker's back. In the "before" picture, the operator lifted and moved heavy containers from an initial location on the floor. In the "after" picture, the containers were set on a table surface further from the floor, thus lessening the vertical lift. Work measurements that involve turning with the heavy object also produce an elevated risk of spinal injury (Kumar, 2002).

NIOSH (1994) developed an equation for computing recommended weight limits. The equation requires several measurements that are multiplied together to produce the result: the weight of the object, horizontal distance of the object's initial position to the operator's body, vertical distance (which is less if the object is positioned about 1 m off the floor), how high the object needs to be moved, turning angle, frequency with which the lift occurs, and the grip strength required to grasp the object. A companion document (NIOSH, 1994) provides numerous and probably familiar illustrations of best lifting practices for minimizing musculoskeletal injuries.

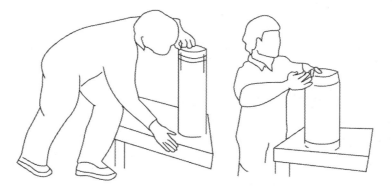

Figure 9.10 Workstation requiring repeated heavy lifting before and after reengineering.

Liberty Mutual Insurance (2005) developed a similar calculation routine where the calculation result is framed around percentile norms for lifting by males and females under a variety of lifting conditions. Their website provides a set of interactive pages for entering measurements similar to the NIOSH measurements. The result that is returned is the percentage of the population who can perform the task. They recommend structuring the tasks to match the capabilities of at least 75% of the occupational population. The website also has facilities for calculating the efficacy of pushing and pulling tasks and carrying heavy objects. The equations have been updated recently (Liberty Mutual Insurance, 2020; Potvin, Ciriello, Snook, Maynard & Brogmous, 2021).

The repetitiveness of a task could place a profound limit on the maximum acceptable effort associated with a lifting task. All other things being equal, the maximum acceptable effort value drops to 50% of peak effort if the operator is under load for 10% of the work cycle (Potvin, 2012). If the operator is under load for 30% of the work cycle, the maximum acceptable effort drops a bit to 45%, but fatigue sets in after one hour.

9.4.2 Walking surfaces

Pushing and pulling tasks often involve carts, so in addition to the strength requirement for setting the cart in motion there are two other sources of demand on the operator. One is the controllability of the cart, such as steering and control of tipping, which are inherent in the design of the cart and its wheels. There are two types of forces involved, *initial force* and *sustained force*. A cart designed for high inertia will require less sustained force but might be more difficult to stop in an emergency. The image of a "freight train running downhill" comes to mind as an extreme example of this principle.

Optimal flooring for carts involves a trade-off between two physiological demands. The concern is not only strength requirements, but disproportional strain on leg and back muscles and the respiratory system that could be induced by different configurations of materials. Maikala, Cirello, Dempsey, and O'Brien (2009) compared the demands for pushing an equipment cart across a plywood floor and a teflon floor; the coefficient of friction for the plywood was 2.6 times that of the teflon. Muscle demand was 26% higher for plywood, but oxygen demand for the initial impulse was 19% greater for the Teflon.

Another type of trade-off to consider when choosing factory flooring is the extent of ordinary walking that one should expect on the floor, and whether the pedestrians would be carrying a load of any size. Slippery surfaces encourage falls.

Walking on overly rough surfaces can have a substantial impact on gait. Railroad workers often have to walk on ballast, which are three sizes of rocks used for drainage in and around the track area. The largest rocks create the roughest surface. Wade, Redfern, Andres, and Breloff (2010) found that ballast sizes affect just about every aspect of gait measurement (see previous chapter). The overall recommendation is to use the smallest ballast on walking surfaces or on surfaces where the operator stands while working to minimize the demands on the ankle and the knee as they coact to maintain bodily stability and to minimize muscle fatigue.

9.4.3 Seating

Kantowitz and Sorkin (1983) noted that there would be no particular seat that would suit everyone, but a few truisms about seating have emerged nonetheless: (a) The majority of the body weight should by carried by the bones in the buttocks rather than by other body parts; (b) the pressure of the thighs against the seat should be minimal; (c) feet should be placed squarely on the floor or a footrest should be used; (d) it should be possible to change posture easily in the seat; and (e) the lower back area should be supported (p. 478). Proper support of the lower back alleviates the weight on the vertebrate by a substantial amount; the precise amount depends on the person–seat fit (Pheasant, 1996) and thus lowers the potential for back pain or pinched nerves after many hours of sitting. The ability to change posture comfortably facilitates the flow of nutrients to the vertebrae, flow of waste out of the vertebrae, and lessens fatigue (Adams & Dolan, 2005).

Chair designs can be evaluated giving users a modest task to perform, such as maintaining focus on a computer monitor and keeping contact with the keyboard, while being asked to change positions in the chair periodically (Bush & Hubbard, 2008). The chair would be equipped with sensors along the arm rest and seat. The user would wear sensors on the neck, elbow, wrist, upper and lower legs. There would be a sensor pad under the chair to capture movements of the chair. The criteria of chair performance would be sustained head, hand, and eye location on the part of the operator, sustained contact with the back of the chair while working or changing position, and the ability to change position comfortably.

There is now reason to think that too much time spent sitting can present health challenges even in the most comfortable seats (Husemann, Von Mach, Borsotto, Zepf, & Scharnbacher, 2009). Employees' work routines should be organized with a mix of standing and sitting and some opportunity or necessity to walk around. Sit-to-stand desks have been developed for this purpose. A review of 47 experimental studies indicated that their strongest impact was on the experience of comfort or discomfort; their impact on actual health outcomes and productivity were minimal (Chambers, Robertson & Baker, 2019). Ten to 30 minutes of standing per hour was the most agreeable balance to research participants working on computer tasks, compared to no standing or standing for longer proportions of time (Barieri, Brusaca, Mathiassen, & Oliveira, 2020). Small adjustments in monitor height are needed to produce the optimal impact (Fewster, Diffel, Kadam, & Callaghan, 2019).

9.4.4 Hand tools

Hand tools convey physical forces from the human hand to the working end of the device. Optimal hand tool designs attend to the forces involved, the shape of the handgrip, and the size of the hand. Consider any handy assortment of handles on kitchen utensils. Note how the shapes of the handles vary from a straight piece of wood to a handle with a greater contour for the handgrip. Some utensils have a soft surface that is compatible with a wide range of handgrips so long as the diameter of the device is not uncomfortable.

Different tools require different handgrip shapes. The fist that grabs the hammer is using the *power grip* (Pheasant, 1996). The curved handle on the pliers facilitates the *torque grip*, which is a motion of grabbing hold of the device and twisting it.

The *precision grip* is used to hold a pencil, pen, exacto knife, or similar device. Pencils have evolved to a standard diameter, undoubtedly with the help of the manufacturers of pencil sharpeners. Slight differences in the contours of the barrels of common pens may add a little comfort for some users. No one knows for sure whether the shapes make the handwriting any more legible, however.

9.4.5 Carpal tunnel syndrome

Many tools require or support repeated handwrist motions. When these tasks are performed all day every day over many years, carpal tunnel syndrome could result. The condition is an inflammation and irritation of the nerves going through the carpal bones of the wrist (Figure 9.11). The condition is painful and the remedy often requires surgery.

It is possible to redesign workstations to lessen the risk of carpal tunnel syndrome. In some cases, it might be possible to design machines to perform the repetitive handwrist tasks. Another group of options is to redesign the hand tools or simply adjust the seating so that the hand extends straight from the arm rather than bending upward from the wrist.

Computer keyboards offer some opportunities for preventative designs. Some users are comfortable with a palm reset that is placed in front of the keyboard; it elevates the palms so that the hand is working at a downward slope relative to the keyboard. Another group of innovations involves keyboards that are split down the middle so that the portions of the keyboard that are operated by each hand are angled at 15°. This orientation minimizes horizontal bending of the wrist, especially for augmented keyboards that

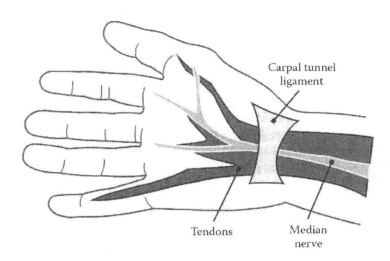

Figure 9.11 The carpal tunnel.

contain many more function buttons than the standard QWERTY keyboard. Ergonomic keyboards are discussed further in Chapter 12.

Vibration is also responsible for some incidences of carpal tunnel syndrome. A jackhammer hitting the pavement in the street has an impact on the street, but not so visible is the wave of vibration through the hammer back through the hand of the operator. If the jackhammer (or similar vibrating tool) does not produce carpal tunnel syndrome, it can produce a nexus of disorders involving damaged nerves, bones, other connective tissue in the forearm, muscles, and joints (International Standards Organization, 1986). Fingertips turning white is a signal that significant damage has occurred.

Bus drivers have an occupational exposure to carpal tunnel syndrome. The large steering wheel requires larger turn motions than does an automobile, and conveys a substantial amount of vibration from the vehicle. Incidence levels of carpal tunnel syndrome among bus drivers is greater for drivers who also have other types of occupational stress exposures (Guastello, 1991, 1992). It is likely that the drivers who were experiencing greater levels of stress were holding the steering wheels more tightly than other drivers, thus transferring the vibrations to a greater extent.

9.5 COMPUTER WORKSTATIONS

The visual display terminals (VDTs) that are common on desktop computers today were first introduced on a widespread level in the late 1970s, particularly in office environments. By the early 1980s, many office workers were reporting eyestrain, body pains, nausea, and even miscarriages that were allegedly due to computer terminals. The devices themselves emitted electromagnetic radiation at levels far below the acceptable levels for common television sets or other appliances. Wattenberger (1980) reported a series of studies that examined what the evils could possibly be. A series of concise studies were conducted with telephone company personnel.

9.5.1 Directory assistance operators

The sample consisted of directory assistance operators, half of whom did their work on computer while the other half used the older paper-and-pencil method. (Computer-based directory assistance functions were just being phased in at the time the study was conducted, so it was not difficult to find samples of operators who were already accustomed to one or the other method.) Each group completed a report of body pains and the Job Descriptive Index (JDI), which is a standardized measure of job satisfaction (Smith, Kendall, & Hulin, 1969) that is still available today in an updated form. The JDI contained five scales for satisfaction with pay, promotion, supervision, coworkers, and the work itself.

Wattenberger (1980) reported significant differences in job satisfaction on all scales except supervision. Greater satisfaction was reported for operators doing the computer-based task. Note that pay, promotion, and coworkers are not directly related to the experimental comparison, which was the difference in the work itself. Of the body parts affected by the differences in work systems, only neck pain was significantly different for the two types of workers. Neck pain was reported by 68% of the computer terminal users, which was a higher rate than that reported by the paper-and-pencil users.

9.5.2 Technical service representatives

Wattenberger (1980) used the same quasi-experimental plan with telephone technical service representatives, half of whom performed their tasks on computer terminals whereas the other half used paper-and-pencil materials. The participants in the study completed the same inventory of aches and pains, and the JDI.

Job satisfaction results once again favored the computer terminal users, but this time the only effect was satisfaction with the work itself. There was no spillover to other aspects of job satisfaction.

Service representatives who used the computer terminal experienced greater physical discomfort in four body areas: headache, nausea, blurred vision, and sore buttocks. The former three symptoms go together as symptoms of eyestrain. The results of the studies on directory assistance operators and service representatives led to an experimental analysis of computer workstation experiences.

9.5.3 Workstation experiment

Participants were computer users from a variety of telephone company job groups. There were two independent measures. One was whether the computer used a dark screen background and lighted lettering (common in those days) or a light screen background and dark letters (similar to contemporary screens for word processors or printed paper). The other independent variable was time of day; the researchers were looking for a circadian rhythm effect. Circadian rhythm is considered further in Chapter 10.

The two dependent measures were a subjective rating of discomfort and the near point. The *near point* is how close to the computer the person must sit in order to be comfortable. Results showed no effect for discomfort ratings, but both experimental factors and the interaction between them affect the near point. The results showed that the near point varied over time of day, and different trends over the day were recorded for light screen and dark screen.

Wattenberger (1980) arrived at several conclusions: (a) In spite of the folklore that was growing at the time, people liked computerized versions of the jobs better than they liked paper-and-pencil versions of the same jobs. (b) There were some true physical discomforts associated with computer workstations that could be isolated from imaginary discomforts. (c) Comfort levels did change in the course of the day, and the extent of change depended on the computer screens being viewed. (d) Reorganizing the workers' tasks so that the workers' entire day was not spent seated in front of a computer screen could reduce discomfort. (e) Discomforts could be lessened by designing and using adjustable computer desks and computer chairs.

Other related findings (Eastman & Kamon, 1976) indicated, furthermore, that the optimal orientation of a computer screen relative to the operator was an upward slope of 15°. This angle would match the natural angle of vision from the opposite direction. The reading target should be located within the range of 15° to 20° downward slope of the eye to minimize discomfort (Rempel, Willms, Anshel, Janchinski, & Sheedy, 2007). Although many studies on computer workstations have been published since 1976, it does not appear that the basic points that were made early on were the least bit shaken (Alyan, Saad, & Kamel, 2021).

9.5.4 The near point

When the early studies on computer workstations were conducted, the typical screen was a dark background with illuminated characters. The characters were the standard ASCII typeface equivalent to Courier 12-point font on today's word processors. Today the white background is the norm in word processing. The type is usually black, and typefaces vary considerably in style and size. Colored type and images are far more common now compared to a generation ago especially on web pages and when working with computer graphics.

According to Gunnar Optics (2022), "70% of U.S. adults experience digital eye strain as a result of increasing use of digital devices" which would also include handheld devices, video games, and similar devices with back-lit screens. Symptoms include: "Dry or itchy eyes, eye strain, watery eyes, headaches [and] blurred vision." Their response was a design for eyewear that wraps around the head to some extent to keep in moisture around the eyes, a lens that filters spikes in fluorescent light sources. The lenses are coated and made (optionally) with progressive magnification to support one level of magnification for looking at a screen and a slightly stronger one for reading from source material that is positioned close to the keyboard.

Research resumed on the best location for the near point in response to the growing number of people afflicted with digital eye fatigue and sore necks in addition. The near point itself represents a balance between sitting close enough to see clearly and minimizing the extra demand on the ocular system that goes with near-field vision. The participants in Rempel et al. (2007) performed 2.5 hours of visually intensive work on a computer while sitting at three distances from the screen; the distances were 46, 66, and 86 cm). The furniture was controlled so that the visual targets would appear within the 15–25° window, and the back of the chair was set at a slope of 110°. Once the set heights was adjusted for the individual participants, they could move their body position within the chair but could not move the chair itself or the computer screen. Dependent measures were several indicators of posture within the chair, head angle, and effective visual distance, with several significant effects associated with viewing distance. The net effect was that the longer viewing distances induced greater fatigue because the participants made more postural adjustments to compensate the distance.

Rempel et al. used standard text sizes found in most business applications (10–12 point type), and noted that longer viewing distances require larger size type in order to resolve visual accuracy than what is commonly used in business environments. Most programs, however, are equipped with a visual zoom feature instead. So, although the stress on the oculomotor system is greater at shorter distances, visual acuity is resolved at shorter distances. Postural and seating adjustments, zoom, and visual correction all offer some degrees of freedom for resolving visual acuity and minimizing fatigue.

9.5.5 Workstations in health care settings

Although the ergonomics of computer workstations had been known for 30 years, the classic issues of body aches and pains arose in the health care industry (Hedge & James, 2012). The new national health care system in the United States, which went live in 2014, required all hospital facilities to convert all their paper and other non-digital records to digital form. As a result, male physicians in one hospital were reporting up to five hours per day working on digital data entry in addition to two hours per day working on a computer in their home environment. Their female counterparts reported up to an hour more per day engaged in the same tasks. Up to 30% reported pains in their (usually right) hands, wrists, forearms, and upper arms, and 50–80% were reporting pains in their upper backs, lower backs, necks, and shoulders.

Lessons from the past strongly suggested that their workstations were not ergonomically suitable. Furthermore, the proliferation of laptop computer designs precluded the 15° rise in the keyboard which affects the angle of the hand on the keyboard; laptops are typically designed with flat-to-surface keyboards, which either instigates or exaggerates the physical discomfort. Two further questions immediately arise: If the physician is spending so much of the workday doing data entry, what is happening to patient care? Why are highly skilled individuals spending so much time on data entry tasks?

Hedge and James (2012) only considered the ramifications of so much typing. The friendliness of the software was not unpacked. At the time, there were at least four different software vendors whose products were targeted to the formatting of a database that would be compliant with national health care requirements, and over 200 software vendors producing reports that needed to be converted in some way or fashion. One would anticipate that any sources of user unfriendliness in the design of the software product can be expected to add to the perceptions of physical discomfort.

DISCUSSION QUESTIONS

1. How would anthropometric data be valuable in virtual reality computer programs? Describe some different types of programs and the data requirements that might be associated with them. (This question will be revisited in a later chapter.)
2. Are there any potential musculoskeletal hazards in the electronics assembly industry associated with jobs that involve coil winding or trimming, circuit board wiring, fastening parts, and packing products? What types of injury potential are evident, and why?
3. Is an adjustable desk really necessary in a computer workstation? If this suggestion were implemented, would there be any side effects or deterrents?
4. Describe some consumer products that could be improved with better attention to anthropomorphic data.
5. Are some consumer or industrial products better suited to right-handed people than they are to left-handed people? How or why would that be the case?

6. Some designs for writing utensils might have placed a value on style and shape over functionality. Consider a variety of designs, and discuss their relative merits. Does the pen slip out of the user's hand or require the user to hold the pen unusually close to the point or unusually further away from the point?

7. The workstation studies were conducted with desktop computers, which typically involve a desk and a chair. What about laptop users? Where do they place their laptops when using them? Is it really on their laps? What is the visual angle to the screen? What distancing do the users regard as comfortable?

Chapter 10

Stress, fatigue, and human performance

10.1 PHYSICAL STRESSORS

Stress, according to the classical definition, is the nonspecific reaction of an organism to any environmental demand (Quick & Quick, 1984; Selye, 1976). This definition is all-inclusive, and suggests that any stimulation at all is a form of stress. Both desirable and undesirable events can produce stress (Holmes & Rahe, 1967); the negative, undesirable consequences of stress are really matters of degree and interpretation.

The physiological dynamics of stress are rooted in the two subdivisions of the autonomic nervous system. The sympathetic nervous system is responsible for bodily arousal when it is activated. Once sufficient energy has been expended, the parasympathetic nervous system is activated and the body relaxes; relaxation is a physiologically active process, not a passive process. Stress reactions occur when arousal is frequent, but there is no opportunity to expend energy at the moment when it is necessary to do so, thus prolonged arousal results. Exercise programs are thus frequently recommended for alleviating the effects of stress in their lives (Zwerling et al., 1997).

There are many types of stress that a person can experience. Although they all have similar effects, it is convenient to classify them according to their sources because the goal is, ultimately, to target and reduce unnecessary stress both in the workplace and in personal life. The four broad categories are physical stressors, social stressors, speed and load stress, and disregulation due to irregular work schedules and sleep loss. The latter has some additional connections to fatigue.

10.1.1 Toxins

Physical sources of stress would include toxins, excessive heat, excessive cold, high altitudes, underwater environments, and noise. Toxins comprise a large and varied category. As a general rule a person is likely to experience the common stress symptoms, but with specific health impairments peculiar to the toxin (Travis, McLean, & Ribar, 1989). Many of them are severe to the point of becoming clear and present dangers more often than other sources of stress. OSHA standards for the safe handling and ventilation of chemicals should be consulted; requirements are specific to the type of toxin involved. Toxins present in confined spaces present additional dangers and challenges as described in the previous chapter.

10.1.2 Extreme temperatures

Excessively cold temperatures in the work environment can impair work performance and cause health problems, notably hypothermia. Nonetheless, appropriate clothing and training to perform the work under cold conditions can counteract the performance effects of cold (Enander, 1989). For less extreme environments, the standard of comfort is a normally ventilated room that is heated to 72°F, although people do express some individual differences in preference for a little more or less heat (Kantowitz & Sorkin, 1983). Turbulent airflow produces the sensation of colder temperatures, or windchill. Windchill is the psychological equivalent of actual outdoor temperature plus turbulence.

Clothing compensates for cold temperatures, of course. Each article of clothing has a number of *clo units* associated with it. A person should wear enough clo units to produce a comfort level equivalent to the standard. Ultimately people pick clothing based on what is in their closets and the known comfort levels associated with those items. In more extreme environments, workers advise each other as to how many layers of thermal underwear and sweaters are needed in addition to the outer garments to get through a night of −35°F with the windchill factor.

Martin et al. (2019) examined 124 studies involving military personnel who were subjected to excessive heat, cold, high altitude, and work underwater. The researchers were particularly concerned with the combined effects of physical and cognitive demands found in military activities. In the case of cold, they found "little impact on tasks of attention ... processing speed ... and auditory vigilance" (p. 1216). Cold temperature produced a negative impact on choice reaction time after exposures of more than two hours. The impact of cold on working memory capacity and executive functions were highly variable, such that the more demanding cognitive tasks were more likely to show impairment.

Gloves are often part of standard personal protection gear as well as cold-weather apparel. Gloves can be problematic because they sometimes interfere with fine psychomotor skill by exerting *sheer forces* between the operator's hand and the outer environment. The choice of glove is a trade-off between psychomotor flexibility, warmth, and protection from cuts and scrapes, and it is possible to purchase gloves that suit a range of possible requirements. For instance, one manufacturer advertised long ago, "Gloves so thin you can pick up a dime."

Excessive heat is a different matter, as there is a limit to the amount of clothing a person can remove. Performance decrements begin to appear at 86°F after 8 hours of exposure, but begin after 45 minutes of exposure to 100°F heat. The critical temperature of 86°F is the point where the body starts to build up heat internally. Ventilation can ameliorate some but not all of the effects of severe heat (Kantowitz & Sorkin, 1983). Current OSHA standards for heat exposure are predicated on the type of industry and the chemical or other hazards that are also present. Two standard responses to extreme heat exposure are to stay hydrated by keeping water available and using salt tablets. The salt helps the body retain water and slows the loss of electrolytes through perspiration.

Hancock, Ross, and Szalma (2007) conducted a meta-analysis of 48 studies published over the previous 60 years on the effects of thermal stressors on task performance. A meta-analysis is a set of statistical techniques that isolates the sizes of the effects of independent variables on behavioral outcomes reported in a variety of studies conducted in different settings with different sample sizes. The use of a common metric of effect size permits generalizations about overall effect size and helps with the search for explanations for the variability in effect sizes that are reported. Importantly, small samples sizes, which are common in human factors research, can produce a challenge to statistical significance when studies are reported individually. Meta-analysis, however, combines all the studies and sample sizes to produce a much larger total number of observations. Hancock et al. (2007) found that the overall effect size for thermal stress on performance was –0.34 standard deviations of the performance metric, which in turn was comparable to an 11% loss in performance overall. Effect sizes varied by whether the individual studies focused on heat or cold and the extremity and durations of the exposures that were studied.

In the case of heat exposure, performance loss increased relative to thermo-normal conditions as the ambient heat became more intense. The drop in performance was not as severe as the increase in internal temperature build-up. Heat exposure affected perceptual performance the most, followed by motor performance; cognitive performance was the least affected of three types of performance.

In the case of cold exposure, performance loss began at 52°F and dropped proportionately with colder temperatures, relative to thermo-normal conditions. Perceptual performance was again the type most affected, but the effect was stronger for the cold (–1.13 SD) than for heat (–0.78 SD). The effect of cold on motor performance (–0.41) was about the same as the effect of heat (–0.78). The effect of cold on cognitive performance actually produced an *increase* in performance (+0.41) instead of a negative effect (–0.23) for heat.

The effect of duration of exposure was not so simple. An acclimatization effect occurs whereby people adapt to the temperatures in much the same way as they adapt to other forms of repetitive or constant stimuli. Hancock et al. found that a cross product of exposure time and intensity was a good predictor of the effect size, however. Martin et al.'s (2019) review indicated that acclimation training for heat and cold appeared to help with performance issues, but the best form and duration of the training remains inconclusive.

The majority of studies on high temperature and stress examined the impact of heat stress on physical work. Using previous publications on the subject. López-Sánchez and Hancock (2019) examined the results of heat stress on cognitive activity and found an inverse power law relationship (Equation 10.1):

$$t = c \left[(T - T_{ref}) / T_{ref} \right]^{-a}. \tag{10.1}$$

In Equation 10.1, T_{ref} is the temperature for normal or comfortable working condition, T is new higher temperature, and t is exposure time before performance declines, c is a scaling constant, and $-a$ is the shape coefficient for the power law curve. Large values of $|a|$ indicate a faster decay time for performance. The analysis and comparison of curves showed faster decline for more demanding tasks; vigilance declined faster than dual task performance, and dual task performance declined faster than simple mental operations. The review study by Martin et al. (2019) also showed that excessive heat had a stronger negative impact on more complex cognitive tasks.

Because of the close connection between inverse power law distributions and self-organizing behavior generally, López-Sánchez and Hancock (2019) conjectured that a self-organizing process is taking place in response to heat stress as the brain regulates and integrates physical conditions and cognitive demands. Comfortable temperature ranges support one type of organized state. High temperatures trigger alternative responses and a reorganization that might not be optimal for cognitive performance.

10.1.3 Extreme altitudes

Studies in high altitudes either required transporting research participants to particular high altitude locations or simulating the altitude effect by fitting the participants with low-oxygen respirators and face masks in an environmental chamber (Martin et al., 2019). The effects of altitude were mixed overall, but when performance deficits did occur, the benchmark altitudes for performance deficits to occur were 3,700 m 4,300 m, and 5,100 m, depending on the task.

The stressors from working underwater arise from higher-than-normal atmospheric pressure and breathing a prepared mixture of air from a tank.

Studies on its effects on cognitive performance have been conducted in the ocean, pools, and environmental chambers (Martin et al., 2019). Most were conducted with experienced divers, but some studies utilized inexperienced volunteers. The effects of depth on performance were variable across a wide range of cognitive tasks. Some performance deficits were observed at 50 m below sea level, and others started to appear at 310 m. Martin et al. also reported studies showing that anxiety underwater affects performance, but only in cases where cognitive load was closer to capacity.

10.1.4 Noise

Noise is another physical source of stress, in addition to its ability to interfere with signal detection and produce hearing damage with prolonged exposure. Noise sources produce arousal, which can be either good or bad, but if the task is already challenging enough, it produces discomfort. High noise levels require the operator to scan the noise for meaningful signals, some of which could denote danger, which means a greater cognitive processing demand is required to separate important signals from irrelevant sound.

The research on the effect of noise on performance produced some generalizable trends (Cohen, 1980; Kantowitz & Sorkin, 1983). First, a good deal of the stress-related impact of noise is simply the annoyance level associated with the sound. People are less annoyed by a noise source if they believe it serves a useful purpose. Second, the noise is more stressful if it is intermittent rather than continuous. Third, noise is more stressful if it is uncontrollable, rather than controllable. Thus, one way to control for the effect of stress is to give those who are exposed to it some modicum of control over it, teach them to look for degrees of control, or remind them of degrees of control that are already available that they might have neglected.

10.2 SOCIAL STRESSORS

Stress can be induced by social sources, and it is convenient to subdivide those into source categories: work related and nonwork related. Work-related social stressors would include role ambiguity ("What am I really supposed to be doing here? No one is making it clear what they want"), role conflict (trying to meet conflicting demands), obnoxious supervisors and coworkers, job insecurity, new job assignments, insufficient authority to perform tasks necessary for the work assignments, and planning for retirement, to mention a few (Holmes & Rahe, 1967; Kahn, Wolfe, Quinn, Snoek, & Rosenthal, 1964).

A personal success can be stressful too. Not all stressors are unwanted events. Stressful events are arousing, which is what makes them stressful. Nonwork-related stressors include illnesses or death in the family, divorce, changing homes or hobbies, change in work patterns of family members, changes in eating or sleeping habits, financial difficulties, the holidays, and Christmas in particular (Holmes & Rahe, 1967).

On the other hand, it is also known that a good social support system can counteract the other forms of stress to produce an adaptive result. The relationship between negative life crises among submariners and illnesses was lower for people reporting stronger social support systems than for people with weaker social support (Sarason, Sarason, Potter, & Antoni, 1985). The same point has been underscored for promoting the well-being of military personnel returning home (Cornum, Matthews, & Seligman, 2011; Reivich, Seligman, & McBride, 2011) and people experiencing severe stress from any life sources (Pincus & Metten, 2010). Engaging social support is one aspect of a broader construct of *resilience*, which is consider further later in this chapter.

Personal social networks, that is, friends and family, are the primary sources of social support. In contained environments, camaraderies within a work group commonly develop around shared objectives, shared challenges, and shared interests that are unrelated to the work environment, such as athletics, music, or other entertainment preferences. In severely stressful and isolated environments, the social cliques sometimes ostracize people who are showing signs of poor adaptation, which can be expected to make matters worse for people who are already having difficulty adapting (Palinkas, 1987).

10.2.1 Crowding and isolation

Crowding and isolation are forms of stress that derive from both social and physical sources. In the case of crowding, work groups in offices show better performance and less absenteeism if there are fewer walls around their personal workspaces, greater distance between one person and another in the work area, and fewer people in the office area overall (Oldham & Fried, 1987).

Note that there is a qualitative difference in the experience of working alone, versus working with one other person, versus working with a larger number of people. There is a bit of folklore that traces back to the British Navy wherein the boss assigns two men to a job no matter how small the job. This writer has seen firsthand that the technique worked particularly well for industrial maintenance and house renovation crews. Although the idea appeared silly to people who encountered it for the first time, they reported when the workday was over that they completed their tasks faster, and they did eventually need to have a coworker assist with some aspect of the assignment. In the case of the house renovation crew, the work for 15 people that would have required 8 hours if done with individual work assignments was done in 5 hours with the two-person method.

The case of working alone rather than with a coworker is only one form of isolation. Much more stressful forms of isolation are experienced in Antarctic winter-overs and outer space travel. Submariners are also isolated from communication with the topside world for weeks at a time, but their daily experience appears to be dominated by the close crowding rather than the isolation. The valuable social support system mentioned earlier requires interpersonal sensitivity and responsiveness on the part of each submariner in order for the crew to remain effective, however (Sandal, Endresen, Vaernes, & Ursin, 1999).

The predictors of psychological well-being among workers in extreme environments do not generalize across all extreme environments. Differences in situations with respect to physical, individual, and social demands affect the profiles of the stress-resistant person (Sandal, 2000). Observations such as these have led to a broader proposal that individual differences that affect performance in PMSs should be given a lot more attention than they have been given traditionally (Szalma, 2009).

10.2.2 Electronic monitoring

Electronic monitoring of employees' work performance became a concern in the early 1990s as a result of widespread conversion of work to computer-based media. Monitoring capability facilitated feedback from supervisors or their avatars. Although, according to principles of basic learning theory, feedback reinforces effective performance, corrects ineffective actions, and raises performance overall, employees saw the presence of electronic monitoring as stressful regardless of whether the substance of the feedback was positive or negative (Aiello & Kolb, 1995). Restricted mobility from being tied to a computer terminal is a source of stress by itself, but it became reportedly worse when monitoring systems were implemented.

Aiello and Kolb (1995) recommended that building a climate of positive social support was essential for minimizing stress and for turning feedback systems into positive influences on performance. Organizational climates can exacerbate stress by encouraging aggressive and oppositional management practices and relationships among coworkers (Van der Velde & Class, 1995). Organizational members respond to a negative social environment with passive-defensive strategies such as angling for approval; acting in conventional or conforming ways when individual thought, action, and expression would be appropriate and valuable; dependence and avoidance.

10.3 SPEED AND LOAD

In the chapter on psychomotor skill and control (Chapter 8), the relationship between errors and RT was observed in situations where work speed was allowed to increase and workload was held constant. Chapter 7 addressed the issues connected to human cognitive channel capacity and the effect of its limitations on speed and accuracy. A classic study characterized what happens when workload and speed are both allowed to vary.

Conrad (1951) made a distinction between speed stress and load stress. *Speed stress* is a reaction on the part of persons working on a task that has the effect of worsening their performance beyond what might be expected from the physical characteristics of the work or equipment involved. *Load stress*, on the other hand, changes the character of the task. As the number of signal sources (e.g., visual displays in the experiments) is increased, more time is needed to make judgments simply because of the greater amount of information that is being processed. Conrad's experimental participants were engaged in a clock-watching task in which they pressed a key as a pointer approached the 12:00 or 6:00 position on any of the clock dials used. In the various experimental conditions, two, three, or four dials were used, and speed was varied. Errors increased as the cross product of speed and load increased.

10.3.1 Working too slowly

Greater than average error rates also occur in low load conditions, and are not readily explained by exceeding limits to channel capacity. Vigilance tasks are common examples. Boredom or stress resulting from boredom is one possible answer. Although the literature that addresses slow-speed conditions is sparse, NDS offers a viable explanation for performance changes in both directions. Consider first the available data on working too slowly.

A physical rehabilitation specialist has a regime for muscle training using relatively standard progressive weight resistance equipment, but with the special regimen of pushing against the weights very slowly (Ross & Ware, 2013; Ware, 2011). The trainee's legs visibly shake while doing the slow push movements in the early stages of training; the initial deficit could have contributed to this effect. The training regime has reportedly a high rate of success. The explanation for the effect is that the neural mechanisms are being forced to find new pathways or firing combinations that are more viable than the initial pathways that are damaged or impaired. The newer and more viable pathways consolidate through further training.

Kremen (1966) studied a tracking task in which the target moved at different speeds and in different directions. Accuracy and smoothness of the tracking motion were the two dependent measures. Increasing the target speed resulted in smoother motions; slow-speed performance was more erratic.

Hausdorf et al. (2001) studied the variation in gait for people walking at their normal pace or at the same, faster, or slower pace that was regulated by a metronome. They were looking for differences in the fractal dimension of the time series of stride intervals under those conditions; fractal dimensions would correspond to levels of complexity and turbulence in a time series that arose from a deterministic process rather than a random process (see Chapter 3). Hausdorf et al. (2001) found that the coherence of the time series dropped and randomness increased under the three metronome conditions. West and Scafetta (2003) developed a mathematical model that emulated Hausdorf et al.'s results; it required two internal control centers, one located in the motor control mechanism and another in the central nervous system. The external driver that produces the speed demand appears to disrupt the correspondence between the two internal systems (West, 2006).

The theory taking shape is that deliberate too-slow pacing disrupts preexisting psychomotor coordination or cognitive automaticity. The further implication is that more than one control center is involved in the psychomotor process, and there would have to be a feedback loop of some sort between them or else they could not coordinate.

Most of the empirical work to date on synchronization within individual psychological processes has been done with psychomotor tasks (Jirsa & Kelso, 2004a, 2004b; Haken, Kelso, & Bunz, 1985; Sleimen-Malkoun, Temprado, Jirsa, & Berton, 2010; Ward, 2004). An application that gained considerable notoriety was a finger-tapping experiment that exhibits the classic synchronization structure very clearly: Experimental subjects tap alternating index fingers to a metronome. When the metronome speeds up enough, a phase shift occurs whereby the two fingers tap simultaneously. Working too slowly would produce the opposite effect: By sliding below the synchronization threshold, the highly connected subsystems become decoupled.

Some studies on thought–action patterns also suggest that decoupling could be involved. Hollis, Kloos, and Van Orden (2009) recommended looking for patterns of engagement and nonengagement of executive control functions while performing a task. The nonengagement conditions could be the result of automaticity setting in, or they could result from *mindlessness* (Robertson, Manly, Andrade, Baddeley, & Yiend, 1997), which is the tendency for the mind to wander while performing the task. It now appears, however, that some experimental tasks that were thought to induce mindlessness were actually producing workload and fatigue effects that resulted from relatively *high* mental loads (Helton, Weil, Middlemiss, & Sawers, 2010). Although the propensity for mindlessness to occur under conditions of *low* cognitive workload was not considered explicitly, Balagué, Hristovski, Aragones, and Tenenbaum (2012) recorded periods of task-relevant and task-irrelevant thoughts while participants performed physically demanding exercises. Their results showed that the patterns of task-relevant and task-irrelevant thought oscillated aperiodically through the work session. The switching between the two types of thought was interpreted as a coping mechanism for the boredom and eventual physical fatigue. Task-related thought appeared to be a part of efforts to reshape the performance of the task and consistently took over shortly before the exhaustion point.

10.3.2 Signal detection tasks

Effects similar to those of working too slowly appear in signal detection or vigilance tasks, and there is evidence favoring the existence of multiple performance processes. Signal detection tasks offer the opportunity for two types of errors, misses and false alarms, that can be biased in favor of one error or another by the rewards or penalties associated with them. Balci et al. (2011) designed an experiment that induced participants to make a trade-off between accuracy and rewards. They found that a bias toward accuracy dominated the early stages of learning and performance, and a bias toward higher overall rewards dominated later on. In this case, the two subprocesses gained coordination with practice. Guastello, Boeh, Schimmels et al. (2012), reported a similar shift in performance strategy in an episodic memory task before and after an hour of work on another task.

In a visual search example, Zenger and Fahle (1997) manipulated load by increasing the search target area to produce a pressure on response time. Missed targets were more common than false alarms, and they were traceable to a speed–accuracy trade-off.

Szalma and Teo (2012) constructed a vigilance task that varied the number of screens containing possible targets and the speed of screen presentation. The found a linear decline in accuracy as a function of speed for low-demand content, but the classic breakpoint was observed as the number of screens increased from one to eight. False alarms declined with increasing speed for the eight-screen condition only.

According to Koelega, Brinkman, Hendriks, and Verbaten (1989), research prior to their study had shown that the density of targets and the memory load inherent in the task affected whether error rates set in during slow or fast working conditions. They assessed performance on four vigilance tasks for speed, accuracy, electrodermal response, and subjective ratings. The tasks varied in memory demand and by type of stimuli. A principal components analysis of the measurements produced two components that indicated that speed and accuracy could result from two different underlying cognitive processes.

10.4 WORK SCHEDULES

The standard work week in the United States is 8 hours per day, 40 hours per week. Employers who need staffing for 16 or 24 hours per day add second and third work shifts. More businesses in recent decades have found the need for staffing 24 hours a day, 7 days a week, and this norm is not limited to the United States. The puzzle is how to define work shifts that produce the best results with the least stress and strain on the workforce. There are some people who prefer evening or night shifts by virtue of activity patterns in their family lives or natural tendencies to be night owls. The need for employees during nonstandard times, however, greatly exceeds the number of people for whom such work arrangements are actually convenient and stress-free.

According to Costa (2003), there are thousands of different work schedules in operation throughout the world. They can be distinguished, however, by several characteristics: (a) the number of work shifts in the course of the day; (b) the length of the workday; (c) the extent to which night work is involved; (d) whether weekend work is part of all work crews' schedule or whether weekend-only crews are involved; (e) whether rotation of work hours is involved and the direction of rotation; (f) the amount of time between changes in work schedules; and (g) whether work starts at different times of day for a particular crew. One common type of rotation is across times of day, such as 2 weeks on the 11:00 p.m. to 7:00 a.m. shift, followed by 2 weeks on the 3:00 p.m. to 11:00 p.m. shift, followed by 2 weeks on the 7:00 a.m. to 3:00 p.m. shift. Other employers might rotate in the opposite direction.

The stress impact of work schedules arises mostly from interference with circadian rhythm, which is a daily biological cycle. Another large component of stress is the interference with the patterns of home life. For instance, workers on the 3:00 p.m. to 11:00 p.m. shift might experience greater stress than those on the 11:00 p.m. to 7:00 a.m. shift because they leave for work when nobody is home and arrive home when everyone is asleep. Those who work through the night may be out of synchrony with their natural biological processes, but they do cross paths in the morning and evening with the rest of the household.

10.4.1 Circadian rhythm

The human body exhibits a natural cycle of waking and sleeping known as *circadian rhythm*. Within the waking and sleeping parts of the day, there are patterns of wakefulness and sleep. The full cycle takes 25 hours if the person is in a contained environment, such as a cave, without contact with the outside world. The cycle is closer to 24 hours long for most people who are exposed to the normal planetary light–dark cycle (outside the polar regions).

The purpose of sleep is to restore the physical and mental well-being of the individual. Prolonged sleep deprivation eventually results in mental imbalances that resemble psychoses, and multi-organ failures. At the same time, sleep disturbances are symptomatic of many forms of mental illness, so it is perhaps better to say that sleep and mental well-being have reciprocal effects on each other. Sleep occurs in 90-minute subcycles, each of which consists of four stages. The four stages range from light and almost awake (Stage 1) to deep (Stage 4) and are characterized by distinct EEG patterns associated with each stage (Andreassi, 1980).

At the end of Stage 4 sleep, the sleeper returns through the other stages quickly to a different form of Stage 1, known as REM sleep, where the acronym stands for *rapid eye movements*. REM is also known as *paradoxical sleep*: REM sleepers look like they are going to wake up because the eyes are darting back and forth as if looking at something, but these sleepers are very difficult to wake if someone should try to do so. REM is the stage in which dreams are remembered most often, although dreams occur in other stages.

Repeated 90-minute cycles exhibit progressively less time in the deepest sleep. The subjective feeling of restfulness is related to the amount of time that one actually spends in Stage 4 sleep. Thus the first sleep cycle is perhaps the most important of all.

Waking periods display different patterns of alertness, body temperature, and adrenaline levels in the bloodstream (Akerstedt, 1977; Colquhoun, 1971; Wojtezak-Jaroszawa, 1978). If the person awakes at 6:00 a.m., alertness is mediocre at first, but increases to its peak by midmorning. The peak may be sustained until after lunch, when the parasympathetic nervous system is doing its job to process lunch. Arousal level reaches a nadir by the later afternoon, but a natural shot of adrenaline around that time improves the alertness levels up in time to drive home from work. Alertness peaks again in the late evening, and tapers off around a normal bedtime of 10:00 p.m.

If the person is staying up all night instead of going to sleep, the alertness level continues to decline until it reaches its lowest point around 3:00 a.m. People who are compelled to stay up all night will report that if they can force themselves through the 3:00 a.m. period, the adrenaline shot that comes in the early morning will carry them through into the next day. A chart of the circadian alertness level appears in Figure 10.1.

Sleep is triggered by natural cycles of melatonin levels in the brain. Wakefulness is triggered by adrenaline and also by light. People who need to force themselves awake during the night are thus advised to turn lights on. People who are trying to sleep in the day should darken their rooms. In the Arctic and Antarctic Circles, the pattern of outdoor lighting where it is night for six months and daytime for six months can pose a serious challenge to sleeping for those who have not acclimated to the environment. In the less extreme subpolar regions, and to a lesser extent in the temperate zones, the short daylight hours can induce *seasonal affective disorder* toward the end of the dark season for some of the residents. Seasonal affective disorder is a mild form of depression traceable to living in darkness for too long; it is a temporary affliction, but it is probably not helping people who might be prone to mild clinical symptoms for other reasons.

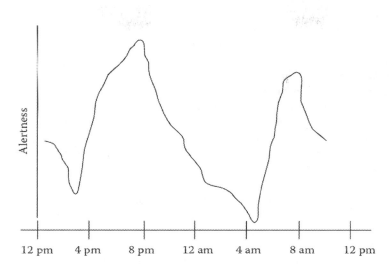

Figure 10.1 Alertness levels during a 24-hour period.

10.4.2 Dysregulation

Dysregulation is a particular form of stress that is characterized by a combination of changes in eating and sleeping habits, and difficulties with them. It is often caused by shift work or irregular work schedules (Depue & Monroe, 1986; Kundi, 2003). Shift work often has deleterious effects on individual work performance because it interrupts or conflicts with the daily biological (circadian) rhythm cycle, which includes the sleep–wakefulness cycle, concomitant brain wave patterns, and variations in body temperature and adrenaline secretion. Permanent night workers have been found to show out-of-phase rhythms, which, in turn, had an effect on their productivity and effectiveness (Agervold, 1976; Malaviya & Ganesh, 1976, 1977). Rotating shift workers express greater amounts of sleep inadequacy, gastrointestinal ulcer, eating problems, irritability, and job dissatisfaction compared to nonrotating shift workers (Tasto, Colligan, Skjel, & Polly, 1978).

Military operations often require prolonged work periods at irregular intervals in order to execute missions or maintain vigilance and readiness under hostile conditions. Although there is a lot of variability as to what a particular exercise or mission would require, military personnel experience shortened or fragmented sleep periods as part of their training and official operations (Naitoh, Kelly, Hunt, & Goforth, 1994; Thompson, 2010).

In addition to the physiological outcomes from the disruption of circadian rhythm, collision with family demands, job demands, social conditions, and individual characteristics also play a role in the amount of strain experienced by a worker (Costa, 2003). Job satisfaction issues and related labor conditions can make a stressful situation worse. For instance, in this writer's experience, some rotating crews in a manufacturing environment reported that they preferred their night shifts to the day shifts because there were fewer interruptions from management that only took place in the daytime. Night shift workers in public transportation reported the opposite experience; all the rowdy and dangerous characters come out at night even though there might be fewer passengers overall.

The social conditions in Costa's (2003) review article included the tradition of shift work that exists in some industries, the supportive nature of the working atmosphere, and the ease of transportation to and from work. The worker's social support outside the workplace is also important. Work–family conflicts vary in size and scope. Workers who do not have families, however, may have social networks that may support some work schedules or run contrary to them.

Individual characteristics include age, gender, general health, personality characteristics, and the extent to which the workers are able to manage sleep at nonstandard times. One important personality characteristic is internal locus of control, which is the belief that good and bad outcomes in life are under the individual's control even though the role of the environment may appear substantial to other people (Smith & Iskra-Golec, 2003). As a general rule, people with internal locus of control are less susceptible to the negative effects of stress than those with external locus of control. Externals believe that good and bad outcomes in life are under the control of the environment and there is not much that an individual can do about them. Although both extremes can be dysfunctional at time, the internals are more likely to search for ways to control their stress sources.

It is not a simple matter to optimize a work schedule for a given situation given the number of variables involved—schedules that might be relatively good in one situation may be relatively poor in another (Costa, 2003). Kundi (2003) offered some recommendations concerning night work scheduling. Whenever possible, the schedule should: (a) minimize the number of hours of night work; (b) minimize the variability of night work hours; (c) spread the night assignments evenly across work crews; and (d) support longer time periods on a given work schedule rather than shorter periods.

10.5 CONSEQUENCES OF STRESS

The consequences of stress are both performance-related and health-related outcomes. The health-related outcomes are both psychological and physiological.

10.5.1 Performance

The effects of stress on work behavior can take the form of errors, accidents, absenteeism, and turnover. Figure 10.2 depicts the effect of arousal on performance (Yerkes & Dodson, 1908). Too much or too little arousal, besides being subjectively stressful, results in suboptimal work performance. Of further interest is that difficult tasks require less arousal from other sources for optimal arousal, whereas easy tasks are best performed with alternative forms of arousal added, such as playing a radio in some situations (McGrath, 1976).

People can cope very effectively with jobs that do not engage the optimal level of arousal. According to Hancock and Warm (1989), the long history of workload–performance research shows that the inverted-U curve in the Yerkes–Dodson law is actually flat on top. When workload gets too far outside the normal comfort zone, people engage in coping strategies to stretch their zone and maintain a steady and effective level of performance at the optimal level (Figure 10.3). Some of the commonly used coping strategies for work overload include off-loading complicated or time-sinking tasks to other people or another time, ignoring social interactions that are irrelevant to the task, using automatic thinking processes and less executive control, and working for greater speed and less accuracy. In the case of work underload, the individual might engage in conversation with coworkers, play the radio, or do something else while the jobs in the low-volume task pile up to a critical mass. The choices of strategies would be limited by what is afforded by or inherent in the work itself.

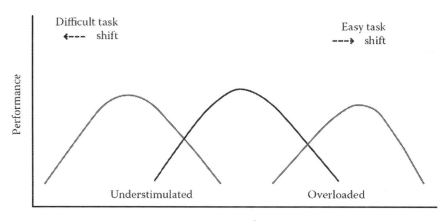

Figure 10.2 The Yerkes–Dodson law for performance as a function of arousal.

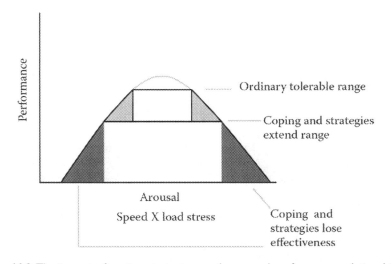

Figure 10.3 The impact of coping strategies on the arousal-performance relationship as described by Hancock and Warm (1989).

When demand exceeds the coping zone in either direction, a sharp drop in performance occurs that Hancock and Warm (1989) characterized as possibly resembling a catastrophe or chaotic function. Numerous investigations of catastrophe functions for the workload–performance relationship were reported subsequently; they are summarized later in this chapter. Augmented cognition and other decision aids are attempts to stretch the effective coping zone. Such help is often paired with decisions about whether to trust the automation system to perform tasks correctly. There are also situations where the software creates more work than it reduces.

One of the more theoretically challenging observations in the stress experiments is the aftereffect of stress phenomenon: Work performance decreases after a stressor has been introduced. When the stressor is removed, however, depressed performance levels continue; the typical effect is shown in Figure 10.4. A viable theory about stress and performance needs to account for the aftereffect phenomenon. The three best-supported explanations for the phenomenon are, according to Cohen—persistent coping, adaptive cost, and learned helplessness.

According to the persistent coping theory, people under stress reorganize their mental efforts such that they are actively fending off the stressful stimulus in the form of a coping mechanism. The key point is that, once the coping mechanism has been initiated, it is difficult to turn it off immediately. The act of coping with the stressor persists after the stressor is removed (Deutsch, 1964; Epstein & Karlin, 1975). Perhaps, at an unconscious level, the person is not sure that the threatening situation has truly ended.

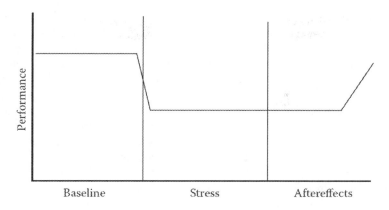

Figure 10.4 Aftereffects of stress on performance.

The adaptive cost theory (Glass & Singer, 1972; Helton & Russell, 2011) holds that the process of adaptation depletes cognitive resources and, as a result, performance remains depressed until the individual recovers or is sure that the "coast is clear" before regrouping resources and ignoring the possibility of stressful stimuli. The depletion of resources, or *cognitive shrinkage*, explanation is consistent with theories of human information processing: People scan the stressful stimuli for information that could be meaningful to their health, safety, or success. By doing so, mental channel capacity is consumed, and the available capacity with which to perform the primary task becomes smaller. Once the channel capacity has been reconfigured to include the stress source, it takes a while to reallocate that channel capacity to the main task.

Stress has a greater negative impact on performance when the stressors are uncontrollable and unpredictable (Cohen, 1980). According to the learned helplessness thesis, people under conditions of uncontrollable stress either learn, or have already learned, that there is nothing they can do about the stressor or its effects on their performance, and therefore they do not try to do anything (Seligman, 1975). Learned helplessness reactions have been demonstrated empirically in experiments with both people and animals. Learned helplessness was a premier theory of clinical depression, which is symptomatically similar to severe stress reactions.

10.5.2 Health

Anxiety is an irrational fear, according to classic clinical psychology. It starts as an adaptive response to real-world events in which a life-threatening event is imminent and an evasive or corrective response is required. Anxiety becomes a clinical condition when the apprehension becomes a steady state of mind that is independent of any actual threats and compromises well-being in some way. Anxiety can be experimentally conditioned in animals and humans (Eysenck, 1997; Wilson & Davison, 1971), and the origins of anxiety through similar conditions is a strong, although partial, explanation for numerous psychological disorders. Anxiety is a nonspecific symptom of many psychological disorders, and could be regarded as the mind's alarm system that goes off when something is wrong. Anxiety is frequently observed as a symptom of prolonged or chronic stress. Indeed, the clinical profile of traits observed for people who have experienced chronic stress exposure are often not readily distinguishable from the personality profiles of people suffering from clinical depression. *Posttraumatic stress syndrome*, on the other hand, is a severe psychological reaction to acute and intense stress; the symptoms can be similar to those of a personality disorder or depression (Glover, 1982; Natani, 1980; Walker & Cavenar, 1982).

Strain is a concept borrowed from material science meaning the amount of deformity caused by stress. The material science concept of elasticity versus rigidity is analogous to notions of psychological resistance to stress. Some personality variables, such as locus of control and Type A personality syndrome affect a person's resistance to stress. The broader concept of buckling and elasticity is less well developed in conventional psychology and elaborated further on in this chapter. For the most part, health professionals use the word *strain* to indicate any of the health consequences or deformities produced by stress.

The physical health consequences of long-term stress include hypertension, heart disease, kidney failure, alcohol and drug abuse, and medical consequences of alcohol and drug abuse. Psychological health consequences of stress include anxiety, impaired interpersonal relationships, impaired decision making, and clinical depression. The onset of many of these health disorders can be traced to background variables and trigger variables (Guastello, 1992, 1995). Hazards in the workplace and physical types of stressors act as background variables. Anxiety and social forms of stress act as trigger variables, with the medical disorder having an acute onset. In order to obtain an understanding of the health impact of stress in a workplace, it is necessary to separate the effect of prolonged exposure to stress from the effect of advancing age, which is another background variable.

10.6 FATIGUE

Fatigue is the loss of work capacity over time. The experience of fatigue can come from any of three sources. The first and most obvious type results from a heavy and fast-paced workload. It is a natural consequence of work and is not symptomatic of a problem in the work environment by itself, except when people are compelled to work too long without enough rest periods. The dynamics of fatigue and performance are explained later in this chapter.

The second type of fatigue is produced by the lack of sleep or interrupted sleep. This source of fatigue is often confounded with fatigue of the first type as a heavy workload day wears on. The same is true for jobs with longer than usual hours of cognitive underload, such as long-distance truck driving.

The third type of fatigue is reported sometimes from people who have not been doing extensive quantities of work, but they have the subjective feeling of tiredness nonetheless. In those cases, the report of fatigue signifies an emotional reaction, which could be related to the workplace, such that the person would like to escape from the work situation.

The decrement in work capacity over time is often inferred by decrements in performance on physical tasks (Kroll, 1981; Starch & Ash, 1917; Weinland, 1927). The *work curve* is a graphic plot of performance over time; it was first introduced by Mosso (1894, 1915). The work curve can be decomposed into two parts: the curve itself and deviations from the curve. The curve denotes a loss of work capacity and decrement of performance caused by extended time on the task (Figure 10.5). The variability from the curve increases dramatically toward the end of the work period and denotes loss of control over one's neuromuscular system (Ash, 1914; Starch & Ash, 1917; Weinland, 1927). Seven types of work curve were identified, those that: (a) dropped gradually; (b) dropped suddenly; (c) maintained stable but high performance; (d) maintained stable but low levels of performance; (e) maintained stable levels of performance prior to, and sometimes after, the output drop; (f) were highly unstable; or (g) increased suddenly (Crawley, 1926; Ioteyko, 1920; Marks, 1935).

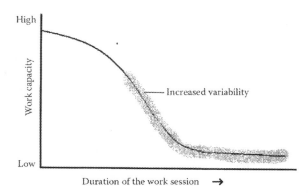

Figure 10.5 The work curve for a physical task showing a decline in work capacity and greater variability in performance as fatigue sets in.

10.6.1 Physical fatigue

Some participants in the earliest physical fatigue experiments showed an increase in strength or work capacity after having been exposed to a physical work set that would have fatigued other participants. Two explanations emerge from the literature: exercise and scalloping. Crawley (1926) found that subjects increased work output in a subsequent test session if the same muscle groups were utilized in a previous session. More generally, exercise is known to increase work capacity by an additional 50% compared to preexisting levels (McCormick, 1976). Crawley also found that individuals who knew their work period would end expended their greatest effort just before the close of the session. Research participants who were not so informed exhibited a more typical work curve.

The additional effort before the close of the session later became known as *scalloping* in learning and behavior research because of the shape of the learning curve that was produced. As an example, pigeons that were trained to expect a reward if they pecked at a key during a fixed time period; the majority of the pecking occurred shortly before the end of the time period. Students "cramming" for an exam is another familiar example.

The physical fatigue that is produced by labor or exercise arises from a drain on food energy in the body generally and a build-up of lactase in the muscles. In current practice, physical fatigue is measured by electromyography (EMG) sensors placed on the surface of the skin in critical muscle areas, which measure changes in voluntary muscle contraction, and accelerometers, which measure muscle tremor (Yung, Manji, & Wells, 2017). Increased eye-blink frequency and duration and postural sway are commonly used measures of whole-body fatigue. The physical aspects of cognitive activity are detected by increased oxygenation in specific brain areas, as measured by fMRI or near-infrared sensors; fatigue is usually detected by increased levels of alpha and theta waves, as measured by EEG sensors (Craig & Tran, 2012; Ogawa, Lee, Kay, & Tank, 1992).

Although cognitive and physical fatigue are produced by relatively independent pathways, there is some crossover that we might experience as "I'm too tired to think straight." The crossover also works in the other direction as people figure out how to regulate their efforts to maintain performance standards (Matthews, Hancock, & Desmond, 2012). We expand on the role of self-regulation later in this chapter in the context of compensatory abilities.

Studies of fatigue most often consider physical and cognitive separately because of the way the research problems are defined. Studies of the correspondence between the two are in much shorter supply because the time periods in laboratory research are usually not sufficient to produce the combination of effects that transpire in a real-world workday. Yung et al. (2017) conducted a study that was designed to work around some of the

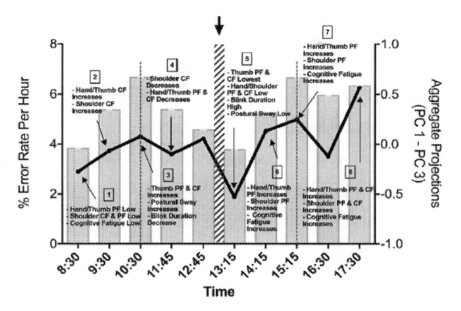

Figure 10.6 Concordance between physical and cognitive fatigue in a daylong laboratory task. EMG scores were aggregated into three components that were represented together in this figure. (Reprinted from Yung et al. (2017, p. 1041) with permission of Sage Publications.)

limitations of the usual laboratory research and investigate the physical–cognitive crossover in fatigue. Their study involved 11 participants who performed a daylong work activity, which was a repetitive micropipetting task in a chemical laboratory environment; they took a 30-minute break for lunch. The participants wore accelerometers and EMG sensors in the head, neck, shoulders, arms and hands in locations specific to the physical task. Performance observations were the end result of physiological and cognitive aspects of the task. The 46 physiological and behavioral measures were aggregated into three composites by principal components analysis. The results showing the correspondence between physical fatigue and performance over time appear in Figure 10.6. Not only did the two types of fatigue track each other reasonably well, they were also consistent with the known circadian trends across the workday.

10.6.2 Cognitive fatigue

The early fatigue researchers reported that cognitive fatigue follows the same essential temporal dynamics as physical fatigue. Work capacity declines after successful intervals of demanding mental work, and recovers after a rest period (Ash, 1914; Starch & Ash, 1917). Another important feature of the early studies, however, was that two human performance phenomena were taking place instead of one. Although the experiments were designed to reduce work capacity over time, which would induce an increase in RT, repeated testing would induce a practice effect, which would have the impact of lowering RT.

The studies of cognitive fatigue were sparse after the Ash (1914) experiments. Two reviews of research surfaced in the early 1980s (Holding, 1983; Welford, 1980). Both reviewers reported that the effects of fatigue on performance and stress were often obscured in the way researchers defined problems and experiments. Some laboratory research, nonetheless, uncovered some interesting phenomena. Bills (1931) reported a *blocking effect* in cognitive fatigue experiments (as in "mental block") where excessive reaction times would occur after prolonged work on an aperiodic basis. Blocking effects became typically defined as an instance of RT greater than two times the mean RT (Welford, 1980). The blocking effect was thought to occur as a means of dissipating the effect of fatigue. There are two critical aspects of the blocking effect that signify fatigue; one is the onset of the block, and the other is the increased number of blocks that appear as the experiment continues.

Although some studies showed that the time needed to recover from cognitive fatigue is less than the time needed to recover from physical fatigue, recovery time could be dependent on the specific examples of mental and physical tasks used in those experiments (Welford, 1980). In some situations, recovery requires a prolonged recovery time, and is not qualitatively different from the aftereffects of stress phenomena discussed earlier, according to Holding (1983). Aftereffects of fatigue often do not transfer to subsequent tasks, and may be related to the repetitiveness of the task and level of arousal or boredom. Perceptual fatigue, according to Holding's interpretation, is strongly related to arousal, and can be manipulated by reward structures. Underarousal has similar effects on stress and performance, as discussed previously.

Experiments on cognitive task performance after prolonged work periods did not consistently result in a decline in performance for all people who were subjected to the same or similar experimental conditions (Dureman & Boden, 1972; Holding, 1983; Poffenberger, 1928). For instance, Poffenberger reported that people who performed prolonged mental work reported feelings of tiredness in much the same way as they do after having performed extensive physical work. Performance on cognitive tasks, however, may remain stable, or improve after prolonged work, in spite of reports of tiredness. Again, practice or automatization effects could counteract other reasons for declines in performance. The differences in cognitive work curves produced by the same stimuli and experimental conditions for different people can be explained by a more inclusive equation and theory that encompasses all varieties of work curves; the theory is explained in the next section of this chapter.

Two other trends in cognitive fatigue research suggest that nonlinear dynamic processes could be involved. First, the physical work curves show greater variability from the curve with prolonged work, and similarly, RT tends to become more erratic after prolonged mental work (Holding, 1983). Second, there is a provocative theory suggesting that at least some forms of fatigue are the result of increased noise in the neural system, rather than the failure of the neural system to transmit signals. The variability from the work curve over time, particularly blocking phenomena and the neural noise concept, strongly suggest that chaotic processes are operating (Clayton & Frey, 1997; Guastello, 1995; Hong, 2010).

10.7 CATASTROPHE MODELS FOR COGNITIVE WORKLOAD AND FATIGUE

Researchers continue to grapple with the relationship between workload and fatigue and whether they constitute one process or two separate processes (Fan & Smith, 2017). The two cusp models separate the two processes, starting with the simple graphic in Figure 10.7. Performance declines as workload increases (or severely decreases) over time. Fatigue produces additional performance deficits when there is extended time on task. A third effect occurs over time, however, which is a practice or momentum effect that results in gradual improvement in performance over time.

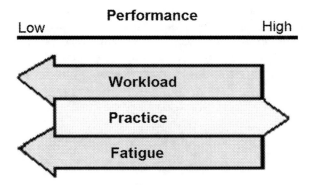

Figure 10.7 As workload increases over time, workload and time on task have separate and sometimes conflicting effects on performance.

10.7.1 Buckling stress

10.7.1.1 Physical demands

The development of the buckling stress model (Guastello, 1985, 1995) began with definitions of stress and strain as material scientists know them, which are a bit different from the psychological definitions, although similarities can be seen. *Stress* in material science is defined as the amount of load per unit area of a test piece of material, usually measured in pounds per square inch. There are a number of stress types, such as bending, pulling, and torque. Given a pulling stress, *strain* is the ratio of the change in length of the test piece to its original length, measured in inches per inch. The *modulus of elasticity* is the ratio of stress to strain. A plot of stress versus strain results in a curve that has both linear and nonlinear segments, with a peak known as *ultimate tensile strength*. In a materials fatigue test, stress is plotted against the number of stress cycles to failure, N, for a given level of stress, where a failure is a break in the test piece. The stress–N curve has an asymptotic lower limit of stress, called the *endurance limit*, which is linearly related to ultimate tensile strength, and indicates the range of stress the test material can withstand before fracture. Some materials increase in strength whereas others decrease with successive stress cycles.

In Zeeman's (1977) application of the cusp catastrophe model (Equation 3.6) to Euler beam buckling, an elastic horizontal strut was given a vertical load and a horizontal compression. The amount of buckling is the deflection of the beam under the vertical load. The system was modeled as a cusp with deflection as the dependent measure (a difference score). Compression, which is the ratio of two times the modulus of elasticity times to length, functions as the bifurcation factor. Vertical load is the asymmetry factor. Buckling occurs at the cusp point, which is the most unstable point in the system.

Euler buckling is actually a group of related phenomena. The situation just described was one in which both ends of the horizontal beam were pinjointed, as shown in Figure 10.8. If both ends were fixed, buckling load is four times that of the pinjointed example. If one end is fixed and the other pinjointed and free to move sideways, the critical buckling load is

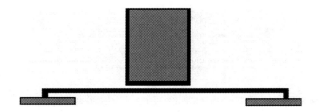

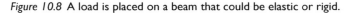

Figure 10.8 A load is placed on a beam that could be elastic or rigid.

one quarter of the value for the pinjointed example. In one of the original problems, the objective was to determine how long one could make a thin pole before it buckled under its own weight. On the other hand, beams that are too short fail by crushing rather than buckling (Gordon, 1978).

The same tests of strength, elasticity, and so forth that are made on building materials can be made on living samples of bone, kangaroo tendon, and butterfly wing. As we shift our attention nonliving systems to humans, the task is redefined from one of measuring the strength of various materials to measuring the strength of composites of materials. The composites can also think, which, in turn, necessitates subdividing the problem into primarily physical and primarily psychological forms of stress.

The statistical model for the cusp that was used in the experiments with this theory is

$$\Delta z = \beta_0 + \beta_1 z_1^3 + \beta_2 z_1^2 + \beta_3[\text{elasticity}]z_1 + \beta_4[\text{vertical load}], \qquad (10.2)$$

where z is a behavior that is measured in normalized form at two points in time with different workload conditions, and β_i are regression weights. The first application of Euler buckling to work performance required a close-up study (Guastello, 1985) of the wheelbarrow obstacle course data from the physical ability project that was mentioned in Chapter 9. In that simulation, subjects pushed a wheelbarrow over a 50-ft course, where the obstacles were traffic cones and barriers on the floor that were designed to simulate railroad tracks. The course was completed three times, first with a load of 100 pounds of sandbags, then 200, and 300 pounds. The criterion was time in seconds to complete the course.

The act of pushing a loaded wheelbarrow over a floor obstacle is similar to the strut that is fixed at one end and free to rock sideways at the other. The worker exerted the compression force; the wheelbarrow tipped when buckling occurred, dumping sandbags on the floor. The worker then had to set the wheelbarrow and its contents upright and continue if possible, thus increasing the performance-time score.

The research participants completed the 100-pound course with a modal response time of 50 s. The RT distributions were bimodal for both the 200-pound and 300-pound courses, with the lower modes at 40 s. The upper modes were comprised of all those who could not complete the course; their score was set equal to the longest completion time for the sample, which was 664 s. Differences in performance-time in the human model replaced strut deflection in the mechanical model (z); data analysis showed the cubic effect (Equation 10.2) which denotes critical change. The asymmetry term was vertical load as in the Zeeman (1977) model. The modulus of elasticity–height ratio (bifurcation factor) was represented by several experimental variables: body fat (as measured by calipers), exercise habits (number of hours of exercise per week outside of work), height, body balance (as measured by a rail walk test), and gender (the sample of participants contained 79 men and 50 women).

10.7.1.2 Cognitive workload model

The cognitive workload model (Figure 10.9) is also analogous to the buckling of an elastic beam (Guastello, 1985, 2003). The asymmetry parameter, which governs the proximity of the system to the thresholds for sudden change, is the workload that is placed on the system or, in this case, the human operators. The bifurcation parameter, which governs the degree of discontinuity, is psychological elasticity versus rigidity, which is a topic of ongoing research

The hysteresis between stable states that occurs in catastrophe models has also been evident in workload research that was designed through a more conventional paradigm. Hysteresis has been reported in other workload research, although without a specific reference to the cusp catastrophe model or its specific forms of analysis (Hancock, Williams, & Manning, 1995; Jansen, Sawyer, van Egmond, de Ridder, & Hancock, 2016; Morgan & Hancock, 2011). The gist of their results runs as follows: Performance drops as a result of adding stress (workload in this case), but remains depressed when the stress is removed. This persistence of low performance is an example of the stickiness associated with the high-error/low-performance attractor in the cusp model. Hysteresis in workload dynamics suggests persistent coping explanation is more consistent with the workload dynamics, whereas hysteresis in fatigue dynamics reflects resource depletion. As an illustrative example of hysteresis in a workload experiment, Jansen et al. (2016) prepared a driving task with a secondary task of responding to police calls on the radio. They had a sequence of low, high, and low workload conditions and reported hysteresis in the form of performance volatility on a memory task and driving speed in the high workload condition.

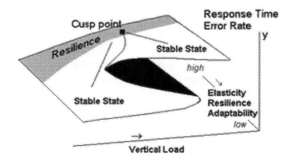

Figure 10.9 Cusp catastrophe model for buckling and cognitive workload.

Hysteresis persisted in the second low workload condition; the task that was more affected depended somewhat on whether the drivers were given an instruction to treat both tasks equally or to give priority to driving. This zig-zagging pattern of performance is an example of movement around the bifurcation manifold. There would be less stickiness if the attractor points were not reached during the fluctuations or if the time frame associated with each measurement were long enough to envelop all the stickiness that could exist.

10.7.2 Elasticity–rigidity

Individual differences in cognitive abilities, personality, or other psychosocial variables make up a substantial portion of the elasticity–rigidity parameter in the workload model. For instance, a person who is rigid in the sense of the trait conscientiousness would likely stick to standard procedures and task scheduling, even when faced with anomalous situations requiring special attention in some way. A more flexible individual would consider rearranging a batch of tasks to allow those that arrived later to move forward sooner if they did not require special handling. A flexible individual might imagine different methods of doing the task that work around the problematic elements. The rigid approach would maintain performance quality and quantity without introducing new forms of decision complexity, which could induce more mental workload. The rigid approach, however, only works up to the point when the anomalies pile up and produce a bottleneck. The more flexible approach finds the best procedure for maximizing throughput in a timely manner without compromising accuracy, qualitative performance goals, or fairness to other people who might be involved.

According to Thompson (2010), resilience consists of both cognitive and emotional elements. Cognitive resilience includes preplanning, rehearsing, focus, dividing tasks, prioritizing, and practice. Emotional resilience would include the ability to recognize one's own emotions and the emotions of others and act appropriately. Furthermore, demands from an onerous physical environment can add to the demands from the cognitive task.

The model shown in Figure 10.9 depicts the broader range of cognitive workload reactions where resilience is comparable in meaning to elasticity in physical applications. Indifference to cognitive workload is actually a response of a stable system that has a sufficient capacity for withstanding peaks in demand. Beyond a critical point, however, performance breaks down suddenly, in which case we have a different stable state that is undesirable. The person or system could make an adaptive response at this point, however, to preserve the level of performance (low response time or error rate). Too much flexibility, however, can be costly to the system and can destabilize it; this point is elaborate further on in this chapter. As Figure 10.9 shows, the resilience portion of the system is located in the part of the response surface that does not contain the stable states. It contains the cusp point instead, which is most unstable point in the system.

Several elasticity–rigidity measurements have acted as bifurcation variables across different types of tasks when using the cusp model for workload. The construct of elasticity as used here bears a close resemblance to the construct of resilience as used by some systems analysts (Hollnagel, Woods, & Leveson, 2006; Sheridan, 2008) and clinical psychologists (Pincus & Metten, 2010). It also bears some resemblance to the construct of *antifragility* (Kiefer & Myer, 2015; Taleb, 2012). Whereas "resilience" is used to describe a person or system's ability to recover quickly from a (highly) stressful situation, "antifragility" describes a person or system that is not seriously affected by situations that would be calamitous for others. Whereas resilience is an adaptive response to something stressful that already happened, an antifragile system is capable of adapting before the event occurs to make its impact as small as possible. It does so by using the available degrees of freedom in a situation advantageously. An antifragile system can be trained or developed with exposure to a broad range of irregularities or random disturbances (or *perturbations*) in the learning environment (Gorman, Dunbar, Grimm, & Gipson, 2017; Kiefer & Myer, 2015; Shockley, Richardson, & Dale, 2009).

A highly resilient state is potentially unstable because it is located at the cusp point, which is the most unstable point on the cusp response surface. Once a resilient effort has occurred successfully, the system needs to lock it into place by essentially adjusting either or both control parameters to steer the system toward a stable attractor. That can be done by protecting workload against extreme surges in the asymmetry variable or by making the resilient response more of a habit than a temporary innovation (bifurcation moving toward rigidity). The known individual differences in elasticity–rigidity fall into four categories—affect, coping, fluid intelligence, and conscientiousness.

10.7.2.1 Affect constructs

As mentioned previously, anxiety is a rational response to real-world threats. The negative consequences of anxiety as a trait are best known for slowing response time, interfering with lucid decision-making (Cox-Fuenzalida, Swickert, R., & Hittner, 2004), and increasing frustration with a task (Rose, Murphy, Byard, & Nikzad, 2002). It can also have a positive impact on performance, however, by narrowing attentional resources to details that other people might miss (Ein-Dor, Mikulincer, Doron, & Shaver, 2010). Its bidirectional influence supports its candidacy as a bifurcation variable in a cusp catastrophe model. A person's level of trait anxiety is more likely to become relevant to performance when interpersonal challenges or physical hazards are present (Guastello Boeh, Schimmels et al.,

2012, Guastello & Lynn, 2014; Naber, McDonald, Asenuga, & Arthur, 2015) or when workload is too low (Vyal, Cornwell, Arkin, & Grillon, 2012). In the latter case, the excess mental workspace capacity is occupied by task-unrelated ruminations. Experimentally induced states of anxiety or negative affect also have a history of inducing rigidity (Liu & Wang, 2014), in contrast to positive affect, which promotes flexibility. Flexibility can turn into distractibility, however, depending on whether the experimental task is goal-oriented or not.

Emotional intelligence (EI) facilitates the understanding of one's own emotions and the emotional messages from other people (Mayer & Salovey, 1997). This understanding can aid the individual in forming appropriate actions in response to emotions. Low EI denotes rigidity in the form of indifference, which could be a buffer against stress effects. When stress gets too high, however, the system can buckle and snap producing any of a number of abreactions if the individual is not aware of his or her own emotional level or those of other people (Thompson, 2010). Ralph, Gray, and Schoelles (2010) likened the abreactions within a complex cognitive–emotional system to that of squeezing a balloon: Squeeze it in one place, and the balloon stretches in another location; squeeze it harder, and it pops.

Empathy is usually defined with both a cognitive and an emotional component. The cognitive component is the ability of one person to take the point of view of another. The emotional component is the ability to vicariously experience the emotions of other people. More contemporary thinking about empathy also contains other facets such as self-regulation of emotion (Gerdes, Leitz, & Segal, 2011; Lietz et al., 2011), particularly when responding to coworkers (Guastello, 2016). Empathy bears some conceptual overlap with EI, and thus it would also act as a bifurcation variable.

Frustration also falls within the affect-related cluster of bifurcation constructs. A frustrated individual could channel the reaction positively to maintain or improve performance. A frustrated individual could also abreact and either withdraw from the task entirely or, in laboratory studies, withdraw psychologically from trying to solve the task-related problems. Studies with the cusp model for workload (Guastello et al., 2014; Guastello, Reiter et al., 2015) measured task frustration with the NASA TLX (Hart & Staveland, 1988).

TLX, which was introduced in Chapter 7, is a set of subjective ratings of workload as defined in Table 10.1. The widespread nature of teamwork in sociotechnical systems in the past few decades has generated some new observations about group workload (GWL). Working with a team produces sources of workload—notably coordination and communication demands—that extend beyond individual workload issues. Thus, Helton, Funke, and Knott (2014) developed a set of subjective workload ratings for GWL that parallel the TLX rating constructs. The GWL ratings are defined in Table 10.1.

Table 10.1 Individual and group work scales

Individual Scales, NASA TLX[a]	Group Workload Scales[b]
MENTAL DEMAND: How mentally challenging (e.g., thinking, searching, deciding) was the task?	COORDINATION DEMAND: How much coordination activity was required (e.g., correction, adjustment)? Were the coordination demands to work as a team low or high, infrequent or frequent?
PHYSICAL DEMAND: How physically challenging (e.g., pushing, pulling) was the task?	COMMUNICATION DEMAND: How much communication activity was required (e.g., discussing, negotiating, sending and receiving messages)? Were the communication demands low or high, infrequent of frequent, simple or complex?
TEMPORAL DEMAND: How much pressure did you feel performing the task because of the pace of the task?	TIME SHARING DEMAND: How difficult was it to share and manage time between taskwork (work done individually) and teamwork (work done as a team)? Was it easy or hard to manage individual tasks and those tasks requiring work with other team members?
PERFORMANCE: How successful were you at achieving the goals of the task?	TEAM EFFECTIVENESS: How successful do you think the team was in working as a team? How satisfied were you with the team-related aspects of performance?
EFFORT: How much energy was put forth to achieve your level of performance on the task?	TEAM SUPPORT: How difficult was it to provide and receive support (providing guidance, helping team members, providing instructions, etc.) from team members? Was it easy or hard to support/ guide and receive support/guidance from other team members?
FRUSTRATION: How discouraged, bothered, irritated, and annoyed were you because of the task?	TEAM DISSATISFACTION: How emotionally draining and irritating vs. emotionally rewarding and satisfying was it to work as a team?

Notes:

[a]   Reprinted from NASA Hart and Staveland (1988); in the public domain.
[b]   Reprinted from Helton et al. (2014); in the public domain.

Some of the studies with the cusp model for cognitive workload included the TLX ratings. Mental, physical, and temporal demands were tested as asymmetry variables, although the experimental manipulations of workload usually accounted for more variance in the behavioral responses, and as a result, those three subjective ratings did not add much to the overall accuracy of the cusp model. Performance and effort demands and frustration did play unique roles as bifurcation variables, however.

Only one group-level study of workload with the cusp model has been published thus far (Guastello, Correro, & Marra, 2019). In that instance, the group averages on the TLX asymmetry variables explained more performance variance than the experimental conditions, and the group averages on the TLX bifurcation variables explained more variance than the group averages on the individual elasticity–rigidity variables. The interpretation is that group members can agree on the workload levels, but the team members arrive at their ratings through different pathways. In the same study, the GWL ratings were all tested as bifurcation variables, but they did not add anything to the predictive accuracy of the model. GWL ratings, however, have been very helpful for explaining synchrony within a team; this topic is considered further in Chapter 14.

10.7.2.2 Coping constructs

One construct of *coping flexibility* is centered on emotional adjustments in the clinical sense of long-term life issues (Kato, 2012). People who have a broader repertoire of coping strategies are likely to be more resilient to stress and emotional hardship. Another construct is oriented toward cognitive strategies such as planning, monitoring, decisiveness, and inflexible responses to changing work situations (Cantwell & Moore, 1996). All strategies would denote rigidity and, like anxiety, could reflect the potential for an adaptive or maladaptive response. Coping has played important roles as bifurcation variables (elasticity–rigidity) in the workload models for a vigilance task (Guastello, Reiter et al., 2016) and an N-back task (Guastello, Reiter et al., 2015); the latter also placed a heavy demand on working memory.

10.7.2.3 Fluid intelligence constructs

Working memory, which limits workload capacity, is part of a broader category of mental operations known as fluid intelligence (Kane & Engle, 2002; Kane, Hambrick, & Conway, 2005). Three such variables have emerged as important for elasticity–rigidity in cognitive workload dynamics: field independence, anagrams, and algebra flexibility. There is a good possibility that there are other fluid intelligence constructs that are yet to be discovered in this context.

Field independence is a cognitive style that allows a person to separate the perception of a central focus point in a visual image from the background imagery. Although field independence (or dependence) appears to be a spatial perception task, this difference among people has further implications for selecting salient information from a background flow of irrelevant information.

People with strong field independence are likely to use more of their working memory capacity in a challenging task (Pascual-Leone, 1970). Stamovlasis (2006, 2011; Stamovlasis & Tsaparlis, 2012) found that field independence functioned as a bifurcation variable that separated high and poor performers' chemistry problem solving. In a financial decision-making task, field dependence was associated with the high-bifurcation side of the cusp response surface for both optimizing decisions and risk-taking choices, and it was also correlated with greater risk-taking overall (Guastello, 2016).

The ability to solve anagrams reflects one of several cognitive measures of creativity (Barron, 1955), which are part of fluid intelligence (Hakstian & Cattell, 1978). Performance on an anagram test played a role as a bifurcation variable in a vigilance dual task (Guastello, Reiter, & Malon, 2016) and a small but consistent role in predicting changes in risk-taking under conditions of fatigue in a financial decision-making (Guastello, 2016).

Algebra flexibility is based on the idea that, in addition to learning the rules of algebra, students should be flexible in their use of algebraic principles to solve problems (Rittle-Johnson & Star, 2009; Rittle-Johnson, Star, & Durkin, 2009; Schneider, Rittle-Johnson, & Star, 2011). The construct takes the form of a test with word problems that require the respondent to compare methods for getting the problem started or organizing the procedure for solving the problem. Algebra flexibility played a salient role as a bifurcation variable in the cusp model for workload using an N-back task (Guastello, Reiter et al., 2015); greater flexibility with algebra was associated with less rigidity under increasing workload and better performance overall.

10.7.2.4 Conscientiousness constructs

Conscientiousness in the five factor model of personality traits (Costa & McCrae, 1992) is a trait that predisposes one to pay close attention to details, rules, and task orientations. The trait also includes planfulness and a lack of impulsivity. It thus implies a type of rigidity (MacLean & Arnell, 2012). In some earlier workload experiments, conscientiousness was measured as a broad trait as it is framed in the five-factor model. In other experiments, it was analyzed as two narrower traits. One of the narrower traits measured conscientiousness in the sense of Factor G on the Sixteen Personality Factors Questionnaire (16PF; Cattell, 1994), which encompasses the predisposition to attend to details and so forth, while the other measured *impulsivity versus self-control*, which would be similar to Factor Q3 on the 16PF. In principle and in practice, it is possible for people to be rigid in the sense of Factor G and flexible in the sense of Q3 (Guastello, 2016). Impulsivity was correlated with higher ratings of TLX temporal demand, and the narrower form of conscientiousness was correlated with higher ratings of TLX performance.

10.7.2.5 Summary of cusp models

Table 10.2 summarizes the results of nine experiments with the cusp model for workload and performance. Each study was designed to tease apart

Table 10.2 Summary of results for cusp models for cognitive workload

R^2 Cusp	R^2 Linear	Bifurcation	Asymmetry
Episodic Memory[a]			
.44	.32	Unknown	Peak load
Pictorial Memory[b]			
.53	.13	Anxiety	Incentive condition
Multitasking at Two Points in Time[c]			
.49	.35	Unknown	Task difficulty
.75	.18	Self-determined task order	Unknown
Vigilance Dual Task, Night Scenario[d]			
.33	.29	TLX frustration	Load increase
Financial Decisions, Optimizing, Altogether and Five Blocks Separately[e]			
.39	.36	Conscientiousness, impulsivity	Work speed
.46	.32	Various	Various
Financial Decisions, Risk Taking, Altogether and Five Blocks Separately[e]			
.25	.25	Work ethic	Load condition
.28	.23	Various	Various
N-Back Task[f]			
.98	.62	Algebra flexibility, TLX performance, TLX effort, TLX frustration, Inflexibility, monitoring	2- to 3-back, TLX temporal demand
Vigilance Dual Task, Day Scenario, Miss Rates, False Alarms[g]			
.39	.11	Field dependence, anxiety, irresolute	Workload direction
.44	.04	Anxiety, inflexibility, irresolute	Unknown
Emergency Response Simulation, Monsters Killed, Game Points[h]			
.51	.30	TLX performance, TLX effort, TLX frustration	TLX mental, physical, and temporal
.00	.00	None	None
Forecasting Chaotic Events[i]			
.56	.14	Forecasting accuracy	
Unweighted average			
.46	.25		

Sources: [a]Guastello, Boeh, Shumaker et al., 2012; [b]Guastello, Boeh, Schimmels et al., 2012; [c]Guastello, Boeh et al., 2013; [d]Guastello et al., 2014; [e]Guastello, 2016; [f]Guastello, Reiter et al., 2015; [g]Guastello, Reiter et al., 2016; [h]Guastello, Correro et al., 2019; [i]Guastello, Futch, and Mirabito, 2020. Table is adapted from Guastello et al. (2020, p. 188) with permission of the Society for Chaos Theory in Psychology & Life Sciences.

the effects of workload on performance from the effects of fatigue. The collection of studies investigated different cognitive operations in order to ascertain the generalizability of the models. Some of the studies used more than one performance measure in some of the studies; readers are encouraged to consult the original research reports for further details. Some of the elasticity–rigidity variables that were discussed above were discovered after the first few studies were completed; as a result, some of the control variables are listed as "unknown" in the table.

Overall, the average R^2 for the cusp models was .46 compared to .25 for linear models containing the same research variables. Thus the non-linear aspects of the model accounted for nearly a third of the performance variance.

The emergency response study (Guastello et al., 2019) was framed around group-level workload and performance. In that instance, the group averages of TLX ratings were sufficient bifurcation and asymmetry variables, in contract to the more specific variables discussed already. The TLX and GWL ratings in that study were correlated with several of elasticity–rigidity variables, however (Guastello, Marra et al., 2017). It appears that team members might arrive at their subjective ratings of workload from different pathways, but the team's collective impression is ultimately the measure that affects the performance outcomes.

10.7.3 Fatigue

Ioteyko (1920) determined a general formula for the work curve that satisfied all varieties of work curve that were known at the time. After a few other interesting studies that intervened over the ensuing 50 years, Guastello and McGee (1987) showed that the effects of fatigue on physically demanding work performance could be modeled as a cusp catastrophe, and that Ioteyko's formula was very close to the cusp catastrophe model, even though it predated catastrophe theory by a half century.

In the cusp model for fatigue (Figure 10.10), the bifurcation parameter is the amount of work done between two measurement points in time, and the asymmetry parameter consists of compensatory abilities or learning conditions. Individual differences in cognitive abilities contribute to the compensatory ability parameter in the fatigue model. The model is again bidirectional. Performance can go up or down over time. Fatigue is typically understood as a unidirectional loss of work capacity over time, yet work that fatigues some individuals might enable others to get "warmed up." Practice and momentum effects would be part of that warm-up process.

Another feature of fatigue that is represented in the cusp model is that there is a substantial increase in the variability of performance that accompanies the net decline in performance over time. The variability is actually hysteresis around the curved manifold in the middle of the cusp surface between the cusp point and the area where the bifurcation effect is the widest (Guastello, 2016; Guastello & McGee, 1987). Fatigue in that context and in the context of the present study reflects the effects of time on task performance and does not assume any additional complications from sleep loss or disturbances. Performance dynamics associated with sleep issues can be studied as cusp functions also, however (Guastello, 2003).

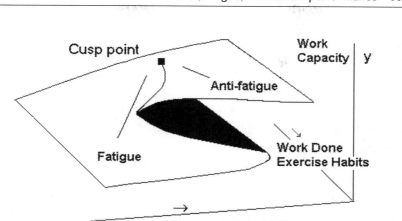

Figure 10.10 Cusp catastrophe model for mental fatigue.

The cusp catastrophe model for fatigue was first developed for physically demanding work in a steel mill (Guastello & McGee, 1987) that required significant amounts of upper body strength. The cusp model for change in arm strength, accurately accounted for the actual data and more so than linear comparison models that were based on the same qualitative variables. Several bifurcation variables were found in the study, some of which had negative regression weights and some had positive weights. Bifurcation variables with negative weights in the model, indicating that the variables promoted fatigue, were the amount of work done, greater amounts of exercise outside of work, labor experience, and higher body fat percentage. Bifurcation variables with positive weights in the model were weight and motivation (subjectively assessed by the experimenters based on the participants' demeanor). Leg strength emerged as a strength compensation variable.

10.7.3.1 Compensatory abilities

The earliest efforts to study cognitive fatigue through the catastrophe paradigm stuck close to the physical labor prototype (Guastello & McGee, 1987). In that case, fatigue was evidenced by a drop in a primary capacity, such as arm strength, and the compensatory ability was a related capacity, such as leg strength. Other compensatory effects in physical labor activities have been reported since that time (Côté, Mathieu, Levin, & Feldman, 2002).

Later efforts took advantage of the developments in cognitive research. Historically cognitive fatigue is incurred by the executive function of working memory (Helton & Russell, 2011, 2013, 2015; Logie, 2011;

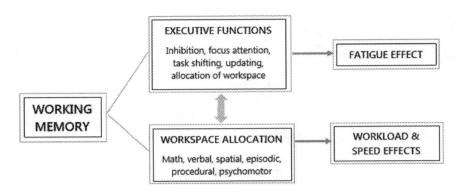

Figure 10.11 Relationships among cognitive workload, cognitive fatigue and working memory functions. (Reprinted from Guastello, Reiter, et al. (2016, p. 516) with permission of the Society for Chaos Theory in Psychology & Life Sciences.)

Thomson et al., 2015). The executive function of working memory allocates mental workspace for multiple cognitive domains to operate or for several pieces of information to be processed at the same time (Oberauer & Kleigel, 2006). Figure 10.11 summarizes the connection between cognitive workload, fatigue, and the primary facilities of working memory. The separation between the functions of working memory and their effects on performance is not quite as clean as it may appear in Figure 10.11. Mental workspace is also allocated to the executive functions themselves, and the resources for executive and more basic functions appear to be drawn from a common pool or resources (Barrouillet, Portrat, & Camos, 2011). As a result, demands placed on executive functioning impact workload in the shorter term and produce fatigue with longer time on task.

The principle of degrees of freedom was introduced in Chapter 7 along with its plausible role in the management of cognitive workload. The degrees of freedom principle underlies fatigue as well. According to Hong (2010), the increase in performance variability that is observed during the low-production phase of the work curve suggests an internal search for a possible reconfiguration of degrees of freedom. There are two plausible directions for such a search. According to the *redistribution principle*, the individual is searching for a lower-entropy means of accomplishing the same task or goal. If a *loss of total entropy* was occurring, however, the individual would not only be trying to regroup internal resources, but also reducing the need to respond to the total complexity of task situation and gravitate instead toward easier task options, changing tasks, slowing the work pace, changing body positions, changing the order of subtasks, simplifying the tasks, or "cutting corners" in other ways to the extent possible. Although it is a question for further investigation to determine the circumstances in which either of the two principles would be more likely to occur, the probable answer is "whatever works" in a given situation, given the affordances that are available.

10.7.3.2 Summary of cusp models

Table 10.3 summarizes the results of nine experiments with the cusp model for fatigue and performance. Each study was paired with a cusp analysis of workload and performance in Table 10.2. Some of the experimental manipulations produced stronger workload effects than fatigue effects. Sometimes fatigue was more prominent.

The statistical model for the cusp that was used in the experiments with the fatigue theory is

$$\Delta z = \beta_0 + \beta_1 z_1{}^3 + \beta_2 z_1{}^2 + \beta_3[\text{work done}]z_1 + \beta_4[\text{compensatory abilities}], \tag{10.3}$$

where z is a behavior that is measured in normalized form at two points in time after a meaningful amount of time on task has transpired, and β_i are regression weights. The amount of work done is a combination of units of work presented to the research participants and the quality or accuracy of their performance during those presentations.

Compensatory ability variables were not identified for some of the tasks, and in some cases one of the experimental conditions produced an opportunity for a compensatory effect. When the compensatory ability variables were identified, they were measures of memory capacity that were not directly associated with the main task or types of fluid intelligence. The cusp models were more accurate descriptions of fatigue and performance effects than simple linear models composed from the same control variables overall. The mean R^2 for cusp models was .41, and the mean R^2 for the best linear alternative models was .23. This comparison also indicates that nearly half of the variance accounted for by the cusp is associated with its nonlinear features.

10.7.4 Multitasking

10.7.4.1 Ordering of tasks

One goal of the multitasking study (Guastello, Boeh, Gorin et al., 2013) was to assess the impact of task ordering cognitive workload and fatigue. Participants were 105 undergraduates who completed measurements of arithmetic, spelling, spatial visualization, and anxiety prior to the main task. The experimental task was a set of seven gaming tasks that tapped different combinations of perceptual and motor ability: accuracy, coordination, memory, an arithmetic–spatial multitask, perception, reaction time,

Table 10.3 Summary of results for cusp models for mental fatigue

R^2 Cusp	R^2 Linear	Bifurcation	Asymmetry
Episodic Memory[a]			
.53	.50	Intervening work done	Peak load
Episodic Memory, Memory Peak, Pictorial Memory[b]			
.30	.16	*Unknown*	*Unknown*
.39	.07	*Unknown*	*Unknown*
.52	.59	Intervening work done	Episodic memory peak
Multitasking: Accuracy Task, Memory Task, Perception Task[c]			
.47	.56	Fully alternating order, self-determined order	Spelling
.53	.33	Self-determined order, intervening work done	*Unknown*
.44	.20	Intervening work done	*Unknown*
Vigilance Dual Task, Night Scenario[d]			
.43	.17	Secondary task completed	Slow speed condition presented first
Financial Decisions, Optimizing, Altogether and Five Blocks Separately[e]			
.56	.24	Speed condition, intervening work done	Field dependence
.24	.21	*Various*	*Various*
Financial Decisions, Risk Taking, Altogether and Five Blocks Separately[e]			
.44	.27	Speed condition, intervening work done	Field dependence, anagrams
.35	.19	*Various*	*Various*
N-Back Task[f]			
.47	.37	Intervening work done, 2- to 3-back load shift	Algebra flexibility
Vigilance Dual Task, Day Scenario, Miss Rates, False Alarms[g]			
.26	.39	Secondary task completed, working in pairs, wearing GSR sensors	*Unknown*
.35	.02	Working in pairs, wearing GSR sensors	*Unknown*
Emergency Response Simulation, Monsters Killed, Game Points[h]			
.49	.10	Group size, number of opponents, experimental session	Fluid intelligence
.59	.04	Group size, experimental session	Fluid intelligence
Forecasting Chaotic Events[i]			
.54	.02	Forecasting accuracy	
Unweighted average			
.41	.22		

Sources: [a]Guastello, Boeh, Shumaker, et al., 2012; [b]Guastello, Boeh, Schimmels, et al., 2012. [c]Guastello et al., 2013. [d]Guastello et al., 2014. [e]Guastello 2016, [f]Guastello, Reiter et al., 2015. [g]Guastello, Reiter et al., 2016; [h]Guastello, Correro et al., 2019. . [i]Guastello et al., 2020. Table is adapted from Guastello et al. (2020, p. 191) with permission of the Society for Chaos Theory in Psychology & Life Sciences.

and speed of response. The multitask required two responses: (a) selecting the one of four spatial stimuli that matched a target shown on the screen, and (b) making a single-digit addition or subtraction calculation. Both answers needed to be entered by a computer mouse within two seconds to be scored as a correct response. Each task was performed seven times, totaling 50 minutes. There were four experimental conditions: (a) the seven tasks were completed in completely alternating sequence, (b) all examples of one task were completed before moving on to the next task, and the multitask was done first; (c) same as previous but the multitask was done last; (d) participants could do the tasks in any order they chose so long as they did seven examples of each task.

Results supported the effect of task ordering, such that allowing participants to choose their own order produced better performance than any of the other three orderings. The flexibility to choose one's task order appeared as a bifurcation variable in the workload model; discretionary task ordering reflected elasticity versus rigidity. Discretionary task ordering also appeared as a bifurcation factor in the fatigue model. Task switching could be an effective means of alleviating fatigue, thus improving performance, but at the same time task switching places demands on the executive functions, thus potentially reducing performance.

There was also an interesting result for performance over time. All seven tasks showed the same temporal trend as shown in Figure 10.12. All bends in the curve were statistically significant. There appeared to be a learning effect at first, followed by a fatigue effect, followed by an apparent shift in cognitive strategy at Time 4, further improvement, then another performance decline. The apparent shifts in cognitive strategy were thought to be a result of reorganizing some cognitive degrees of freedom associated with the task.

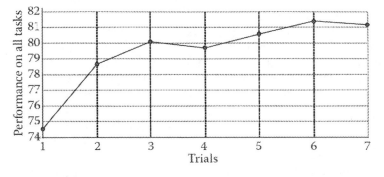

Figure 10.12 Performance on seven tasks across four experimental conditions. (Reprinted from Guastello, Boeh, Gorin et al. (2013, p. 35) with permission of the Society for Chaos Theory in Psychology & Life Sciences.)

10.7.4.2 **Vigilance dual tasks**

The two cusp catastrophe studies with vigilance dual tasks (Guastello, Malon et al., 2014; Guastello et al., 2016) approached the multitask aspect of cognitive load from a different yet commonly used perspective. The situation was designed to be analogous to the work of a hospital nurse who monitors security cameras in several patients' rooms while performing other tasks. The situation also captured the work of someone who monitored building security cameras while performing other tasks; the first study involved a nighttime scenario, and the second involved a more difficult daytime scenario. In the nighttime scenario, the vigilance task was composed of a sequence of virtual reality images from a security camera that was monitoring a building when no one was supposed to be in any of seven different rooms. An intruder, who was designed to look threatening, appeared on a random basis in 10% of the frames. When the participants spotted an intruder, they rang a desk bell. The experimenter had a script of the timings for the appearance of intruders and marked whether the participant produced a correct hit, a miss, or a false alarm. Because false alarms were relatively rare compared to miss errors, only the miss errors were analyzed.

For the secondary task, participants completed a 300-piece jigsaw puzzle. The puzzle task contributed one score, which was the total number of pieces assembled at the end of the 90-minute session.

There were three experimental manipulations. One was the speed of presentation of the security camera frames; there were three speeds comprising a repeated measure. The second independent variable was whether the participant experienced changes in speed from slow to fast or from fast to slow. The third independent variable was whether the participant worked on both tasks while working alone or working along with one other participant.

In the cusp model for cognitive workload, the asymmetry variable was *load direction*, which was the change in miss rates on the vigilance task between slow to medium speed, slow to fast, or the reverse conditions. The model was supported further with NASA TLX ratings of frustration as the bifurcation variable. There were no additional effects from EI or conscientiousness. People who reported being frustrated by the two-task assignment had either a high or low error rate compared to less frustrated people. The experience of frustration sometimes acts as a cue for people to expend the right amount or type of effort to be successful at the task; sometimes it dissuades people from engaging in the task any further. In essence the results were consistent with the theory with the value added that frustration had the opposite effect of EI.

The fatigue model was supported with the amount of work done (number of pieces assembled) on the puzzle (secondary task) as the bifurcation variable. The asymmetry variable was whether the participant started with the fast condition first or the slow condition first. The increase in misses was greater if they started with the slow condition. Instead of an ability variable working here, the operating variable denoted a situation where the participants developed two strategies for managing the two tasks such that performing in the slow condition before the fast condition produced stronger fatigue.

The vigilance dual task with the daytime scenario was more difficult. Instead of targeting an intruder carrying a weapon and running through an empty building, the daytime scenario targeted an intruder carrying a briefcase and walking through rooms that could be occupied by people who were supposed to be there and moving around (Figure 10.13)

Some of the participants in the daytime experiment who were working in pairs also wore galvanic skin response (GSR) sensors. It was hypothesized that wearing GSR sensors would contribute to workload as an effect of electronic monitoring. The results showed, however, that wearing the sensors did not contribute to workload beyond simply working in pairs, but they did make a significant contribution to fatigue. The contribution to fatigue was in the bifurcation variable for "work done."

10.8 OTHER STRESS AND PERFORMANCE DYNAMICS

This section of the chapter describes two more catastrophe models connecting stress precursors to performance outcomes. The essential points leading to the development of these models—anxiety and performance and diathesis stress are considered next.

10.8.1 Arousal, anxiety, and performance

The catastrophe models describe sharp declines and improvements in performance as a function of stress and other variables. The size of the performance changes is governed by a bifurcation parameter. The proximity of the system to a point of change is governed by the asymmetry parameter. As in the case of any other cusp catastrophe model, there may be more than one research variable associated with either control parameter. Note that "arousal" in this context is not the same as "anxiety." The level of excitement associated with arousal often only becomes anxiety when it interpreted as a negative experience or a sign of an impending negative experience.

The anxiety and performance models were framed around sports performance (Hardy, 1996a, 1996b; Hardy & Parfitt, 1991). Although Hardy (1996a, 1996b) did explore catastrophe models that were more complex, the cusp model that is shown in Figure 10.14 appears to be the theoretical resting point. Physiological arousal, as measured by heart rate, contributed to the asymmetry parameter. Cognitive anxiety contributed to the bifurcation variable. Self-confidence also functioned as an asymmetry variable.

Figure 10.13 Target intruder, sample of a target stimulus, and sample of a nontarget stimulus in the vigilance dual with daytime scenario. (Reprinted from Guastello, Reiter, et al. (2016, p. 516) with permission of the Society for Chaos Theory in Psychology & Life Sciences.)

As physiological arousal increased, the golf player was more likely to move into the manifold region of the behavior surface. Anxiety would have the effect of bringing out the best or the worst in a player, as observed in the basketball study. Self-confidence would bias the player toward higher levels of performance; self-confident people can experience the same sources and levels of arousal as other people, but their interpretation of the arousal does not allow them to be overwhelmed.

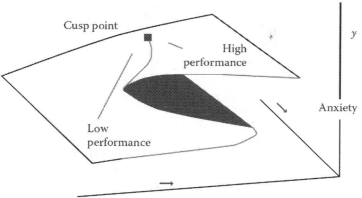

Figure 10.14 Cusp catastrophe model for the effect of anxiety and self-confidence on sports performance.

The foregoing relationships are consistent with standard findings in social psychology (summarized in Aiello & Kolb, 1995) regarding the effect of an audience on performance in sports, public speaking, or other activities. If the performers are confident of their abilities to perform, the added arousal generated by the audience produces better performance. If the performers are not so confident, the presence of the audience has a detrimental effect. The negative effects of anxiety on performance are stronger when the task requires a heavy load on working memory and has its most pronounced negative impact by disrupting automatic processes in the executive cognitive functions (Baumann, Sniezek, & Buerkle, 2001; Sartory, Heine, Mueller, & Elvermann-Hallner, 2002; Williams, Vickers, & Rodrigues, 2002). Anxious performers tend to refocus their attention on tasks that involve lesser amounts of executive functioning (Williams et al., 2002).

Anxiety can also arise from a long-term personality trait that could have genetic origins; in that case it would not be just a transient condition (Eysenck, 1997; Leary & Kowalski, 1995). It can have both positive and negative effects on performance under stressful conditions (Ein-Dor et al., 2010; Guastello & Lynn, 2014; Matthews & Campbell, 2009). The negative impact of stress is that it leads to broken concentration, response delays, distortions of judgment, and errors; in this context it might be regarded as a resilience deficit. The positive impact is that it is sometimes associated with greater vigilance and attentiveness to small cues that could signify a large impact. By having both positive and negative roles, anxiety should function as a bifurcation variable. Additional control variables were considered in the experiments with the cusp model for buckling stress reported later in this chapter.

10.8.2 Levels of performance

Some researchers have reported three stable states of work performance, rather than two, however (Guastello, 1981, 1987, 1995; Kanfer, 2011). The three states are: (a) not good enough to meet the performance criterion; (b) good enough; and (c) a higher level of excellence that represents one's personal best in the matter. Three-state systems can result from substantial individual differences in ability, motivation, and differences in the surrounding organizational context that affect motivation within the organization. The motivation component requires two control parameters, one for extrinsic forms of motivation and one for intrinsic sources. Taken together the three states and four control parameters comprise a butterfly catastrophe model (Guastello, 1981, 1987, 1995). Motivation in this context is both positively and negatively valenced; an oversupply of negatively valenced experiences would be interpreted as "stressful" in the negative sense by most people.

Three levels of functioning are often (but not always) associated with entire jobs rather than specific tasks, and usually require longer time horizons to unfold rather than what is usually the case in contemporary experimental analyses of fatigue. Vestiges of three states can be seen in Figure 10.12 for the multitasking experiment, although the lowest level of performance does not appear to be stable.

10.8.3 Diathesis stress

Diathesis stress models attempt to explain why the effects of stress take the form of one somatic symptom in one person but a different symptom in another. The explanation is that there is an underlying vulnerability in a system, or a hierarchy of system strengths and vulnerabilities, and the consequence of stress affects the weakest element in the system. Thus, a weakness in the stomach area could predispose a person to eating and digestion problems, but a cardiac weakness in someone else would manifest as a different symptom.

Figure 10.15 depicts a diathesis stress model for the effect of shift work as a stressor on the effectiveness of a manufacturing organization (Guastello, 1982). The organization was a precision printing organization in which the workforce worked rotating shifts; this is the underlying source of vulnerability. The stressors took two forms. One was cognitive demand as the bifurcation parameter. Some of the printing jobs (woodgrains utilizing four to eight colors in a rotogravure process) required relatively short print runs, and some required very long press runs. The printing cylinders wore down during the longer print runs and the ink flow changed by subliminal amounts over short periods of time. Changes in ink colors that resulted from

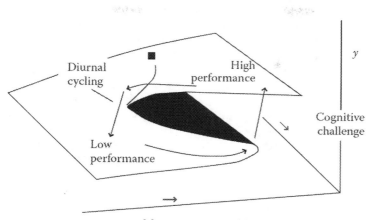

Figure 10.15 Cusp catastrophe model for diathesis stress showing the exacerbating effects of cognitive demand and management-induced stress.

the cylinder wear produced color-matching difficulties and considerable waste of materials and labor in the process. The other source of stress came from management influences. The asymmetry parameter in Figure 10.15 represented some management personnel and relationships over a 13-week period, which were ultimately for the better.

The dependent measures were changes in material waste and machine–labor time for two versions of the same print run done on two different shifts, for example, morning versus afternoon or afternoon versus night. For short runs, which were less demanding, there were small fluctuations across shifts, but the more demanding runs produced larger fluctuations. In this situation, the morning and night shifts were equivalent in performance, but the drop in productive efficiency occurred in the afternoon (second shift) when it did occur.

During times of negative management conditions, the flux in productivity was anchored at the low efficiency mode of the cusp response surface. During the last weeks of the data series that involve the positive management conditions, the flux in productivity was anchored at the high efficiency mode of the cusp response surface. The largest fluctuations in efficiency occurred during the weeks between the two endpoints. The management solution to the production efficiency problem was to retain the new operations management procedures and to place a limit on the length of a print run so that they could reengrave the cylinders before the color-matching crisis was likely to occur. Their customers were given price incentives to place their job orders in batches of workable size.

The diathesis stress principle generalizes a bit further. It basically consists of an ongoing source of variability in performance, which would be the fluctuations induced by circadian asynchrony. The workforce was sufficiently practiced at keeping the impact of those fluctuations to a minimum, but the effect loomed in the background of the system's activities waiting to be activated by new stimuli. The new stimuli consisted of a background variable and a trigger variable. The background variable was the management situation which went from not-so-good to something much better; this was the asymmetry variable in the model. The trigger variable was the cognitive demand of the task, which could be traced to a quantifiable feature of the task, and for which the machine system was partially responsible; this was the bifurcation variable in the model.

The occupational health effects of stress mentioned earlier are another example of diathesis stress dynamics (Guastello, 2013). The concept of diathesis stress dynamics originated in behavioral medicine and thus meanders outside the scope of human factors.

DISCUSSION QUESTIONS

1. How would you design a stress management program based on the occupational stress effects described in this chapter?
2. Describe a strategy that you would use to evaluate some choices in nonstandard work schedules. What criteria would you use in an empirical study? (Hint: How is this question similar to, or different from, evaluating a human–machine system as discussed in previous chapters?)
3. What new information about the control of stress effects comes uniquely from the nonlinear cusp models?
4. What does the set of cusp models for stress and work performance tell us about the nature of stress?
5. How would you evaluate mental fatigue in modern technological work?
6. What are some other cognitive strategies a person might use to buffer the effects of a challenging mental workload?
7. For all the public health messages that continue to circulate telling people what they should eat or not eat, do or not do, or the help they can get to minimize health disorders, have any of those messages tried to cure people of "workaholism"?

Chapter 11

Occupational accidents and prevention

11.1 CAUSES OF DEATH AND INJURY

The word *cause* is a treacherous word because it requires a human evaluation of the outcomes produced by a complex system. Much of the information on causes of death, especially from past decades, is only available from death certificates, and the designation of *cause* on certificates is inevitably predicated on the status of medical knowledge available at the time of death. The task of defining a cause of death requires a gross simplification of the process leading up to many *proximal* causes of death. Thus a bit of skepticism is warranted when interpreting alleged causes that were recorded at any one time in history.

Social psychologists have specifically studied the attribution of cause, which is the process by which ordinary people view situations and assign cause to either the primary actor or the immediate situation. To make a long story short, one typically chooses between a dispositional attribution and an environmental attribution. The former assigns cause to some characteristic of the person who is the accident victim or something the person did to promote the resulting event. The latter examines the forces at play in the environment and how those forces may have shaped a particular outcome and the victim's apparent actions while the event was unfolding. Humans tend to have biases in their causal attributions that place greater weight on dispositional explanations rather than on environmental explanations, especially when the available evidence is insufficient for determining which explanation is more appropriate. In other words, when in doubt, there is a tendency to blame the victim. This bias should be avoided or prevented as a general rule, but it happens nonetheless. An accident investigation is usually required, however, to uncover a deeper explanation for how the primary actors ended up in the critical situation.

DOI: 10.1201/9781003359128-11

11.1.1 Death statistics

With the foregoing caveats in mind, we can still see some broad trends in causes of death. In 1900, accidents were the seventh most frequent cause of death after influenza, tuberculosis, gastroenteritis, heart disease, stroke, and kidney disease (Sternberg, 1995). By 1986, the accidental death rate was cut roughly in half from 72.3 deaths per 100,000 population to 39.5, but it was ranked fourth of all causes of death (National Safety Council, 1989) after heart disease, cancer, and stroke. As infectious diseases became curable, other causes of death moved up the ranks. The same rankings persist in more recent data based on 34 countries (National Safety Council, 2002). As of 2010, the top four causes of death in the United States were still the same. As medical technology made progress in preventing or slowing down the death rates from the top three origins and improving longevity, Alzheimer's disease rose to the sixth most prevalent cause of death after respiratory diseases (Alzheimer's Association, 2010; Centers for Disease Control, 2007). It used to be said that a person died *with* Alzheimer's disease and not *from* it, but as longevity increased it was possible to say "from."

The ranking persisted until the start of the COVID-19 pandemic in March, 2020. The COVID-19 pandemic produced widespread disruptions in employment, travel patterns, and health outcomes. According to the National Safety Council (2022) and the Centers for Disease Control (2022), COVID-19 skyrocketed to the number 3 cause of death in 2020 after heart disease and cancer. Accidents remained in fourth place. Stroke and respiratory diseases dropped to fifth and six places, respectively. The top five causes of accidental (preventable) death from all origins combined were, in order: opioid overdose, suicide, motor vehicle crashes, falls, and gun assault.

Heart disease, cancer, and stroke are stress-related to some extent; age and genetic dispositions play major roles also. A good deal of the stress in life is occupational in origin, and links to major medical outcomes have been documented (Guastello, 1992, 1995; Landsbergis, Schnall, Schwartz, Warren, & Pickering, 1995). Greater incidences of cardiovascular disease have been associated with a three-way interaction among male gender, job strain, and age in years over 40. Job strain in this context was defined as a combination of high job demands with little decision control (Landsbergis et al., 1995). It is noteworthy, however, that the occupational conditions that could lead to cardiovascular disease could also lead to other medical disorders or psychological health problems, as discussed in Chapter 10.

11.1.2 Occupational accident trends

Of all the accidental deaths recorded in the United States in 1988, motor vehicle (nonoccupational) accidents accounted for 47% of incidents. Occupational accidents (including motor vehicles for transportation workers), home, and public accidents accounted for 11%, 23.4%, and

18.8% of accidental deaths respectively (National Safety Council, 1989). The occupational death rate of 4.3 cases per 100,000 working people is only one third the rate reported in 1928 when specific statistics first became available. It is not clear at first blush, however, how much of the progress at the societal level has been due to safer work practices versus shifts in technologies introduced for reasons other than safety, industry composition in the economy overall, or the exportation of hazardous work has been to other countries. Outsourcing the recycling of electronic waste is a relatively new example (Côté, Gravel, Gladu, Bakhiyi, & Gravel, 2021).

The median occupational death rate over 40 countries was 4.2 per 100,000 in 1998 (National Safety Council, 2002). The rate for the fourth quartile (top 10 countries), however, ranged from 6.5 to 31.0 deaths per 100,000. Interestingly, the later statistics for the United States showed a drop in the rate of motor vehicle accident death to 43% of all accidental deaths, and an increase in the occupational death rate to 16% of all accidental deaths. The occupational death rate for the United States had also increased to 5.0 per 100,000 by 1998, but tapered down to 3.1 per 100,000 by 2019 (National Safety Council, 2022). The total quantity of occupational accidental deaths dropped to 10% in 2020, which corresponded to a 9% reduction of total employment hours caused by COVID-19.

Although deaths represent a high watermark that is helpful for understanding accident trends, nonfatal injuries are both a major concern and the criterion used most often for evaluating hypotheses or the efficacy of prevention strategies. If a particular type of accident cannot be eliminated altogether, it could be reduced in severity. There is a 2000-fold scale factor between deaths and disabling injuries just reported, and large multipliers also exist between disabling injuries and injuries with no lost work time. Heinrich (1931) reported that for every nonfatal accident, there are, in turn, approximately 10 near-miss encounters; this was perhaps the first report of a scaling relationship. The multipliers have been recently updated to show that for every work-related fatality, there are 30 lost workday cases, 300 reportable injuries, 3,000 estimated near-misses, and 300,000 estimated at-risk behaviors (Gnoni, Andriulo, Maggio, & Nardone, 2013). Although the specific scaling factors vary across industries and over time, the general principle is that there is an exponential distribution of severities, and a number of factors could be involved between the near-miss and the actual accident. Some analysts, such as the Federal Aviation Administration study near-miss accidents in hopes of preventing real ones.

Figure 11.1 illustrates the most disabling occupational injuries, based on cost information available from insurance company sources (Braun, 2008). The MSK group contains musculoskeletal injuries incurred through over-exertion or repetitive motions such as carpal tunnel syndrome. The falls include falling to the same level or a lower level of a structure. Striking includes incidents where the individual was either struck by an object or

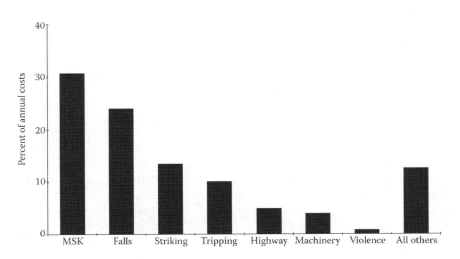

Figure 11.1 Most disabling injury groups by type, based on U.S. insurance data from 2005.

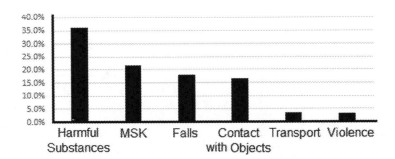

Figure 11.2 Most disabling injury groups by type, based on data from the National Safety Council (2022) for the year 2020.

struck against an object. Machinery accidents involve being caught in machinery somehow. Other groupings should be self-explanatory.

Figure 11.2 illustrates the most disabling occupational injuries in 2020, based on frequency counts from the National Safety Council (2022). The most frequent group, hazardous substances and environments is composed mostly of COVID-19 deaths plus others that would have been part of "all other causes" in the 2008 insurance data. Falls include slipping and tripping accidents. Contact with objects combines striking and machinery accidents. Transport includes highways plus low-incidence transposition accidents such as trains, airplanes, and water transport. Violence includes altercations with animals, which is a well-known agrarian hazard.

11.2 STRUCTURAL RISK MODELS

Structural risk models have been widely offered as heuristics for understanding accident occurrence. Structural risk models vary in complexity and can be ordered as: single-cause mechanisms, chains of events, factorial approaches, process-event sequences, fault trees and Petri nets, multiple linear models, deviational models, and catastrophe models (Benner, 1975; Guastello, 1989, 1991; Rowe, 1977).

11.2.1 Individual accident proneness

The concept of accident proneness first appeared in the 1920s when insurance statisticians discovered that approximately 90% of the industrial accidents involved only 10% of the people in the workforce. That finding led to the premature conclusion that those 10% were chronically doing something wrong, and they were thus labeled *accident prone*. The label provided an illusion of explanation for the mysterious probability structure. It was shown later, however, that the 90%–10% finding could occur by chance if one assumes that a Poisson statistical distribution, rather than a Gaussian (or normal) distribution, generates accident incidence rates (Mintz & Blum, 1949). With the change in the assumed statistical distribution, several data sets no longer showed abnormal accident frequency rates. Those that continued to show deviations from Poisson expectations could not be interpreted as evidence of individual accident proneness; environmental causes could be responsible just as readily.

Poisson distributions have become very useful for problems involving unusual rates. For instance, if the average computer hard drive crashes after of 84 months of service, how likely would it be for a drive to crash after 48 months? Even if we are crashing drives faster than average, how unusual is it for us to be crashing drives this fast? For an accident problem, a similar question might be: If our industry incurs 7 injuries per 100 workers per year, how likely is it for 1 person to incur 2 of the injuries?

Exponential distributions are also useful for studying accident data. In an exponential distribution, a zero frequency of an event will be the most common observation (statistical mode), one event will be the next most common, and so on (Figure 11.3). For distributions where the overall incidence rate of an event is low, the shape of the Poisson distribution can be approximated by the exponential distribution (Evans, Hastings, & Peacock, 1993). The exponential distribution has an interesting property whereby the mean is equal to the standard deviation in theory. Thus, when some engineers want to apply a "safety factor" to a system, they might be prepared for incidence rates of the mean times six. "Six sigma" is a popular criterion that indicates that the system would respond correctly to all but 0.25% of situations.

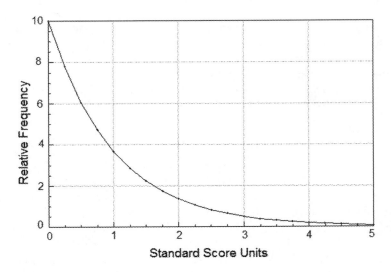

Figure 11.3 Exponential distribution.

In spite of Mintz and Blum's (1949) conclusions, the subsequent 40 years hosted a parade of individual variables studied for purposes of advising employers how to avoid selecting accident-prone job applicants. The prevailing concepts in the past 20 years have centered on impulsivity, personal or social maladjustment, and alcohol or drug use. These personal attributes of people will be considered later in this chapter under the section on accident-prevention programs.

11.2.2 Single-cause models

Single-cause models take the form, "If condition X is present, then Y will occur." Nothing especially complicated is taking place in the modeling sense. Some examples might be: "The spinal cord injury was caused by diving into shallow water," "The train wreck was caused by a faulty connection in the signaling system," "The deaths in the underground mine fire were caused by a failure to develop an evacuation plan," and "The worker was knocked unconscious by a wrench that fell from an overhead platform."

Although these explanations for events look convincing and concise, the pictures of the events become a bit more complex when we ask additional questions. In the case of the diving accident we might ask, "Was it a pool or outdoors?," "If it was a pool was there a deep end?," "If there was no deep end, was there a sign saying 'No Diving'?," "Could the diver see the sign, the deep end of the pool, and so on?," "If not why did he dive?," "Was the diver familiar with the body of water?," and "Was alcohol involved?" As

Table 11.1 Example data for machine systems and accident outcomes

	Accident	No Accident	Row Total	Odds
System 1	2 (5.48)	40 (36.52)	42	0.119
System 2	13 (9.52)	60 (63.48)	73	0.217
Column Total	15	192	115	

soon as we address these questions we have the potential for multiple single causes, chains of events, factorial models, and other structural varieties.

A significant correlation coefficient, or one of its nonparametric substitutes, does not determine causation; it only determines an association between two variables. Although it is usually the case that a proximal cause will produce a correlation with accidents, correlation is not sufficient to support causation. To determine causality, the researcher must manipulate the independent variable experimentally or else use a quasi-experimental research design.

Single-cause models are often expressed as *odds ratios*. An odds ratio only requires a dichotomous predictor, a dichotomous outcome, and some frequency counts. Consider an example: Flying metal debris often causes lacerations to people in a radius of a few feet from the metal-cutting work. Suppose we observed that work sites with one type of metal-cutting tool and machine equipment produced more reportable lacerations than work sites with a different set of tools and machines. We would set up the data as shown in Table 11.1. The independent variable is Machine System 1 versus System 2. The dependent measure is the number of people who suffered no lacerations versus those that did.

The first step in the analysis is to determine whether there is an association between the two variables. For this purpose we could use the chi-square (χ^2) test with the Yates correction for continuity. However, because the expected frequency for one of the cells is very low—and in many situations the expected frequencies could be even lower—the likelihood χ^2 would be the favored choice of statistic. (The likelihood χ^2 can be used for problems with larger subsamples also.) The expected frequencies of accident events are shown in Table 11.1 in parentheses. The expected frequency is the row total times the column total, divided by N = number of observations for this problem. The χ^2 statistic is computed as:

$$\chi^2 = 2 \ \Sigma \ Ob \ \ln \ (Ob/Ex) \tag{11.1}$$

where Ob = observed frequency for a cell, Ex = expected frequency, and the summation is over all four cells (Howell, 1992). The χ^2 statistic for this problem is 5.83, which is statistically significant at $p < .025$. (A p value of <

.05 is sufficient when testing only a small number of potential single causes with a sample that is representative of the intended population of events and people.)

Now that an association between the two variables has been established, the next step is to produce the odds ratio. The odds of an accident with System 1 would be 5/42, or 0.119. The odds of an accident with System 2 would be 13/60, or .217. Take the ratio of the two odds, and we can state that the odds of an accident in System 2 are 1.82 times as great as the odds of an accident with System 1.

Odds ratios can be misleading when taken out of context. It is one thing to say that the odds of an event are "twice as likely" when the base-rate chance of occurrence is 0.1, but quite another when compared to a base rate of 0.000001. Orders of magnitude matter.

Another alternative structure for a single cause model would be a *break-point* relationship between the hypothetical cause and effect. In a break-point relationship, the amount of X must increase up to a critical threshold before any Y can occur (Rowe, 1977). This relationship is essentially non-linear, and implies a transition from a stable safe state to an unstable state. The speed–accuracy trade-off (Chapter 8) is a familiar example.

11.2.3 Multiple-cause models

Single causes can expand into multiple-cause models as soon as multiple Xs are found to be associated with the unfortunate Y. The logic then becomes, "If X_1, X_2, or X_3, are present, either alone or in some combination, then event Y will occur."

The diving accident scenario that was introduced earlier is based on some real events. A drought in Wisconsin in 1988 resulted in severely low water levels in many public waterways. There was in fact a spate of 11 spinal cord injuries resulting from diving into shallow water. Branche, Sniezek, Sattin, and Mirkin (1991) assembled a control group of 22 men who matched the 11 injured in several ways: They lived in the same counties of Wisconsin, used public water facilities that summer, and were between the ages of 18 and 45.

Ten independent variables were tested for comparison between the injured group and the control group. The injured were 12 times more likely to have entered the water from a dock or pier, 10.8 times more likely to have dived rather than jumped, 7.5 times more likely to have dived into water less than 5 feet deep, 6.0 times more likely to have used medication on their last water recreation, 4.3 times more likely to have been unfamiliar with the body of water, 4.0 times more likely to have used alcohol on their last water recreation, and 2.9 times more likely to have a high school diploma or less education.

The statistical analysis does not assume which variable or condition preceded another, but the risk factors themselves might suggest a logical pathway. At the end of the analysis, however, all the hypothetical risk factors could be channeled into two important proximal ones: the water was too shallow and the person dove in from a dock or pier.

11.2.4 Domino models or event chains

Domino models are chains of events. The Chicago fire is a classic example. The cow knocked over the lantern, which ignited the hay, which ignited the barn. The fire from the barn spread to the next house, then to the next house, then to the next house, until it burned the last available building. Note the critical importance of each step in the process on each subsequent step (Figure 11.4).

Some multiple-car collisions are examples of event chains. Imagine a line of cars stopped at a red light. A new car joins the pack from the rear, but its driver fails to apply the brakes soon enough. The new car hits the one in front of it, forcing that car into the one in front of it and so on. The chain of collisions might be broken if there was enough space between the cars, the drivers who were about to be hit had their feet firmly on the brakes, or one of the vehicles was large enough to absorb enough shock without being forced forward.

Another type of traffic event chain might start with the lead car of a moving pack. The lead car hits a surprise patch of ice and spins out of control, whacking the next car in line. The third car takes evasive action but slides into a ditch while doing so.

Short chains of events are often inherent in occupational accidents. In behavior modification approaches to workplace safety, management introduces a plan to recognize and reward specific safe behaviors, such as

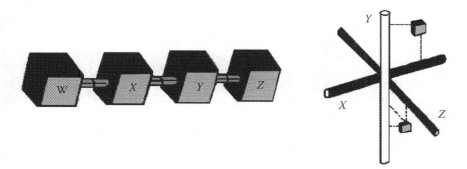

Figure 11.4 Visual comparison of event chains and factorial models of accident causation.

wearing protective eyewear. The plan produces a higher utilization of protective gear, which should produce fewer injuries to the eyes.

Reniers and Dullaert (2007) developed a computer program that would uncover the potential for domino scenarios in a process industry. In chemical processing, for instance, a chemical accident of one type leads to additional accidents, making the end result a lot worse compared to what might have been the case if the chain were interrupted after the first event. The analysis requires a database of probabilities of events and collateral events. It calculates how much danger is exported to adjacent events given that a source event occurred. From that point it is possible to calculate the odds of particular domino scenarios by cross multiplying the odds of events within a hypothetical chain. The underlying logic is not very different from that used to construct fault trees, which are considered here a bit later.

11.2.5 Factorial models

Single causes can expand into factorial models as soon as multiple Xs are found to be associated with the unfortunate Y. The logic then becomes, "If X_1, X_2, or X_3, are present in some combination, then a known level of Y will occur." Here we are saying that an accident is explained by a combination of risk factors. More commonly, however, we are saying that a work unit's accident rate is explained by a weighted combination of factors, such that:

$$AR = b_0 + b_1 X_1 + b_2 X_2 + \cdots + b_n X_n. \tag{11.2}$$

Equation 11.2 is a multiple linear regression equation where AR is the accident rate, b_i are regression weights, and X_i are the independent variables. In many cases, however, better results might be obtained by converting AR to $\log_{10}(AR + 0.05)$. A visual comparison of the chain of events and factorial models appears in Figure 11.4.

Consider an example of an accident analysis that was conducted in this fashion. The situation involved 658 employees at a secondary metal fabrication facility (Guastello & Guastello, 1988). The employees completed a paper-and-pencil survey that covered a number of variables related to safety management, different kinds of stress, an d different kinds of hazards. There was one question added to the end of the survey, which asked how many accidents they were involved in during the past year. Note that "involved" denotes not only actual injury, but includes workers who may have played some role in a collective task in which someone in the work area was injured. The involvement rate in this situation was about double the OSHA-reportable accident rate for the same time period.

There were many results from the survey that were useful to the safety management objective at the plant. Three variables were isolated with a stepwise multiple regression procedure, where individual accident involvement was the criterion. The factorial model for the location, however, was composed of three variables: anxiety levels, the number of environmental hazards (from checklists), and danger levels. Questions regarding danger levels indicated the level of severity of potential injuries in a work location.

The coefficient of multiple regression $R = .35$ ($p < .001$), indicating that the three variables accounted for 12.3% of the accident involvement variance (R^2). The regression weights indicated that anxiety accounted for approximately 10% of the accident involvement variance, and the other two variables accounted for an additional 1% each. Although these relationships are small, they are within the expected range for factorial investigations of accidents. Nonetheless, there was a lot of unpredicted behavior variance. Statistical analyses stronger than multiple linear regression are probably needed to account for it, which is part of the motivation for the other accident models that are considered next.

11.2.6 Process-event sequences

The simplest form of event sequence model accords less attention to causes and more attention to the outcomes leading up to an accident. The nuance here is that an accident is a process, rather than a single discrete event. Surry (1969) conceptualized the accident process as a hazard build-up cycle. At first, the workplace is safe with no uncontrolled hazards. As people start to work, however, tools are left out in workspaces, and different people enter the workspace to do different things with different tools and equipment. People and objects move around and make opportunities to bump into each other. In other words the entropy builds up along with the hazards. Eventually hazards accumulate to a critical level when an accident occurs.

An intervention based on the hazard build-up cycle would emphasize training for good factory housekeeping. Other possible forms of training would center on the best care for tools, and procedures that would minimize the acceleration of the hazard build-up. Workers should learn to recognize the build-up cycle and to spontaneously intervene by reorganizing their workspaces for safer operation.

11.2.7 Fault trees

A fault tree is a diagram that represents a sequence of events. The sequence could represent a process or a chain of events. The distinguishing characteristic of a fault tree is that it contains one or more branching functions that denote choices, options, or opportunities for multiple outcomes. A fully defined fault tree would contain estimates of probabilities associated with each branch at a node.

The fault tree in Figure 11.5 is a reinterpretation of the Sunshine Silver Mine disaster (Denny, Gilbert, Erdmann, & Rumble, 1978). The disaster started several floors underground in a trash heap in an obscure part of the working cavern. The ventilation system was inadequate, and caverns filled up with smoke. Not everyone was able to detect the smoke before it blocked their exit to the nearest escape route. Analysis showed that the mine did not have a working evacuation method, so the miners had to figure the way out extemporaneously. Those that survived to the next set of options faced insufficient elevator space and respirators that often did not work. If the devices did function properly, many other miners did not know how to use them properly. The net result was 91 deaths out of 173 miners within an hour.

It is also possible to construct a fault tree prior to a calamity. If the possible nodes where an error can occur can be mapped out in advance, even if the error rates are not known, then it would be possible to insert diagnostics into the system that will alert operators or system controllers to impending errors.

Figure 11.6 depicts the core elements of a fault tree whereby events and sub-events have probabilities (p) associated with them. There is an initial condition that triggers possible combinations of outcomes. At each juncture, there is an opportunity for a risky situation to continue or terminate. Event 3 in Figure 11.6 defines when the accident or injury actually occurs or a recovery is possible. The odds of an accident would be $(p_1) \times (p_{2a} \mid p_1) \times (p_{3a} \mid p_{2a})$. Breaking the sequence down into steps illustrates points of possible recovery as well as a picture of how it happened.

A closer look at Figure 11.6 reveals that the hypothetical calculation just made here constitutes the odds of a particular domino sequence occurring. The essential difference between a fault tree and a domino structure, however, is that multiple overlapping domino structures could exist, and "danger units" could spread in multiple directions from the source. Fault trees, in contrast, tend to address fixed event structures that could be prevented from reaching criticality if certain yes/no gates were open or closed. In some simpler contexts, the distinction between the two types of processes is not blatant.

11.2.8 Flow charts and Petri nets

The social science literature is speckled with models of phenomena that are depicted as configurations of boxes and arrows. The contents of the boxes could indicate objects or events. The arrows convey a vague notion of causation. If the flow of arrows heads in one direction only, the model could be tested using linear structural relational analysis. That type of analysis produces patterns of linear correlations such that higher correlations are interpreted as stronger causal links, as if the correlations resulted by a

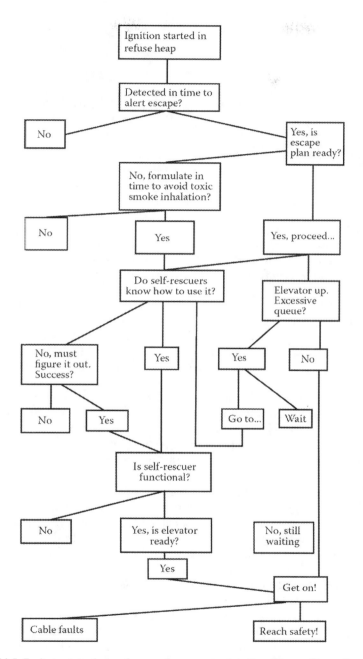

Figure 11.5 Fault tree analysis of an underground mine fire. (From *Chaos, Catastrophe, and Human Affairs: Applications of Nonlinear Dynamics to Work, Organizations, and Social Evolution* (p. 210), by S. J. Guastello (1995), Hillsdale, NJ: Lawrence Erlbaum Associates. Copyright 1995 by Lawrence Erlbaum Associates. Reprinted with permission.)

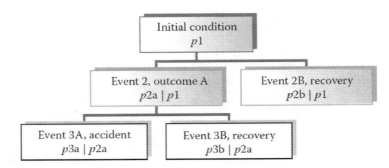

Figure 11.6 Segment of a generic fault tree annotated with conditional probabilities associated with events.

causative pattern. The analysis would also ascertain whether the patterns that were predicted within a hypothetical model were stronger than links that one would obtain from alternative patterns of correlation.

Petri nets also use boxes and arrows and often resemble fault trees. Petri nets, however, define system elements as the flow of information among subsystems of person–machine networks. *Marked Petri nets* are embellished with big dots in the boxes that indicate that a parcel of information has reached a particular box or subsystem (Peterson, 1981). Marked nets are used in industrial engineering where some form of person–machine communication is taking place.

Love and Johnson (2002) used a Petri net to display the conditions leading to the Clapham Junction train wreck in the United Kingdom. An improper repair on a signaling system gave several trains the wrong cues leading to a collision between two of them. A third train crashed into the wreckage of the first two. Four other trains passed through the same junction and could have easily been involved in the multitrain collision had the timing of their arrival been different. Further investigation into the event uncovered that the technician responsible for the faulty repair had been working excessive hours because there were too many problems to fix in too little time and without enough qualified help. The event resulted from fatigue on the part of the technician, something that was not defined in the Petri net model. One might drop backwards in time a step further, however, and ask why management allowed the situation to develop as it did.

Advances in computer technology have made it possible to develop simulations of accidents in animation form (Johnson, 2002). One way to use the technique would be to vary system attributes, for example, the speed of the vehicles and the timing of possible evasive maneuvers, and determine visually if a particular event would have taken place under specific conditions. Such simulations may be worth the proverbial 1,000 words for conveying to nontechnical audiences, such as courtroom juries, what actually happened in a given situation. On the other hand, the asset of this method is also its limitation: Erroneous assumptions that are built into the scenarios in the simulation would also be easy to understand and remember and could be more convincing than a more accurate explanation that is conveyed in a less dramatic form.

11.2.9 A complex and circular causation network

Figure 11.7 is a substantive model of industrial accident causation that organizes numerous qualitative research findings on the linkage between human and environmental characteristics and occupational accidents with connections among stress, anxiety, errors, hazards, perceived danger, beliefs about accident control, and errors and accidents. The definitions of stress in its many forms and anxiety were presented in Chapter 10. Correlations between anxiety and accidents are known (Guastello, 1993), although the relationship appears to be bidirectional. Leary (1990) identified two paths by which anxiety could lead to a behavioral disruption. In the first case, a

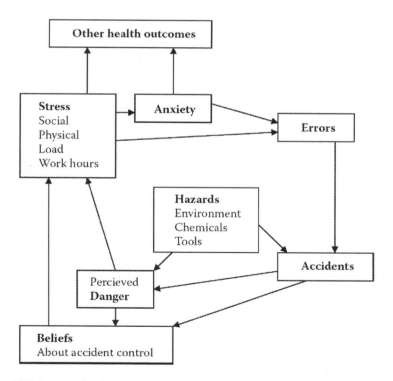

Figure 11.7 A network of psychosocial factors in the accident process.

person's cognitive evaluation of a situation, here a dangerous one, leads to anxiety, or the excessive fear that something bad will happen. As a result, actions are not taken that could have been taken because the person could not decide what to do. A second path is one where a person engages in a behavior that would ordinarily be appropriate, but anxiety intrudes and interrupts the behavior sequence. The interruption may take the form of an inopportune hesitation, or of dropping everything and making a quick exit to go do something else. Thus, people climbing a mountain or working on a tall building tell each other not to look down; the arousal level produced by the height and perception of the possibility of immediate danger could result in a behavioral interruption at a very bad time.

Accidents, or even some near misses, could easily trigger elevated anxiety on the part of the worker (Guastello, 1995). Even if the person involved was not personally harmed or as badly harmed as others, the effect of seeing what happens to others could become traumatic. What we have essentially established are two mechanisms by which stress can be regarded as contagious. In one pathway, there is within-person build-up caused by repeated exposures to stress and personal accident involvement. Each accident or serious near miss contributes to elevating stress beyond the threshold at which anxiety sets in. The environmental cues that trigger anxiety can be exacerbated by the individual's initial level of anxiety, and coping strategies could control the effect (Chapter 10). In the second contagion mechanism, an accident experienced by one person produces stress to a coworker. Thus there is transpersonal stress build-up. The role of anxiety in safety climate or culture is discussed later in this chapter.

The next two loops in Figure 11.7 express how danger and beliefs about accident control are connected to the circular stress–anxiety–error–accident cycle. Hazards in the environment are perhaps the first prerequisite for any type of accident process. If we all worked in rubber rooms, there would be little point in discussing these concepts. Although they make direct contributions to the accident process, hazards also lead to perceptions of the level of danger, and that level of danger modulates the person's stress level. People also habituate to stress and danger levels, which could be a benefit in one sense, but a limitation in another if it means that the individual becomes oblivious to the hazards that need to be negotiated (Wilde, 1988). Hazards can be further separated into their sources of origin, such as tools, chemicals, or environments where fires and explosion occur.

Some people believe that the good and bad things that happen to them in life are the results of their own behaviors, whereas others believe that most events in their lives a matter of luck. This polarity is well known as the *internal* versus *external locus of control* concept (Rotter, 1966). Several attempts have been made to link a variation of the locus of control construct to accident involvement. Some people believe that accidents are controllable whereas others believe that accidents are matter of bad luck. The concept of beliefs about accident control was first introduced as an *outcome* of accident involvement (Guastello & Guastello, 1986), although other researchers have used it as a concept of individual accident proneness (Arthur & Doverspike, 1992).

Each connection in Figure 11.7 presents an opportunity for safety management. The concept of safety climate was first expressed by Zohar (1980) who was investigating safety practices, and workers' views of those safety practices, that distinguished factories with good safety performance from those with poor performance. Attitudes toward the organization's safety program and its effectiveness, worker training, availability of needed tools and personal protection equipment, and the foreman's attentiveness to rule violations, all served to distinguish high- and low-performing groups.

11.2.10 Cusp catastrophe model of the accident process

The cusp catastrophe model is a risk analysis structure that can be used to describe a single accident event or a system of individual or group accident experiences. The strength of its predictive value resides primarily on empirical analyses done with occupational samples. Extensions to the interpretation of a single accident can be made from the collective results. An accident investigation could be framed around two questions: What were the background conditions or variables? What were the trigger variables?

The cusp catastrophe response surface appears in Figure 11.8 and is labeled for the general accident model (Guastello, 1989, 1991, 1992, 1995, 2003). Change between the two behavior states in a cusp model is a function of two controls, asymmetry (a), and bifurcation (b). The shaded areas between the two stable states in a cusp catastrophe are a region where very few points fall. That vacant region contains the *cusp point* at its extremity. The cusp point is the most unstable point on the surface; see Chapters 3 and 10 for further explanation and other examples.

There are two groups of variables in the accident model that correspond to the two control parameters. The asymmetry parameter consists of environmental hazards. The bifurcation variable consists of several types of stress or stress-reducing variables that are collectively labelled as *operator load*. Operator load includes various forms of stress such as work pace and other contributions of safety management; physical stressors such as heat, cold, noise, and crowdedness; and social stressors such as poor job security, changing job assignments, and difficulties with superiors or coworkers. Work group size also contributes to the bifurcation parameter in work groups, wherein the accident rates of small groups are more unstable than those of larger groups (Guastello, 1988, 1989). On the other hand, small groups also have access to locally stable low accident rates.

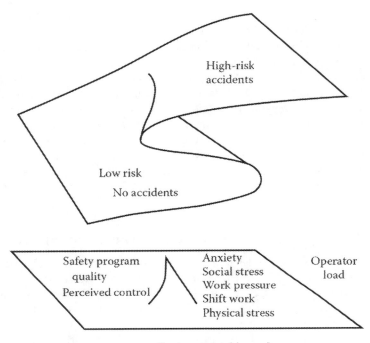

Figure 11.8 The cusp catastrophe model for occupational accidents. (From *Chaos, Catastrophe, and Human Affairs: Applications of Nonlinear Dynamics to Work, Organizations, and Social Evolution* (p. 219), by S. J. Guastello (1995), Hillsdale, NJ: Lawrence Erlbaum Associates. Copyright 1995 by Lawrence Erlbaum Associates. Reprinted with permission.)

The process for a single accident works as follows. The system begins at a stable state of low risk. Risk remains low while ambient hazards and load variables slowly change. Once the control variables reach a critical point, however, there is a sudden change to a high risk where an accident occurs. Subcritical risk levels are shown on the response surface also. The process for group-level accident rates works in a similar fashion. At the group level, however, we can specify returns to lower accident rates when the values of the control parameters have receded sufficiently far.

Empirical tests of the cusp model have been made with industrial work units (Guastello, 1989; Guastello & Lynn, 2014), bus drivers (Guastello, 1991, 1992), and hospital personnel (Guastello, Gershon, & Murphy, 1999). The critical variables were measured by a survey that was completed by employees. The variables that are specifically relevant to a work environment would vary with different situations, although the general themes still apply.

Table 11.2 summarizes the three accident situations and their formal constraints. The strongest group dynamics appear in industrial settings, which are relatively closed systems. Employees tend to work together in an intact unit. The output of one person affects the others. Outputs include contributions to the relative safety of the work environment, and atmosphere of stress, anxiety, and confidence in safety management. Bus drivers, however, work mostly alone in a relatively open system, although there should be some radio contact with a central controller who can initiate help where needed. In fact, the primary recommendation for this group, which was adopted, was to provide a better radio and response system for drivers in emergency situations. Health care employees usually work in cross-functional teams that are reconfigured many times per day according to the needs of patients and the flow of patients during emergencies. They represent an example of metastability, which is characteristic of a CAS.

The accident rates in any business' experience sometimes vary sharply over time. The cusp model would explain the variation as change around the bifurcation manifold. Reason (1997) also observed the oscillations as well as the two stable states (Figure 11.9). His explanation was that the safety management factor was actually changing, such that it sometimes favored low accident rates and less productivity. At other times management favors faster production, which promotes more human error and accidents. Management is likely to gravitate toward higher risk and higher production after it has experienced an extended period of lower risk, lower production just because it loses the memory of what could happen if it speeds up the work processes.

The model requires that the dependent measure (accident count or rate) be measured at two points in time. In the industrial setting, the dependent measure was the work groups' accident rate based on their OSHA-200 report at two points in time. For the bus drivers, accidents were measured as self-reported counts at work and self-reported counts of accidents in personal automobiles; this difference would convey the occupational risk exposure controlled for a base rate of driving accidents for that driver. For the health care setting, the two dependent measures were self-reported individual accidents of all forms and self-reported incidents of blood-specific events such as needle sticks and sharps. In the health care setting it was possible to predict the general accident rate from blood-specific incidents more accurately than the other way around (Guastello et al., 1999).

11.2.11 Complex dynamics, events, and deviations

A multilinear event series (Benner, 1975) is a compilation of multiple simple-event series and fault trees. The premise is that a complex manufacturing system is composed of activity patterns that occur simultaneously. In principle, the workflow patterns have been designed by management to

Table 11.2 Comparison of accident models and situations

Model Property	Manufacturing Facilities	Public Transportation	Health Care
Group dynamics	Employees tend to work together in an intact unit. The output of one person affects the others. Outputs include contributions to the relative safety of the work environment, and anxiety, land confidence in safety management	Employees usually work alone, but there is some radio contact with a central controller. Hazard levels change as bus moves through different neighborhoods and as traffic situations change. Stress, anxiety, and confidence in management are shared by virtue of common experiences	Employees usually work in cross-functional teams that are reconfigured many times per day according to the needs of patients. Stress, anxiety, and confidence in management are shared by virtue of employee cohesion and common experiences
Type of system	Relatively closed	Relatively open	Relatively unstable
Type of observation pairs	OSHA-200 accident reports for work groups measured at two points in time	Self-reports by individuals for bus accidents in personal automobiles	Self-report by individuals for occupational accidents generally and for blood-specific risks (needle sticks, sharps)
Environmental hazards (asymmetry)	Tools, toxins, poor housekeeping, crowded walkways, ratings of danger severity	Unruly and potentially violent passengers, need to reprimand passengers for radios, smoking, etc. Hazard levels vary by shift.	Verbal abuse from coworkers and patients. Exposure varies by professional group
Cognitive load (bifurcation)	Physical and social stressors, safety climate, beliefs in accident control, work group size, amount of time spent working on recommendations from the ergonomic report	Social stressors (job insecurity, changing assignments, difficulties with coworkers or management).	Shift work, safety climate, depressive symptoms, work pace, physical stressors (noise, heat, cold)
R^2 cusp model	.42	.63	.75
R^2 best linear alternative	.05	.26	.07

Source: From "Nonlinear Dynamics, Complex Systems, and Occupational Accidents," by S. J. Guastello, 2003, *Human Factors and Ergonomics in Manufacturing 13*, p. 297. Copyright 2003 by Wiley. Reprinted with permission.

maximize work output and to minimize accidents. When patterns of activity fall out of sequence, or outside the toleration limits for those sequences, risks escalate.

Kjellén and Larsson (1981; Kjellén, 1984a, 1984b) extended the multilinear event concept by introducing the concept of deviations. The conceptual model assumes that there is a normative and functional work pattern with which to begin. When deviations from norms are introduced, they are carried through the system from subprocess to subprocess and their impact magnifies. At some point in the magnification process a burst of energy is released that suddenly transforms the system into the second stage of the accident process. During the second stage, personnel have the opportunity—maybe—to take evasive action either to prevent the accident itself, thereby transforming it into a near-miss event, or minimizing the damage. Large accumulated deviations may interfere with successful evasive maneuvers. The third stage of the process is the actual delivery of harm to the employee; by that time, the event is final.

Sources of deviation in a work setting could be associated with the flow of materials, changes in personnel assignment, flow of information, equipment anomalies, intersecting activity patterns, and environmental disturbances (Kjellén & Hovden, 1993, p. 421). Kjellén and Hovden presented an example of how a confluence of deviations described a construction accident: A regular worker was out sick and was replaced by an apprentice. The task of the day was to erect a beam. A crane that would have been very useful was needed somewhere else. There was ice on the beam. The worker erected the beam manually, but it was crooked, so he walked out onto the beam to realign it, slipped on the ice, and fell to the floor below, breaking a rib and puncturing a lung.

The deviation principle in accident investigation is similar to the deviation principle when ensuring quality in manufactured goods. Kjellén and Hovden (1993) observed through experience, however, that the standards against which one might compare deviations are more rigorously defined in product quality management and permit fewer sources of deviation. Operationalization of the model requires detailed information sources. Both issues, in their opinion, explained the limited uses of the deviation principle in actual practice. Larsson (1993) observed that the deviation models should include cognitive and decision-making aspects of the work process that could describe the onset of some forms of deviation and correct others.

11.3 GROUP DYNAMICS, SAFETY CLIMATE, AND RESILIENCE

Although accidents happen to people who are often carrying out activities on their own initiative, occupational accidents are group phenomena to varying extents, as mentioned earlier in conjunction with Table 11.2. This section of the chapter expands on the group dynamics leading to accident occurrence or control, and the concepts of safety climate, safety culture, and resilience.

11.3.1 Group dynamics and complex technologies

The contrasting group dynamics for industry, transportation, and health care situations indicate one source of self-organization. The group work environment involves numerous local interactions among the individual employees and management representatives. In the course of an industrial group task, the employees coordinate their actions with each other, and wherever possible watch for hazardous acts of others that could affect themselves personally, or warn others when a hazardous situation is developing.

In transportation situations, the information flow among workers is much weaker, and is likely channeled through a central controller. The majority of safety-related decisions are going to be made on the spot by the driver in response to momentary changes in conditions.

The health care situations often involve cross-functional teams of nurses, physicians, and technicians. Emergency (stress) levels can vary at a moment's notice. In hospital emergency rooms, it appears that a substantial number of adverse outcomes can be traced to communication failures among personnel, particularly when the criticality of the situation spikes (Garosi et al., 2020; Gorman et al., 2020; van der Oever & Schraagen, 2021; Xiao, Patey, & Mackenzie, 1995) and the problems or tasks increase in complexity (Perry, Jaffe, & Bitan, 2022). Communication patterns tend to change when criticality escalates, although the nature of the patterning depends on the communication demands of the situation and roles played by the interlocutors. Analyses of transcripts of team interactions showed that, for a sample of 54 surgeries in two hospitals, 63% of surgeries had deficiencies in communication in the form of communication equipment failures, protests, and irrelevant conversations (Garosi et al., 2020). Collecting knowledge did not appear to be widely problematic, but there were numerous communication gaps related to the integration of knowledge (Perry et al., 2022).

Coordination training among team members will help any one particular person integrate well with others. This is an important point if a smooth assimilation of a new employee is desired, which is often the case. Experimental studies on group coordination indicate, however, that coordination will become chaos when more than 50% of the component employees are not familiar with each other (Guastello, Bock, Caldwell, & Bond, 2005; Reason, 1997). Group coordination can occur without any management help (Guastello, 2002), although placing "coordinator roles" within large work groups (Comer, 1995) could facilitate coordination. The optimal ratio of workers to safety coordinators appears to be about 15:1 (Guastello & Guastello, 1987b).

Computer-driven automation in manufacturing and elsewhere can have a favorable impact on safety insofar as a greater physical distance is placed between the operator and mechanical forces that induce injury. The questionable side of progress, however, is that the trend in automation is to place greater cognitive loads on the operators while reducing physical loads (Eberts & Salvendy, 1986). To complicate matters further, the operators work through computer software that is intended as a virtual representation of real systems. Unlike virtual reality software, however, the fidelity of cause and effect is not present in the software. Air traffic controllers, for instance, must detect signals on a screen, maintain radio contact with many aircraft, and remember where they put the airplanes. Consider this scenario reported by NASA researchers for aviation safety:

> In 1991 a tower controller at Los Angeles airport cleared a Metroliner to position on a runway and hold for release while she instructed other aircraft to cross the other end of the runway, a common practice. Several unavoidable delays occurred, and the Metroliner was not conspicuous in the twilight with poorly designed airport lighting. The controller forgot that she had not released the Metroliner to take off, and she cleared another aircraft to land on the same runway, which it did, crashing into the Metroliner.
>
> *Stone, Remington, & Dismukes, 2001, p. 1*

Coordination and safety in complex systems are considered in greater depth in the next chapter.

11.3.2 Safety climate

The understanding of the group phenomenon known as *safety climate* is starting to undergo some development and with some added complexity. The original construct of safety climate (Zohar, 1980) captured a social or organizational climate that was characterized by the adoption of safe operating practices and workers' evaluations of those safety practices. Safe and unsafe work environments could be distinguished, all other things being equal, on the use of personal protection equipment, vigilance on the part of workers and management, the safe behavior of employees, and good housekeeping.

The concept of climate was not far removed from the broader concept of organizational climate (Schneider & Reichers, 1983) that was prominent at the time for distinguishing the social atmospheres of organizations. An organizational climate describes patterns of interactions between the individual and organization as a whole, and can distinguish organizations in much the same way as personality distinguishes individuals. Some of the classic distinctions in climate include humanistic and participative versus authoritarian and exploitive (Likert, 1961; McGregor, 1960) or trait-like constructs such as level of achievement orientation, degree of risk taking, level of individual autonomy, and warmth and support (Saal & Knight, 1988). The more specific notion of safety climate can be expected to combine with other aspects of the organization's climate to produce a more unique experience regarding safety matters (Wallace, Popp, & Mondore, 2006). Ironically, the same background climate and administrative policy type that supports a positive climate for safety also supports other aspects of organizational performance (Bitire & Chuma, 2022; Herzberg, Mausner, & Snyderman 1959).

Although the original concept of safety climate is closely tied to the experiences of workers on the shop floor, safety climate, like the broader concept of organizational climate, is strongly affected by policies of upper management. Management policies affect the relative priority of work speed and profits versus safety, the choice of accident-prevention programs, the efficacy of those interventions, equipment maintenance policies, and adequate maintenance of equipment (Reason, 1997). Kjellén and Hovden (1993) advocated the development of safety information systems to assist with greater top-down control over deviations in work processes and to identify sources of unwanted novelties that build up into hazardous circumstances.

Bottom-up dynamics also influence in the production of a climate. Climate levels within a work unit are usually thought to be similar to the perceptions of safety climate in the organization as a whole (Zohar & Luria, 2005). Greater variations in climate occur in work groups, however, when the supervisors have greater autonomy for setting work pace or enforcing standards. The level of participation in safety functions is recognized as an important facet of climate and contributor injury levels, and researchers advocate strategies for participatory ergonomics (Halpern & Dawson, 1997; Jensen, 1997; Moore & Garg, 1997; Neal & Griffin, 2006).

In spite of the strong conceptual case for the role of management in safety climate, the correlations between measures of safety climate and actual individual accident involvement or group accident rates are not always consistent (Clarke, 2006; Fullarton & Stokes, 2007; Johnson, 2007). According to Clarke's (2006) meta-analysis, the corrected mean population correlations (ρ) between safety climate with safety compliance and safety participation were .43 and .50 respectively. The ρ between compliance and participation with accidents and injuries were .09 and .14 respectively, indicating a substantial attenuation effect between employee actions and actual outcomes. Because the lower bound of the 90% confidence intervals was still positive, those four values of ρ are considered generalizable, however. Safety climate actually showed a stronger relationship to injuries ($\rho = .22$, based 28 studies and a total $N = 17,695$), but did not generalize because the lower bound on the 90% confidence interval was negative and an insufficient amount of the variation in correlation coefficients could be explained by artifacts. The confidence intervals indicated that 10% of the relationships between climate and accidents were likely to run in the counter-logical direction, and a thus substantial amount of situational specificity was occurring overall.

Several writers suggested sources of situational specificity for safety climate, such as technology (Zohar, 2003), beliefs about accident control (Guastello, 1989, 2003; Neitzel, Seixas, Harris, & Camp, 2008), stress and anxiety levels (Dunbar, 1993; Gandit, Kouabeanan, & Caroly, 2009; Guastello, 1989, 2003; Hellesoy, Gronhaug, & Kvitastein, 1998; Kerr, McHugh, & McCrory, 2009; Nagashima et al., 2007; Parkes, 1998; Quick, Murphy, & Hurrell, 1992; Sauter & Murphy, 1994), organizational size (Fullarton & Stokes, 2007) and organizational subunit size (Guastello & Guastello, 1987b; Guastello, 1988), and other variables that are part of the broader spectrum of organizational climate (Wallace et al., 2006). Anxiety as a safety climate variable is considered next.

Anxiety is often observed or experienced as a consequence of job stress. Job stressors variously include work load and time pressure, stressors with social origins such as job insecurity, or those with physical origins such as noise, heat, cold, shift work, and physical danger sources (Quick et al., 1992). It could be transient, arriving and disappearing as the stress sources come and go, or it could be longer term effect. Longer term effects might reflect personality traits, but they could also reflect long-term job exposures, particularly if anxiety levels in the work group are high. Some plausible pathways for anxiety to disrupt behavior were discussed previously. Although anxiety has origins that are not simply reducible to job stress, stress management programs have been effective for reducing accident rates by an average of 15% (Murphy, 1984; Murphy et al., 1986), and by implication reducing state anxiety levels as well.

More recent studies, however, have reported positive effects of stress on performance, with the interpretation that coping mechanisms appear to be playing a decisive role in channeling arousal into greater focus on the task instead of transforming it into disruptive anxiety (Gandit et al., 2009; Matthews & Campbell, 2009). Others have observed that anxiety can serve as a positive force in a group because it results in greater vigilance over risk factors (Ein-Dor et al., 2010). Oddly, the extant literature on safety climate, anxiety, and accidents has not articulated a role for environmental hazards in conjunction with the psychosocial variables.

11.3.3 Cusp model for safety climate

In light of the aspects of safety climate that had not been sufficiently addressed, the cusp catastrophe model mentioned earlier was considered once again, this time with specific attention to the safety climate concept (Guastello & Lynn, 2014). The instability that is intuitively associated with an accident is actually located in the *change* between the stable states.

The bifurcation manifold is actually a *pattern* of instability. The respective roles of asymmetry and bifurcation variables correspond to what early epidemiologists called background variables and trigger variables, respectively (Fine, 1979; Guastello, 1995). Hazard levels would be the background variable or variables.

Safety climate and anxiety were the two primary bifurcation variables of interest in this study. If the safety climate is poor or indifferent, a time series or distribution of events would show work groups split between those with lower and higher rates. The outcome that prevails would depend on the hazard level. Positive scores on safety climate favor low-accident conditions, but although they are desirable, they are not necessarily stable. Positive safety climate can be disrupted by a new hazard or other aspects of operator load.

Reason (1997) noted that work pace by itself can have what amounts to a hysteresis effect on accident rates; *hysteresis* is the irregular oscillation between stable states that is often observed in catastrophe phenomena: An organization might make concerted efforts to reduce accidents by adjusting the work pace demands. Eventually, however, the organization starts to demand higher production output and as a result the accident rates start to zigzag up and down until the control point lands on "up" (i.e., higher risk) as depicted in Figure 11.9. Although Reason did not invoke catastrophe models as part of his exposition, the movement across the bifurcation manifold is essentially what he was describing.

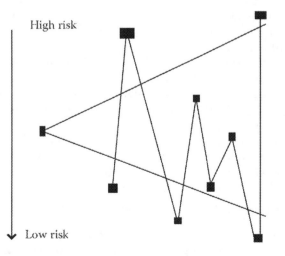

Work pace speeds up over time

Figure 11.9 Hysteresis and risk levels.

Anxiety displays the attributes of a bifurcation variable: The high end of the scale predisposes the individual or group to performance disruptions that could be precursors to accidents or it could predispose the individual or group to closer monitoring of the environment for threatening signals. Low anxiety takes the individual or group away from either extreme, but the absence of anxiety does not generate a locally stable outcome.

Catastrophes are often the result of phase shifts (Chapter 3), which are critical moments when a situation emerges or self-organizes to produce a result that is not readily reduced to elementary contributing causes. Systems tend to self-organize when their internal entropy levels are high, where-upon events take on a structure without any input from outside the system, such as a manager. For instance, Kjellén and Hovlan, (1993) noted several accident theorists have focused on the build-up of energies in the system that begin with deviations in procedures or materials, carry through further parts of the work process, and have gone out of control by the time the energies transform into an injury to an operator.

Situational specificity plays a substantial role in self-organizing processes (Hazy, Goldstein, & Lichtenstein, 2007). Although a process may take on a generalizable structure, the specific variables that are salient in the environment will shape the particular outcome. For instance, if one were to include safety climate as it is generally known as part of a broader array of constructs that are potentially operative—safety *management*, stress from different sources, anxiety, perceptions of danger levels, safety locus of control, and physical hazard inventories—different variables would be more pronounced in some situations than in others. Differences in the tight and loose nature of the work groups' boundaries also affect the outcomes; contrasts between steel mills, transit operation, and health care environments have been noted previously.

The investigation (Guastello & Lynn, 2014) was defined broadly enough to cover the probable range of possible contributing psychosocial and environmental elements and to anticipate that some contributing elements could be more salient than others when viewed across situations. In an illustrative example, 1,262 production employees of two steel manufacturing facilities completed a survey that measured safety management, anxiety, subjective danger, dysregulation, stressors and hazards. Instead of using intact survey scales, however, the survey items were factor analyzed to produce four scales that reflected the ongoing dynamics of the facilities: safety management, which was close to the original definition of the construct, anxiety, a cluster of hazards associated with lifting, and a cluster of hazards associated with the use of construction tools. Safety climate and anxiety were tested as two bifurcation variables. The two hazard groups were added together to form a single asymmetry factor. Age and experience were combined into another asymmetry variable. Nonlinear regression analyses showed that the accident process was explained by the cusp catastrophe model; the accuracy of the cusp model ($R^2 = .72$) exceeded that of the next best log-linear model ($R^2 = .08$), which was composed from the same four variables. All four control variables played a statistically significant role as expected.

Of further interest, the older and more experienced people were involved in more of the accidents over the same three-year time period. Ordinarily the opposite is true. The interpretation was that the more experienced workers got the more challenging assignments and were thus the ones who were more often exposed.

The R^2 coefficients for the cusp models appearing in Table 11.2 along with the R^2 values for the next best linear alternative measures can be combined with the results of the safety climate study to assess the relative efficacy of the cusp model as a theoretical approach for accident risk modeling. The average value for the cusp models was .63, and the average value for the comparable linear alternatives was .12. The cusp showed an advantage of 5.25:1 in terms of accident variance accounted for by the model.

11.3.4 Safety culture

The concept of safety climate was eventually transformed into *safety culture* in the semantics of safety officers. The transition took place in the wake of the Chernobyl nuclear reactor disaster in the Ukraine in 1987, when it was noticed that different societies had different views of what constituted a risk and what the risk was worth in terms of societal benefits (Douglas & Wildavsky, 1982; Pidgeon, 1991). Policymakers in the European Union (EU), meanwhile, are currently grappling with the task of developing uniform occupational health and safety standards (OHS) for EU employers. Perhaps the most fundamental problem facing transnational policymaking is the extent to which program policies should balance uniform compliance, national interpretation, employers' discretion, and worker participation. Next, there is a need to balance physical environment and process standards. Each of the constituent countries has its own history of labor and safety litigation. These differences affect the form of the general solution and the norms of compliance (Frick, Jensen, Quinlan, & Wilthagen, 2000).

The agglomeration of organizations through buyouts produces situations where multiple approaches to OHS can exist within one organization. Confusion or diffusion of responsibility could result. For instance, for Swedish organizations, if one organization buys out another, the responsibility for OHS still belongs to the unit that was purchased where the dangerous work presumably is being carried out. Here the OHS objectives could be in conflict with other questionable directives from the holding company's management. The exportation of dangerous work to countries outside the European Union, where norms and expectations for OHS transfer greater risks to the workers, presents more opportunities to exploit workers or diffuse responsibility for OHS. The growing size of the secondary labor market in the United States and European Union also reflects more opportunities for organizations to shirk OHS responsibility by outsourcing to less advantaged operators.

OHS management within any one organization has often stood outside the purview of most management functions. There is now considerable agreement that OHS objectives must be integrated into the mainstream of an organization's business concerns. The implementation of OHS programs is thought to benefit from worker participation initiatives, and policymakers at the EU level are placing a strong emphasis on employee participation components to OHS.

In spite of the emphasis on participation, however, which is usually associated with the more progressive forms of management, there are strong subcurrents of authoritarianism in management that result in blame-the-worker policies. Wockutch and VanSandt (2000) presented several contrasts in high-profile OHS plans in the United States (DuPont) and Japan (Toyota), although they noticed some curious similarities also:

> When there is an oil spill in a US factory we erect a cage around the spill, whereas in Japan they put up a sign warning workers to be careful. The critical fact from this perspective is that in neither place is the oil spill cleaned up.
>
> *p. 372*

11.3.5 Swiss cheese model

Even a well-designed system can have sensitive areas where safety risks can get through. The goal of safety management is thus to find those sensitive areas and provide additional defensive strategies that will limit the flow of risks. Reason (1997) likened the situation to a series of slices of Swiss cheese that are arranged in a staggered fashion like the squares depicted in Figure 11.10. Although other OHS concepts emphasize the role of the individual, management is ultimately responsible for investigating risks and incidents and setting priorities as to what constitutes a risk that requires attention.

Reason (1997) suggested further that some of the fluctuations in an organization's accident rate could be traced to waffling in priorities that are expressed by various policies. Improper maintenance of systems may be at least as important as the actions of the human at the machine interface; the analysis of several major disasters such as Apollo 13, Three Mile Island, Bhopal chemical plant, the Clapham Junction train wreck, and several airplane incidents all showed significant contributions of poor maintenance or maintenance policies.

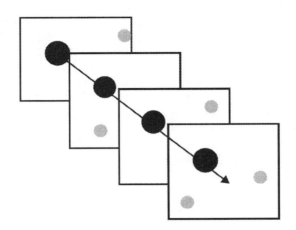

Figure 11.10 Swiss cheese model of risk management.

11.3.6 Resilience engineering

The concept of resilience in human factors and ergonomics (Hollnagel, Woods, & Leveson, 2006; Sheridan, 2008) poses questions such as: How well can a system rebound from a threat or assault? Can it detect critical situations before they fully arrive? Building resilience into a system requires more than the analysis of accidents in hindsight or implementing a right-minded plan. It involves regular monitoring of the system for its proximity to critical situation, anticipating possible critical situations, and taking action to adapt as necessary.

The stress–strain diagram shown in Figure 11.11 depicts the placement of resilience properties within the broader scope of the system's functioning (Sheridan, 2008; Woods & Wreathall, 2008). Strain is an independent measure. In the low-to-moderate regions of strain, the outcome is a fairly consistent increase in the system's ability to meet demands of various types. Beyond a certain point, however, more strain is producing diminishing increments of capability in what the theorists call the "extra region." At the extreme of the extra region the system faults. The spot marked "A" would be a good place in the process to introduce an adaptation that would stretch the capability of the system.

Successful adaption to possible risks begins with a calibration of possible risks according to their level of severity and probability of occurrence. In a traditional risk matrix, the outcome categories would range from negligible to marginal, critical, and catastrophic. In a resilience-oriented risk matrix, the outcomes would range from negative (catastrophic in the extreme) to neutral in the middle, to positive at the high end. The second is the probability of an event, ranging from rare to routine. The task is then to assess all types of performance events for the sources of variability or flexibility in the system. Insights regarding the events to be avoided can be forthcoming from understanding the processes that produce unexpectedly positive outcomes (Hollnagel, 2011).

The definition of a preemptive strategy requires a good sense of how a CAS operates (Dooley, 1997). One feature of the CAS involves the use of sensors that take the form of information gathering capabilities that can inform organizational members of changing events in the outside environment or internal operations. The most useful information inflows result in the system-level equivalent of situation awareness (Chapter 7). Woods and Wreathall (2008) characterized this form of situation awareness as *calibration*, which is to know where the system is located along the stress–demand function in the face of changing circumstances.

A second feature of the CAS, which is not independent of the first, is to have a clear sense of how the system is organized for work and information flows. This idea is similar to Level 2 situation awareness (Chapter 7). Here one should look for *functional resonance*, which is how the variability associated with one part of the process crosses to the adjacent parts of the process and travels through the system (Hollnagel, 2012; Leonhardt, Macchi, Hollnagel, & Kirwan, 2009). It would be then be possible to pinpoint bottlenecks and non-resilient or nonadaptive features of the process. The bottlenecks are essentially those associated with parallel cognitive processes (Chapter 6). Non-resilient or nonadaptive features of the process often involve automated functions which, traditionally, permit little variation in the process and output and thus work very efficiently for a limited range of circumstances. A resilience event would occur when the human intervenes to make an adaptation that either extends or overrides an automatic process after detecting that a supercritical event was on the verge of occurring. Adaptive automation processes (Chapter 2) can be interpreted as attempts to build a modicum of resilience into a PMS.

The third important feature is the production of effective adaptive responses. Recognition-primed decision-making was suggested as one medium for doing so (Woods & Wreathall, 2008). Some adaptations are more expensive than others regarding the cognitive resources required; for instance, an adaptation made by a work group requires a greater expenditure of resources than an adaptation by a single individual. The adaptations themselves are essentially the dynamics of self-organization all over again (Wears, 2010).

As with other self-organized systems, there is an optimal level of variability associated with high levels of performance (Leonhardt et al., 2009; Schuldberg, 2015); the variability has been associated with pink noise in other contexts. Although there is a folk tendency to think of high performance as synonymous with consistent performance, the relationship is not

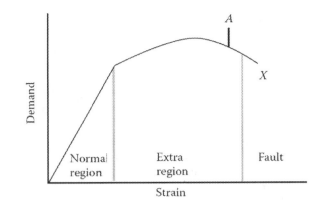

Figure 11.11 Stress–demand resilience function.

really true. A certain amount of variability in the system is needed to generate an adaptive response. The grain of truth that remains in the folk logic is that high performing systems, although variable, spend comparatively less time in the very poor regions of the performance spectrum (Guastello et al., 2013).

Before departing this topic, it would be useful to close two more loops in the connection between resilience engineering and other concepts already covered. First, the "extra zone" in Figure 11.11 is comparable to the adaptive range in workload response described by Hancock and Warm (1989), and the fault region of the demand curve corresponds to what happens when the adaptive range is exceeded. Second, the construct of resilience in work systems corresponds to the rigidity–elasticity principle in the cusp model of mental workload (Figure 10.9). Once again, rigidity has the benefit of controlling system variability up to a point, after which a decisive fault can be expected. Elasticity and resilience, on the other hand, span the region of the surface that contains the cusp point, which is the most unstable point on the surface. It follows that too much flexibility, or resilience capability, can make a system unstable in the long run. Here one should look for features of the system that use too many degrees of freedom (or control) to get a job done well enough when fewer degrees of freedom would produce a result that is just as good.

11.4 ACCIDENT-PREVENTION PROGRAMS

In spite of all the effort and expense that goes into accident-prevention programs each year globally, there is scant information available on the relative merits of the known accident-prevention strategies. Decision makers are thus destined to make important decisions based on unreliable or disorganized information. The lack of sufficient program evaluation information was first identified as a problem in the early 1990s (Guastelo, 1993), and subsequent researchers have called attention to the lingering deficit (Lehtola et al., 2008; Pederson, Nielson, & Kines, 2012; Shannon, Robson, & Guastello, 1999) which still persists.

This section of the chapter fills that information gap by identifying the salient program types, and by compiling conclusions about how well they worked to reduce accident rates. The information sources were scientific journal articles that were identified through hand searches of the literature, computerized literature searches, and other articles identified through leads found in primary sources. Reports were also solicited through announcements that were published in relevant journals in the mid-1990s. As it turned out, the vast majority of material that was generated from the announcements was published somewhere; there was very little material hiding out in file drawers or fugitive literature. Occupational vehicular accidents were included in the present review, but nonoccupational vehicular accident studies were omitted. Because theories and technologies evolve and improve over time, this review was limited to techniques that were tested since 1977; earlier studies were included to the extent they pertained to techniques that were still under investigation. The review that follows was built on a review that was first published in *Safety Science* (Guastello, 1993), two interim updates that were compiled and presented at conferences (Shannon & Guastello, 1996, 1997), and direct searches of later literature.

There were ten broad categories of intervention: personnel selection variables, technological intervention, behavior modification, poster and promotional campaigns, safety committees, medical management, near-miss accident reporting, comprehensive ergonomics, other management interventions, and governmental interventions. Studies were included in the analysis if they contained actual accident rates or counts as the dependent measure. As a result, studies that used safe or unsafe behaviors as their criterion were not included in the computation of effect sizes in Table 11.3 because of the strong attenuation effect between unsafe acts and actual injuries as reported earlier. Also, studies that expressed the results in monetary costs of accidents were not included because there are too many variables related to health care costs, especially rising costs, that only become involved after the accident occurs. The concern here is, in contrast, for the means of preventing the occurrences.

A measure of effect size, D, was computed for all studies listed in Table 11.3:

$$D = [(R_{pre} - R_{post})/R_{pre}] - [(S_{pre} - S_{post})/S_{pre}]. \qquad (11.3)$$

Table 11.3 Comparative evaluation of safety programs

Program Type	No. Coeff.	D(%)
1. Personnel selection	70	4.8
Personal maladjustment	21	5.7
Social maladjustment	8	6.4
Alcohol use	10	0.8
Drug use	4	0.0
Cognitive variables	24	6.5
Nonwork accidents (lifestyle)	3	0.0
2. Technological interventions	12	54.4
3. Behavior modification	11	53.1
4. Poster and promotional campaigns	3	−1.0
5. Safety committees	11	33.7
Quality circles	1	20.0
Safety committees, generic	3	36.0
Discussion groups	3	17.5
Joint labor-management committee	2	55.0
Two-group review routine	1	10.4
6. Medical management	6	26.5
Stress management and exercise	2	15.0
Employee Assistance Programs	2	32.5
Medical management	1	29.4
First aid training	1	34.7
7. Near-miss accident reporting	2	0.0
8. Comprehensive ergonomics	9	53.1
Musculoskeletal injuries targeted	2	77.5
All others	7	46.1
9. Other management interventions	6	55.0
Philosophy + behavior modification for management	1	30.0
Autonomous work groups	1	161.1
Self-regulation	1	46.0
Management-leadership training	1	71.5
Multifaceted safety campaign	1	−2.6
Multifaceted safety + management	1	24.0
10. Governmental interventions	29	9.7
Finnish national program	2	18.3
OSHA onset, regulations, inspections	5	2.6
OSHA industry-targeted inspections	12	18.9
Increase in workers' comp costs	3	−46.3
Workers' comp, experience method	3	25.0
Universal precautions for health care	1	0.0
Australian regulatory system	1	15.0
Constructions fatalities, USA	2	26.8

R_{pre} and R_{post} are the accident counts within a given period of time before and after the intervention, respectively. S_{pre} and S_{post} are the raw numbers of accidents before and after the intervention observed for the control group. In some studies the dependent measure was expressed as a rate. Because an accident frequency count is equal to the accident rate times the number of exposure units, a comparable D can be obtained by inserting a rate in place of R_{pre}, R_{post}, S_{pre}, and S_{post}. Unlike the evaluations of other types of safety programs, the personnel-selection studies were based on correlation coefficients. The translation of correlation coefficients into a measure comparable to D is explained later.

11.4.1 Personnel selection

Personnel selection is a group of strategies that is directed at picking the right job applicants. Ability and personality tests of various sorts are commonly used. For safety applications, the objective is to identify job applicants who are not likely to have workplace accidents. Almost all personnel-selection studies were based on correlations between an employee's safety record and how they scored on tests of interest.

For correlational studies that were taken from the initial review (Guastello, 1993), the correlations were transformed to percentage increase in relative efficiency by using utility expectancy tables. The tables required two additional parameters: the selection ratio and the base rate of success, which were set at .20 and .70, respectively. A selection ratio of .20 signifies that the organization can afford to hire only the most promising 20% of their job applicants. A base rate of .70 is equivalent to defining success on the job as an individual not incurring an accident in four years of exposure, based on the average accident rate of 8.30 for all private sector U.S. industries as reported by the National Safety Council (1989). For correlational studies dated after the first review, utility was calculated using Jarrett's (1948) continuous utility formula with assumptions of an exponential distribution, a selection ratio of .20, and a base rate of .70 were retained.

Six types of variables for personnel selection have been tested: personal maladjustment, social maladjustment (which would include external beliefs about accident control), alcohol use, drug use, cognitive variables, and reports of nonwork accidents. Eight coefficients for personal maladjustment were available from the initial review. Arthur, Barrett, and Alexander (1991) produced seven, and six more were reported by Barofsky and Smith (1993), Jacobs, Conte, Day, Silva, and Harris (1996), and Lardent (1991). These variables included stressful life events, anxiety, a measure of distractability, trust, tensions, insecurity, even temperedness, concern with being on time for work, other life events, and regard for authority.

Seven coefficients for social maladjustment were available from the initial review. Jacobs et al. (1996) added another test of the safety locus of control variable. Arguably, there is a fine line between personal and social maladjustment. Nonetheless the average (unweighted) effect sizes for those two groups of variables were 5.7 and 6.4, respectively. In other words, it appears that it is possible to reduce accident rates by 5.7% by using personal maladjustment variables in a selection context.

The studies of maladjustment variables, however, were always postdictive in nature, rather than predictive. Maladjustment could just as easily be the result of working too long or not enough in poorly controlled environments where safety management was ineffectual. Even if the correlations between individuals' personality characteristics and their accident rates were to be regarded as if they were as good as predictive values, personnel-selection techniques rank among the least effective methods of accident control of all the available options.

Self-report inventories or assays of blood samples were used to measure alcohol use. If such measures are to be used to select personnel, it would be tantamount to assuming that alcohol use *off* the job was associated with accident involvement while working, which is not easy to support. The eight studies reported in the earlier review (Guastello, 1993), and the two studies uncovered since that time (Dawson, 1994; Holcom, Lehman, & Simpson, 1993) show very little association between the two variables. The average effect size for this personnel-selection technique is 0.8%.

Drug use has been measured by self-report inventories or urine tests. Urine samples are usually tested for marijuana and cocaine, and the percentage of people testing positive for only marijuana ranged from 67% to 93%, depending on the industry (ACLU, 1999). Once again, if such measures are to be used to select personnel, one would need to verify that drug use *off* the job affects accident rates at work. The two studies reported in the initial review, and the two studies uncovered afterwards (Feinauer & Havlovic, 1993; Holcom et al., 1993) found no association between the two variables. Manufacturing organizations that adopted urine-testing programs were found to have obtained no benefit from the program compared to other organizations in the same industries and geographic region (Feinauer & Havlovic, 1993). The effect size for this personnel-selection strategy is 0.0%.

Cognitive tests appear to have a better track record. Auditory selective attention tests showed positive results in one study (Arthur, Barrett, & Doverspike, 1990). More broadly defined tests of selective attention showed positive results in 13 other occupational samples (Arthur et al., 1991), and tests of field dependence and field independence showed positive results in nine occupational samples (Arthur et al., 1991). General cognitive ability did not show a relationship to accidents in a large sample of bus drivers, however (Jacobs et al., 1996). The average effect size for the 24 correlations was 6.5%.

The last group of personnel selection studies was reported by Salminen and Heiskanen (1997). Three extensive stratified samples of Scandinavians responded to a survey in which they were asked about their involvement in work accidents, home accidents, and public situation accidents. The samples were taken in 1988, 1990, and 1993. The hypothesis was that there would be such a relationship, which would in turn mean that accident proneness could be traced to a person's lifestyle. The average correlation between work and other accidents was .02 in each sample, thus rendering an effect size of 0.0%.

The emphasis on individual-level explanations for occupational accidents (and illness) continues to be fueled by claims that the same people who are involved in accidents are also those with higher absenteeism rates. Verhaegen (1993) reported, however, that no such correlation exists in an ergonomically suitable environment. Furthermore, it is only when the environment is ergonomically degraded that one might observe correlations between personality measurements and accidents; an ergonomically sound workplace would leave no room for such individual differences to influence risk. According to Quinlan (1988) the unwarranted individual focus is an outgrowth of practices in industrial psychology, dating back to the scientific management era, where psychologists are called in (and paid) by management to solve so-called work problems. The right answer for the client was one that blamed the victim in some way. Such viewpoints overlooked collective causes of accidents and management's role in them, particularly when the pace of production was stepped up beyond safe working limits.

11.4.2 Technology interventions

Technology interventions may be broad in scope, such as introducing robots to perform dangerous work (Karwowski, Rahmi, & Mihaly, 1988), redesigning an entire manufacturing facility (Harms-Ringdahl, 1987; Kjellen, 1990; Mohr & Clemmer, 1989), outrigging loggers with improved personal protection gear and machine guards (Klen & Vayrynen, 1983), or introducing vessel-tracking systems into a maritime operation (LeBlanc & Rucks, 1996). Technology interventions could also be more event-specific, but nonetheless valuable: preventing needlestick injuries in hospitals with a new needle cap design, using visible tent stakes on camping expeditions, redesigns of oil-drilling equipment, rollover protection for farm vehicles, and introducing a barrier between train tracks and a roadway (Zwerling et al., 1997). Of all the occupationally traceable deaths to astronauts reported by Peterson, Pepper, Hamm, and Gilbert (1993), 47.1% of the known causes have been eliminated by improved equipment or materials, such as the O-rings in the construction of the space shuttle. The average unweighted effect size for these contributions is 54.4%.

Before continuing further, it is noteworthy that robot systems have the potential to reduce injuries, but they also have the potential to create new hazards because of the fast and forceful motions of the machine parts. Relevant remedies include emergency kill switches, radar detection of humans in the work envelope, and the placement of control stations for the humans well outside the machine's work envelope (Karwowski et al., 1988).

Comparable values of D for this group of technological interventions are not available, however.

A new genre of technology intervention involves software solutions for accident prevention, and a crop of examples came to the foreground recently in conjunction with fatigue-related accidents. It was noted in Chapter 10 that fatigue, the loss of work capacity over time, can arise from working too long on a mental or physical task or sleep deprivation. The latter case has been a concern for accident prevention in the transportation industries— long-distance truckers, railroad workers, and airline pilots—but also health care settings where emergency response and extra-long work shifts required to respond effectively to critical situations are the norm. One technology for addressing the matter is the use of "universal" work-schedule controls that require operators to have sufficient time between shifts to get enough sleep so as not to incur sleep debt (Balkin, Horrey, Graeber, Czeisler, & Dinges, 2011; Dawson, Noy, Härmä, Åkerstedt, & Belensky, 2011; Tucker & Folkard, 2012). Schedules are based on algorithms that combine wake– sleep cycles and body temperature cycles in different ways to predict suscep- tibility to fatigue events, slow response time, and errors. At present there is no strong evidence that one algorithm is better than another for this pur- pose. The effectiveness of algorithms for schedules is limited by whether the individual actually obtains the necessary hours of sleep when there is opportunity to do so.

The translation of fatigue conditions based on sleep sufficiency into accident rates requires the analyst to separate sleep deprivation, time on shift, and workload effects such as traffic density (Gander et al., 2011). In addition, there is the observation that errors could be more likely in the early part of the shift and not necessarily toward the end. Furthermore, although errors can be serious enough, translation into accidents is mitigated by all the usual factors. As a result, the evaluations of sched- uling algorithms that could provide effect sizes for Table 11.3 have not been forthcoming.

New technologies that are being explored for motor vehicles include EEG monitors, percent eye closure with eye tracking performance monitors, and adaptive workload systems. Concerns are intrusiveness, data loss in the online hook up, calibration, ability of the technique to make predictions of near-future driving quality based on the metrics, and conflicts between readings of fatigue events compared to non-fatigue driving errors (Balkin et al., 2011). For instance, alert drivers make more micro-corrections for road curvature for maintaining lanes than fatigued drivers, while fatigued drivers make more macro-corrections; yet a macro-correction could be a legitimate response to a road hazard such as a bad move by another driver. The efficacies of these techniques are considered promising by their proponents, but not yet established. A lot will depend on what decisions are made by, or based on these new data collection devices and what actions are taken as a result. A similar strategy for fatigue has been advanced for airline pilots (Williamson, 2012).

A different type of technological intervention takes the form of software to assist management in allocating safety resources (Shakioye & Haight, 2010). The thesis is that risks are dynamic over time, and not all safety management concerns can be attended simultaneously. Thus, it would be advantageous to predict risky events in the near term from the allocation of time spent on different classes of activities, and forecast which combination of activities would be most effective in the future time interval. The four classes of activities incorporated into the model were: (a) safety awareness and motivation; (b) skills and training; (c) tool and equipment design and implementation; and (d) other equipment-related activities (p. 47). The central algorithm was a 15-variable polynomial function, which was calibrated on 30 weeks of incident and safety resource allocation data. It was not clear whether the incidents were all reportable injuries, or included near-miss events. The R^2 for the model was .70 predicting incidents; R^2 adjusted for the number of variables in the model was .36. The model recommended that 11% of person-hours (a forestry operation) be allocated to safety activities in order to meet the organization's safety targets. There was no particular claim that the tool provided a reduction in accident rates when used in this fashion. The study did indicate, however, that the safety targets could be achieved with a 4% reduction in person-hours invested in safety tasks by using the activity-switching recommendations from the program.

11.4.3 Behavior modification

Behavior modification programs are based on operant learning theory. Safety officials, who could be work group supervisors, identify specific desired behaviors, such as wearing one's personal protection equipment properly, or compliance with particular safety rules. The identified behavior could be the number of reportable accidents themselves. The group of employees would then receive feedback on their behavior, often at the group level, with the goal of 100% rule compliance or zero accidents for the work group in a unit of time. Feedback is often presented in the form of highly visible posters; informational feedback acts as a form of reward to reinforcement in these circumstances. Some implementations of the program may also include a lottery for prizes, wherein only employees in units that have met the goal receive chances to win. Other versions provide censures for employees who repeatedly violate safety rules. McAfee and Winn (1989) published a compilation of programs and types.

The initial report on the effectiveness of behavior modification programs for controlling actual accidents (Guastello, 1993) showed an average effect size of 38.6% for six implementations. A seventh program that had been in place for over ten years in a uranium mine brought the average effect size up to 59.6%. Since that time four more reports have been collected (Gregerson, Brehmer, & Moren, 1996, giving two examples; Martinez & Brito, 1994; Zwerling et al., 1997). The average effect size for the 11 studies was 53.1%.

11.4.4 Poster campaigns

Safety posters are frequently found in industrial sites that typically give generic advice to employees to be mindful of safety matters. Better posters will give specific safety information. In the previous report, one safety poster program had a net positive effect on workplace safety, whereas another actually had a negative impact. In the latter case, the control group improved its safety performance faster than the poster group. Another similar case of negative impact was reported by Gregerson et al. (1996). The average effect size for the three poster studies is thus negative, −1.0%.

11.4.5 Safety committees

This category is new since the previous report, although safety committees of various sorts have been used in industry since time immemorial. One report characterized a safety committee as a *quality circle* (Saarela, 1990). A quality circle is a committee of employees who perform similar types of work who meet voluntarily to solve product quality, productivity, and cost-reduction problems. Past research has shown that the quality circle technique has been successful in those objectives in addition to improving the quality of work life and reducing absenteeism (Marks, Mirvis, Hackett, & Grady, 1986). Several results for broader varieties of safety committees have been reported (Havlovic & Feinauer, 1990; U.S. Department of Labor, 1988; Zwerling et al., 1997). The variety with the greatest power to enforce recommendations appears to be the joint labor-management committee. The next most influential is the employee safety committee. The least effective of the known varieties is the discussion group.

Menckel, Carter, and Hellbom (1993) reported an unusual variety of group procedure that involved two decision teams. One group was responsible for investigating hazardous situations and soliciting corrective ideas. The other group was responsible for effective implementation of corrective procedures. Another underlying goal of the procedure was to develop safety knowledge and prevention skill among a wider range of employees at the site. In addition to the effect size of 10.4% for accident reduction, Menckel et al. also reported a 35% reduction in accident severity for this procedure.

When all varieties of safety committee are considered together, their average effect size is 33.7% based on 11 reported outcomes. The highest marks to go the joint labor-management committee with a 55.0% reduction in the accident rate.

11.4.6 Medical management

This category started with two studies of stress management and exercise programs in the first review. Since then four new reports have surfaced on the effectiveness of an employee assistance program (EAP; Nadolski & Sandonato, 1987; Wickizer, Kopjar, Franklin, & Joesch, 2004), medical management (Zwerling et al., 1997), and first aid training (Zwerling et al., 1997) in reducing accident rates.

As a general rule, EAPs address a wide range of psychological issues, of which stress and anxiety are a part. The study by Wickizer et al. (2004) was one of the five programs from the construction industry evaluated in Lehtola et al. (2008). It was actually characterized as a drug-free workplace program that included, "a formal policy; drug testing; treatment; worker assistance; education of workers, supervisors, and managers; information; education' facilitation (financial incentives); and enforcement (drug testing)" (Lehtola et al., 2008, p. 80). By all appearances it was an EAP added on top of a drug testing program; the EAP feature was necessary to be compliant with the provisions Americans for Disabilities Act (ADA) of 1990 for the proper response to employees with substance abuse problems. Given that alcohol-related problems greatly outnumber drug-related problems in the general population and fall under the purviews of ADA and EAPs, it would follow that most of the effective interventions for individuals were alcohol-related. There is a stark contrast in effectiveness for this program ($D = 25.5\%$) and the effectiveness of drug testing at the point of personnel selection. This observation brings us to the broader issue that complex interventions by definition have a lot of parts, and some parts could be producing a much larger portion of the positive effect than other parts.

The medical management program that Zwerling et al. reported was designed to help employees make self-directed improvement on their personal health issues. One would guess that such an intervention had an impact on stress and particular health issues that affected fitness for work; further research would be needed to determine how the management of particular health affected any known intermediate variables that are already known to affect accident rates.

First aid training has an interesting effect on reportable accidents. If more employees knew how to respond to low-grade occurrences and had kits available, then off-site medical attention would not be necessary so often. By definition, the incidents would not be reportable to OSHA.

Hurrell and Murphy (1996) noted that stress-related illnesses comprised more than 11% of worker compensation claims during the 1980s. They regarded stress management programs as "secondary interventions." A primary intervention would alleviate the source of stress in the workplace, rather than shift the burden of responsibility to the employee. Nonetheless, the medical management group of interventions produced an average effect size of 26.7%.

11.4.7 Near-miss accident reporting

The near-miss reporting program is based on the principle that for every real accident that occurs, approximately ten near misses have also occurred. By investigating 11 times as many incidents, a greater number of ideas for preventative measures can be generated. Two examples (Carter & Menckel, 1985; Menckel & Carter, 1985) from the earlier report (Guastello, 1993) did not result in any reduction in accident rates. The probable explanations for the lack of results might be traced to the quality of corrective ideas and whether the best ideas were actually implemented. On the other hand, the Federal Aviation Administration routinely examines near-miss air disasters for possible explanations and prevention of actual disasters. The same is true for chemical process industries (Gnoni et al., 2013). The probable difference between a successful and mediocre near-miss program is the extent to which action is taken on the basis of the findings from the near-miss incidents. Although new strategies for the design and implementation of a near-miss analysis program have been reported, no new evaluative statistics are available yet that produce a measure of D.

Perhaps one of the barriers to success for near-miss accident reporting programs is the tendency for individuals who evaluate risks and outcomes to equate a near-miss incident with a success, rather than a system failure (Dillon & Tinsley, 2008). Similarly, managers of projects with a serious near-miss incident are also regarded as approximately as successful as managers of similar projects without a near-miss; both are regarded as distinctly better than managers of projects with an actual failure. Part of the explanation for this thinking, according to Dillon and Tinsley, lies in the gap between statistical odds of failure and subjective odds of failure. In one of their experiments, participants were given stated odds of event and asked to indicate whether they would engage in the target activity, for example, driving through an area where a hurricane is expected to hit relatively soon. Those who were also provided information about near-miss events gauged the situation as not so dangerous.

11.4.8 Comprehensive ergonomics

Comprehensive ergonomics programs involve a full-scale assault on any and all human factors and ergonomics problems in the workplace. Each organization would have its own list of problems requiring attention. Problems may be identified through use of a survey that is filled out by employees in all work locations (offices are typically excluded), or identified by safety engineers or outside auditors who make a thorough search for potentially hazardous situations, especially where human factors could be involved. Some organizations do not use an audit; instead they implement every program they possibly can, which, with a little luck, could cover a substantial range of human factors or ergonomic situations and hit some important targets. Studies involving OSHA inspections are placed in the category of governmental interventions.

The first report of this series on accident-prevention programs (Guastello, 1993) identified three examples of comprehensive ergonomics programs, although the specific contents of one of the interventions was unclear. Four new examples have surfaced since that time (Doos, Backstrom, & Samuelson, 1994; Helander & Burri, 1995; Lipscomb, Li, & Dement, 2003; Macek, 1987). The latest arrival (Lipscomb et al., 2003) was an evaluation of a program that was motivated by compulsory legislation directed toward preventing falling injuries in the construction industry. Because the program covered a range of activities usually associated with comprehensive ergonomics, it was included in this category.

Musculoskeletal injuries comprise a large portion of total occupational injuries as noted in Figures 11.1 and 11.2. Not surprisingly, some comprehensive ergonomics strategies have targeted musculoskeletal injuries, and two new reports are included in a new subcategory in Table 11.3. Moore and Garg (1998) reported a "participatory ergonomics" program in a large meat-processing industry, which historically has been one of the most hazardous industries in the United States. The particular facility had a baseline rate that was 20% less than the industry average at the time, and a lost-time incident rate that was 25% higher than the industry average; musculoskeletal injuries were mostly in the subgroup of lost-time injuries. The program itself involved auditing, targeting, and correcting as many sources of musculoskeletal injury as possible. Employees were involved in all phases of the program, and an ergonomics specialist was primarily responsible for its design and implementation. The interventions included the invention of new patented mechanical devices for certain meat packing operations. The program began in 1986 in which there was an immediate drop of 50% of lost-time injuries relative to 1994 norms; there was a 17% drop (possibly spontaneous) from 1994 to 1995. Lost-time injuries dropped to an average of 16% of baseline for 1991–1993. Thus the D value for this program was 76.0.

A similar report centered on machine sewing tasks (Halpern & Dawson, 1997). Again musculoskeletal injuries, a substantial portion of which are repetitive motion injuries, were targeted. The program had most of the features of the meat packing program, but with regular use of force measurements and discomfort surveys and without specialized machinery. Ultimately there was a 79% reduction in lost-time injuries recorded for 1995–1996 compared to the baseline of 1992–1993 years. There was also an 11.1% reduction in all Workman's Compensation claims, which included the nontargeted sources of injury.

The average effect size for the seven examples is 46.1%. It is noteworthy, however, that the effectiveness of a comprehensive ergonomics program is limited to the extent that the organization actually carries out the recommendations from its diagnostic analysis. Depending on the complexity of the list of recommendations, organizations might need more or less time to complete the improvements or for the improvements to realize their full effects.

In what could be an interesting twist on the themes of comprehensive ergonomics and usability, some insurance company representatives advocated accident prevention through better product designs (Braun, 2008). The former emphasis on "human error" as the culprit has apparently run its course and the sources of human error are now receiving attention. The history of automobile features was given as an example of what could be possible in other settings. Although new product designs might be characterized as technology interventions, the source of design ideas, however, might reasonably emerge from information uncovered by comprehensive ergonomics programs.

11.4.9 Other management interventions

This group of interventions was directed at improving management effectiveness, rather than the quality of person–machine or person–environment interactions directly. Woodhill, Crutchfield, and James (1987) reported a form of behavior modification whereby management's bonuses were tied to their safety performance. The program was coupled with a widely disseminated corporate philosophy that placed precedence on workplace safety. A learning theorist would probably remark here that the performance–bonus link was responsible for the vast majority of the accident-reduction results, and that philosophy exercises might help managers understand more clearly the contingencies of reinforcement; it is doubtful the philosophy alone would have had much impact.

The autonomous work group intervention (Trist, Susman, & Brown, 1977) had a substantial impact on safety; accident rates improved 161.1%. This uncanny level of improvement was the result of substantial progress in the intervention group combined with a deteriorating safety record in the control group. In an autonomous work group intervention, work that was once organized as a fixed sequence of specialized processes, as in an assembly line, is now given over to a group. The group establishes its own sequence of people and tasks, including the supervision or managing functions that used to be allocated to a supervisor. In other words, management was more or less removed from the situation. Rees (1988) reported a self-regulation intervention that seemed to fit the autonomous work group definition also.

Fiedler, Bell, Chemers, and Patrick (1984) found a substantial safety advantage from an intervention designed to enhance managers' leadership skills and assign managers to work groups that were better suited to their management style. The control group did not show any such progress. Although the specifics of their leadership training are outside the scope of this book, it would be fair to say that there is an alternative explanation for the results: A good many workers might have been simply happy to get rid of their managers and start the relationship over fresh with a new one. A careful test of this particular interpretation has not been reported, however. Altogether, the average effect size for the four studies in the category of other management interventions was 77.2%.

A new addition to this group of interventions (Spangenberg, Mikkelsen, Kines, Dyreborg, & Baarts, 2002) was a "multifaceted safety campaign ... including attitudinal and behavioral aspects (e.g., newsletter, best practices, safety inspections, financial safety award, themes on injury risks); information; facilitation (feedback); enforcement (inspection)" (Lehtola et al., 2008, p. 80) that was directed toward construction workers in Denmark. The safety record three years after the intervention was regrettably a bit worse than the record in the three years prior to the intervention. Another new addition, which also entangled multiple intervention types plus other non-safety management changes (Bunn, Pikelny, Slavin, & Paralkar, 2001, reported in Robson et al., 2007) produced a D value of 24.0 for all reportable incidents.

One could regroup the five studies into two groups: the autonomous work groups that involve the elimination of management roles (average $D = 103.6$) and the three that target the existing management (average $D = 33.0$). Ironically, the values of D favor autonomy versus management. The number of reports is relatively small, however, and a considerable amount of variance in the effectiveness of either subgroup of strategies has yet to be recorded.

11.4.10 Governmental interventions

A total of 29 intervention results were obtained in this category. Two concerned initiatives of the Finnish (in Guastello, 1993) and Australian (Gun, 1993) governments, which showed a net positive effects. Most of the reports pertained to some aspect of OSHA. They included the onset of OSHA in 1970 (Butler, 1994), which actually met with a slight increase in nationwide accident rates, inspections for compliance with regulations (DiPietro, 1976; Mendeloff, 1979; Robertson & Keeve, 1983; Smith, 1976, 1979), and attempts to influence accident rates by various manipulations of worker compensation costs to the employers (Bruce & Atkins, 1993; Butler, 1994; Robertson & Keeve, 1983). Specific regulations were evaluated in two cases. A law requiring rollover protection equipment on farm vehicles was particularly effective (Zwerling et al., 1997); because this intervention involved a specific technological requirement, the results ($D = 92.0$) are classified with technological interventions. OSHA's Universal Precautions for preventing needlestick and sharps injuries in hospitals (and other exposures to blood-borne pathogens) had no effect on reportable needlesticks, sharps, or other injuries (Guastello et al., 1999).

OSHA's purview has historically paid little attention to human factors issues other than establishing allowable limits to noise and heat. Most of the inspections focus on structural issues such as proper electrical service for machine, the presence of machine guards in all the right places, and the availability of proper personal protection equipment. OSHA has expanded its purview into ergonomics-related consultations and inspections (OSHA, 2002) on an advisory rather than an enforcement basis in 2001.

Two further additions to the category resulted from legislation designed to prevent fatal falls in construction (Derr, Forst, Chen, & Conroy, 2001), and fatalities in the excavation phase of construction (Suruda, Whitaker, Bloswick, Philips, & Sasek, 2002). The rate of fatalities from falls dropped 3.5%, and the rate of fatalities in excavation dropped 50.0%.

11.5 EMERGENCY RESPONSE

The mainstay of what we know about emergency response comes from public situations that have experienced earthquakes or other natural disasters. The essential aspects of emergency management are considered next with the understanding that organizations experience internal emergencies also, and that emergency response is usually part of someone's job and one that has a great deal of hazardous exposure.

11.5.1 Hazard perception

In an occupational setting, one of the important differences between experts and novices is that experts know how to recognize signs of a hazardous condition, and novices often attend to all the wrong cues. When the tsunami struck Southeast Asia in late 2004, the unusual behavior of the water, as seen from the shore, gathered a lot of attention, but too many people had no clue that the receding water would be followed by a huge rush of water over the shoreline very soon afterwards. Figures 11.12–11.14 are photographs taken by an anonymous photographer, which were circulated through the internet without further credit. Some important group dynamics were taking place in the photos. In Figure 11.12, the crowd is watching. In Figure 11.13, some people are backing away, while other are left behind. In Figure 11.14 most of the crowd is running away, but some people insisted on staying and watching what would happen next; they found out. One can only wonder about the effectiveness of the umbrella.

Although it was a post hoc interpretation of events, it was possible to interpret them nonetheless as a result of a cusp catastrophe model (Guastello et al., 2008). The two stable states were "continue gawking" and "running away," and there was a discontinuous change between the two states. The asymmetry parameter was knowledge of the situation; some people had prior knowledge of what a tsunami would look like when it unraveled, while

Figure 11.12 Risk perception of the crowd before the tsunami reached the shore. (Reprinted from Guastello et al. (2008); photograph in the public domain.)

Figure 11.13 Risk perception of the crowd when the tsunami reached the shore. (Reprinted from Guastello et al. (2008); photograph in the public domain.)

Figure 11.14 Risk perception of the crowd after the tsunami reached the shore. (Reprinted from Guastello et al. (2008); photograph in the public domain.)

others were either uninformed or did not make the connection between what they saw and any public service announcements that they did receive.

The bifurcation factor is social facilitation. The very presence of a crowd affects judgment. One might respond quickly if other people were not around, but seeing how other people were responding affected the responses of the target individual. As we saw in other arousal–performance dynamics, the presence of other people is arousing also, and predisposes people to one or another response very distinctly. Thus in the presence of danger signs, groups are slower to react than individuals; the non-reaction by some delays the reaction by others.

For the more common types of emergencies, citizens with telephone access call 911 operators (112 operators in Europe). The callers are not experts and are not familiar with everything the emergency operator would need to know to initiate an effective response. Experts viewing photos of accident scenes spend more time viewing critical information points than novices (Prytz, Norén, & Jonson, 2018). The operator at the receiving end of the call would need to guide the caller, who could be under a high level of stress, through key questions.

11.5.2 Time ecologies

Emergency response systems, like other types of public policy, operate on multiple time horizons or *time ecologies* (Koehler, 1999). At the slowest time horizon, senior management is identifying and interpreting risks of an outbreak of a natural or other type of disaster. The time horizon is occupied by foresight and action planning and could extend for many years. Sociopolitical systems that fail at this level are seriously impaired when an actual disaster strikes, and their focus of attention shifts to the more immediate time horizons (Comfort, 1996; Pauchant & Mitroff, 1992; Reason, 1997).

The management of hurricane Katrina when it struck New Orleans in 2005 serves as an example (Cigler, 2007; Derthick, 2007; van Heerden, 2007). The city had been warned for many years by the Army Corp of Engineers that the levees were in danger of breaking, but no action had been taken. It was only when the hurricane was within striking distance that city officials told the public to leave town as soon as possible. There was no evacuation plan, however, and those who had their own transportation had to work through traffic congestion in the best way possible. Many people were left behind in the city, and bus transportation out of the most affected areas was uncoordinated at best. Those who were moved were taken to a convention center in the city, which became severely overpacked. The inside of the center was also uncontrolled, turning it into a pressure cooker for theft, rape, and assault, according to news reports.

To continue, the midrange time horizon initiates when the disaster actually strikes. The horizon for rescuing people from an earthquake region is about five days (Comfort, 1996). The majority of people rescued who survive are rescued within the first two days; the odds of survival given rescue decays sharply afterwards. Meanwhile, all the shock elements of unplanned physical locations, time of day, availability or impairment of medical or transportation resources, fires and explosions, and a generally fast-changing situation require instant adaptive responses.

The chaos of the situation should be taken literally, with sensitivity to initial conditions figuring prominently in how the events unfold (Farazmand, 2007; Koehler, 1995, 1996). Many, sometimes hundreds, of formal and informal organizations and citizen groups mobilize their capabilities over the short time horizon (Comfort, 1996; Morris, 1906; Morris, Morris, & Jones, 2007). Coordination among them is especially challenging; complex systems in a high level of entropy can produce surprises of their own (McDaniel & Driebe, 2005; Sellnow, Seeger, & Ulmer, 2002). The skill for managing chaos is thought to be in short supply in the general population of management personnel (Guastello et al., 2021a, 2021b). Nonetheless, Morris et al. (2007) gave high marks to the U.S. Coast Guard and U.S. Air Force for their coordinated actions in the Katrina disaster, as did Morris (1906) to the mayor of San Francisco for managing that historic earthquake.

While the activities at the midrange time horizon are getting started, events are occurring at the scale of hours and minutes. Koehler (1996) emphasized the critical and problematic nature of timing at this level of activity. For instance, one decision maker can ascertain that a hospital emergency room has a certain amount of carrying capacity at a particular moment, and then dispatch some casualties to that hospital. By the time the casualties arrive, other decision makers had the same idea and dispatched more casualties to the same location, producing a bottleneck. Other critical events are connected to the discovery of new casualties or the prevention of concomitant disasters, such as fires in the wake of an earthquake, or the change in the path of a forest fire caused by a sudden shift in the winds. Human communication and the physical movement of people and equipment are not always fast enough to compensate. The psychological representation of time by disaster respondents and victims is strongly constricted to the demands of the present moment. The ability to see the future, even in the short horizon of a disaster response, is greatly impaired.

11.5.3 Situation awareness and sensemaking

Situation awareness is usually regarded as a *process* of perception and interpretation of events that can be assisted by technology, rather than a particular outcome (Durso & Sethumadhavan, 2008; Endsley, Bolté, & Jones, 2003; Wickens, 2008b), as discussed in Chapter 7. Communication technologies are vital to an effective ER. For instance, satellite photography can provide ground personnel with information about which areas are affected to what extent and which escape routes are operable and which ones are not. Cellular phones can keep emergency response personnel closely connected—until the batteries die.

Sensemaking (Weick, 2005) places joint emphasis on the process of gathering relevant information and the cognitive integration process that occurs shortly afterwards. Expectations that are only based on known information can produce some automatic actions that misfire if the interpretation of the situation is wrong. Preparedness for the unknown, surprising, or emergent events could produce more advantageous results. The Center for Disease Control's correct diagnosis of West Nile Virus was only obtained after the Center became aware of lab tests that did not fit the original hypothesis and new information about West Nile Virus that was not previously on record. The arrival at the correct diagnosis was facilitated by coordinated communications among the responding agents. In fact, coordinated communication efforts are critical to sensemaking in dynamic situations (Baber & McMaster, 2016). Information can become siloed in governmental administrations so that one agency or task force is unaware that they collected or stumbled upon pieces of information that could be critical to another unit's operation.

The Center for Disease Control's experience raised the issue of how best to prepare for an emergent disease epidemic or bioterrorist attack. One does not prepare for the new disease exactly. Rather one prepares a reasonable *strategy* for finding out what it is and formulating a coordinated response with relevant agents. In the case of the COVID-19 pandemic, the coordination efforts extended to the World Health Organization. Meanwhile, widespread distributions of vaccinations were made possible in about a year after the announcement of the pandemic because scientists had already gained some knowledge of viruses that shared some similarities with COVID; testing for effectiveness and safety took a substantial amount of time.

ER decisions are also dynamic in nature: The decision made at one moment affects the options available soon afterwards. Dynamic decisions made by teams can be studied using microworld simulation software (Cooke & Gorman, 2013), which can also be used as part of a training program. There are other group dynamics to consider in conjunction with emergency response, notably group size and coordination, with are considered in Chapter 14 along with other practical implications of complex systems.

The principles of resilience and adaptive responses introduced earlier in this chapter are highly relevant as well. An effective emergency response requires an interplay between well-rehearsed scenarios and responses and recognizing when the original plan does not cover what is happening at the moment. For instance, in the first responses to an earthquake or massive flood, the first steps are to shut off the gas lines to prevent fires, shut off electricity from the power plant feeding into damaged transformers, and limit telephone communications in and out of the affected area. The next steps are to identify and interrupt domino sequences and launch rescue teams. Other task forces are then simultaneously working on restoring or importing essential services. The chaotic elements of the disaster present a challenge to RPD, but a wide repertoire of the latter is essential. For this reason, wise municipalities in high-risk areas stage practice drills.

DISCUSSION QUESTIONS

1. Consider a situation where someone dove into shallow water and suffered a spinal cord injury as a result. Construct a chain of events leading up to this incident. Can this same situation be conceptualized as a factorial model? How about a fault tree?
2. Consider the Claptham Junction train wreck. Do Petri nets convey anything differently from a standard fault tree analysis? Is one method more informative than the other in this situation?
3. Retrace the major disasters at the Three Mile Island nuclear facility or the chemical factory in Bhopal. What errors occurred at the human–machine interface? Was stress a factor in any known way? Did management contribute to the problem in any way? Hint: It would be necessary to look up key documentation on these incidents. Reason (1997) recommended Kemeny (1979) with regard to Three Mile Island, and Meshkati (1989) with regard to Bhopal.

4. In January 1986, the space shuttle Challenger exploded shortly after liftoff, killing all persons on board. The technological culprit was a failure of O-rings, which were simple sealing devices in hydraulic equipment. Was that the only safety factor that was involved?

5. In February 2003, the space shuttle Columbia exploded when it reentered Earth's atmosphere after a 16-day mission. NASA and other investigators are trying to figure out why the incident occurred. What can you uncover about the accident in the categories of technology or engineering, operator error, stress, or management error? It may be desirable to revisit this question again after reading the final chapter on human factors in outer space.

6. The current culture of technologists often takes the position that the vast majority of accidents or other unwanted events are the result of human error, and therefore new forms of automation should be developed to minimize human input. Is this a wise perspective?

Chapter 12

Human–computer interaction

12.1 THE CHANGING NATURE OF THE INTERFACE

Developments in computer technology have changed the nature and scope of the human–machine interface itself. As usual, HFE efforts lag behind the development of products, the technology changes before HFE research has run its course on a generation of equipment, often rendering the HFE research results pointless. For this reason, Newell and Card (1985) encouraged human factors specialists to become involved in the early stages of the design processes much more often than they had been before.

In the conventional machines that were discussed to this point, the control and the display are two separate entities, even though they are often juxtaposed on a control-display panel for optimal use. In the earlier phases of the computer's commercial existence, computers were heavy on controls and control sequences, light on displays and display intelligibility, and abominable for contemporaneous juxtaposition. When the VDT or monitor became the primary display medium, it became a large part of the control medium as well. Think about it. We can move a mouse that moves a pointer—a virtual finger—over a picture of a button. The button activates when we click the mouse. Touch-sensitive screens allow users to skip the mouse and poke a finger at the display screen.

The human–computer interface has changed so substantially since its inception that the changes formed the proverbial big circle. The first interface was actually an entire room that had portions of the computer equipment lined up against the walls that surrounded the human. Built in 1946 as the first prototype of an automatic computational machine, it surrounded the team of people who were operating it (Heppenheimer, 1990). That was a lot of hardware for a device with a memory of 1 kilobyte!

The operator set myriad controls on the Electronic Numerical Integrator and Computer (ENIAC) every time a new task was required. The concept of programming had not been introduced yet. The operator consulted an abacus or a sliderule during the control process; the handheld calculator did

DOI: 10.1201/9781003359128-12

not exist either. By the 1970s, the handheld calculators that became widely available had computational memories larger than ENIAC.

In the post-ENIAC era, the interface eventually became a desktop device of one sort or another. By the 1990s, the entire computer and display device had become small enough to carry in a briefcase. Some computer devices can be worn in a shirt pocket—wearable computers. Cell phones are ubiquitous examples that combine the telephone, e-mail access, and access to a wide variety of computer programs and popular websites. Biomedical engineering continues to develop implantable devices such as heart pacemakers, neuro-regulators that can predict and prevent epileptic seizures, and prosthetic devices that can operate from the host's intention as we would move our natural limbs (Nathan, Guastello, Prost, & Jeutter, 2012).

Identification chips can be inserted into pets containing the pet owner's identification and the name and location of the veterinarian who installed it. The goal is to be able to identify the dog and its owner if the dog were lost; most people who find a lost dog dial the phone number on the dog's human-readable ID tag. Efforts have been made to make chips that can be used to actually locate the dog, but they seem to need an energy source that has not been designed yet. Students in some school districts are required to use ID tags that have an identifier chip. One school district in Texas was requiring high school students to use them so the district can see who is present in school and who is not, crack down on truancy, and improve their state funding levels, which are predicated in part on student attendance levels (Huffington Post, 2012). The daily tabulations would interface with a telephone call list that counselors would use to ascertain the whereabouts of missing students. Students who do not comply with the ID program would not be allowed access to common areas such as the cafeteria or the library.

Yet other combinations of technologies can once again surround the operator—VR chambers—although with very different purposes and results compared to ENIAC. Another way of surrounding the operator is with the use of sensor textiles. Gloves have been designed that contain electrodermal sensors in the fingertips to detect physiological arousal (Valenza, Lanata, Scilingo, & De Rossi, 2010). Gloves are more portable and versatile than the exposed sensors that have been used historically in electrodermal response (also known as GSR, see Chapter 10) studies. Valenza et al. have, so far, been able to isolate four NDS patterns of electrodermal response that supposedly correspond to personality patterns; a generalizable theory about those relationships is not yet available, however. The sensor gloves were also characterized as a step toward smart textiles that could read biometric information from the wearer. The possible uses of biometric clothing and sensor gloves are not yet clear, nor is the impact of washing machines or dry cleaning on the lifespan of the product.

12.2 CONTROLS

12.2.1 Keyboards

The QWERTY keyboard was initially designed for use with the first typewriters in the late 19th century. The first typewriters were mechanical devices that bore a strong similarity to the mechanism of a piano. Pressing a key would activate levers. At the end of the last lever was a die that was cast with two characters that we know as uppercase and lowercase (capital and small letters in most cases). The die would strike a cloth ribbon that was loaded with ink. The ribbon was positioned just in front of the paper on which the print was going. When the die hit the ribbon, ink from the other side of the ribbon would impress on the paper. The paper was manually fed into the carriage portion of the typewriter. The carriage was fitted with grippers for the paper, adjustments for single and double spacing, margins, and tabs. The shift key on the keyboard would connect to the carriage and raise it up so that the hammer with the die would strike the uppercase letter, rather than the lowercase letter.

The origins of the QWERTY keyboard are not historically clear in the sense that there were any human factors experiments to rely on, but some HFE sensibilities were apparently involved. In one saga, the concept of the typing phenomenon would involve the use of all eight fingers and the thumbs on the space bar. In another saga, all the letters of the word *typewriter* appeared on the top row of letters as an aid to potential salesmen who wanted to show off the device without actually having any typing skill. The third saga seems to be the one that is the most strongly corroborated: The QWERTY design was actually intended to *slow down* the typist so that the letters would not be typed faster than the capacity of the report action of the mechanism. If the typist typed faster than what the machine could handle, the keys would jam and a mess might appear on the typing page.

Typewriter mechanisms eventually improved for faster report times. A prototype keyboard was introduced around 1915 in which the letters were rearranged to maximize left-right alternations between the hands, thus allowing faster typing, especially for two-fingered typists. The new keyboard arrangement did not become popular. The user population had already become accustomed to the QWERTY arrangement, and relearning time was prohibitive. Thus, the preference for QWERTY persists today even though the lever mechanisms behind the keyboard have disappeared.

When electric typewriters were introduced in the 1950s, they replaced part of the lever mechanism with an electrical contact. The intended result was to reduce the physical forces needed from the typist, thus improving typing speed, and to deliver an even impression on the paper from each key-stroke. Anyone who bought got a surprise, however. The keys on the first models had insufficient resistance to them, and the keys would be depressed at undesired moments. If a typist lingered on a key a fraction of a second too long, or hit a key a little too hard, multiple *keeeeeeeeeeeeeeystrokes* would appear suddenly. The remedy was to reintroduce a little more resistance to the keys and to give additional resistance, if not to stop altogether multiple automatic keystrokes. The exceptions were glyphs that the user might actually want to strike repeatedly such as *X*, periods, hyphens, and underscores.

Sometimes, designers forget history. A colleague bought a 2010 model of a popular brand of laptop computer. He was well accustomed to laptops. Imagine his surprise and frustration when he got the *keeeeeeeeeeeeeeystroke* effect.

Error correction on a typewriter was simple for decades: Do not make them, or you will hate yourself afterward. Corrections were made with erasers. The skill was to erase the unwanted characters without ripping a hole in the paper, and to retype the new characters in exactly the same position in which the originals had been placed. The paper feed and adjustment mechanism—and the person using it—had to be well controlled, or else the corrected text would not fit into the erased space exactly as intended. Correction fluid and correction tape were eventually invented, and they were very popular.

The next level of evolution for the typewriter took the form of the IBM Selectric®, which was widely adopted for use in commercial offices in the 1970s. The Selectric was noteworthy for several features (besides weighing a proverbial ton). A revolving sphere that contained all the type characters replaced the hammer-and-die mechanism. Spheres were interchangeable, and the common preferences were regular versus italic type, larger versus smaller size, and math symbols. In other words, the office typist was now able to change fonts. The Selectric was also the platform for building the first word processors; we return to word processing later on in this chapter.

Ergonomic keyboards were first tested in the 1920s, then forgotten, revisited in the 1960s, then revisited again in the 1980s when the reports of carpal tunnel syndrome and other musculoskeletal injuries were accumulating among people who spent the better part of their workdays entering data (Rempel, 2008). The concept is that hands, which are attached to arms and shoulders, approach a keyboard at a 25-degree angle. Thus, one feature of the ergonomic keyboard was to split the keyboard accordingly as shown in Figure 12.1a and 12.1b. The design in Figure 12.1a included a rest pad for the palms of the hands and some elevation in the back of the keyboard. The design in Figure 12.1b splits the keyboard with some control over the separation angle and more control over the vertical angle. As a result, the wrist pad could not be attached, but it is included as a part that can be freely positioned on one's desk.

Some performance studies for the earliest designs showed fewer typing errors with the ergonomic keyboard compared to the normal straight keyboard, but typists remained more comfortable with the straight keyboard

Figure 12.1 Keyboard designs: (a) and (b) are ergonomic designs, (c) is a side view of keyboard supplied with a desktop computer, and (d) is the keyboard portion of a laptop design.

for many years, and many still are. Performance studies on the newer series were showing that performance was actually better on a standard keyboard, but the gap in performance disappears after an hour. Ratings of pain and discomfort with prolonged use favor the ergonomic design, however. Ergonomic keyboards are seldom included as standard equipment at the time of purchase, but they can be purchased separately.

The keyboard in Figure 12.1c is a side view of a standard desktop keyboard. It has two increments of elevation that are manually operated, the keyboard is not split, and the palms rest on the desk in front of the front edge so that the thumbs naturally fall on the space bar. The keyboard in Figure 12.1d is a standard laptop configuration that has no rear elevation. It is expected that the two hands approach at a 25-degree angle, but the only place to rest the palms is on the keyboard surface to the left and right of the position pad. Some thumbs might align with the space bar, but for other people, the thumbs drop further down around the position pad and click-buttons, causing unwanted clicks. This could be another situation where the protection against the accident operation of controls could be improved.

12.2.2 Keypunch machines

The QWERTY keyboard, now augmented with specialized controls, was harnessed for computer operation during the 1950s in the form of a keypunch machine. Computer programming and operation at that time usually did not allow the operator to give commands to the computer directly.

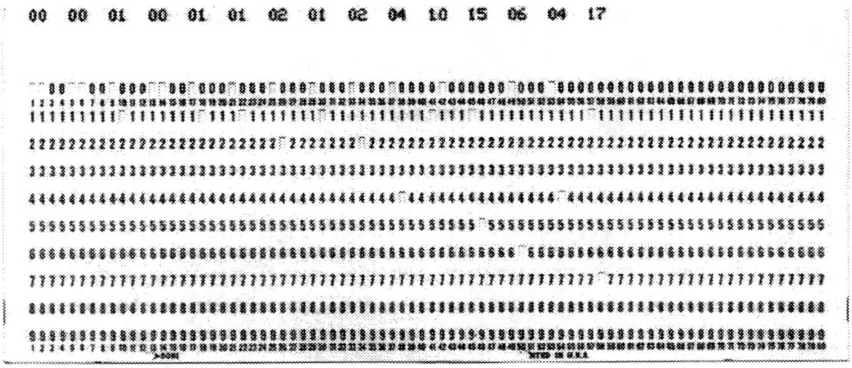

Figure 12.2 Computer punch card.

Instead, the operator had to prepare a stack of punched cards to execute the programs. The cards were prepared with a keypunch machine.

An example of a punch card appears in Figure 12.2. Each card had the capacity for 80 keystrokes, and represented a short line of type comparable to what appears on our computer screens today. The operator loaded a set of new cards into a bin that was located on one side of the machine. The feeder mechanism moved a card into a position where typing and punching would occur. The typing operation would result in a set of holes in the card that would eventually allow electrons to pass or not pass through each particular location. This binary code corresponded to what was printed in human-readable characters at the top of the card. When the card was finished, the operator pushed the release button, and the finished card would move to the collection bin.

We should pause to reflect on what was taking place. Typing errors occurred often, but the only way to correct them was to eject the card and catch it before it reached the collection bin, or to mark it for later removal. As the keypunch machine aged, the mechanism would go out of alignment, and some of the cards would be punched a little off target. It was not always possible to detect an off-punch card until the batch had been executed.

Once a set of cards was finished, the operator took the stack of cards to a device known as a *card reader*. The top card identified the user and the start of a new job. If the card readers were located in inconvenient and cramped spaces in the computer facility, someone could bump into someone else who was holding a stack of 500 cards. The cards would go flying and needed to be reassembled in the correct order.

Was the deck of cards reassembled correctly? Were the error cards removed? Were there any lingering errors on any of them? Were any off punch? The users would know the answer to these questions in about 45 minutes. A typo that we would discover in less than a minute today, and one that could be fixed in a few seconds, would require as much as 45 minutes to reach the user in the form of an error notice on printed output. The reply time could be as short as 15 minutes during times of low computer traffic volume. Thus, a culture of people evolved who did their work at 3:00 a.m. in order to work more and wait less.

12.2.3 Numeric keypads

The computer keyboard has a numeric keypad sitting on the right. The numeric keypad was introduced during the 1950s, and it is strikingly similar to the keypad on a touch-tone telephone. The human factors experiment behind its design (Deininger, 1960) was described in Chapter 2.

Notice that the arrangement of numbers on the numeric keypad places the 1-2-3 row at the bottom whereas it appears at the top on a telephone (Figure 2.3). During the early days of computer programming when the numeric keypad was first introduced, a large amount of typing consisted of series of 1s and 0s. If the 1 and the 0 were placed close together, it was easier for the keypuncher to twiddle two keys. The widespread use of touch-tone phones occurred many years later as binary typing became less common; thus, confusion of population stereotypes was not a problem.

12.2.4 Membranes

Membrane keyboards were first introduced in the early 1980s. They were piloted for inexpensive home computers and several commercial applications. The first QWERTY membrane keyboards were flat surfaces. One only had to tap the spaces that indicated the letters, and thus, minimal physical forces were needed. The attempt to improve speed, if not also to make an inexpensive device, backfired as users made more errors because they were not used to the lack of tactile feedback that lets them know when they have hit a key properly. Manufacturers then responded by making ridges in the plastic membrane to signify the location and space associated with a key. The additional feedback seemed to help, but not enough to replace the button-type keyboards.

Membrane keyboards found a few homes in applications where keypads were needed, but the kind of typing that lends itself to the QWERTY configuration did not make the transition to membranes very well. Figure 12.3 is a membrane keyboard from an electrical power plant operation that was installed around 1990. Note the absence of a QWERTY pad. The key panel to the right is an alphabet pad, but it appears that only single letters were needed to input data. The next generation of power plant equipment

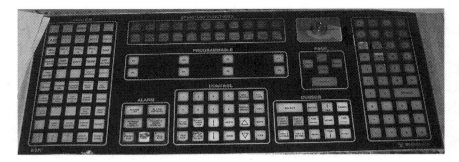

Figure 12.3 Membrane keyboard, 1990 vintage.

from the 2002 era reverted to a standard desktop-style keyboard with a QWERTY configuration.

Membrane keyboards are easier to clean than the standard variety. Thus, membranes have survived on the interface of the cash registers in fast-food restaurants where the environment has cooking oils in the air. In the same application, however, the touch of one button for *cheeseburger* is all that is necessary to sell a cheeseburger; the price is already programmed into the machine. Menus and prices need to be updated periodically, however. Thus, the relative merits of the membrane have to be understood in the context of the control operations that are required.

12.2.5 Positioning devices

The mouse, the four cursor arrows on the keyboard, trackballs, and joysticks are perhaps the four most commonly occurring positioning devices. As for which was the best device, the answer is, once again, to pick the right tool for the right job; a situationally specific study may be needed. All other things being equal, Fitts' law seems to apply well to the response times associated with positioning movements, and the mouse is fastest for most tasks (Card, English, & Burr, 1978; Landauer, 1987).

To complicate matters a bit, the motor skills of the user can affect the speed and accuracy of the use of any particular device. Adults who learn to use the mouse for the first time in their lives (there are not too many of them anymore) experience some lack of control or jitter in their mouse placement; they might prefer the predictability and simplicity of pushing the buttons that correspond to the arrow keys. For the accomplished users, however, the mouse is usually the preferred device, and may be the only way to activate icon controls on a program. The arrows, on the other hand, are good for moving the cursor a small number of discrete steps in the four orthogonal directions.

A quick reminder of Fitts' law would be helpful here. Movement time is a function of the size of the target, larger is better, and the distance traveled to reach the target. For mouse and trackball operations, the trajectories all fall in the same 2-D plane of the computer screen. For touchscreen interfaces (see below), the movement is more similar to the original context of reaching a button on a control panel from by someone who is positioned outside the plane of the control panel.

A trackball is essentially a mouse that has been built upside down. It sits in one place on the desktop, and the users roll their fingers over the ball to move the cursor on the screen to the desired point. The trackball would have an enter button that corresponds to the left-click on the contemporary mouse. The trackball was introduced as an alternative to a mouse for a desktop computer that was placed in a small space where there might not be enough space to accommodate a mouse (Shneiderman, 1987). The trackball requires fine motor control over the fingers as opposed to gross motor movement of the arm. The reach for the enter button was probably more convenient for the user with the mouse as the left-click button is located under the index finger. As additional uses for a right-click button were introduced, the mouse gained superiority because its other control is conveniently located under another finger, and the operators can maintain their grasp while operating the buttons.

On the other hand, the trackball design may have found a home on most laptop computers. There is no mouse because there is no desktop to roll it on. Instead, there is a little red nubbin tucked in the middle of the keyboard with enter buttons to the left and right of the space bar. The nubbin was replaced by a touch-sensitive pad (e.g., Figure 12.1d) with two buttons underneath corresponding to the left-click and right-click control movements. Wireless mice became available, nonetheless, for people who still preferred them. Users' preference can probably be gauged in terms of the numbers of laptops and desktop computers sold every year. There is obviously a trade-off taking place among the conveniences of one feature over another.

The joystick originated in equipment where a 2-D controller was needed, such as an airplane or a stick shift in an automobile. Joysticks also showed up in arcade games and remote control toys where the control of the object in a desired direction was not complicated by a positioning device on a computer screen: Move the joystick, see the results, and click the button on top to enlarge something. See Figure 8.14 for an example of a positioning device for a remote control security camera.

12.2.6 Touchscreens

Touchscreens bring the simplicity of a single touch to the display, rather than the keyboard. Here we have an example of the display being the control panel simultaneously. In the example shown in Figure 12.4, the user only

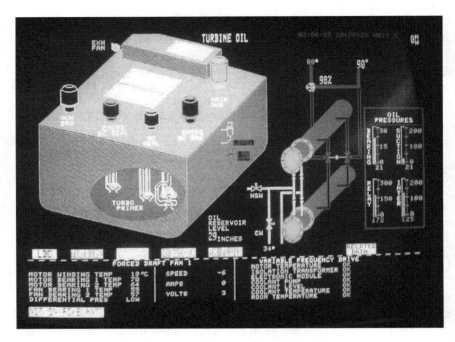

Figure 12.4 Touchscreen control with graphic user interface, 1990 vintage.

needs to touch an item that is depicted on the screen, and the screen changes to produce a new image of what is inside the original small target space.

Although some touchscreen applications have endured since they were introduced in 1990s, others have been replaced by more conventional keyboard and mouse configurations. The touchscreen seems to have reached its limit in safety-critical environments when it induced too many chances for error by hitting the wrong target or the amount of pressure on the screen not being sufficient to produce a response as soon as it was needed. Another limitation is that the amount of time the operators had to stand or sit by the screen with their hands and arms in the air ready to poke a target produced fatigue that was not generated by keyboard and mouse controls.

12.2.7 Styli

A stylus has the look and feel of a pen. It operates when it is touched to the screen, but it allows for more control precision. It was a device of choice in the graphics industry years ago because it capitalized on the artists' control precision with a handheld drawing tool, yet allowed for data entry. Fingertip touchscreens, on the other hand, appear to produce their best

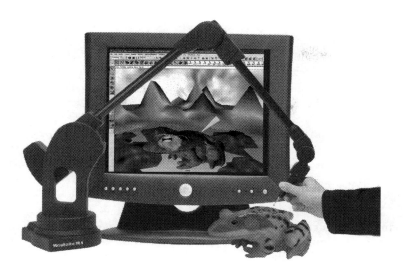

Figure 12.5 Microscribe® Digitizer used to encode a 3-D object. (Photograph copyright 2005 Immersion Corporation. All rights reserved. Reprinted with permission.)

results when the targets are large and few in number. The equipment shown in Figure 12.4 was eventually replaced by a mouse-driven system.

Figure 12.5 illustrates the use of a stylus for 3-D data entry. Here the operator wants to encode the shape of a model frog into an animation or a VR program. The frog will be used as an object within the program thereafter.

12.2.8 Gestural interfaces

A gestural interface allows the operator to command a machine with a wave of a hand. The first examples of a gestural interface were introduced in the 1920s with a musical instrument, the *theremin*, which was named after its Russian inventor Leon Theremin (Pinch & Trocco, 2002). The theremin was a box-shaped device with metal plates on its upper surface. The metal plates emitted an electronic beam above the plates. The motion of the hand interrupted the beam, and the interruptions were translated into sound. One hand controlled pitch, and the other controlled volume. The theremin did not gain a great deal of popularity when it was first invented, but it was featured in the soundtracks of some B-grade horror movies in the late 1940s and early 1950s. It experienced resurgence in the mid-to-late 1960s when it was discovered by some musical ensembles specializing in the psychedelic genre with electronic sounds (most notably Ultimate Spinach, Lothar and the Hand People, and the Beach Boys' hit "Good Vibrations").

Although the unique fluid sound of the theremin was very attractive to experimental musicians and virtuosos did exist, the overall lack of tonal control was a serious limitation for professional musicians who wanted to adopt the new sound relatively quickly. Most theremins sold during the 1960s were manufactured by Robert Moog, an expert performer himself, who eventually combined the theremin's electronics with new controllers and a piano-type keyboard to make the famed Moog Synthesizer.

One motivation for developing the gestural interface in the late 1980s was to assist people who were working in challenging environments, such as Mars, so they could operate their equipment without having to remove the glove portion of their environmental suits. They would not have to fiddle with the complication of low gravity when turning knobs or throwing switches.

A more viable application of the gestural interface concept is in VR systems. The system requires the use of a glove that contains a system of sensors, as in Figure 12.6. Moving one's hand in the air will interact with a virtual object, which the computer will interpret according to the laws of

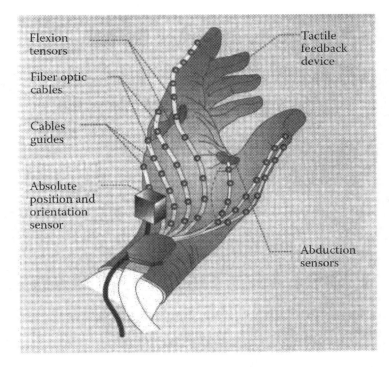

Figure 12.6 Data glove for VR applications. (Copyright 1996 by Walter Greenleaf, Greenleaf Medical. Reprinted with permission.)

physics—or not as the designers prefer (Greenleaf, 1996). The basic limitation of the gestural interface is that the hand motions must be executed with stereotypic precision. The motions for a specific command must be executed exactly enough so that the machine recognizes the command and distinguishes it from other gestural commands.

The gestural interface and the touchscreen have been combined in the operation of the current generation of smart phones. The most common motions are the flip-up, down, left and right to scroll, and squeeze the thumb and index finger inward or outward to zoom in or out. The fingers maintain contact with the screen thus circumventing the need for a sensory glove and reducing the degrees of freedom in the motion that the device has to read. The stereotypic smartphone motions seem to have been learned easily. There have been informal reports, however, of users accidentally activating a program or deleting something while making the scroll motions. Designers might want to consider how the newer devices could be requiring comparable advances in the protection of controls from accidental operation.

Another application for gestural interfaces is to manipulate very large, wall-sized displays. The operator makes a stereotypic motion to capture a section of the image and move it from one place to another. The real-world applicability is somewhat limited as is the repertoire of motions that can operate such a system. The 3-D mouse discussed later in the chapter appears more versatile for the manipulation of large images.

12.2.9 Mobile devices

The laptop computer and smartphone are two of the most commonly used mobile devices. There are others, however, such as devices used by paramedics and those in specific industries where, on the one hand, observation and recording of information from the work context is a substantial part of the operator's job. So far, any ideas in this chapter concerning the usability of any of these devices have been confined to control operations in a relatively stationary context. Mobile devices, by definition, are meant to be used as the operator moves around, however. Thus, usability testing for the controls, displays, and program layouts needs to take context into consideration.

One of the biggest advantages of the mobile devices is also one of its biggest limitations: They are small. How many controls can be reasonably condensed into a small space? One rule of thumb is to organize the screens or control panels to accommodate the 95% percentile male finger. The frequency with which this rule is actually observed in mobile designs is an open question.

Baber (2008) took the position that virtually every task performed with a mobile device is a dual-task problem. The operator is walking, paying attention to other things and people, and possibly manipulating other devices simultaneously. Thus, the designers need to visualize all the contexts in which the device might be used as well as the features and operation of the device. How does the effectiveness of the operation compare at a quiet desk with office lighting to a train station with noise, varying lighting, and other people present? How does the operation of the device affect the other activity? For instance, people walking on treadmills walk 30% slower while using a personal data assistant than walking without the device. Are the pedestrians on a busy street as attentive to traffic as they might be otherwise?

Individual differences among users can be expected, some of which could be extreme. For instance, a colleague reported that he acquired an interest in cell phone use while walking and its effect on gait when he saw a young woman walk into a public fountain while using a cell phone.

12.2.10 Gaze control

Chapter 5 explains the uses of eye-tracking for determining the patterns of attention to portions of a visual display. Researchers are now experimenting with eye-tracking, which can be used as a form of control. In close-space interaction, gaze could be used as a substitute for a mouse and keyboard for people with motor impairments (Koesling, Zoellner, Sichelschmidt, & Ritter, 2009), as part of a brain-control interface apparatus (Pasqualotto et al., 2015), or as part of VR apparatus used in conjunction with a gestural interface (Deng et al., 2017). In principle, the technology could be adapted to longer distances of operation where the person or robot is too far away to reach by pushing a button or a touchscreen. The eye-tracking sensors, once it is calibrated to the individual, can pinpoint the position on the screen that corresponds to something the user wishes to control.

The limitation of gaze control, however, is that the amount of visual attention to a location does not necessarily mean that the user wants to interact with the location. The user might instead be monitoring the display without seeking any changes or solving a problem. This phenomenon has become known as the *Midas touch* problem. Some of the sources of error associated with gaze control have been localized to complexity of background imagery and background motion (Caroux et al., 2013), and overall size of the targets on the display screen (Niu et al., 2021). Targets should be larger if they fall outside the optimal range of field of vision.

An alternative control design registers gaze position but does not activate the control until the user blinks. Koesling et al. (2009) compared gaze and blink controls with research participants performing a typing task on a virtual keyboard. The blink system resulted in shorter response times and fewer errors. The subjective reaction from the participants, however, was that the gaze control *felt* like it took less effort and worked more accurately. The implications for fatigue with longer term eye-blinking have not been ascertained.

12.3 MEMORY ENHANCEMENTS

Somewhere between the controls and the displays were innovations in computer memory and programming. Note the shifts in the allocation of functions between person and machine as these innovations occurred.

12.3.1 The word processing challenge

The first dominant silicon form was the mainframe computer that worked on a timesharing system. All users would submit their jobs, and the system would allocate chunks of time to each job up to a working capacity limit. The theory of fixed human cognitive channel capacity was analogous to this principle of time sharing.

For systems that worked on punch card operation, word processing as we know it today was not a viable option. Electronic word processing required devices that were dedicated to single user tasks and allowed for a quick viewing of the results from keyboard entry. One of the first models for the office worker (as opposed to folks in the printing and publishing industry) was the MagCart® typewriter by IBM. MagCart was built from a Selectric unit, and it had a small memory device that sat on the desk and plugged into the main unit. Typists could now save, retrieve, and reuse typewritten documents. The memory mechanism recorded the information on plastic cards that contained a magnetic strip on one side—not unlike the credit card, which had already become popular in the mid-1970s. They were very expensive relative to their times, and did not gain much popularity before the advent of personal desktop computers in the early 1980s.

Type is not print. Type allocates equal spacing to each letter, number, or non-alphanumeric character. Print, on the other hand, allows variable spacing for narrow and wide characters. Different fonts often have different character spacing requirements. The word processor as we know it still had not come to life. The printing industry could purchase a transition technology in the form of the IBM Compositor® for a mere $10,000 new. The Compositor was an overgrown Selectric® that had a working memory of 8,000 characters. The operator could use print fonts, and could change the font within a document by inserting stop codes through special keys on the keyboard. As the Compositor replayed a stored document (on paper, not on a VDT), the replay would stop where indicated, and the operator would change the font. The fonts were designed as removable mechanical balls. Additional memory like those that accompanied the MagCart® typewriters could be attached.

The Compositor was only capable of font sizes up to 12 points. For larger type, one needed to use photographic enlargements or press-on type that was set by hand. The organization of pages containing graphics and text had to be composed by hand in an operation known as *keyline paste-up*. Here we encounter the original meaning of the word processing verbs, *cut* and *paste*. "Cut" really did require scissors or a razor blade, and "paste" really required glue or hot wax.

12.3.2 The desktop computer

The Apple Corporation gets most of the credits for introducing and popularizing the personal computer. According to Turkle (1984), the personal computer was motivated by the needs of folks with engineering backgrounds who loved computers and wanted their own, but who obviously could not afford a mainframe, did not have space for it in their homes if they could afford it, or could not use their employers' equipment for the tasks they had in mind. The mainframe user who was used to the norms of speed and computational power may have regarded the personal computer as a bit silly. The original varieties had no memory storage capacity. The device was operated using two floppy disks—one contained the program and the other contained the source data that the user wanted to operate on and the eventual output file. Hard drives were not part of the desktop computer until IBM introduced the 286-series computers.

Floppy disks in those days were made out of soft material and were indeed floppy; they had a capacity of 0.75 MB. They were eventually upstaged by the next generation of disk, which by then was only used for storage, had a harder plastic shell, and had a capacity of 1.25 MB.

The 286- and 386-series computers were still slow. A colleague who required massive number crunching in his work timed a job performed on a 386 with the same job performed on a mainframe to which he connected by telephone linkup. He reported that the mainframe was 100 times faster than the desktop when operated after midnight.

Substantial improvements in the user's experience occurred when Apple introduced the Macintosh circa 1988. It performed the basic word processing functions that we know today and contained some capability for fabricating graphics. It was a windows-based system, rather than a one-screen system. Microsoft Corporation introduced its version of a similar interface for IBM platform operating systems in 1993 and called it *Windows*.

Desktop computers, in spite of their early primitive nature, introduced three new concepts in human–computer interaction: the CRT display system, the concept of *user friendly*, and the metaphor in programming. By 1981, the vast majority of keypunch machines had hit the junk pile, and mainframe systems were equipped with CRT displays. The user could now see the input and the output without having to go through the print-and-wait process. A 45-minute exercise in error correction was now reduced to about 2 minutes. The time required to figure out what to fix and how to fix it is another matter entirely.

User friendly quickly acquired a lot of meanings, including but not limited to the advantages of a well-organized CRT display system. The broader use of the phrase indicated that the program and hardware support were adapted to the needs and preferences of the user, rather than requiring the user to learn and think like a computer and adapt to its limitations. In other words, any ideas that were inspired by good human factors thinking could qualify as steps toward user-friendliness.

The concept of user-friendliness eventually morphed into *usability*, which was the recognition of the role of human factors and ergonomics by designers of products of all types. The concept extended beyond human error and performance into matters of convenience versus annoyance (Reiss, 2012). The most annoyed user of anything in the present era is probably not subjected very often to the limits of the machine's or system's capabilities, or its inability to bridge the communication gap between the human and the rest of the system. The gap often occurred between the user's goals, knowledge, and experience and the deficits in the product design. Convenience matters as well as functionality, and systems are often designed around the convenience of the producer at the expense of user frustration.

One persistent gap between the human and the machine was the effort that it took for the human to understand enough of the intricacies of a computer program to operate it. Programmers came to rely on metaphors for this purpose. By adopting visual concepts that were already known to the user, the program could bridge the transition between population stereotypes from the concrete world to analogous functions in the virtual world. The *cut* and *paste* verbs are examples. Another example is the use of the desktop and file folders in current operating systems. Graphics programs use metaphors that echo physical artwork tools such as pencils, pens, paintbrushes, and paint buckets. In other graphics programs, a box has a complex set of meanings. Although some metaphors might only be meaningful to people who are entrenched in a particular line of work, the metaphor approach has been effective for conveying the operation of program features to the intended users.

12.3.3 Menu structures

In the pre-windows interfaces, menus of actions or files would be hierarchically organized. The broadest category would constitute the top-level directory. Groups of more specific operations would appear once an option from the top-level directory was selected. Tertiary and lower order groupings were possible. A limitation of these menu systems, assuming that they were logically organized, was that if the users did not reach the desired item on the first try, they had to use a control to back up to a previous menu level and try another option. Users needed to remember what they saw on the earlier screens and probably write down the pathway for later use. A significant advance in usability was to provide a toolbar with the top-level menu listings and pull-down menus; thus the user can take a quick peek at the options to see if the desired elements are present, and try several pull-down menus in rapid succession.

A strategic design question is how to organize the program elements under the first-level menu options (Landauer, 1987). The principle of semantic categories would be applicable. A sample of probable users should be given a list of functions and asked to sort them into categories. One can then compile a matrix showing the number of times each program element was categorized with each other element. The elements should show clusters that were meaningful to potential users, and the clusters should be distinctly different in meaning.

In the early days of interactive systems, the advisement was that the number of clusters should remain small, and seldom more than seven in number, in keeping with the principle of the magical number 7 ± 2 (Galitz, 1993). The idea turned out to be convenient for systems that could not load up quickly due to limited processing power, but not optimal. The act of searching through menus turned out to be more similar to scanning an array of information for a target (Norman, 2008). The range 7 ± 2 might be good for describing how many chunks of information people could keep in their short-term memories or how many categories they tended to impose on a system of elementary ideas, but scanning a menu for the correct item is a different mental operation. Menu organization can have its best impact on work speed and error rates by organizing information into meaningful categories and spatial arrangements. If a hierarchical menu is needed, search time is much faster if all levels of the hierarchy are visible at once. The practicality of this recommendation could be limited by the size of the list of options, but it is effective when it is feasible.

12.3.4 Data storage capacity

The primary advantage of a hard drive compared to its predecessor desktops is that it can store all the program files and files produced by or for the programs without reloading with every use. The storage capacity is divided between two modes, passive storage of things not in use and random access memory (RAM), which is the space needed to operate the programs. The function of RAM is equivalent to that of working memory in humans. The hard drive was essentially a recovery of the function of the mainframe computer, but without the total processing capacity, at least for another decade. The main disadvantage of the hard drive is also that it can store everything. When it crashes, everything is potentially lost. Thus, the culture of users reminds each other to back up files.

As hard drives grew in capacity, so did the programs and files they made. Preservation required media with larger capacity, thus the zip disk and zip drives appeared. The zip disk had the look and feel of a floppy disk, but was a little larger and had capacities of 100 MB and 250 MB. Zip drives were prominent for only a few years before they were replaced by burnable compact disks (CDs) with 800 MB capacity. Within just as short a period of time, computers were no longer equipped with floppy or zip drives as standard equipment; the user had to acquire peripheral devices, which plugged into a Universal Serial Bus (USB) port, to continue to use the stored material. When people started to make video products on their computers or just wanted to use their computers as video players, burnable digital video disks (DVDs) were introduced. DVDs have roughly five times the storage capacity if used for non-video files.

The limitation to the first generation of burnable CDs was that once they were burned it was not possible to add or remove files, even if there was a lot of unused storage space remaining. With rewritable CDs, it was eventually possible to add, overwrite, or delete files.

The next major advance was the jump drive (also known as the flash drive or data stick). Smaller than a pack of chewing gum, it contained all the drive mechanism and storage space necessary to plug into a USB port and store files of any format. Capacities of 2 GB are common, and much larger sizes are available today. One can add, replace, and delete files with all the convenience of the earlier floppy disks. People found that they could store all their important files on just one or two of them. The greatest asset of the jump drive was also its greatest limitation: Large amounts of information could be stored on one of those little things, and just as easily lost or chronically misplaced, so was everything on it. Would it be advisable to back up the jump drive on an ordinary computer?

The history of storage devices highlights a tension that existed between the user's needs and manufacturers who were preoccupied with introducing new products. Some capabilities were gained, lost, and regained as designs iterated. Users wanted to store data, retrieve it, organize in a way that was personally meaningful, perform tasks of growing levels of complexity with it, share the results with friends and associates, and maintain control over what they thought they owned.

12.3.5 Clouds

The industry is now encouraging individuals and businesses to use their cloud storage, which is actually a "farm" of computer servers controlled by the service provider. This innovation in computer memory seems to be making the full circle back to the early 1980s with devices that had no hard drives like smartphones and tablets (Duan, 2013). Smartphones have the convenience of being able to deliver Internet, including movies and television programs, in a handheld device. All the content is stored on a provider's website and downloaded as needed. The tablet is mostly an overgrown smartphone with a larger display that is more suitable for reading text or watching video content. Another advantage is that all the responsibility for updating and backing up files was put on someone else. The new trade-off is between convenience and privacy.

The cloud system together with the absence of a hard drive works for people whose computer usage is limited to sending brief text messages, making phone calls, or using corporation data sources that they have no intention of altering. Cloud systems for home entertainment purposes do not operate very differently from turning on broadcast television and radio. Rather than programs always being broadcast at times specified by the supplier, the user can watch whatever whenever, halt a download or stream, do something else, then go back to the show without missing anything.

The conveniences afforded by streaming video sources used to be one of the main reasons why people bought videotape machines—record the show, play it later, start and stop at will, and fast-forward over the commercial messages. Although programming the videotape machine to turn on and record when the owner was absent was troublesome for users, that did not prevent some households from accumulating three or four of them. Videotape recorders and DVD players were also good for owning a library of favorites that could be played repeatedly at will. Now all those annoyances can go away, the user can pay per view, the control of the content resides solely with the supplier, and the user does not actually own anything that could be resold.

Cloud facilities have become attractive to some users for storing and backing up files. The downside to some, however, is that suddenly some other agents have access to all of one's material, private information, intellectual properties, and so on. Some people value their privacy more than others, and some people hedge their uses of cloud systems by maintaining some control over some of their material and not uploading everything. Some individuals and businesses would not use them at all.

The cloud systems are actually making two full circles. Some tales from the trenches would be useful here. In the early 1980s, Yours Truly visited a major transportation company who, in the course of things, expressed their excitement about their new computer system that stored all their shipment information, much of which was geographic in a satellite system. Actually, this was not much different from communications satellites that facilitated long-distance telephone, emergency response, and so forth. The access to the database was done through a telephone-based system that was equipped with accelerators to accommodate all the data flowing back and forth. "Do you have the information backed up in case the satellite goes out?" "No need, the technology is very sound." Uh-huh. During that time period, one military and political concern was for building the "Space-based Defense Initiative," which addressed in part killer satellites that would cuddle up to an enemy's satellite and explode itself, or perhaps launch an anti-satellite attack from earth.

I never heard of anything bad happening to the transportation company's system (good thing), but not everyone in the early 2000s was so fortunate. The setting was a university clinic for psychological services that used digital video recordings of patient–therapist sessions for training purposes. In some systems for training, the sessions are saved on DVD and do not leave the premises in any way. In this one particular case, however, the clinic adopted a system of storing the recordings on a cloud server. There was a huge storm and the server went out. The clinic lost access to its files for three weeks.

12.4 VISUAL DISPLAYS

As with conventional machines, the first major design questions were to decide what should be displayed. Questions about how logically followed after a sufficient time lag. Here we capture the broad strokes of evolution from the first error messages to the GUI.

12.4.1 Error messages

In the days of the early computer languages such as FORTRAN, the first step that the computer would take in processing the program was to run the program through a compiler program. The compiler would determine whether the grammar, syntax, and specification of commands (and spelling) were correct. If an error were detected, the output would state the program line number containing an error. During those times, the programmers were mostly from a population of experts. Thus, a small hint concerning an error would be enough to alert the programmer as to what needed correction. Novices and students could become terribly frustrated, however, and they would lose a great deal of time trying to determine what the error message implied.

Eventually, the normative error documentation produced more straightforward explanations of what was wrong. Better error messages would state some possible corrective actions that the user (no longer a programmer) could try. Here we saw the essence of the smart display. Instead of simply reporting what offended the compiler, the display framed the information in terms of what the user wanted to know or do. This technique eventually carried over to displays that might appear on computer-based or conventional machines: The displays would frame their information in terms of what the user should do and not simply report the system status.

Sometimes, what you see really is what you get from a computer program. The disappointing results of a control operation might not need any explanation. For these moments, the *undo* function solves the problem. This is yet another innovation made possible by the concept of short-term memory.

The computer can hold the last version of the product, for example, a photograph that is going through an enhancement process, and revert to the original version when *undo* is clicked. It can also permit a *redo* command. Undo functions should be easily accessible in almost any type of program.

12.4.2 Screen organization

A user-friendly organization for a single screen of information draws on the best wisdom of conventional machine control and display panels and that of good text layout. Elements should be organized by function, sequence of use, frequency of use, and whether the elements are used together to perform an operation. The relative size of elements matters. Elements should be large enough to be seen from the normal viewing distance and to form an easy target for a pointing device if a control operation is associated with the element. On the other hand, the operator should not have to scroll up and down a screen to any great degree to complete the work associated with the screen. There is obviously a balance that needs to be struck with regard to the size of elements and the choice of elements to put on a particular screen.

Elements that are used less often for particular circumstances can be aggregated to a second screen or a pop-up window that can be opened and closed as desired. The logic for organizing a multiscreen system is not especially different from the logic for organizing a menu of options that was sketched previously.

The combination of computer technologies over the years has been successful in reducing the morass of control and displays for a power plant control room, such as the mid-1970s vintage example in Figure 8.19, to a set of six monitors plus some ancillary items in the 2003 system in Figure 12.7, left panel (only four of the six monitors are shown). One of the displays shows the alarm status of sensors throughout the system. If one of the line entries is interesting, the operator only needs to point and click on the line and a detail of the sensor's history pops up in a window as shown in the right panel of Figure 12.7. Note that the information can be presented graphically in a line chart or in a table of numeric values. The graphic and tabular representations obviate the need for the types of historic displays that were used in decades past.

In spite of all good intentions, designers sometimes miss something important such as relative demand. Figure 12.8, left panel, is showing a critical system stability parameter. The operators can only see what is going on when they get close to the screen and if they know what to look for. The big secret is shown in the right panel of Figure 12.8, which is a photographic enlargement of a portion of the main screen. The operators do not want the dot to cross the left curve (which is red in the original). Hopefully, some other element of the system is calculating the proximity of the point to the curve and will display an alarm condition when the point crosses the curve.

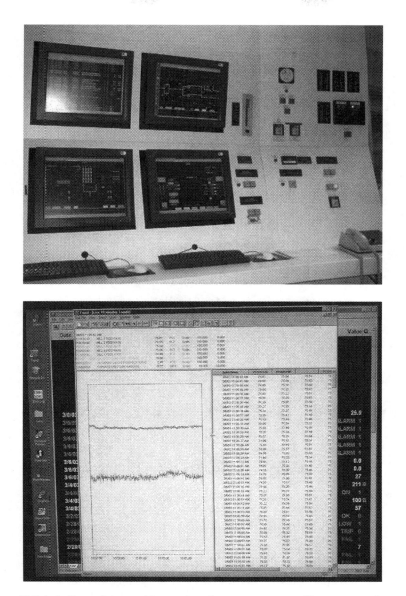

Figure 12.7 Left: Four of six graphic user interface screens controlling a power plant with a few ancillary controls and displays, 2003 vintage. Right: Screen 5 allows for clicking and expansion into a historic display.

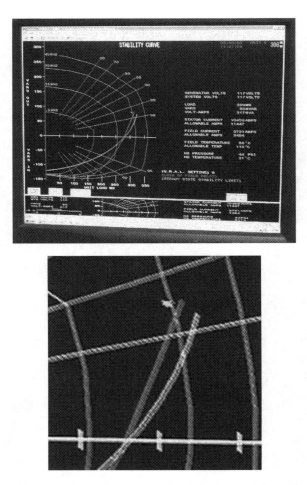

Figure 12.8 A problem in demand representation. Left: Screen 6 from the power plant control system shows a critical system parameter. Right: Close-up of the same screen shows where the critical item is located.

Screens pop up when the system needs to convey a message to the user about an aspect of the system status or a recent command that was just given. Where should a message window be placed relative to everything else on the screen? According to a study by Haubner (1992), the message windows should always appear in a fixed location rather than a variable location. In less critical cases, a system might allow the user to move the pop-up window to a convenient location on the screen while working back and forth between the main window and the pop-up. Pop-up messages that pertain to something critical, however, should not allow the user to move the window or otherwise continue working until the issue has been resolved.

Human–computer systems are not as closed and self-contained as they once were. The Internet access is usually running constantly. Sometimes, the only way to use a program is to go to a website; it cannot be downloaded for private use with any degree of convenience. The Internet connection means the outer world has a connection to the operator as well. Pop-up windows are widely used for instant messaging, advertising, and, unfortunately, the insertion of malware. Security software and Internet service providers once provided pop-up protection, until they started using it for their own purposes. If anything, there is a regular cat-and-mouse game going on whereby the security system figures out how to counter the new threat and the threatening agent figures out how to bypass the security system. The security systems also have their own series of pop-up messages to deliver.

The typical user responds negatively to pop-up windows, ranging from mild irritation to strong annoyance, and quickly develops the heuristic of ignoring all pop-ups. Bahr and Ford (2011) made three recommendations to potential senders of pop-ups: (a) When the user is busy, do not interrupt. (b) Save an intrusive pop-up only for emergencies. (c) Present subtle alerts that the operator can go see later, rather than alerts that grab full attention.

12.4.3 Graphical user interfaces

Graphical user interfaces (GUIs) are a form of screen organization that is, or was, intended for large database and control units. Accommodations for larger file sizes were often needed when they were first introduced as a standard practice. There is a hierarchical organization of information that is implicit in such a design. The first-level screen depicts the entire database at an overview level. The graphic screen shown in Figure 12.4 is an example; the operator was working on a second-level screen at the time the photograph was taken. The operator would then click (touch) a desired component in the big picture, and a second screen would appear showing the contents of the specific area. The operator could then click an object on the second-level screen to zoom deeper into the information base.

Another example of the view-click-zoom architecture (Shneiderman, 2000) appears in geographic databases that could contain marketing and demographic information or locator information. The top-level graphic would show a recognizable map, for example, of the United States. The user would click a state and see the details at a statewide level, click a county to see what is going on there, pick a city within a county, and so on.

12.4.4 Use of color

The computer offers the temptation to apply elaborate (and sometimes tasteless) color schemes because of the low costs associated with using color on a screen compared to print. The issues associated with color in conventional interfaces apply just as well to the computer interface: Higher contrast provides more information than lower contrast. Here *contrast* refers to both luminance and chromatic distinction.

According to Galitz (1993), the empirical studies were divided as to whether color enhances performance. Extreme contrasts or too many different colors in a display can cause discomfort and distraction from the content. In some cases, the users thought their performance was enhanced by color, but it really was not. Galitz advised that the number of colorations be kept to less than 10 and closer to 4.

Perhaps the critical question for computer interfaces is the same as it is for conventional machines: Does color add information, or is it frivolous and decorative? A productive use of color in a GUI would be to aggregate items by function or other relevant classification category and then color the backgrounds to signify where one category ends and the next begins.

A good example of adding high-contrast color to convey information can be found in systems that display biological information. A cell that is depicted in its original colorations would be of low contrast, and the elements of the cell would not be easily distinguished. Exaggerating the contrast of original elements' colors helps somewhat. Systems produce much more readable results, however, using *pseudo-coloring*. With this technique, the original elements are identified by the system, then displayed using colors that are easily distinguishable to the human although they might bear little resemblance to the original specimen's colorations. Pseudo-coloring is used routinely in PET and fMRI brain images. Brains do not really light up in the way that the pictures on the screen might suggest.

Geographic information systems (GIS) are natural applications for color graphics. For instance, the database of national medical data would be organized around a map of the country. The country would be segmented into irregular shapes corresponding to states and counties (e.g., in the United States). Each entry for a county (or other smallest unit) would contain a piece of categorical data, for example, prevalence rate of a medical condition. The rate scale would be broken down into 4 to 9 categorical levels, each of which is assigned its own color patch. The designer can then choose from a palate of colors to indicate the categories (Brewer, 2006; Brewer, Hatchard, & Harrower, 2003).

The question of how to choose the color palate could be answered better with some organized human performance research. The best suggestions at the moment are as follows: For a bipolar scale, the color series could range from dark blue at one pole to bright red at the other pole, with the other chromatic colors in between. Medical imaging such as the fMRI makes use of this scheme, so it would have some familiarity in other applications. For incidence rates from 0.0 to 1.0, a cascade from light to dark is a popular choice that can be found in chromatic sets, such as shades of blue, shades of green, and so on. Keeping the colors for one display within a chromatic range would be useful when separate maps for several related variables are composed. The user would be able to remember "the red map" or "the yellow map" as a navigational cue. If the categories were really qualitative categories that did not have an underlying continuous measurement connecting them, then a set of colors that jumped chromatic ranges might be less misleading than shades of brightness; aesthetics might interfere here, however.

Just as there are limits to the number of colors a person can reasonably interpret in a display, there is also a limit to how small or finely grained the colorized data should be geographically. Data that are too fine-grained with color changes can lose the visual clarity of an underlying pattern such as a source point or a contour. For this reason, virtual cartographers recommend the use of any of several statistical spatial smoothing strategies (Oliver & Berke, 2010; Waller & Gotway, 2004). The data entries can usually be expected to contain an element of error variance, either in the observed number or in the spatial location to which the number should correspond. Smoothing (in one computational scenario) would involve calculating cascades of medians across a geographic contour (e.g., from one county to the others adjacent to it). Waller and Gotway noted that viewers of GIS data tend to regard the numerical or categorical values to be as accurate as the geographic markers on the map: "In short, a good map of bad data often seems more believable than a bad map of good data" (p. 69).

Aesthetics and color can play a role in the attractiveness of the product—any product—to the potential user (Ji, 2013). This is a marketing issue rather than a human factors issue because it does not affect the joint function of the person and the machine. It might attract enough attention in a (virtual) store, however, to induce a customer (user) to spend some time trying to interact with it or think about purchasing it. If the display hit the viewer at an emotional level in a positive way, all the better. Marketing people might call this phenomenon "usability" too because it affects whether the user *wants* to use to, even though the functioning of the system could be mediocre.

12.4.5 Pop up and wait

In spite of advances in processing power and speed, it sometimes takes non-zero time for the computer to complete an operation before allowing the user to continue working. In these circumstances, it is advisable to use a pop-up window that shows the progress of the operation (Galitz, 1993).

Two iconic glyphs are the hourglass and the little blue spinning circle, either of which just sits on the screen until the operation is complete. Another type of design is the slidebar that changes color from left to right as the operation progresses; the color bar is usually accompanied with a verbal message indicating the percentage of the job that has been completed at a given moment. These wait signals explain to the user why nothing appears to be happening, and dissuades the user from interpreting the nonresponse of the system as a crash.

12.5 AUDITORY AND MULTIMEDIA DISPLAYS

12.5.1 Speech interfaces

Speech interfaces might be used to give the computer commands or as part of a text-to-speech (TTS) display. The computer's speech should be clear and natural, with the intonations and inflections that correspond to normal speech (Van Nes, 1992). One important goal is to hold the listener's attention, and monotonic droning does not do any better here than in real-time speeches.

Longer passages of text should be accompanied by pause and repeat controls. Some applications might benefit from a text or visual outline of the material to assist navigation through the material. One way to enhance navigation is to use quite different voices, usually male and female voices, for different parts of the text sequence. The shift in voices should correspond to a shift in the content of the message.

Some research has been done on the use of voice annotation to make edits to a word processing document. Van Nes (1992) reported that users making the edits preferred a voice system by 2 to 1 when giving instructions to the computer or other humans, but the text annotations were preferred by the people who actually had to make the changes in the document.

12.5.2 Earcons and spearcons

The diminishing size of the GUI on cell phones and other handheld devices has prompted the development new sonification techniques to compensate for the visual restrictions. *Earcons* (Blattner, Sumikawa, & Greenberg, 1989) are brief auditory displays that consist of a few musical notes. They are meant to be as memorable as visual icons, but without the literal connection between the sound and the referent that a true auditory icon would have. On the other hand, they require sound designers to come up with the intuitive quality that makes a memorable earcon. Earcons are meant to be hierarchically structured to convey first- and second-level menu items; different timbres would be used to separate the second-level categories. A limitation of the earcon concept, however, is that a set of sounds cannot be readily augmented if new menu items are introduced later on. There is no formu-laic theory behind their design; their evolution is several steps behind that of the visual icon in that regard. Although a few such earcons might serve the purpose of conveying status information to the user (e.g., "Windows has finished booting up"), the viability of earcons as actionable objects is another matter entirely.

Walker et al. (2013) thus introduced a new sonification system that purports to be more functional than earcons. Spearcons are TTS conversions that have been sped up so fast that they are not readily recognized as actual speech. Frequency patterns underlying the phonemes are preserved, none-theless, so there can be a familial resemblance between similar but different menu items such as "save" and "save as." In a series of experiments, Walker et al. compared the user responses to auditory icons plus TTS, earcons plus TTS, spearcons plus TTS, and TTS alone. In the former three conditions, the icon, earcon, or spearcon was presented first and followed by the TTS; if one design was better than another, the user should be able to respond faster and more accurately without having to hear the whole TTS that transpired in real time.

In two experiments where it was possible to develop full sets of auditory icons to represent objects, earcons clearly produced the poorest results for speed and accuracy; the other three modalities were tied. A third experiment involved learning representations of 30 nouns or cell phone operations at two levels of hierarchy; participants using spearcons outperformed those using earcons in both cases.

Walker et al.'s fourth experiment involved learning and recognizing a set of outdoor objects or events such as buildings, intersections aids, obstacles, plants, usable objects (e.g., public phone, or garbage can), and landmarks. This time, it was possible to create a hybrid set of true auditory icons and earcons and compare the set against spearcons. Spearcons produced better performance.

The final test was a comparison of TTS alone and spearcons alone. The task was to navigate a cell phone menu at different levels of depth. Spearcons produced the short set of response times at all levels of navigation depth by a clear margin. The spearcon technique could accommodate the widest range of objects or menu items, and were best for speed, accuracy, and learning compared to auditory icons or earcons.

12.5.3 Animation and hypermedia

Animation became available in user interfaces along with digital video and audio clips. Systems that used any combination of these display techniques became known as *hypermedia*. The human factors studies that were reported often compared combinations of media techniques for their impact on human performance. Many of the applications were centered on educational products, where system performance was measured by student learning. Some of the commercial applications, such as those that appeared on websites, were evaluated in terms of the amount of time it took for a visitor to find something, and how much time they spent with the site overall.

According to a review article by Betrancourt and Tversky (2000), the data concerning the effects of animation in a variety of instructional program types did not unequivocally support or refute the merits of including animation as an instructive device. In another analysis of the problem, Dehn and van Mulken (2000) reviewed some evaluations of programs that contained an animated character that was apparently giving a narration, and reported only a small effect favoring the use of the animated character compared to narration alone. Later studies involving the same comparison reported no loss in learning by deleting the animated character (Craig, Gholson, & Driscoll, 2002; Mayer, Dow, & Mayer, 2003).

Three other themes seem to have emerged. The first is that animation would show an advantage when the goal is to portray something that actually moves, such as ballet steps or the operation and repair of a mechanical device. In those cases, animation could be better than video, particularly when the sequences are brief (Koller, 1992; Sukel, Catrambone, Essa, & Brostow, 2003). Animation has shown an advantage in teaching computer algorithms, which might require procedural steps (Catrambone & Seay, 2002).

The second theme is the user's working memory capacity. Too many audio or visual effects transpiring at once may capture global attention, but the viewer's attention becomes split between too many channels of incoming information to be able to attend to the important material properly (Mayer et al., 2003; Mayer, Heiser, & Lonn, 2001). This rule applies to extraneous music, or to text that scrolls in addition to the narration; both elements should be avoided.

The third theme is that animated programs and hypermedia generally offer an advantage for students with poor reading skills or for whom English is a second language (Henao, 2002; Liu, 2004). Future research might be more definitive, however, as to whether text continues to have an advantage for the learners' retention of details as opposed to the basic ideas of the content. Students with better reading skills do not appear to gain any advantages from the parallel animation. The quality of the animation, text, or other hypermedia element may be confounding variables worth considering in future human factors studies on this topic (Schmidt, 2001).

12.6 THE INTERNET AND THE WEB

The Internet and World Wide Web (WWW, or what some users refer to as the World Wide Wait, or simply the Web), introduced some new themes in the human factors of system operation that were not hitherto explored in human–computer interaction. This section begins with a brief history of the Internet and the Web. The Web-related topics considered in this chapter are: search engines, information foraging theory, navigation, and web page construction and usability. Other distinctive human factors issues related to the Internet are considered in Chapter 14 in the context of collective intelligence.

12.6.1 Origins

The Internet started in 1969 as a U.S. Department of Defense enterprise to keep the critical government functions operating during times of emergency, for example, nuclear disasters when military control centers planned to operate underground. The system was known as ARPANET in those early days. The network around the United States evolved quickly, and it included physicists (e.g., from national laboratories) and other interested parties in the scientific research community (Gilster, 1993).

Access was initially restricted to large computer sites, all of which existed for noncommercial purposes. Some large sites eventually sold services to qualified users such as educational institutions who could connect their personal computers or mainframes. In time, end users could connect to their mainframe from home by dialing a phone number through their computer's modem. By 1983, ARPANET evolved into USENET, while the military operations channel bifurcated away from the main system. Numerous networks of computers were evolving during that time throughout the world. Eventually, electronic protocols and address systems were worked out so that all systems could communicate. The irregular size and organization of networking systems still persists today.

Perhaps the most used communication systems were electronic mail, which was transmitted in plain text that looked like typewriter type (ASCII characters); file transfer protocol (FTP); and search engines such as GOPHER and VERONICA. FTP systems were particularly useful to physicists who wished to disseminate their laboratory reports to the interested community quickly, rather than having to wait the better part of a year for the journal publication system to complete its course. Computer scientists used FTP to pass computer programs around also. The anonymous FTP sites were constructed to allow anyone access to the material. The search engines were developed to search the Internet for any documents on specific topics that might be held on any of those sites.

FTP and search systems were command driven. If a search produced some results worth reading, the user could enter a keystroke and go to the computer containing the file, then use the FTP commands to retrieve the file. Files that were prepared in ASCII were readable on screen. Anything that was word processed (mainframe computers often offered this capability to their users, especially when mathematical characters were involved) or contained graphics was available in postscript format. Postscript is a data-encoding system that makes desktop printers and laser printers apply dots of ink to exactly the correct positions on a printed page. Postscript files could not be viewed on the computer screen, however. The portable document format (PDF) that is in widespread use today is essentially the postscript format with human-readable features added on. Eventually, interface software evolved and insulated the users from having to know the commands, and they could point and click instead. The residual typing only existed to insert the address of a destination computer in the initial part of the dialog. In fact, this transition was just another example of the replacement of command-driven computer control by point-and-click interfaces.

The network providers essentially regulated Internet traffic until it crossed a critical hub that was regulated by the National Science Foundation. The norms for use prior to 1997 were that a user could transmit information that could be of interest to other users, including information about items for sale, but general commercial use was not permitted. Private organizations could acquire access and a dot-com address, but they were subject to the same norms of use. An announcement of an item for sale was considered information when it was first transmitted, but it was considered advertising (and not allowed) on second and subsequent transmissions. It was doubtful that any Internet police would actually crawl out from the telephone jack and spank anybody for advertising, but the user-reader population at that time was so steeped in the norm of noncommerciality that they would admonish users who broke the norm. The norm was predicated on another norm, which was the unfettered transmission of dubious information, and advertising (which later became known as *spam*) sucked up precious transmission bandwidth on the network and storage space in individuals' mailboxes while it was waiting to be discovered and deleted.

In 1997, the United States liberated the use of the Internet for commercial purposes. Commercial purposes meant the sale of services, technology deployment to increase bandwidth, and lots of spam, all of which proliferated. Software became easier to use and friendlier to millions of computer novices. Some users found it to be more fun than shopping through mail-order catalogs. In any case, the user population changed considerably from the culture of users that inhabited the Internet in the 1980s and early 1990s.

Some aspects of the Internet fall under the legal auspices of the FCC, particularly when security and other issues affecting U.S. citizens are involved. Inasmuch as a great deal of Internet traffic occurs outside the United States and across its boundaries, there is no systematic regulation in place. Security concerns in the United States since 2001 led federal government agencies to monitor e-mail sent across the U.S. border. Reading and sorting the huge quantity of legitimate e-mail to find a suspicious item requires—you guessed it—autonomous search engines that use strategies similar to the ones the humans use.

The Web was first launched in 1989 as a system for connecting the file libraries of computers into one huge library of material (Mayhew, 2003). Its technology is an outgrowth of the previous generation of search engines. Color document display and easy access made the Web's contents and utilization thereof especially attractive, particularly in the wake of the sudden influx of new commercial users after 1997.

Although the Web has the convenience of a huge library with searchable contents, its contents are in no way organized except by means of a user-defined search and the initiatives of content providers to supply links to other sites; the system is self-organized in this regard. Mayhew (2003) noted, and all users should beware, that web contents are unselected, unedited, and in no way imprimatur with any professional credibility simply because they are on the Web. Credible and conscientious people and organizations do produce reliable material of course, but the burden of responsibility is on the user to determine what is credible.

12.6.2 Search engines

Web users could be looking for something specific or exploring (*surfing*) for entertainment, or a bit of both at a given time. The user's first probable step is to search the Web for sites pertaining to a topic of interest. The user's skill for defining a search will often affect the results, especially when topics are complex or keywords have multiple interrelated meanings and uses. The available search engines, nonetheless, will respond to whatever keywords are entered by the user. The naive user may require some trial-and-error experiences that the expert user can sidestep. Having some prior knowledge of one's topic area is also helpful.

Search engines first became available in the late 1970s as a means for searching through technical databases such as *Psychological Abstracts* or *Chemical Abstracts*. This author remembers the Lockheed Dialog® search system vividly; it was an interactive search engine that could be used on virtually any such database of technical abstracts. The data sources containing the abstracts corresponded to periodical print publications produced by professional societies. The databases still exist and thrive, although today access is more convenient in online form compared to a hand search of printed documents. The databases are equipped with their own search engines that work like Dialog®.

An abstract is a carefully written preamble to a technical journal article that captures the essence of the article's contents. The entries in the databases contained the full bibliographic citation of the article and the abstract, and each entry was tagged with an index number. A user searching for articles on a topic would search the keyword index in the publication for index numbers that corresponded to the keyword, then look up the abstract numbers. Each volume or issue of the abstract publication had to be searched separately to see what was published and indexed during a specified time period. This strategy became unwieldy with the technical information explosion that occurred during the 1970s, if it was ever wieldy beforehand. The automatic properties of a computerized search that covered all the issues of the database in one fell swoop were highly desirable.

Literature search engines were, and still are, based on set theory. A user would enter a keyword, and the system would cough up the number of database entries that were found, and give this set a number. The user could then enter a second keyword, and a different set would be generated with a set number. At this point, the user could specify *and* or *or* to combine the two sets into the union or intersection of the two sets. The result would be designated as a third set with the total quantity of qualifying items designated. The user could proceed at this point to view the abstracts, and then make personal selections of which items should be pursued for the intended purpose. The user was advised to be aware of synonyms for keywords, or for technical words that were close in meaning. For instance, *human factors* and *ergonomics* have different shades of meaning that might be relevant to some people but not for others; someone searching the literature might do well to put an *or* between those two keywords before continuing the search.

Literature search engines evolved slowly. At first, they were only available to library staff in most universities. Eventually, when all university faculty had desktop computers and the proper connections from their offices, they could all do their literature searches by themselves quickly. *Psychological Abstracts* and many other databases are now accessible from a web-based technology that is operated by different software companies, which offer additional controllers for limiting searches, expanding to broader topic areas based on keywords selected, and packaging up the selected items for the user.

The contemporary forms of the source databases are often augmented with additional web links that would allow the user to retrieve the actual documents. This feature is limited by contracts and agreements with the original publishers, and whether the library that hosts the search engine also has a paid subscription to the targeted source. The phrase *free access to information* was originally intended to mean "easy to acquire because you can find it and retrieve it," but that meaning was too often conflated with "you do not have to pay for it one way or the other." There is of course a copious supply of information that is not restricted by a paywall. The cost of producing the items is paid by the authors and sponsoring organizations.

The Dialog® concept was built for systems that had direct connections to the database providers' computers, usually through a TELNET connection.

Searching Internet sources, on the other hand, required a different strategy. The GOPHER system allowed the user to browse the file indexes of computers that were connected, see menus of available materials, and retrieve (download) them. This was more user friendly than requiring that a TELNET address be known, remembered, and typed next to a cursor. (We all kept little cheat sheets in our desk drawers with favorite access coordinates.) GOPHER allowed the start of a navigation process, but with all the menu trees that looked alike on the users' screen, the users had to remember where they were located at any point in cyberspace; thus in response, GOPHER offered a bookmark capacity. The set theory method of narrowing did not apply, but some capability of shaping a search strategy was made possible by VERONICA (Very Easy Rodent-Oriented Netwide Index to Computerized Archives; Gilster, 1993).

There are some contrasts between the search strategies intended for a fixed database of published scientific documents and those available on the Web. The contents that the engines search are the contents of the websites. They place a priority on the home page of the website, and there does not appear to be any identifiable consistency to the depth of processing within a site. Although they are built on the GOPHER and VERONICA strategies, they are limited by the file structures of the websites themselves. Goodies that are buried too deeply past the home page will be missed by the rodent.

The user enters one or more keywords at once in the dialog box, and the system responds with web addresses and a brief description that comes from the website's text. The search request automatically connects multiple keywords with an *and* gate. The results are presented in an order selected by the search engine that reflects the probable relevance of the site to the user's stated request, starting with those that meet the *and* criterion. They are followed by results produced if only one of the search terms were used (the *or* criterion). Searches of this type pull up more irrelevant sites than a strictly set-theoretical search. Although this strategy may facilitate browsing by a user who is not pressed for time, it can be daunting to see the note saying that 5,600 sites were collected in response to a query, especially if nothing resembling the goal of the information search shows up on the first two or three screens of material.

Academic search engines produce results that are almost always formatted the same way—title, authors, publication source, abstract—organized in the user's choice of ascending or descending chronological order. Web search engines do differently. They produce page titles; authorship information is random unless an author's name was a keyword, a line or two of text containing the search terms, and the uniform resource locator (URL). Although the search engines respect the *and* gates to some extent, the algorithm prioritizes results by how many links there are to and from the target site. If one were to search for a site by name perhaps because the URL is not known, the target site might not come up on top the search. Instead, one might obtain many sites with "news about" the target. Many of the intruding sites are there for advertising and marketing purposes. Sites with strong commercial interests often pay for web services that magically enhance their placement within a search or position their ads near the top of the search results page.

12.6.3 Information foraging

The behavior patterns that people use to find information bear a substantial resemblance to the patterns that animals and our prehistoric ancestors used to find food (Pirolli, 2007). The food supplies are self-organized into ecological *patches* that support the growth of the food source. Some environments have more patches, and some patches are more profitable with respect to the amount of food or information it can deliver relative to the amount of work it takes to collect the information (food) and find one's way to the patch itself.

The forager's behavior is shaped by a certain need for efficiency in terms of both effort and time constraints. (Less time spent starving is better than more time starving.) Whenever possible, the forager goes to the most productive-looking patch first, exploits it until the law of diminishing returns sets in, then moves on to another patch. Recalling Tolman's (1932) principles of expectancy, valence, and instrumentality, the rat that has learned the maze knows where the cheese is. The rat goes to the place where it has learned that the largest cheese exists with the greatest likelihood. If it finds none, it moves to the place in the maze (patch) with the next-highest cross product of valence and expectancy.

A forager who is new to a particular information landscape (maze) might not know where the good patches are, but could stumble on a good one early on. In Pirolli's (2007) theory, the forager sets out on a random walk to find patches and picks one that looks good relative to the information search (dietary) goal. When it is time to move, however, the forager is unlikely to know where to go next, but can rely on *distal stimuli* that suggest, based on past experiences with other information mazes, which places to try first. The guess might be accurate or not; one can visualize a contribution of signal detection theory here whereby the forager determines which patches are correct hits, false alarms, and opportunity costs. The profitability of a foraging expedition is a function of the number of patches one can visit, which is inversely related to the time spent finding the sites to visit.

Opportunity costs take the form of diverting time and attention to bogus leads when they could be better spent by exploiting an information-productive patch. That brings us to the logistics of junk mail. To paraphrase one of Pirolli's examples, imagine that a business person receives job orders (where the real profits are located) in the mail, but the mail is also overwhelmed by advertising. The advertising itself could contain something of possible small value, but the job orders are clearly large and positive. When the rate of real job orders is high, the time spent on sifting junk mail is not worth the effort. When the rate of real job orders ebbs to a low, however, the sifting junk could be worth the time to process it. Of course, this parable assumes that the individual has only two possible tasks: the real job in one possible form and sorting junk. The business reality often includes maintenance tasks that have to be done eventually and actions that could lead to new business. The consequences of not doing one or the other could very well trump the utility of any junk mail.

Information is hierarchically organized for two reasons. The first is that organization adds value to a morass of relatively raw data by showing relationships among ideas that usually range from broad to specific. The second is robustness; changes in information at the lower levels of the hierarchy have little impact on the broader concepts until a critical mass of change is reached. Information is organized around hubs and authorities. Just as airlines target certain cities to be their more central locations through which flights pass in and out to go somewhere else, some information centers serve the purpose of drawing people to a well-groomed collection of links to other related valuable resources so that the forager wants to go there first and then choose a direction for further foraging. In the scientific literature, the review study, which pulls together results from primary research and extracts themes going on, usually trumps other single articles. It is a likely read for exactly that reason plus it serves as an inventory of all the meaningful studies that were published before it; the forager can then follow particular themes backwards in history as necessity seems to dictate at the moment.

The medium for measuring the influence of a journal article is the number of citations to that article found in later articles. The analogous metrics for websites are the number of hits to the home page and the number of links from the site to other places and the number of links. The distribution of node sizes within a network is distributed as an inverse power law, with a large number of small sites and much smaller numbers of larger sites (Barabási, 2002).

Information foraging theory (Pirolli, 2007) also captures several mathematical–statistical phenomena that were supported by studies of actual web user experience. Of interest, the number of sites a person visits in the context of a search also looks like an inverse power law with many searches ending at one site, but with a long-tailed distribution ending around 30 site visits. If the connection between power laws and self-organizing phenomena has any generalizability, it would appear that not only do sites self-organize in their information content, but also networks of sites do as Barabási, (2002) indicated, and the information forager is also following a non-random pathway. Pirolli (2007) characterized the search in progress as "getting the scent" for the information target. Once the foragers get an idea of where the goodies are located or supposed to look like, they follow the trail and ignore distractions as well as possible.

Some facets of human cognition that could draw people to a particular website. To start, the user should have a good mental model of the structure of information that exists on a topic in the cyberworld. The site designers do not have much control over this mental model, although they might consider how they should make their sites more responsive to users' mental models. This challenge requires knowing something about the user population and its diversity of experts, novices, and casual travelers. It may be instructive to find out what keywords the searchers might use, and then see how the website might fare in the listing of search engine results under those keywords. One trick that some designers use is to position a group of plausible keywords in white type in the white space on the home page. The user does not see the scramble of words, but the search engines find them.

12.6.4 Navigating the site

Navigation in the conventional sense involves finding one's way through a physical space from one point to another. One begins by finding landmarks and orienting toward or away from them. As learning of the space improves, it is possible to specify routes from one point to another and to select among alternate routes to reach the destination. The same principle applies to website navigation, although the physicality of the analogy is a bit weak.

For sites that are rich in content, options, and things to do, the merits of the site can be overshadowed by the sheer complexity of the site's architecture. A standing recommendation is to reduce complexity whenever possible (Whitaker, 1998). In practice, this means designing the site with a simple linear layout of the main sections of material. Two landmark structures are the home page button and the top-level menu of options. Both should be handy to grab for a user who might either be getting lost or choosing to leave one area and look at another.

Once a user has finished reading the material or is otherwise at the bottom of a file, a choice of a button for "top of file" and "home page" is often good enough. Designers should also consider whether some material that was initially planted in one area of the website might be useful to someone exploring a different area. Links across the site should be strategically located. Files that are not available, however, or are inadvertently moved during an update, should not be linked. Check the integrity of the links.

Some site designers see an advantage to linking to other sites on related topics, whereas other designers do not. One motivation for providing the links is to position the site as a place to find links and give it an appearance of centrality. Perhaps the goal was in fact to provide a set of links to a specialized group of users. There are two downsides to external links. First, sites that are all linkage and no unique content will not receive many return visits; users will bookmark their favorite links and just go there. Second, links to outside sites exist at the mercy of other site operators. Sites are erected, torn down, or moved at unpredictable moments, leaving the site with the links in a situation known as *linkrot*. The solution is simple: Check the links regularly and erase those that are no longer operative. The wisdom from building architecture applies here: Less is more.

Links that are really active should be identified as such. The common designations are to underline the words that comprise the link or to change color and font style. In some sites, there are only mouse-overs, which the users might not notice unless they moused over everything (Flanders, 2016).

12.6.5 Web pages

The very existence of a website known as www.webpagesthatsuck.com is a telltale sign that design atrocities happen every day. Its author, Vincent Flanders (2016), contends that the best way to teach good web design is through the example of bad designs and then figuring out what can be done about them. Problems include "accessibility, usability, and regular design problems including general ugliness." The site features bad designs that are reported by web surfers regularly in a column called, "The Daily Sucker." The site also features the top 10 atrocities of web design for a given year. The problem of internal and external linkrot is alternatively known as "Mystery Meat Navigation." Other common flaws are poor graphic design and typography, bad color contrasts and juxtapositions (color might not add information, but it can be a severe detractor), and the use of flash graphics that have no function beyond dizzying the viewer.

Some aspects of good webpage design rely on perceptual psychology and graphic artistry. Bertin (1981) proposed a scheme for evaluating graphic designs that carries over well to webpage design, according to Czerwinski and Larson (2003). The seven aspects of a graphic design are form, color, size, textures, orientation, value, and position. The Gestalt laws of perception (Chapter 5) are good guides for construction and perception of form, whether we are considering the form of a particular image or the formatting of smaller objects within a larger display.

Color has been considered already in the context of human–computer interfaces, and the same sensibilities apply here. *Value* in this context pertains to the attention-getting properties of an image element, by virtue of color brightness, graphic style, or other techniques use of animation. By the same token, too much flash on an object relative to its actual importance to the page undermines the goal of moving viewers to where they want to go because what they see overwhelms what they need to see (Cloninger, 2002).

The appropriate size of objects is relative to the other objects that must appear on the same screen. Larger controls or displays designate relative importance. At the same time, there will be lower limits at which small objects will be missed because they are too small or because they are overshadowed by larger ones.

Orientation refers to the positioning of an object relative to other objects in the display (Bertin, 1981). Some questions to consider: Is it on an angle? Is it close to or away from another object? Is it positioned so that it conveys depth? Are the positioning cues meaningful or conveying pointless information?

Textures refer to the level of detail on the surface of an object. A relatively detailed photograph juxtaposed to an area of large text conveys a distinction in texture. Texture attracts attention to the image as a whole, assuming all components are meaningful, and breaks up the monotony of solid text.

Position refers to the specific location of an object on a page. The upper left and lower right corners command the greatest attention from the viewers (Czerwinski & Larson, 2003). Advertisers might want to place their ads in those positions, but the host site might want those spots for their own purposes. A pattern that evolved in the last 20 years is to place the site's identification banner at the top of the page. A high-priority (highly paid) advertisement would appear underneath it, often with moving pictures. This juxtaposition could mislead viewers to think that the ad is part of site, although this bit of confusion dissipates within the second or two to recognize and ignore it. The primary site content then follows underneath, occupying two thirds of the horizontal space. The last third on the right would contain more links to other related items on the site, followed by advertising below.

Czerwinski and Larson (2003) noted that cognitive variables that may have one form of relevance when they are considered in isolation in an experiment have a different impact when combined with other variables in a web experience. The parameters of graphics are one example. Another is the number of menu options. Although the magical number 7 ± 2 seemed to be a good rule of thumb for determining the number of menu items, using only the magic number of menu items when there are many items that need to be included at the second and third layers only presents confusion to the user as to where to find a particular object. Czerwinski and Larson obtained better results with a layout containing 16 first-level menu items under which the organization of final objects was logical from the viewer's point of view.

Animation is frequently part of the web experience. The points raised earlier concerning animation in programs would apply here. The cognitive load issues have emerged in studies of commercial websites as well. Hong, Thong, and Tarn (2004) reported that site visitors did manage to find flashing items more quickly, but there was no visible impact of flashing on recall. The differential demand associated with flashing items should be reserved for the most important aspect of the site page, if it is to be used at all. Flash diverts attention from other items on the page. Flash and animation are not always in the site's best interests. Commercial sites that take too long to load up due to pointless high-bit-density graphics, or for any other reason, are going to lose customers or viewers (Karayanni & Baltas, 2003).

12.6.6 Interactive pages

Some interactive interfaces are intended to be analogous to paper-and-pencil business forms that are already in use (Galitz, 1993). Indeed some people might prefer to use the paper method instead of the computer interface. Tax forms would be an example. There is utility to some users to log into the Internal Revenue Service and enter their personal data there; their tax reporting requirements are simple, the information they need arrives early, and they need the refund money as soon as possible. The other 20% of taxpayers are more comfortable filling in their forms on paper and then mailing them in; their tax reporting requirements are complex, the information they need takes longer to arrive, and they are either in less of a panic to get their refund or they expect to pay out additional taxes, and they have security concerns with the Internet-based system. Either way, structuring the computer interface the same way as the paper form would make use of the users' expectations (population stereotypes), and it would also facilitate the work of people who use the data. Note that there are two sets of users in this application.

When designing a computer-based form from scratch, Galitz (1993) recommended that the form be devised so that it requires very little formatting on the part of the user. If a decimal point is part of a requested number, the decimal point should appear automatically in the correct place so that the user is not required to type it. Drop-down menus for fields for portions of address such as U.S. states are handy. They typically insert the two-character postal abbreviation instead of the full spelling of the state. Not only does the drop-down menu replace keystrokes, it also puts data into a storage field that can be accessed and reformatted to produce mailing labels (if needed). Drop-down menus are usually hidden under a downward arrow, and they do not interfere with the layout of the form when they are not needed.

Data-entry fields should be labeled. As with labeling controls on a conventional machine's interface, the labels should be consistently located relative to the empty field. Locating the label below the data-entry box is the least popular option, however. The ground rules for producing a questionnaire are virtually the same as those associated with paper-and-pencil surveys; thus, the reader is referred to that body of literature. Ultimately, the goal is to compute everyone's responses as well as collect them, so asking the right questions in the right manner is important.

A crucial situation occurs when an Internet sale transpires. Vora (2003) introduced the concept of friction-free purchasing. The analogy is that when we go to a store, we spend time driving through traffic, finding a parking space, looking for the items, comparing alternatives, standing in line at the cash register, and driving home. On less successful occasions, we return the item to the store and buy something else instead. These are all types of friction the purchaser experiences. A good sales site should make it easy to find the shopping cart, investigate the selections, monitor what is in the cart, remove things from the cart, and make the final purchase. The action sequence for the final purchase should be insulated from advertisements, customer surveys, and other distractions. Importantly, Vora reported that about 80% of online shoppers shop that way because of time constraints, and about 50% shop online in hopes of finding good deals for less money. A site can gauge the effectiveness of its shopping system by the number of abandoned shopping carts, which indicates purchases that were initially intended but ultimately not completed. The shopper should receive a confirmation that the order was received and some information about shipping time and how to get customer service if necessary. Ultimately, the process should instill trust and confidence in the vendor and the vendor's computer system.

12.6.7 Responsive design

In the early decades of web browsing, sites were accessed by computers with display screen that all fell within a limited range of sizes. With the advent of tablets and cell phones with access to the Internet, the old pages did not fit well. *Responsive designs* were devised to solve this problem. The paradigm shifted away from the page as the unit of content to using blocks of different sizes. The effect can be visualized best from an ordinary computer screen. Watch what happens when the user narrows the size of the window: The blocks reorganize on the screen so that the content is completely visible at a good size for reading and everything that was meant to be there is visible instead of hanging off the end of the screen.

An alternate design is to make blocks that only occupy the central half of the computer screen, leaving left or right quarters empty. The simplification is that blocks do not need to change position. It wastes a lot of real estate, but probably works well for users whose only computer-based tasks are done with small-screen devices.

An intermediate design is the wiki format. The main text appears on the central half of the screen. Links for internal navigation or external sites appear on the left, and graphics appear on the right. It is built in blocks that change position with screen sizes, but the standardization ensures that the repositioning of blocks will remain consistent.

12.6.8 Extreme graphics

To some aficionados of the graphic arts, website designs are often uninspired, although they would get a free pass on their aesthetics if they were actually functional. There are artist enclaves, however, who recognize the website as a unique medium of expression and are actively experimenting with graphic design techniques and styles for this medium. Graphic designs can collide mercilessly with usability, however (Cloninger, 2002). Although it is not the most extreme style ever offered, the *Mondrian style* features page layouts that are organized around rectangles of different sizes and colors. Select one, and the screen reorganizes, treating each new page as a pop-up window. Unfortunately, anyone trying to link to an internal page would be unable to do so. There was also the problem of having too many open windows that could overload a viewer's system eventually. There are also styles that are radical in the graphics, but very predictable in their actual functionality. Some of the radical styles found uses in product sales, fashion displays, or showcasing artwork.

The *pixilated punk rock style* was perhaps the most extreme style, as most of its distinguishing features were intended to rattle and disorient the visitor. Some of its antics include fake malfunctions, keyword searches that produce non-sequitur results, missing or misleading labels for links, random links, excessive use of animation and audio noise, unusual image cropping, image enlargements that took the viewer all the way to the pixel level, full-screen images with no visible navigation clues, unlabeled mouseovers where clickable links might ordinarily be found. The pixilated punk rock style does not appear to have found any applications other than as a performance piece that compels a designer to question the purpose of building a site:

> [P]ixelated punk rock is a usability expert's worst nightmare. These sites give their visitors a false sense of control, all the while manipulating and coercing them at every turn. But wait a minute: Haven't I just described most modern advertising in a nutshell? ... Should user empowerment and usability be the goal of an e-commerce web site? Do you really want to encourage surfers to wander around your online stores willy-nilly, reading this FCC-mandated earnings disclosure or that unmoderated product review? Instead, don't you want to funnel them as quickly as possible to that bright, candy-like "order now" button?"
>
> *Cloninger, 2002, p. 119*

The quick visit with extreme graphics leads to another point: Context matters. The design of the site, the user population, and the intended tasks all affect whether sites successfully provide information resources in the sense of the global library, product information and interaction leading to sales, or carrying out an automated government service such as renewing an automobile registration (Bennett and Flach, 2011). There is also the question of what else the users were trying to do when they jumped on the web to look for something. The example of walking while using a cell phone is perhaps just one example of a task being really part of a dual task scenario.

12.7 VIRTUAL REALITY

VR technology is an outgrowth of numerous artistic and technical developments from prior decades—theater, cinema, perspective drawing, pinball machines and arcade games, stereoscopic photography, 3-D auditory displays, and animation, not to mention all the computer programming skills and hardware that make it function. Burdea and Coiffet (2003) define VR as "a high-end user-computer interface that involves real-time simulation and interactions through multiple sensorial channels" (p. 3). They made an additional point of distinguishing VR from intermediate technologies such as those where a virtual component is superimposed over a view of actual reality, or *telepresence*, in which images from remote cameras are assembled into a portrayal of the actual reality. The former type of system does not qualify as VR because the actual reality still peeks through; the experience is not really simulated. The latter type of system may provide a breathtaking viewing experience, but it does not allow for interaction between the viewer and the program.

VR is characterized by two themes: *immersion* and *interaction* (Burdea & Coiffet, 2003; Pimental & Texiera, 1995). Immersion of sorts occurs when we read a fascinating book or watch a good movie; "engagement" would be a better word. In VR, however, immersion means that the system user becomes part of the program. *Interaction* means that the user can introduce actions to what is essentially a movie script, and the script responds. The program is prepared for the user's presence, and can do things to the user (in virtual form) as well.

The VR concept became popularized with the applications to games and entertainment. After all, how else could we stretch the baby's toes so that she'll be easier to grab, ride a buggy through the deep ocean without an environment suit, or play croquet with the boss' head (Kelly, 1995)? The rules of reality can be used, modified, or totally rewritten.

In spite of the potential for science fiction experiences, VR exists to solve problems in real life. Practical applications include medicine, industrial skill training, working in extreme environments such as outer space, hazardous waste removal, and military training (Bergamasco, Bardy & Gopher, 2013; Burdea & Coiffet, 2003; Pimental & Texiera, 1995). The practical applications require a maximally close parallel between the rules of the virtual system's operation and the rules of real system operation. Some of the more reality-grounded rule groups include the angles of incidence where a human user makes physical contact with, or uses weaponry against, an agent (or avatar) or other elements in the program and vice versa, and the duration, force, and implications of a physical contact. The combination of sensory experiences should provide enough sensory input, and input of sufficient quality to give the operator a sense of "being there." A large number of sensors would be needed to permit the types of changes in perceived locations of sights and sounds that occur as the individual moves through the environment. The system should be transparent in its delivery, meaning that the equipment generating the virtual experience introduces as little additional input as possible so that the human experiences the virtual world and not the machine (Bergamasco et al., 2013).

Several human factors issues became apparent in the development of the new controls and displays for VR equipment. Some problems were obviated with better hardware. VR has also opened up two new areas of psychological study—haptic perception and emotional programming.

12.7.1 Helmet and wall displays

If the distinction between a display and a control became blurred with GUIs, the two become more intertwined in VR. The user becomes part of the display in many instances. Furthermore, the concepts and system logistics are not limited to one user and one machine.

The basic VR helmet is an extrapolation of the heads-up display. The design can take the form of an intermediate technology, such as the superimposition of diagrams and reference material over a real visual image of a piece of equipment that an operator wishes to build or repair. It becomes VR, however, when the actual reality is blocked out, and all that one experiences is the content of the simulation. The discoordination between movement and the corresponding visual input is the primary source of simulator sickness.

The graphic content of the helmet display was limited by the graphics-processing capability of the computers to which they were connected. The original equipment used wire frame images. Pimental and Texiera (1995) remarked that if our regular vision were that poor, we would be declared legally blind in most states. The technology evolved to accommodate solid box diagram images, so that a wall could appear solid rather than transparent. In another leap forward, surfaces of walls and other objects could acquire the level of texture that distinguishes a bookshelf from a solid blob. The resolution of 2-D images is expressed *as* $N \times N$ pixel density; 3-D images are calibrated in polygon density. Higher graphic realism is related to the polygon density.

The VR helmet is equipped with visual tracking devices. The image presented to the helmet display depends on the orientation of the head, just as it would in real life. Figure 12.9 depicts a classic example of the control-display relationship. The pilot turns his head toward the direction of a

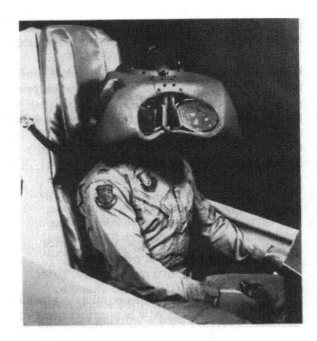

Figure 12.9 The helmet-controlled aircraft artillery aiming device. Photo by Wright-Patterson Air Force Base. In the public domain.

target and squeezes the control, and the airplane artillery fires directly on the target. The early equipment had a significant delay between the moment of the head movement and the corresponding display to the viewer. It was difficult to walk or navigate through the virtual world when the visual feedback, which is instantaneous in the real world, was delayed.

VR helmet designs are still in use, but they are being replaced by two other forms of displays. One replacement design resembles a set of eyeglasses with the exception of a small processor that sits on the nose grip. This lighter weight equipment is particularly adaptable to situations where the viewer is engaged in a real-world task, such as surgery or biomedical viewing of regions deep inside the human body. The virtual material comes up to the screen when the user turns the display on. The annoyance should not be much greater than that associated with bifocal glasses, at least in principle.

Another type of display format is the wall-mounted viewing screen. This class of displays can be as small as three standard computer screens—minus the space associated with the frame between them—or as large as a cinema screen. The full-wall screen can be a fairly good substitute for the VR chamber, but the immersion is not as complete, all other things being equal.

Figure 12.10 Wall display style VR. (Reprinted from Katsavelis et al. (2010a, p. 168) with permission from the Society for Chaos Theory in Psychology & Life Sciences.)

In this context, the user enters a room that is equipped with all the visual, auditory, and tactile display elements and interacts with a program that surrounds the user in a literal sense. The older systems required a network of projection cameras that cast portions of the image onto the correct locations in the room. Computers can now generate visual displays in convincing formats onto overgrown desktop displays that are mounted on a wall.

Wall-mounted displays still require a visual tracking device if a helmet is not used. The distance of the human to wall needs to be computed in order to display the images within the resulting field of vision. A field of vision of 37° appears to be the desirable target presently (Burdea & Coiffet, 2003). Figure 12.10 shows a wall display that is part of an experimental setup for studying gait and its correspondence to visual cues (Katsavelis, Mukherjee, Decker, & Stergiou, 2010a, 2010b). The objective of the system is to build better methods for rehabilitating people with impaired mobility caused by stroke or other medical conditions. The visual images are driven by cameras that watch the person on the treadmill, which is programmed to deliver specified routines. The images are stereoptic, and the person wears 3-D glasses during the experience. The images that are projected emulate the experience of working through a real environment. A similar setup is has been built to train athletes to row in a team (Bergamasco et al., 2013).

Wall-mounted displays can also be used for more conventional purposes such as showing multiple media in windows for groups of viewers simultaneously. Miniature desktop versions are currently available. The miniature contains the spatial equivalent of three larger sized VDTs without the frame boundaries between them. Ideally, the multiscreen display systems could substitute for wall-sized versions of the same content during the design process.

12.7.2 Glove and body controllers

The data glove shown in Figure 12.6 is used in conjunction with a head-mounted display. The user could view objects within a 3-D space, reach out, and grab an object and move it around; the results of those actions would show up on the visual display. With this basic equipment package, a user could interact with a program displaying the current medical condition of a human body, take virtual surgery actions, and see the results.

The data glove design expanded to data suits. In one illustration, two users could put on the suits and move their arms and hands, and the actions of one person would show up in the display of both of them. To make matters more amusing, the program portrayed the two users as lobsters in an aquatic environment. Although the demand for equipment that would allow us to become lobsters is probably still small, the technology advance was primarily the interaction of two or more users in a system simultaneously. Tactile feedback from the data glove or body suit was eventually added (Pimental & Texiera, 1995). This class of devices is essential, nonetheless, for the immersion of a user in a VR chamber, in which the equipment must track the position of bodies and their detailed movements (Burdea & Coiffet, 2003).

Figure 12.11 shows an individual operating a small robot (with limited functionality) using VR-type equipment in addition to conventional tools. The goal of the experiment (Jankowski & Grabowski, 2015) was to determine whether the additional devices were a help or a hindrance to robot operation. There were three experimental conditions: The simplest used an LCD monitor and joystick. The second level used the same tool plus stereo vision produced with stereoscopic glasses. The third added head-mounted VR vision, motion tracking, and a data glove. The VR-enhanced interface produce the best task completion times of the conditions. Stereovision was in second place in most cases.

12.7.3 Anthropometric issues

If the hand is too small relative to the glove, the sensors in the glove start to wrinkle over each other, producing a dilapidated image. The solution, albeit expensive, is to prepare gloves in different sizes.

The manipulation of objects in a virtual environment is limited by the reach of the person's arm or hand. This is probably not a major problem when using a finished program; in fact, it would be realistic if physical reach did limit the user's actions. It creates a bit of a problem during the programming stages, however, especially when a wall-mounted display is depicting

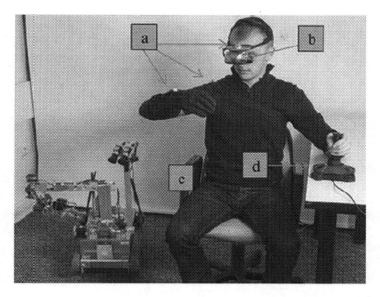

Figure 12.11 VR interface utilizing (a) motion tracking, (b) a head-mounted display, (c) data glove, and (d) joystick. (Reprinted from Jankowski and Grabowski (2015, p. 884) with permission of Taylor & Francis.)

large objects. The cubic mouse (Figure 12.12) was developed as a response to the situation. The cubic mouse facilitates a combination of pointing, capturing, and rotating in 3-D space, and moving objects in the virtual space using six action buttons.

12.7.4 Haptic perception

It is one thing to hold an object, and quite another to let it slip out of one's hand or squeeze it to death. Systems without the proper tactile feedback are susceptible to these failures. The human haptic system is sensitive to temperature, pressure, and texture. Temperature is encoded into the tactile delivery system with heat sinks (Wesler, 2004).

The pressure and texture functions are currently running into the limitations of the human sensory system. For instance, two stimuli of 1 ms duration must be separated by 5.5 ms in order to be perceived as two separate stimuli on the same location. Two successive stimuli on different locations must be separated by 20 ms in order for the operator to detect which stimulus arrived first (Jones, 2001).

We can "see" our way around in the dark if we use our hands to follow walls and furniture. On a good night, we can find the flashlight in the drawer

Figure 12.12 The cubic mouse from Fröhlich, Plate, Wind, Wesche, and Gobel (2000, p. 12). (Copyright 2000 by IEEE. Reprinted with permission.)

when the power goes out, also by feeling our way around. The psychological determination of 3-D objects requires a combination of pressure and texture sensations. The latter is generated by vibration. The psychological theory that underlies how texture and vibration sensations are coordinated into perceptions of objects is inconclusive. In addition, the information that contributes to the perception of an object involves not only the pressure and texture information, but also the movements that the person used to arrive in contact with the object (Smith, 2001).

"He doesn't know his own strength." Real-world clumsiness can be complicated another notch in the virtual world. Haptic feedback systems need to be adjusted for the strength of the user, and clear rules for this aspect of human–machine coordination are still in the making. Systems that provide feedback that is too strong can harm the operator. Operators who are too strong can damage the equipment. These design issues become complicated further if the system is meant to be used under conditions of stress (Smith, 2001; Wesler, 2004).

12.7.5 Training systems

The fidelity of perceptional phenomena, subjective awareness, and performance or system response are extremely important in training system. The classic rule for transfer of training is that training is optimal to the extent that the stimuli received, behavior produced, and the rules of reward, cause, and effect are exactly the same between the training situation and the real one. Learning that is dependent on a feature of the training system that does not have a counterpart in the real world, or is too different from that of the real world, could transfer negatively (Bergamasco et al., 2013). Sometimes, it is better to leave a stimulus feature out of the training environment altogether if it is too dissimilar to the real condition.

In other cases, where the training situation utilizes hints and prompts for the beginner, the support feature can be phased out of the experience in much the same way as training wheels are removed from a child's bicycle. Another support structure for learning involves the use of accelerators, which control the speed of the activity. An example would be juggling very slowly at first so that the learner can grasp the phasing of motions.

Numerous studies have examined the effectiveness of extended reality equipment for training purposes. *Extended reality* would include augmented reality, mixed reality, and full-fledged VR (Kaplan et al., 2021). The premise is that transfer of training would be greater if the simulated environment was closer to the real environment than conventional learning situations such as classrooms. Kaplan et al.'s meta-analysis of 24 studies showed no consistent benefits in training transfer and immersion from the extended reality techniques. A remaining benefit, however, is that virtual media can be used for training in situations that would otherwise be too dangerous for the trainees. The effect of VR on cognitive workload is similarly inconclusive (Lackey, Salcedo, Szalma, & Hancock, 2016).

12.8 EMOTIONS IN HUMAN–COMPUTER INTERFACES

Emotions have only played a small role in human factors until recently. Historically, the concept entered the scope of human–computer interaction in the mid-1980s under constructs such as computer anxiety. During that epoch, many organizations were introducing new computer systems to offices and as part of automation systems for production. Some employees responded favorably to the changes if they saw them as an opportunity for a career advance or a skill enhancement, while others were less happy about the idea of their skills becoming obsolete. The obtuse computer command systems and unfriendly interfaces in vogue at the time could not have helped matters.

12.8.1 Stress and anxiety

The roles of stress and anxiety in work performance and system safety were described in Chapters 9, 10, and 11. Adaptive interfaces (Chapters 2 and 7) detect stress on the part of the operator and alter the appearance or configurations of displays and controls in response (Klein, Moon, & Picard,

2002; Hudlick & McNeese, 2002; Khalid, 2004; Schmorrow & Stanney, 2008). The machine system would use the emotional information to alter the information load experienced by the operator and engage additional levels of automatic decision processing. At the present time, adaptive interfaces can respond to stress levels that are determined by physiological indicators of cognitive workload, but they do not make any particular use of the colorful range of human emotions that usually come to mind when humans interact with each other. Practical applications are limited by the current status of psychological theory that connects cognition with emotion and by measurement techniques that are responsive momentary changes in emotion or mood.

A completely different role of emotions shows up in VR applications for curing phobias. The premiere approach to curing phobias relies on a process called *systematic desensitization*. In this process, a person is given increasing exposure to the source of their fear, for example, heights, spiders, or flying. Some of the exposures are mental imagery exercises during therapy, and other exposures are homework assignments such as to go visit one place or another. VR systems have the potential to deliver systematic exposures in the office environment. For instance, a patient can put on the helmet and take a virtual elevator ride instead of driving around town to find suitably incremental elevators. Flight simulators lend themselves to therapies for fear of flying. Again, it is too soon to make pronouncements with great assurance, but evaluation studies on record indicate that a VR-enhanced therapy can be more effective than traditional means alone (Anderson, Rothbaum, & Hodges, 2003; Wiederhold & Wiederhold, 2003). VR systems for social phobias, on the other hand, place a greater demand on the realism of the humanoid avatars within the program and the emotional expressions that they exude (Rizzo, Neumann, Enciso, Fidaleo, & Noh, 2001).

12.8.2 Emotions as information

The clinical application brings us to some additional roles for emotions in VR. A VR system is more convincing to the user if it gives the user a sense of presence. Presence is currently an active research area, and is thought to be related to the realism of the animation, graphics, and scripts. Part of the realism, however, involves the extent to which the user's emotions are invoked during the virtual experience, and the similarity of that experience to real-life emotional triggers (Riva, Davide, & Ijsselsteijn, 2003; Slater, 2004; Bergamasco et al., 2013). Realistic emotions triggered within the user should be met with compatibly realistic responses from avatars; users appear to be more fond of empathetic avatars than self-centered ones (Brave, 2003; de Rosis, Pelachaud, Poggi, Carofiglio, & De Carolis, 2003).

Although the technology of speech synthesis and speech recognition have improved greatly since they were first introduced, naturalistic interactions between people and machines are limited by the machine's ability to recognize laughter and other subverbal and nonverbal communication elements such as vocal inflections and the facial expressions that usually accompany them. This aspect of human–machine interaction is thought to be critical for the development of household robots or other personal assistance robots.

Some advances in laughter detection have been reported. In one example, multimedia data from an online conversation that were analyzed for bursts of laughter were isolated in 200ms clips along with the visual elements (Scherer, Schwenker, Campbell, & Palm, 2009). A machine learning program (see Chapter 13) was trained to recognize bursts of laughter with 87% accuracy. At present, it is unknown how well the machine learning would transfer to the recognition of laughter by people outside the training video. Eventually, it could be possible to design the robot to engage interactively with the humans in a naturalistic fashion.

If laughter recognition is a challenge for the household robot, the recognition of other human emotions is no further out from the starting gate. There are six types of emotions that register in human facial expressions that are recognized in virtually all cultures: fear, happiness, sadness, disgust, surprise, and anger. Classification algorithms are starting to become available (Schels, Thiel, Schwenker, & Palm, 2009), but the limits of generalizability and potential for comfortable interaction still apply.

A similar research program is underway to program robots to recognize and make communicative gestures such as a hand wave (Sadeghipour, Yaghoubzadeh, Rüter, & Kopp, 2009). Humans make these gestures in different ways, but with the same meaning. The robot needs to recognize the class of gestures and possibly use them in conjunction with speech programming to facilitate natural communication between the person and the machine.

DISCUSSION QUESTIONS

1. At least one human factors engineer was known to remark, perhaps sardonically, that every human action and ability can or will be reduced to two motor skills—one is point and the other is click. How often is this statement true, in your estimation? Do situations exist where it is not true at all?
2. Where in a human–computer interface would Fitts' law apply?
3. What other principles of controls, displays, or psychomotor skill from earlier chapters apply equally well to the human–computer interface?
4. If you look at the file directory on your computer, you will see mention of an A drive, a C drive, and perhaps drives D, E, F, or G. What happened to the B drive?

5. A pundit once said, "Every year the cost per bit of information goes down, but the cost per bit of useful information goes up." Explain what this statement means.

6. Many innovations in computer-based technologies occurred for reasons other than human factors issues, but human factors improvements occurred nonetheless. How many such improvements can you find hidden in the evolution of computer systems?

7. Consider the technology innovation of installing identification chips in pets. Does this technology actually solve a problem, or is it excess baggage that only increases the cost of living or doing business? Are there any other technology applications that you would consider excess baggage or unnecessary complications?

8. Consider the use of identification chips in students' ID badges by the school district in Texas. Does tracking students' whereabouts on a continual basis serve any educational purpose? Does it really solve any perceived problems? Are there any downsides to the system, either functional or social?

9. Consider Bahr and Ford's (2011) recommendations for the use of pop-up windows. Are they likely to solve problems with pop-ups as the user usually experiences them? What is the likelihood that advertisers or anyone else will follow them? Why?

10. Go to the U.S. Postal Service site, www.usps.gov, and figure out how to apply for a permit to mail a monthly magazine at a discount rate and what forms need to be filled out. How long did it take you to get this information (assuming you were not interrupted by people and events that are unrelated to this task)?

11. Describe a plan for evaluating the effectiveness of a VR system. Hint: Try answering the question more than once with different VR applications in mind. What aspects of system performance would you consider here that might not have been considered?

12. If a VR or other system is designed to manipulate the user's emotion levels, what would be the implications for the design of voice-operated controls?

13. Participants in human factors studies that involve wearing EEG helmets or sensor gloves may have been willing because of the novelty of the idea, the premise that the results might assist them in their work in some way, or perhaps because they were being paid for the participation. What do you suppose would be their reaction to the idea of developing monitoring systems that were to be worn by employees everywhere so that employers could monitor their levels of alertness throughout the day?

14. Let's consider some critical incidents in Internet sales: (a) Tell us about a purchase transaction that you made that went very smoothly. (b) Tell

us about another purchase that felt like a complete abomination. What was different about the two experiences?

15. Is the concept of usability that has been assumed so far throughout this book really adequate? What would be some implications of reframing the concept to "usability for how long?" Can you think of any devices that were *too* useful? (Hint: In *whose* opinion would they be too useful?).

Chapter 13

Programming, artificial intelligence, and artificial life

13.1 THE EVOLUTION OF PROGRAMS

The concepts of computer programming, artificial intelligence (AI), and artificial life emerged in close succession, thanks largely to the work of John von Neumann. As we saw in Chapter 12, the first computer required a human to set a myriad of knobs and dials to correspond to a desired computational operation. The advent of computer programming allowed the user to specify and implement most of the control operations in a fast and flexible fashion (Heppenheimer, 1990). In doing so, however, there became a separation between the control operations required to *set the functions of* the machine, and the control operations required to *operate* the machine. For instance, typing text through a keyboard interface is substantially different from what a computer programmer did previously. The programmer specified the rules by which keystrokes converted into text that appears on the screen—or on paper if one were to use additional control actions for producing a print.

Von Neumann contributed his mathematical skills to the Manhattan Project, which operated from Los Alamos National Laboratory, to build the first atomic bombs. In those days there was a job category called the *computer*, which was a human who made computations. Computation requests were submitted from the laboratories that collected data to usually two people who were told only what the computational goals were. They were not told what the experiment was about or how it was supposed to turn out. Modern scientists would recognize this work technique as a *double-blind strategy*. The disjoint between the lab and the people making calculations ensured that errors from biases in judgment would not creep into and affect the results or their interpretation. The two computers were required to agree, ensuring the desired level of reliability.

The Manhattan Project was a historical landmark and the beginning of a modernizing society where numerous high-speed calculations were needed every day. Computers thus gained early notoriety for their ability to process large quantities of numeric data. The non-numeric information-processing activities in later generations of technical systems were also known as *computations*. The same holds true for the information-processing activities of living systems.

Von Neumann's vision for the programmable computer extended far beyond the ability to make numeric computations. Some of his more poignant mathematical proofs led to his proclamation that all life could be expressed as the flow and processing of information (Levy, 1992). It would thus be possible to express all activities of life forms in this fashion and, in essence, create models of life in a silicon environment. Hence we arrived at the concept of *artificial life*.

There was a large gap between the goals of artificial life and the technical capabilities of the early 1950s. Thus progress had to occur in modest incremental forms. *AI* became the bite-sized piece. Can we use computer programs to emulate human thought processes? In doing so, we should be able to speed the process, enhance the reliability, and expand the scope of what humans can accomplish. Thus computer scientists became very interested in scientific advances in human cognition, and psychologists became very interested in computer-based methods for testing theories about human thought.

The next sections of this chapter explain the major increments in the development of computer programs, AI, and artificial life. The overriding principle is that machines have gradually evolved to mimic human thought, and thus the information-processing and control action sequences that humans might employ. Granted, the mimicry remains crude.

13.1.1 Conceptual levels of programs and systems

The dissociation between the programmers' tasks and the users' tasks evolved gradually as programming techniques became more sophisticated. The most elementary form of program was the binary code. Binary code consisted of strings of digits 0 and 1 that specified whether an electron pulse was allowed to pass through a particular location at a particular time. Programming of this nature was obviously very alienated alienated from the goal of the computational activity.

A major leap forward occurred with the development of programming languages such as FORTRAN, COBOL, and Basic. These elementary languages bore a strong resemblance to human language syntax, although there were important differences. Commands must be specified exactly in a programming language, whereas human language allows us to construct many possible statements that have the same meaning, often with different word choices. In fact, this distinction between programming and human languages gave rise to a new form of program, the natural language processor, many years later.

The elementary programming languages, nonetheless, allowed the programmer to specify numerous binary commands through the use of familiar-enough words and base-10 digits. The 80 keystrokes that were applied to a punch card corresponded to a series of binary operations, thus producing the pattern of holes that were punched in the card. Elementary programs allowed the end user enough flexibility so that the user could input different sets of data, which would always be recognized properly if they were entered correctly. The same operations would always be executed, and the same form of output would always be received.

Entering data correctly was a challenge for the typical human who was accustomed to a little tolerance for typographical errors. A keystroke out of place in a computer program could often shut down a program's operation with a fatal error of some sort. The errors were often difficult to find. For instance, numeric 0 was not easily distinguished from capital O by the naked eye. Thus the character Ø came into common use. Similarly, the lowercase *l* was not distinguishable from numeric 1; in fact the two characters were made by the same ley for lowercase *l* on most typewriters in use through the 1970s. By then two distinct characters *l* and 1 appeared on a regular typewriter because of the need to distinguish the two characters in the computer environment.

Human factors played an important role concerning the error messages. In order to write in a program language such as FORTRAN, the programmer interacted with another piece of software called a *compiler*, which recognized the proper specification of FORTRAN statements, and spit back error messages. It took some time before sets of interpretable error messages were forthcoming from compilers. The lack of awareness of human factors issues was not entirely a result of ignorance or indifference on the part of the system engineers. Computation time and computer memory were costly; although it was often possible for the early technologists to imagine error-checking devices, the computer memory required to execute the software was prohibitive.

People who first acquired their programming skills in a FORTRAN-like environment were inclined to think of the data and the program as separate entities. Although this distinction has practical value at times, once of the advances in computer science came from the recognition that there was no fundamental difference between the data and the program. This realization led to the subsequent advances in programming techniques.

The next level of advance beyond the program language itself was the entity that might be called a *shell program*. In essence we would have a program that was large and complex such that it could give the user many possible options for data input, analysis, and output formatting. The user would then have to choose the desired options for a given task. The user would enter commands with the exactitude that was required of program language statements.

The statistical analysis programs that were in common use up until the late 1990s were examples of the shell program. A user had to write a little program that told the big program what to do. The end user was essentially writing program rules that affected the use of rules in the main program; the end users' "little programs" are described later in this chapter as *metarules*.

Symbolic programs allow the programmer or program user to identify a chunk of program rules, treat it as an intact object, and move it around the program at will. This group of capabilities allows us to compose a bitmap picture, pixel by pixel, compress it into another form of graphic object that is smaller in size, such as a .GIF or .JPG graphic, and move it around the file that we are using to prepare a document. The object-oriented programs and programming languages are essentially what allow us to cut and paste most anything in a document that we are composing; this capability is perhaps taken for granted by modern computer users.

13.1.2 Conceptual levels of system design

Computer-programming capabilities evolved from the bottom up, but the typical system designer organizes the design process from the top down. The highest level of conceptualization, that is, that which is most intrinsically human, is to imagine the purpose of the computer program, the range of capabilities that it should include, and the users who will benefit from such a program.

System designers have often relied on metaphors between some familiar real-life activity or situation and the representation of those entities in their programs. Hence we call the first screen on Microsoft Windows a *desktop*. The verbs *cut* and *paste*, which are common in a variety of word processors, originated with standard operations in the printing industry, where *cut* involved the use of rulers and razor blades and *paste* involved a hot wax machine.

The second conceptual level of system development is the *semantic* level. This level of design specifies the operations that can be executed by the system, how they are to be organized, and how they relate to the experience of the user. The third conceptual level of system development is the *syntactical* level of analysis, which compiles the specific control operations that a user must execute in order to make a particular operation occur.

Finally, the fourth conceptual level of system development is the *device-specific* aspects of the program. Ideally, these are not experienced directly by the user, except perhaps during the initial setup of the hardware and software. These features of a program recognize particular pieces of equipment in the system and make the necessary transformations to the processed information so that the end result will appear properly on the screen, the printer, or other device. The manufacturers of peripheral devices are in turn prepared to make their devices operable by as many varieties of computer operating systems as possible.

13.2 ARTIFICIAL INTELLIGENCE AND EXPERT SYSTEMS

Several principles that drove the theory behind AI programs are a particular class of applications known as expert systems. Here we encounter the work of Von Neumann, Gödel, Turing, and Simon. Expert systems exist in several levels of sophistication, which are elaborated later.

13.2.1 Some basic principles

13.2.1.1 Gödel

Gödel produced two ideas that haunted the development of AI applications. His incompleteness theorem (1962) proved that, given any rigorous system of logic, there is at least one legitimate logical statement within the system that cannot be reached from inside that system of logic. Thus there was no such thing as a perfectly closed logical system. At some point there is an element of information that must be supplied from the outside, hence the system user always has a role. This logical leak has become known as a *back door* to the program.

The applicability of Gödel's (1962) theorem to practical AI programs rested on an assumption that all knowledge can be expressed as a set of logical *if-then* statements. This assumption carried no intrinsic distinction between knowledge of simple facts, the kinds of mental operations by which conclusions are drawn based on those facts, and a given problem, or for that matter, the kinds of induction that are involved in creative problem solving. If, on the other hand, we could identify forms of knowledge or mental operation that could not be framed as *if-then* statements, then we would have an unprogrammable situation altogether.

13.2.1.2 Turing

Back doors often have something in common; the missing knowledge link often involves knowledge of the system itself. The Turing test (Turing, 1963) specified a hypothetical procedure, by which one could discern whether a system was truly intelligent or just artificially intelligent. Two curtains would be set up in a large room. Behind one curtain would be a computer. Behind the other would be a human. Both entities would receive a question through a mail slot in the curtain. Both would type their responses to the person who asked the question. If the question were sophisticated enough so that neither the human nor artificial source could answer, the human would reply that he or she did not know the answer, whereas the artificial source would respond in gibberish.

Thus an intelligent system required, at the very least, a modicum of self-awareness. Eventually an element of self-awareness was introduced by the error diagnosis in the compiler program. Programming finesse eventually integrated the error reports into the bodies of shell programs and latter-day programs. Computer systems eventually acquired diagnostic programs (essentially expert systems) to help repairers make repairs, and help programs for the users. Did these advances bring self-awareness to the system? Not by a long shot, but they were useful steps forward from the vantage point of both theory and usability.

The concept behind the Turing test plays an interesting role in some current studies in person–automation interaction. In experiments that use the *Wizard-of-Oz* methodology, research participants interact with an entity that is not visible, but communicates according to a rigid script that portrays one or more features of *human–automation interaction*. The human "behind the curtain" is portrayed to the participants as an actual computer program. In some cases, participants' behavior is then compared with behavior occurring with human–human interactions.

Turing also contributed to the development of a branch of mathematics called *symbolic dynamics*. Suppose we had a sequence of objects or events, and gave each object or event a letter code $a \ldots z$. We might possibly encounter a sequence such as:

$$F_1 = (a,b,c,a,b,c,d,e,d,e). \tag{13.1}$$

The mathematical analysis would allow us to extract sequences from Equation 13.1 so that

$$F_2 = (f,f,g,g,). \tag{13.2}$$

We obviously get from F_1 to F_2 by allowing $f = \{a, b, c\}$, and $g = \{d, e\}$.

Real problems are seldom this simple or obvious. Nonetheless, the math eventually facilitated some more sophisticated operations. For instance, we could identify a sequence of program code, make an operation on it to change it in some way, and put it back into the program where it originated. We can write rules for how these program changes would take place. In doing so, we also arrive at an important insight: There is no fundamental distinction between program and data. The combination of both can be called *program* or *data* interchangeably. All that matters is that symbols that are found somewhere are manipulated by rules from another source, and the rules can be manipulated by other rules in indefinite recursion.

13.2.1.3 Von Neumann

Von Neumann and Morgenstern (1953) provided some substance to the AI objectives through the development of the economic theory of games. A game consists of a set of strategic options, each of which delivers *utilities* to a player. A utility is a generic form of value. The utilities associated with each option, however, depend on the actions taken by one or more other players. Games may be strictly competitive (noncooperative), cooperative, or mixed-strategy games that involve competitive and cooperative options.

In a noncooperative or *zero-sum* game, the utilities of the possible outcomes for one player are opposite the utilities for the competitor. One player's winnings are the other player's losses. Optimal decision-making is thus one where one's outcomes are maximized and the other's outcomes are minimized; this is the *maximin principle*. Players do not typically share information in a noncooperative game, and binding agreements are not possible. For competitive games, there is at least one strategic condition that is a stable expression of the maximin outcome (Nash, 1951), known as a *Nash equilibrium*.

Although it is possible to conceptualize a one-person game, where one player works against the environment, or against a slot machine, the more interesting games are interactive and require at least two players. Thus game theory considers two-person, three-person, and *n*-person games. The simplest games for two players are the two-by-two games, where each player has two options.

Of the many ways of configuring two-by-two games, some games have *trivial* outcomes. Trivial in this case means that the best option for one player is always the same option, no matter what the other player does. In practical applications, a trivial game might not be so trivial. The received wisdom over the millennia is not to underestimate one's opponent, which is generally good advice. On the other hand it is important not to overestimate the opponent's capabilities. In a game of bluff, the power of one player resides in the perception of that power, rather than in the substance of that player's capabilities. When such games are in effect, it helps to simply go through the list of options: If A does this, what is the best option for B? If A does something else, is the best option for B still the same? Is there anything A could do that would change the best option for B?

Competitive game theory principles extrapolate to complex games such as chess. Chess has numerous options for each player that change with every play, and players attempt to control their opponents' options while maximizing their own. The utility of each option depends on what the opponent could do once the play is made. Chess became the prototype computer gaming application in the 1950s. A system user could play against the computer, which acted as a highly skilled player.

A cooperative game is structured in such a way as to provide maximum returns to both players if certain joint plays are made. In those cases information sharing is strategic, and binding agreements are possible (Zagare, 1984). The Prisoner's Dilemma game has become particularly dominant in the game literature because of its capability of producing both competitive and cooperative results. It is thus regarded as a mixed-strategy game. Strictly cooperative games are explored further in Chapter 14 in conjunction with the coordination of work teams.

13.2.1.4 Simon

Rather than explore all varieties of games here, it would be more helpful to refocus on what game theory contributed to human–computer interaction. Interacting with a suitable computer program would have considerable benefits that would not be obtained by acting without one, all other things being equal. Simon (1957) introduced the problem of *bounded rationality*: Although people are capable of rational thought and select the best options for their own interests when faced with a decision, there is a limit to the amount of information that they can process well and explore all possible options. Too much information at once or flowing too fast compromises the human's ability to produce optimal decisions. The decision maker resorts *satisficing*: Examine what is available and take the best shot. It might not be optimal, but it only needs to be good enough.

Simon advocated the use of AI to augment human computational capacity in light of this pervasive problem. He and numerous coworkers soon began developing computer simulations for complex economic events and organizational behavior (March, 2002). Industrial systems simulations involve the modeling of a combination of technical and human contributions (e.g., Forrester, 1961). A useful simulation of a human–technical system, however, would require knowledge of the humans' decision-making process. This requirement extends to a more general problem of how to extract knowledge, code it into a program, and validate the program's output. The topic thus turns to expert system architecture.

13.2.2 Expert system architecture

An expert system is an interactive computer program that functions like a human expert on a particular set of problems. Expert systems might be built to diagnose failures in a computer system or automobile, select wines to accompany various menus, determine preferable courses of medical or psychological treatment, predict outcomes of military or economic scenarios, or help develop recipes for commercial food products (Binik, Servan-Schreiber, Freiwald, & Hall, 1988; Fox, 1984; Klahr & Waterman, 1986). In addition to diagnosis, these expert systems often suggest remedies or series of tests to be performed in order to continue the query to a final solution. Expert systems were first introduced as knowledgeable coworkers for the human expert, but in practice they often serve as replacements for human expert.

Expert systems consist of three main components: the database, the inference engine, and the interface. The database consists of conventional spreadsheets of numbers (e.g., means, standard deviations) and facts and interrelationships such as "Condition A is correlated r with event X." The inference engine is the set of logical rules that the human would use to solve a problem. The interface allows the operator to query the expert system and to obtain answers (Denning, 1986). In the more sophisticated interfaces, a natural language processor that interprets the meaning of the question rather than only the array of specific words bolsters the dialoging program, which facilitates the query. Thus, persons may ask questions in their own language style without having to remember exact commands in order to operate the program.

13.2.2.1 Algorithmic systems

Inference engines can be sequenced on an evolutionary scale from relatively dumb to relatively intelligent. At the most elementary end of the scale, the rules of inference are simply algorithms. An algorithm is a specific set of computations and logical rules that are always performed in the same order. They can produce limited forms of output and respond to limited arrays of questions. Nonetheless, algorithmic systems can accomplish a lot, especially where they model relatively repetitive human decision processes that have only a few if any variations of procedure.

One illustrative use of an algorithmic expert system is the interpretive report generated by a computer-based psychological or medical test (CBTI). Algorithm-based CBTIs require five basic components: (a) the reading of responses to questions; (b) the compilation of test scale scores; (c) the use of prediction equations for additional scale scores; (d) the library of interpretive statements; and (e) the invocation of rules to trigger appropriate interpretive statements. The algorithm set would include equations drawn from the publisher's database, but also could contain equations supplied by the system user. The most basic interpretive rules would divide a psychological measurement into high versus low, or high, midrange, and low score ranges and trigger one interpretation for each grouping of scores. Another basic type of rule triggers statements based on configurations of high, midrange, or low scores for two or more variables by using *and/or gates*. The algorithmic interpretive program would conclude with print commands to format the interpretive statements chosen by the rule system.

Many useful algorithms have been derived through statistical bootstrapping of decision makers' interpretations of multiple pieces of data (Brunswick, 1955; Carroll, 1987; Meehl, 1956). This approach to rule-making is called *policy-capturing regression.* The first step in the process is to consult with the experts to determine what cues they rely on for the decision and how those cues are presented or measured. Next, the procedure requires a set of decisions made by the humans; the decision outcomes are correlated with the preconditions to create a multiple linear regression equation:

$$Y = B_0 + B_1 X_1 + B_2 X_2 + \ldots + B_n X_n,$$
(13.3)

where Y is the outcome and B_i are regression weights, and X_i are the cue measurements.

For the bootstrapping portion of the procedure, a new set of decisions is collected for which the level of accuracy is known, for example, a medical or psychological diagnosis. It is then possible to compare the humans' accuracy with that of the statistical model that was built from the earlier sample. It also becomes possible to test the generalizability of a statistical model to wider ranges of data sources where similar decisions need to be made.

Meehl (1956) discovered that once an equation had been identified in this fashion, the computer's continued use of the equation for making decisions with new examples produced more accurate results than the human decision makers. This was an important discovery because the humans initially generated the entire idea of how the decision should be made, but the statistical rule based on their judgments did a better job. Fifty years later, a meta-analysis (Karalaia & Hogarth, 2008) of 239 studies for which bootstrapping data were available showed that the algorithm produced a mean advantage of 10 correlation points compared to the humans. The advantage was smaller, and sometimes negative, if the outcome measurement was noisy (less reliable) and when the judges had more experience with the decision. The humans' accuracy was prone to be lower when they (or the statistical analyst) employed a linear decision rule instead of a nonlinear rule. Nonlinearity in this context could mean a nonlinear function of the types covered in Chapter 3, or it could be a hierarchical decision rule, such as the type often found in rule-based systems.

Decisions could be made for just about any topic where knowledge is needed, but extra care is required when the decisions are made about humans, such as who to hire for a job or who qualifies for a bank loan. Decisions made by the humans could contain unethical (illegal) biases against women or ethnic minorities; a substantial stream of litigation around adverse impact has transpired over the last 50 years. Discriminatory decision rules should not be allowed to transfer to an AI program (Zhang, Khalili, & Liu, 2020). Statistical procedures exist for rooting out adverse impact that could be embedded in a decision rule (Guion, 1998), and it would be wise to use them in many contexts.

13.2.2.2 Rule-based systems

The second stage of evolution for an inference engine is the *rule-based system*. It is distinguished from an algorithmic system by the use of *metarules* that govern the execution of other rules. Rules and materials may be deductive, inductive, prohibitive, permissive, obligatory, causal, or probabilistic. Some rule systems allow both forward and backward searches of the logical steps. The following discussion of rule spaces, chaining strategies, and classifications elaborate these points.

As an example, consider the task of choosing a wine. One could start with any one of three questions, depending on which one plays the biggest role in the individual's wine choices: Do you prefer red, white, or rosé? Do you prefer sweet or dry? Or what food are you pairing it with? From there one could proceed to narrow the choices by asking the other two questions, followed by more questions specific to wine taste (e.g., oaked or not) or food (e.g., spicy or not). One finally arrives at a small grouping of wine choices that fit the final description. Sometimes it is not necessary to go through all the questions to arrive at a choice. This is an example of a hierarchical decision model and one that is governed by a metarule to get it started—asking about color, dryness, or food pairing—and perhaps another later on in the sequence.

A rule space is a set of rules that is structured by a set of logical interrelationships. The rule space would also be characterized by various attributes of rules such as a rule's rigidity versus flexibility, its specificity or generality, and its precision or vagueness (Sridharan, 1985). Flexibility would be governed by metarules, often of the branching type. A rule is more flexible to the extent that its execution depends on a greater number of metarule arguments holding true.

Precision or vagueness would affect the order of rule execution, where precision usually takes priority over vagueness. A precise rule would take the form: "If test score $F < Y_2$ and $F > Y_1$, print statement ~." A vague rule would take the form: "If 6 or more of the preceding 10 rules are false, go to next rule." The former rule specifies exact conditions for using S_1. The latter is defining a default condition that should be executed when any of several possible conditions exist. The vagueness of the vague rule example would be increased to the extent the "preceding 10 rules" are also vague.

Specificity or generality also governs the order of execution of rules. The order of execution is also determined in part by whether the nature of the decision problem requires a depth-first, breadth-first, forward-chaining, or backward-chaining strategy. Chaining strategies reflect a search from specificity to generality or vice versa.

Hierarchical decision rules are not fundamentally different from a decision tree like the example of the Sunshine mining disaster in Chapter 11. LaFond, Roberge-Vallières, Vachon, and Tremblay (2017) published an illustrative example of a naval air defense decision that required classifying a signal appearing on a radar screen as friendly, hostile, or neutral. They reported that "the decision tree models derived from the participants' judgments actually yielded a larger proportion of correct classifications (98.51%) on the task than did their human counterparts (95.1%) ... which translates to more than 1000 errors avoided" during the course of the experiment (p. 131). Although the 3.3% advantage seems small, any one of the 1,000 errors could be very deadly. For high-stakes decisions, the remaining 500 decisions still need to be reconciled.

13.2.2.3 Chaining strategies

Figure 13.1 describes a set of rules, such that a square or circular node in the diagram represents each rule. The rules are organized at three levels of specificity. A depth-first forward-chaining rule order begins with a general rule, then moves forward toward a specific rule. A depth-first backward-chaining rule order begins with a specific rule or rules, then moves toward a general principle. For breadth-first strategies, the decision strategy begins with an execution of rules at a given level of specificity, then moves toward generality or specificity depending on the nature of the query. In principle, the program is equipped with metarules to execute the basic rules in either a forward, backward, depth-first, or breadth-first direction. The choice of direction would be dependent on the type of question being posed to the expert system.

The squares in Figure 13.1 represent objects within a classification structure. Note that an *object* can be a physical object that is known to the program as a value of a variable, such as "Occupational Group = 7." A physical object can be known to the system as a vector of parameter values, in which case it is also a *frame*; frames are addressed in a subsequent section of this chapter. In the information sense, however, an object can be a single value or a frame as just described, but it can also be a block of rules (or program statements) or a graphic that appears on the screen. Informational objects can be moved around the program and utilized multiple ways.

The object squares in Figure 13.1 are all drawn in black except for one to denote a rule search in progress. The white object might light up in a depth-forward search that is executed in response to the query, "Find an object that is a member of A' and B' sets" (more than one object in Figure 13.1 would satisfy this requirement). Alternatively, the white object may be the subject of a depth-backward search such as, "What are the origins of object C?" The white object could also light up in response to a breadth-first query, "Find the object with the most ambiguous classification certainty value."

The normative leadership model (Vroom & Jago, 1988) serves as an illustrative example of a depth-first forward-chaining rule structure. In the model, the manager proceeds through a series of yes-no decisions regarding how to delegate a job decision; the manager's decision rules are ordered

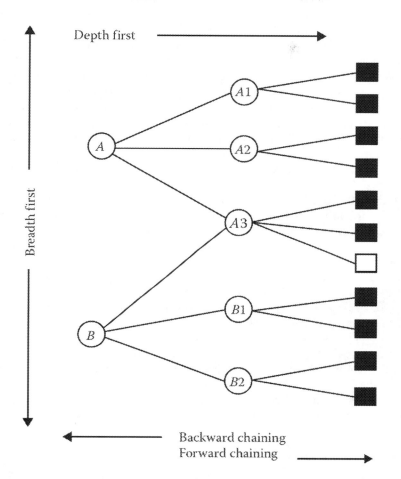

Figure 13.1 Forward, backward, depth-first, and breadth-first frame search strategies.

in priority. The pattern of *yes* and *no* responses results in the most appropriate delegation strategy for the set of facts surrounding the decision and the work group. A backward-chaining order of the same rules begins with the type of delegation strategy that the manager chose, and then proceeds to reconstruct the pattern of yes-no decisions that could have led to a specific delegation strategy.

13.2.2.4 Classification structures

Classification structures represent another type of rule-ordering format. An interpretive strategy for a medical or psychological test may require that a pattern or configuration of measurements define group membership, where the groups are hierarchically organized. Figure 13.1 could also describe an arrangement of ten objects that are organized into five categories, which are, in turn, organized into two broader categories. One might develop a rule system that classifies a case, C, according to a specific-to-general search strategy, a general-to-specific search strategy, or a commonality strategy. Commonality search strategies seek to determine what C has in common with the categorical classifications in the program; it is essentially a breadth-first strategy. In any classification structure, it is necessary to set a criterion for deciding whether two items are the same or different, or for whether an example is truly a member of a class. For some problems, such as the classification of biological specimens into species, the rules of membership are clear. For psychological or behavioral problems, however, the boundaries of classification are often fuzzy or probabilistic (cf. Zadeh, 1981). Hence, decision makers use hedge phrases such as "it is likely that," or "the certainty level is." A database coupled with rules derived from discriminant analysis, regression, or cluster analysis can increase the precision of some types of classification rules.

A *Horn clause* is a rule that defines a negative instance of group membership, for example, "If C scores high on the anxiety scale, C is not a psychopathic deviate." Horn clauses can cause complications with rule structures because they often introduce contradictions. Sharpening the precision and ordering of existing rules within a rule space (Buntine & Stirling, 1988) can minimize complications. An individual rule is overly general if its application causes the selection of a negative instance of a prototype (Buntine, 1987). With psychological or behavioral data, overgenerality would be a matter of degree and would depend on the base rates of category membership.

13.2.2.5 Frame-based systems

The third stage in the evolution of inference engines is the frame-based system. There are two uses of the word *frame*. A frame can refer to a set of parameters in the database that describe an object in the system. Each parameter in the set would constitute a meaningful piece of information that could be used to place an object into one category or another. A frame can also refer to a prototype in a classification structure; it could reside in the database or in the rule structure. In both usages, a frame is analogous to a chunk of data in human skilled memory (Linde, 1986). For instance, one might write rules that invoke separate characteristics of a prototype or a possible example of a prototype, or one might chunk all characteristics of an example or a prototype into an intact object. That intact object can be manipulated within the program. Some frames represent prototypes (abstract principles) in the system, whereas others represent commonly used deviations from prototypes (examples). Each frame has a number of properties that may be modified or canceled to produce a new and more appropriate result. When frames are added or modified, the rule system is structured to utilize the newly acquired information.

That brings us to a riddle: "What animal is large and weighing over a ton, gray in color, with big floppy ears, a long nose that drags the ground, likes to eat hay and peanuts, and lives in the trees?" Answer: An elephant—I lied about the trees (Brachman, 1985). The conceptual problem is to determine when we can add characteristics to a prototype or deviate from the prototype and still have an example of the prototype.

The use of rules for classification purposes constitutes a rudimentary frame-based system. Frame-based systems build on rule-based system design. Once the profile of scores has been compared with a library of example prototype profiles and the best match has been selected, the program would do more than simply generate a generic interpretation. It could also determine deviations from the prototype and retain the new data for purposes of self-modification of the program's rule structure. Frame-based systems are prone to a frailty in that there is a fuzzy cutoff point beyond which the prototype-deviant profile of scores no longer resembles the prototype frame (Brachman, 1985). The deviant profile might then resemble another prototype or no known prototype at all. Consequently, a frame-based system should contain rules that impose limits on prototype deviation for identifying true prototype matches.

A frame-based system would not only contain a large stock of frames, but also would allow users to add or modify frames. When constructing frames, it is important that the rules used to define frames be relevant to the interpretation objectives in question (Linde, 1986). If an interpretive strategy is based on causation data, more frames are required to the extent that the causation mechanisms are not precisely known (Pazzani, 1987b).

In addition to classifying profiles against prototypes, frame-based systems have other potentialities. First, they could describe the probability of a case belonging to a prototype represented by a frame by drawing on knowledge generated through statistical procedures such as discriminant analysis or cluster analysis. Second, they could retain case data for possible building of new frames in a recursive system. Third, they could prescribe a series of operations or activities required to transform an example into a frame prototype by means of a backward-chaining rule complex. Fourth, they could predict or explain new behaviors from preexisting knowledge and rules. The interface for a frame-based system should also allow the following interactions: (a) to enter new rules; (b) to select parameters that govern the execution of various rule groups; (c) to compose a frame; (d) to recall and modify a frame; (e) to compare frames; and (f) to execute the sorting of new cases against prototype frames.

13.2.2.6 Example spaces

An *example space*, which is a collection of prototype frames in a frame-based system, needs to be supported by several types of rule groups. These would include rules to assess sameness or difference, rules that set criteria for fuzzy classifications, rules that allow for transformations of prototypes, and rules that set limits on allowable deviations from prototypes.

Another type of rule in frame-based systems is the *open-textured* rule. Unlike other types of rules for frames, where a prototype is defined and examples subsequently provided, open-textured rules define a concept in terms of a set of examples. Prototypes are created by a partial backward search. A new item is accepted as a member of a set if it is similar to any one of the existing members of the set. Open-textured rules are typically fraught with hedge phrases such as "due care," "reasonable length of time," and so on (Sridharan, 1985). Open-textured rules often appear in legal decision-making and allow precedent decisions to impact on subsequent decisions. The relationship between precedents and decisions is an example of a recursive process.

13.2.2.7 Recursive systems

Automatic revision of the rule set is intrinsic to the fourth stage of evolution (Brachman, 1985; Sridharan, 1985; Thomas & Gould, 1975), the *recursive frame-based system*, whereby the expert system can learn through experience by collecting new frames, and modifying its rule system. The principle of recursion lies at the heart of a self-teaching expert system. A recursive program is one that can call on itself as a subroutine and execute a routine that changes its own programming.

A recursive system has the capability of updating and revising its database and rule structure. In a recursive frame system, a rule is conceptualized as if it were a frame. Rules have properties such as the contents of an *if* clause, content of a *then* clause, the presence and number of and/or gates, the elements of any computations that are involved, the confidence associated with a particular rule, whether a rule is a simple rule or a metarule, and so forth. These frame-based rules would then be read and manipulated by a new set of recursion rules. A schematic of a recursive frame system appears in Figure 13.2. As mentioned earlier, recursive systems capitalize on the conceptual breakthrough in AI that there is no fundamental difference between a program and data (Turkle, 1984).

The left side of Figure 13.2 describes the non-recursive frame-based system discussed up to this point. Program sections for the recursive inference engine and for modifying system processes on the right side of the diagram describe the new elements. There, algorithmic rules, metarules, and frames for prototypes are read as data. The recursive inference engine then

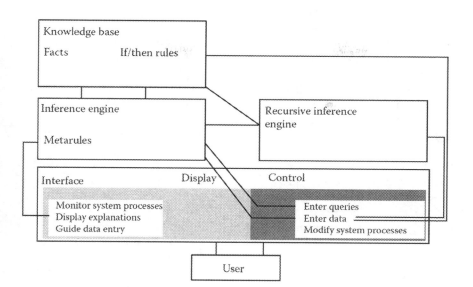

Figure 13.2 Expert System architecture with recursion component.

executes rules for rule modification in Step A. Next, the core program in the left section of the diagram is called as a subroutine or symbolic object (Step B). More rules are then executed whereby the new rules are integrated into the original program (Step C). There should be a check for logical inconsistencies at Step C, and flags to the user at Step D to indicate where new library statements are needed. Learning can take place in a rule-based system. The system would contain rules to retain new data and to modify the knowledge base accordingly. A rule-based system with such a capability is structurally more advanced than ones without it, but not as advanced as a recursive frame system. In the latter, both the knowledge base and rule structure for making inferences and decisions can be modified.

13.2.2.8 Interface requirements

Up to this point, no distinction has been made between interactive and noninteractive expert system programs. The distinction is important because rule-based expert systems are equipped with interfaces by which the user may query the database and invoke generic tasks, such as to provide interpretations of scores for various purposes, to classify, or to make recommendations. The need for an interactive mode is especially relevant where the decision goals require a substantial amount of customization or situation specificity from one decision event to another. The interface should allow the user to expand the database, modify rules, or reflect different decision priorities.

A new level of interaction occurs in the interface between the recursion rules and the user, in terms of control functions and displays needed to operate the system. The recursion rules would emulate the strategies the user would invoke to modify the program manually. The user may desire the option to invoke or not to invoke the recursion subsystem, or to enter or modify any criteria for enhancing the program.

The prototypic expert system was designed as *an assistant* to human experts that were (a) grappling with specific and high-stakes decisions, and (b) capable of operating the program and utilizing its recursive features. The essentials of AI have grown new tentacles into AI-infused consumer products with which the end user has little expertise or capability of to contribute, except perhaps the skill required to get the AI program to produce results. The system programmers manage the database and its updates. The influence engine is shaped by the population of users' queries. The updates are also drawn from the users' queries and behaviors such as watching a movie or giving up in apparent frustration with their interaction with the program. The interface is confined to the users' TV remote controller or cell phone, and it can be operated by keypad, voice interactions, or a combination of both.

Amershi et al. (2019) reported that errors in usability are widespread due to designs that do not adhere to interaction principles that were established for more traditional products. They studied the features of 20 of the most widely used AI-infused products and arrived at 18 recommendations (p. 3):

1. Make clear what the system can do.
2. Make clear how well the system can do what it can. The first two points reflect the *transparency* of an automated system, which in turn has a strong influence on the user's trust in the system.
3. Time services based on context: How quickly does the user need to know the updated information, as when driving on highways?
4. Show information that is relevant to the specific situation.
5. Match relevant social norms of conversation and referents.
6. Do not reinforce biases toward stereotypes by gender, age, ethnicity, or anything else.
7. Make the user's role easy to invoke a command.
8. Support efficient dismissal of the avatar.
9. Support easy corrections of commands or queries.
10. Offer suggested options instead of autocompleting a message from a user.
11. Explain why the system did what it did, for example, travel route choices made by the system. This is another example of transparency.
12. Remember recent interactions with the user. They provide context.

13. Learn from the user's behavior so that the system can personalize the interaction to the greatest extent possible.
14. Minimize disruptions in the dialog with the user when updating databases and query protocols.
15. Encourage granular feedback, such as the user marking something as important, preferred, or a "favorite."
16. Convey consequences of user actions, such reactions to advertisements that were generated by the system in response to a recent search or other action by the user.
17. Allow the user to control what information the system can collect and operate on at a global level.
18. Notify users about changes. Notifications are currently rendered by pop-up windows. Ironically, there was a time in the early 2000s when one of the favorite controls available to a user was to stifle pop-up windows. Pop-ups are no longer as discretionary as they once were.

Mueller (2020) also emphasized the importance of teaching users what the system can and cannot do. In the case of picture recognition and classification, humans can finesse distortions like shadows, incomplete images, and occlusions and classify the objects according to more than one schematic. The AI system is often limited in both capacities. The uninformed user tends to expect that the computer can think like a human until disappointing results occur.

Another type of example occurs in verbally controlled interactions with the AI product. Avatars like Siri and Alexa are capable of responding to a limited range of commands initiated by the human. Although advances in natural language processing allow the human to ask questions or give commands using different words with similar meaning, the computer cannot carry on a conversation; the social aspects of human–human conversation are generally lacking. An analysis of transcripts of human interactions with a digital assistant (*Wizard-of-Oz* methodology) while driving (in a simulator) showed that the humans' utterances contained many social elements of conversation like turn-taking during the question and answer sequence, phrases that reflected understanding ("uh-huh"), vague language (e.g., indefinite pronouns that a human would understand in context), hedge phrases (e.g., "just a little bit"), and polite responses (Large, Clark, Quandt, Burnett, & Skrypchuk, 2017). The next generation of AI product would need to evolve to recognize and respond to those types of utterances. Effective responses from the avatar would need to be responsive to the context of the interaction to some extent.

13.2.2.9 Smart integrated displays

The smart displays encountered in Chapter 5 originated with the principles of AI, and it follows that they should play a role in the displays generated by

AI-infused products. The smart display as we (should) experience it today builds on lessons learned concerning principles of visual perception, recognition primed decision making, ecological psychology, and the fact that we *interact* with displays more so than ever before.

According to Bennett and Flach (2011), our dyadic model of the person–machine interface (Figure 1.1) should be modified into the triadic arrangement in Figure 13.3. The new element is the system of meanings associated with the task environment. The operator is not a passive viewer, but is instead trying to accomplish a task and is looking for salient visual cues, or affordances, that signify opportunities to take one sort of action or another. Importantly, control actions are limited by operators' mental models of possible situations; one needs to understand what one is looking at.

Situations change over time, a point that is addressed by NDS theory specifically. Bennett and Flach argue for including recognition of system states as part of the knowledge base that should in turn be built into smart displays for a better representation of the situation for greater improvements in situation awareness. The NDS connection leads to five implications for display design: (a) How many system states are meaningful, and what are they? (b) What variables affect change in the situation? These would be control parameters of some sort. (c) How do control parameters affect the change in the system state? (d) What are the critical values of the control parameters that are associated with the change in situation? (e) What are the boundary conditions that induce limits to problem solving or situation awareness, or might contribute to a reorganization of the situation? In more colloquial terms, when do we have a major game change when the rules of engagement change distinctively?

The systems of meaning that make the PMS triadic are manifest in three levels of abstraction. The most abstract level ("highest" in the jargon of information technology) involves computation of the mental processes that are part of problem solving. AI or expert systems are essentially *computing* mental processes. The second level is *representation* of the results of the computation in a form that conveys maximum information to the viewer. The third level of abstraction occurs at the hardware level, which is deliberately out of the end user's view, while the user focuses on what is being represented.

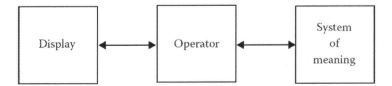

Figure 13.3 Triadic person-machine system.

For an illustrative example, reconsider the discussion question from Chapter 3: The Air Force wants to know whether a new pilot–cockpit interface will help fighter pilots destroy enemy aircraft in air-to-air combat situations. How can signal detection concepts be used to assess pilot performance (in a simulator)? The first step is to frame the performance question as a signal detection problem. Five dimensions are involved: three for space and two for movement. The target aircraft in the dogfight not only occupies a 3-D space while standing still but can move north-south, east-west, and vertically. The predator aircraft can also move in the same space. The goal is to move so that the target falls within the firing envelope of the predator's guns. In days gone by, flight motions were calculated by the pilot in the moment. An intelligent aircraft, however, would be equipped with sensors to detect the location of the target, convey the location information to the predator pilot in a form that immediately results in a strategic movement. When the target is computed to be within the envelope it would represent this status in the form of an audio signal and a flashing light. The predator pilot could then push a button and—boom.

The triadic principle of display design generated the concept of the integrated display (Bennett & Flach, 2011). If there are several pieces of information that contribute to the meaning of situation, the gestalt laws of perceptual grouping would apply: If the whole were greater than the sum of the parts, the goal is to design a display that would convey the emergent condition in a holistic fashion in one integrated graphic. Figures 5.21 and 5.22 are examples of integrated displays.

13.3 ARTIFICIAL LIFE

Artificial life denotes a range of computational techniques that emulate an aspect of living systems. Examples include neural networks, autonomous agents, cellular automata, genetic algorithms, and evolutionary computations, and agent-based modeling more generally. The former two items are considered in this chapter. The other items are discussed in Chapter 14 on complex systems.

13.3.1 Neural networks

Whereas the expert system attempts to emulate complex cognitive processes, the neural network is better suited to visual pattern learning and motor response (Chandrasekaran & Goel, 1988) or simple psychophysical judgment tasks (Vickers & Lee, 1998). The neural network programming concepts were initially intended to emulate actual psychological processes. Several possible architectures have been developed over the years, but most have shown to be equivalent with respect to performance accuracy (Greenwood, 1991; Vickers & Lee, 2000). However, see Minelli (2009) for a review of the primary mathematical models of real neural behavior. Importantly, the *binding problem* in neuroscience, which is the set of rules by which the firing patterns of neurons correspond to what we experience as a thought or information, has not been solved.

A neural network attempts to emulate four basic actions of self-regulating and adaptive systems: discrimination, recognition of identity, adaptation, and memory (Vickers & Lee, 1998). The system distinguishes one stimulus pattern from another. It must recognize an incoming stimulus as a match for one that is already known. It must produce suitable responses to incoming stimuli as well as change its responses to changes in the stimuli. Memory is needed for all the foregoing functions, as well as a means for moving information in and out of it.

The neural network architecture requires a sensory subsystem and an effector subsystem. Accumulator functions, which underlie both subsystems, are program units that keep track of the number of times a stimulus has been encountered and associated with a response. Accumulator functions represent the learning curves that are typically found in human learning experiments. They may be statistically represented as cumulative normal distributions in some neural network systems, whereas other systems may define learning curves as deterministic functions that resemble the formation of an attractor in the nonlinear dynamics sense (Bar-Yam, 1997; Vickers & Lee, 1998).

Real neuronal firing patterns differ for novel stimuli compared to familiar stimuli. Once a stable recognition-and-response mechanism has been formed, it can then remain stable or shift in response to new stimuli during its virtual life span. Neural network programs thus undergo a training process whereby numerous stimuli-and-response combinations are presented and the program accumulates an association between them. Training may be unsupervised or supervised. In supervised learning, the human operator is trying to make the program learn specific connections and perhaps specific probabilities associated with them. These objectives are accomplished through the preparation and organization of the training stimuli. In unsupervised learning, there is a chance that the system would not learn the intended objectives, and there is a chance that the system might learn something unexpected, which could be a welcomed result or not, or just plain wrong (Greenwood, 1991).

Virtual associative memory functions allow the program to recognize an incoming pattern as an example of a prototype, compare prototypes, and automatically distinguish new prototypes. For these memory processes, each stimulus-and-response pattern is defined as a group of informational elements. A set of cognitive rules are needed within the program to allow the comparisons of stimuli and the execution of modified actions (Vickers & Lee, 2000).

Within the memory banks, the informational elements act like little hooks whereby elementary units of similarity are allowed to accumulate; the memory is now making new associations. Additional rule groups are needed to keep track of the new associations and execute modifications to the stimulus–response patterns. If one were to activate a particular proto-type (frame) one could readily call up adjacent frames that had some par-tial association with the target frame. To prevent possible confusion in the system, the system would need to be equipped with both activation and inhibition mechanisms, like real neurons, to permit the activation to spread to the desired extent (Kohonen, 1989).

Vickers and Lee (2000) identified a speed–accuracy trade-off in neural networks that bears some resemblance to the speed–accuracy trade-off in human psychomotor processes. They assessed the accuracy of a neural net-work program for a simple psychophysical judgment as a function of the number of learning exposures. Their experiment also contained a manipu-lation that required the program to deliver judgments with high or low confidence. Low-confidence judgments actually produced higher mean proportions of correct responses compared to high-confidence judgments. The speed–accuracy function for the low-confidence judgments showed a sharper S contour than the function for high-confidence judgments.

Bar-Yam (1997) and Sprott (2013) identified a phenomenon of neural networks that bears some resemblance to stochastic resonance in human psychophysical judgment. It appears that the presence of some noise in the stimuli allows the neural network to make adaptive responses more readily than what would occur if noise-free training stimuli were used. According to Bar-Yam, the absence of noise allows the program to lock into a stable stimulus–response attractor quickly and inhibit the program from leaving the attractor. If noise is introduced during training, the primary stimulus–response attractor may take more trials to acquire, but the program will have also learned to respond to variations in the stimuli.

13.3.2 Machine learning

Artificial neural networks have found their way into computational machines that mine data of many different types and look for patterns. In the broader range of possible applications, there is not particular need to represent a specific neural theory of computations; the only goal is to find answers to questions that usually have little to do with neural activity. Other algorithms with similar objectives have evolved, notably k-nearest neighbors, Random Forest, Naïve Bayes, support vector machines, and extreme gradient boosting (Braarud et al., 2020).

Machine learning algorithms have found their way into the human factors toolkit for problems such as classifying driver distractions (McDonald, Ferris,

& Wiener, 2020) and classifying workload in a nuclear power plant based on aspects of speech (e.g., number of syllables, pauses within utterances, rate and duration of speech) within a three-person team (Braarud et al., 2020). The Random Forest procedure offered modestly better accuracy than the other five computations in both applications. Although the computational details of the algorithms are beyond the scope of this book, the contribution of machine learning to human factors and other problems are worth watching in the near future.

13.3.3 Autonomous agents

An *autonomous agent* is a piece of programming that recognizes target stimuli and effects actions when the stimuli are encountered. What makes them autonomous is that they can be loaded into a program system and allowed to do their jobs without direct control of the humans. One might then send an agent out to a group of websites, look for targets, and report something back to the human who initially sent the agent. The agents are sometimes known as little robots, or *bots*.

There are different levels of complexity for the design of autonomous agents. The simplest level would be written as part of a software upgrade package. The agent would look for files with particular names, determine if those files need to be changed, and execute the replacement of one file with another. The same technique can be used for questionable purposes, however. Some of the so-called updates are not intended for improving the user's functionality at the time of entry, but to read information attached to the files and determine how the user is using files and programs. The updating routine often occurs automatically whether the user likes it not; sometimes the visits are discretionary.

The second-level agent can do some web surfing on its own. Web search engines send agents out regularly to patrol websites regularly with the objective of making a response to a search inquiry that is as complete as possible. The agents might fetch information only when explicitly asked to do so, or they might be sent out on a daily basis to update the master system (and human user) on topics that were requested on a continuing basis. Agents inevitably interface with other users' systems, which they can only do if their presence is unnoticed or recognized as benign. Otherwise the targeted system has a security breech, which can be a huge problem.

The F-Finance program (Moemeng, Gorodetsky, Zuo, Yang, & Zhang, 2009), mentioned later, uses agents to gather data from real-world trading sources for testing investment algorithms. Program trading itself has been around since the mid-1980s: Agents look for situation that match a trading target and execute purchases and sales. The agents can create a problem for each other however. The stock market crash of 1987 resulted from trading program reading each other's trades; a pattern of sales was detected, then the drops in prices following the sales triggered more sales at a loss. There were no fundamental problems with the securities themselves or the market conditions as a whole. The Securities and Exchange Commission responded by imposing an agent of its own: If the Dow-Jones index drops more than 100 points in an hour, all trading stops to give the automated investors a chance to reset their parameters.

Figure 13.4 makes reference to several databases involving pictures, fingerprints, medical services, animal health, and travel patterns. Each type of database is formatted differently and limited in scope. The process of extracting and collating data from disparate sources can be accomplished by using multiple autonomous agents that mine the individual databases and converse with each other to produce results by combining and integrating their findings (Cao, 2009). The agents work in a *swarm* configuration (see Chapter 14) where agent-to-agent controls and displays keep the agents' information collection coordinated.

Swarming agents may be the means for getting around the dotted line of privacy and security in Figure 13.4. Users supply medical information to physicians because it is essential for the medical service and different information that is essential to travel services to airlines. Airlines and medical facilities state that they guarantee privacy over the records so that anyone who is not a legitimate part of the service provision does not have access, and they will not sell the information to outside parties. Marketing and sales

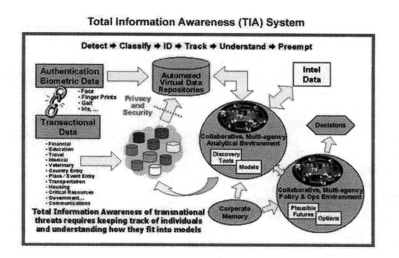

Figure 13.4 Overview of a suggested national security information and surveillance system. (From U.S. Department of Defense, 2002; in the public domain. DARPA's Total Information Awareness Program Homepage, www.darpa.mil/iao/TIASyst ems.htm).

operations are large sources of demand for information. Separate autonomous agents, however, can be designed to enter each database legitimately and get out quietly, then collate all the information and pass it on to entities that do not have a legitimate access to the information. The agents would need to be recognized as friendly.

A similar situation can arise in conjunction with the social networking sites. Facebook reportedly has a billion or more users worldwide, most of whom upload pictures of themselves and other people to share with their "friends." Facebook, in its early stages of development, put a tag on each picture that was uploaded so that it would be possible to deploy face recognition software to every picture to catalog who was in it. According to Sengupta and O'Brien (2012), Facebook received a lot of pressure from investors to sell picture data to increase profits, and from regulators in European telecommunications agencies to stop tagging pictures altogether because of privacy laws there. This type of software for mining and collating data has been in use outside of the United States and European Union for monitoring political gatherings and possibly preempting protests and insurrections (Fuchs, 2012).

13.4 LARGE-SCALE INTEGRATED DATABASES

13.4.1 What is possible?

Figure 13.4 depicts a hypothetical architecture for a large-scale integrated system designed by the U.S. Department of Defense (2002). The draft includes many possible data sources that can be integrated and cross-linked to identify international terrorist threat risks. Note the combination of pictorial information (e.g., faces and fingerprints), movement data (gait), and links with other government and private agencies with which the potential terrorist might have interacted. A system like this one would benefit from contemporary computer languages that translate other human and computer languages and information modalities into other modalities. The web page with the Total Information Awareness system is no longer available, and one can only guess why. The model, nonetheless, stands as an illustrative example of what someone could imagine for a massive database.

The arrows and connector graphics indicate information flows among the major components of the system. The little drums to the right of the list of databases shown in small type are icons for a database that date back to the pre-GUI days. It is not clear what the crossbar labeled "privacy and security" really means; it could mean that security protections need to be in place, or it could mean that a privacy protection barrier exists and somebody needs to get an "arrow" through it. Similarly, the chain with the broken link does not convey anything specific, but it could mean that if a person could not be identified through physical recognition, then it would not be possible to link to their other data sources.

The question about "privacy and security" leads to another issue concerning medical information that appeared in the list of databases. The Total Information Awareness system would require at minimum a massive digitalization of all medical records. The impetus for digitizing them started as part of the national health care system in the United States that was introduced in 2009. In theory, if all records were digitized in a central database, then physicians who were working on a patient could see the results of tests ordered by other physicians and not duplicate efforts by ordering another test. In practice, however, not all tests are ordered to produce the same information for every patient, and the life expectancy of test results could be very short. The efficacy of using digitization to eliminate redundancy is thus not as great as it might appear at first blush. Furthermore, the access to patients' medical records is controlled by Federal regulations and limited only to the patient, the patient's designated agents (often family members), and the service providers specified by the patient. One political scientist observed, "In today's mostly paper-based health records system, privacy is protected largely by fragmentation, inefficiency, illegibility, and general chaos" (Rothstein, 2007, p. 487). Digitization of medical records could be the first arrow to leak through the dotted line.

The two pictorial bubbles in Figure 13.4 signify the location of the inference engine, but the substance of the inferences was not specified at the time the drawing was produced. Data mining and machine learning techniques would be the likely tools for getting it started. The interface for the Total Information Awareness System would be complicated if it were to allow humans to enter a variety of information databases and monitor their coordination. The operation of the inference engine would require several points of human entry.

13.4.2 Database construction

Scientific research usually starts with a research question, organizes a database that would permit answering the question, and performs the pertinent statistical analyses needed to draw the appropriate inferences. One common variation is to build a large database that contains more data fields than necessary to answer a limited set of questions, but is available for later use to answer other questions that might be investigated later on. Another variation is to start with a question and look for a database that could be "out there" that could be used to answer the question. In the first scenario, there is a close alignment between the question and the data that are supposed to generate the answer. The second case is an answer that is looking for a question; researchers generating the questions would need to ponder whether the database is constructed closely enough to the way they would have done if they started from scratch. Are the constructs what they are looking for? Are the constructs measured in such a way that does not compromise the interpretation of what they analyses are supposed to mean and what inferences they are supposed to deliver?

"Big Data" sets are often constructed with other users in mind, and their usability can be disappointing. There are (at least) three broad challenges to the construction of so-called "big data" sets: (a) The database should contain variables that are measured in the formats—for example, categorical vs. numeric, number of decimal places, and temporal structure (seconds, minutes, weeks, months)—that would be most useful for answering intended questions. (b) The database should be constructed to support not only the types of statistical analysis that could be of immediate interest to one user group, but also a wider range of analyses that might be imagined in the future since the architects of the database are not necessarily the future users. (c) The preparation of the optimal database would have domain-specific requirements (Sangali, 2018). Financial forecasting, climate change, and human neurological activity data could share some common principles, but the three domains are not going to conform to a single mix-and-do recipe.

Although the problems of "big data" might not appear much different from those of "small" data, the problems become larger when automated data collection and analysis run roughshod over the requirements for proper statistical analysis of *any* type and ignores the users' requirements to extract meaning from the analysis (Secchi, 2018). Data preprocessing becomes more demanding when the research objectives involve nonlinear analysis or large quantities of time series (Guastello & Mirabito, 2018).

The measurement structure for a time series database requires that one establish an optimal level of time granularity in the measurements (Bettini, Jajodia, & Wang, 2000). A time granule is the smallest amount of time in which a meaningful event could transpire. Small granules can often be combined into larger ones (business days into weeks). The time granules for one variable might need to be consistent with those of other variables for some types of analysis. Observations might be partially overlapping or discretely segmented, depending on whether artificial smoothing is desired. A well-chosen temporal structure, which is often predicated on sampling rates would impact the choice of the *optimal lag time* that should be adopted in the time series analysis. The source data could be under- or over-sampled. Ideally, the time lapse between two consecutive observations should reflect the natural or physical amount of time required for a value of x at time 1 to transform into x at time 2 (Theiler & Eubank, 1993). Optimal time lags should be deliberately investigated from a perspective of naturally occurring change processes prior to the main analyses rather than relying on automatic computations (Guastello & Mirabito, 2018, p. 457–458).

13.4.3 Data mining

In the expert systems discussed up to this point, the data could consist of the primary (or raw) observations such as those the research scientist would analyze, or it would contain facts that have been distilled from a primary source into IF/THEN statements. *Data mining*, however, involves applying statistical analyses to large primary databases without a great deal of regard to any a priori research questions or preferences for particular inferential statistics used to answer those questions (Hancock, 2012). Mining programs read the data fields of a database and execute whatever computational or statistical analysis they know how to do. Although their broad-based analytic strategies can be focused or redirected by the human end user, the overall goal is to set them into automatic operation to produce anything they can find: emerging categories of objects or events, predictive relationships between one thing and another, trends over time and forecasts, and so on.

Data mining has become increasingly common now that huge databases are being generated every day by social networking sites, medical record-keeping in hospitals and clinics (Epstein, 2010), travel arrangements, every commercial site where people buy products online (Li, Ogihara, & Tzanetakis, 2012; Ohsawa & Yada, 2009), finance and securities transactions, (Kovalerchuk & Vitysav, 2000) and web search results whenever anyone searches the web for information. Big Science projects such as the Human Genome Project that commenced at the end of the last century to map the entire human genes and chromosomes, got what they wished for: voluminous data that needed some automatic processing for the human investigators to figure out what any of it means for medical practice (Ao, 2008). Furthermore, these databases are updated on a continuous basis, thanks to all the individuals in society who visit physicians, websites, use online banking facilities, and enjoy the convenience of seeing their friends and posting pictures of their lives online instead of remembering and storing their information on their own computers or other means of private control.

Although these databases exist for one explicit purpose, nothing prevents their use by entities mining data to extract patterns and produce "models." There is also the cat-and-mouse game of internet security and malware— viruses, worms, Trojan horses, phishing expeditions, and so forth—that are designed to disrupt the flow of legitimate transactions across websites or commit outright fraud. Thus there are data mining strategies designed to identify the prototypes and flow of malware (Aggarwal & Yu, 2008; Masud, Khan, & Thuraisingham, 2012).

Data mining programming continues to evolve meanwhile. The more automaticity that is allocated to the programs, however, the greater the potential for surprise to the operator. Some surprises could be beneficial, but there is a greater potential for meaninglessness if some filters are not applied by the humans operating the program. The program F-Finance, for example, offers an interesting combination of classic expert system architecture and data mining. The program allows financial traders to insert their favorite algorithms for picking or selling securities into the inference engine (Cao, Wang, & Zhang, 2004; Moemeng, Cao, & Zhang, 2008). The program launches a data collection, applies user-defined algorithms, and reports back on the accuracy or productivity of the algorithms.

In the interim, it is fair to state that the main contribution of data mining to the broader scope of AI is in the production of data and material for inference engines. There has been little in the way of unique advances for controls and displays or related aspects of user interaction, possibly because the community of users is still preoccupied with the veracity of the algorithms and not so much with usability issues. Nonetheless, there is some potential for new developments:

> [H]uman-computer interaction issues in [distributed data mining] offers some unique challenges. It requires system-level support for group interaction, collaborative problem solving, development of alternative interfaces (particularly for mobile devices), and dealing with security issues.
>
> *Moemeng et al., 2009, p. 56*

13.5 VALIDATION ISSUES

Having described expert system architecture, it is now possible to assess the probable sources of validity and invalidity. Validity issues pertain to the knowledge base, the rule structures, and the final report that is part of the human–computer interface display.

13.5.1 Knowledge base validity

The cornerstone of validity for expert systems lies in the extraction of knowledge from subject matter experts (Culbert, Riley, & Savely, 1989; Gruber & Cohen, 1987; Pazzani, 1987a; Vale, Keller, & Bentz, 1986). Even if the experts agree, the information they provide for the knowledge base may not be optimal. Research on "illusory correlations" (Chapman & Chapman, 1967, 1969) suggests that there are often systematic errors in experts' judgments of the relationships between predictor variables and outcomes. These relationships might not exist at all, might exist to a lesser degree than was previously thought, or might be in the opposite direction than what was thought. Sometimes the facts of a situation can change. For instance, a study of eminent clinical psychologists found that they performed only slightly above chance (Little & Schneidman, 1959) in their diagnoses of

clinical disorders. Moreover, extensive research on clinical versus actuarial prediction demonstrated that actuarial predictions can be superior
to clinical predictions (Goldberg, 1968, 1969; Karalaia & Hogarth, 2008;
Meehl, 1954). On the other hand, collective knowledge base in that area has
improved in the last 50 years.

System planners need to work closely with subject matter experts to
identify the information that should be included in the expert system,
to determine the level of specificity of that information, to ascertain the
structure of the information, and to collect the specific facts that need to
be included. In the case of psychological knowledge, experts can and do
rely on published research for CBTI content, but a substantial amount of
psychological research is not in a form amenable to CBTI rule building
(Lanyon, 1987). Although the validity of the psychological concepts may
have been established in carefully controlled situations, the validity of new
generalizations made by CBTI system developers may be questionable and
have often gone untested (Lanyon, 1984).

13.5.2 Expert knowledge space

The first questions to address are how to choose the human experts and
how many should be involved. The solutions depend on the structure of
the knowledge base. The first system design task is to identify the range of
problems the designer wants to solve and to specify what subproblems, or
elements, are involved (Koppen & Doignan, 1990). Next, experts need to
be queried regarding which elements are closely connected to other elements. If the experts know how to solve element A, can they also solve B?
Experts would be given a list of element pairs in the form of a survey or
interview and asked, "If an expert can solve A correctly, then is solving B
correctly likely, based on the specific knowledge required to solve A and
B?" Elements are thought to be quasi-ordered such that element B may
build on A, but that sort of directionality is not required. Each resulting
aggregate of elements is thought to represent a knowledge space. Having
identified the knowledge spaces, the next task is to identify human experts
for each space.

Koppen and Doignan (1990) provided proofs concerning the separ-
ateness of knowledge sets, and addressed to some extent the logistics of
collecting data from experts about the structure of knowledge. When
perceptions of similarity are at issue, psychologists commonly resolve
them with multidimensional scaling, which is a close relative of factor analysis. Multidimensional scaling would process a matrix of similarities of
knowledge elements to produce a small and efficient number of underlying
dimensions, like the color wheel in Figure 4.8.

13.5.3 Extraction of knowledge

Having identified the knowledge questions to pose to the experts, the experts are asked to solve the problems and to reconstruct the mental steps required to solve the problems. The think-aloud technique (Chapter 7) would be helpful here. Ericsson and Simon (1980) cautioned that much of the verbal data obtained from the experts can be easily influenced by the instructions they are given regarding how to complete the tasks and reporting how the problems were solved. Different types of probe requests make different demands on memory or require additional processing for the experts to make a response. The excess cognitive load demand required to verbalize the thought process would slow the verbalization process, but not necessarily to affect the accuracy of the report.

Verbalization methods have been both concurrent and retrospective in nature (Ericsson & Simon, 1980). In concurrent methodology, experts are asked to think aloud as they solve the problem and are given free rein to structure their verbalizations as they find comfortable. Alternatively, the experts may be asked for specific information at various points in the process. In retrospective methodology, experts may be asked to report their solution strategies after each problem, or after a set of problems. Newell and Card (1985) adopted this approach in their think-aloud technique.

A programming problem arises, however, when experts generate different diagrams, different directions of information flow for otherwise topologically equivalent diagrams, and different odds for each linkage. Rege and Agogino (1988) developed an algorithm for synthesizing such conflicting diagrams that could be useful in developing interpretive reports. The problem could be reduced substantially, however, by compiling the knowledge base from empirical research, whereby the judgments of many experts (clinicians) are bootstrapped over a wide enough knowledge space.

Militello and Klein (2013) encouraged system designers to identify the most challenging problems that their expert sources have ever tried to solve along with the solutions the experts invoked. The challenging problems tax the underlying expertise more so than the routine problems. The more routine problems and solutions would come to the foreground as the big problems are analyzed.

If the most challenging tasks are not already known or easily forthcoming, the *critical incident technique* (Flanagan, 1954) would be a good place to start. The core procedure is simple: Experts are brought together to swap stories about challenging situations, what they attempted to do, and what happened next. The emphasis in the narratives is on the recounting of events, rather than drawing conclusions about why things happened a particular way. Someone would need to transcribe the stories as they are recounted; the transcriber should prioritize objectivity over editorial comments or inferences.

People with less than maximum expertise should also be included in the discussions because their solutions might *not* be what the best experts would have done. This is important information too. The analysis of narratives is essentially qualitative as the events are sorted into meaningful categories; this phase of the procedure might not be a simple task and the data analysts try to decipher what were the most salient aspects of the situation that contributed to the outcome.

Extraordinary events are more likely to be remembered and recounted, which make the critical incidents technique ideal for finding the situations where maximum expertise is of interest (Guion, 1998). Routine events are less memorable, but one moderator of the session can probe for them. The critical incidents technique has been very useful over the years for diagnosing and understanding performance issues in a variety of contexts. The technique has undergone some enhancements over the years as well, in which researchers apply a brief questionnaire or checklist to the incidents; the content of the checklists also varies by context (O'Connor, O'Dea, & Melton, 2008; Renshaw & Wiggins, 2007). In the case of expert systems development, the focus should be placed on the types of decision goals and strategies used by the experts (Militello & Klein, 2013).

13.5.4 Decision validity

The first questions and strategies were encountered already regarding knowledge base validity and the accuracy of experts' decisions. Experts' decisions could be rendered as a single multiple regression statement like Equation 13.3. For that type of rule, we would compare R^2 for the equation based on the experts' decisions against the real answers to an independent sample of examples that had outcomes determined by independent sources. Updates and intended improvements to the expert system can and should use R^2 as a benchmark for evaluating the level of improvement. If the first version of the expert system showed an improvement beyond the judgment of the human experts, the update should reflect a wider margin.

13.5.5 Signal detection technique

Dichotomous judgments are susceptible to two types of error: misses and false alarms. A decision rule that is rendered by an expert system should be evaluated in terms of both types of errors along with the accuracy of correct hit. The human judgment that usually underlies the computer code makes a trade-off between false alarms and miss errors due to the (dis)utilities associated with each type of error.

Lehner (1989) presented a method based on signal detection theory for assessing the incremental accuracy of an expert system without having to use human experts in the evaluative process. For each decision point or *node*, the accuracy of the rule is compared against the baseline accuracy rate. The baseline accuracy rate is the rate at which the decision could be guessed correctly by chance alone. For instance, if the odds of an event A is 60%, then a person or program could be correct 60% of the time by guessing that all outcomes are event A. One would expect a program to do better than that. The difference between rates is d'. There was no rule, however, for aggregating the d' coefficients over nodes, although one could foresee the use of combinatorial statistics for producing a system accuracy rate based on multiple judgments, each with their own baseline rates.

13.5.6 Meta-interpretive reliability

Some expert systems, such as those associated with some medical or psychological tests can contain up to 20 or more measurements that are interpreted in combinational form, either through a linear regression equation, such as Equation 13.3, or more complicated means involving metarules. Their final results take the form of a verbal report that interprets the measurements; the CBTI mentioned earlier would be a class of examples. The value of a good measurement system and rule structure can be demolished by a sloppy report. Lanyon (1987) identified three levels of validity assessment that would apply to this class of AI products. The first is the statement level of analysis whereby each statement in a report is evaluated for correctness. The second is the narrative level of analysis where the validity of the global impression of a person created by the set of interpretive statements is assessed. The third level pertains to the correctness of decisions made by the expert system as interpreted by the end user of the system. Several types of validity are elaborated here.

Errors in the knowledge base or inference engine will result in interpretation errors. Additional sources of error may arise at the interpretation level of analysis in the form of poorly crafted statements, contextual influences that arise when statements are organized into reports, and overgenerality caused by high base rates for the interpretive statements. In light of these potential sources of error, reports from expert systems that are delivered in text (e.g., CBTIs) be evaluated at both the statement and global levels (Guastello & Rieke, 1994; Lanyon, 1987).

Much of the error in an expert system report may be traceable not only to the rules, but also to the process by which data and rules are transformed into words, words are compiled into reports, and reports are read and understood by users. *Meta-interpretive reliability* describes the reliability of that process. The concept is based on the Shannon–Weaver communication model, along with classical and generalizability theories of test reliability (Endres, Guastello, & Rieke, 1992). Meta-interpretive reliability is the degree to which a human expert who is reading a product from an expert system can reconstruct the original data that the program relied on to produce the report. Endres et al. also found that the humans' accuracy in reconstructing the original data pertinent to one part of the report was influenced by nearby text that was generated from different combinations of measurements, and rules for inference.

13.5.7 Barnum effect

The overgenerality that arises from high base rates for interpretive statements in a CBTI is known as the *Barnum effect* among personality theorists (Baillargeon & Danis, 1984; Furnham & Schofield, 1987). The problem compounds for people who read the interpretations of their own personality test score reports because the Barnum effect is greater where the interpretive statements are favorable or flattering to the examinee. In response to the potential for distortions of ratings from the Barnum effect, Guastello and Rieke (1990) developed the *full Barnum* experimental design, to assess the validity of a CBTI at the statement level of analysis. The analysis partitions CBTI accuracy judgments into two quantities: variance associated with the Barnum effect, and discrimination validity. *Discrimination validity* is the relative amount of variance associated with the CBTI's ability to distinguish individual examinees. The design also allows for the assessment of the influence of context on ratings of accuracy for individual statements.

The full Barnum experimental design is based on a two-way analysis of covariance with one repeated (within subjects) factor and one fixed factor. For the repeated factor, examinees are presented with two interpretations of their test scores; one interpretation is a real report, and the other is a bogus report composed of interpretations based on mean scores for the sample of examinees. Examinees are asked to rate each statement from each report for accuracy. Ratings of statements, which are made in terms of percent accuracy may be aggregated (summed or used separately) over the entire report or sections of the report. The mean accuracy rating for the bogus report is the amount of accuracy associated with the level of Barnum effect in the typical report. The discrimination validity of the CBTI is the difference between mean accuracy ratings for the real and bogus reports. The between subjects factor may be a counterbalancing effect such as whether the real or bogus report was rated first. The covariate is reserved for a property of the report that could influence ratings of accuracy, such as an index of favorability or length of the report.

In the full Barnum design, some interpretive statements can be found in both real and bogus reports. Ideally, these statements should be rated as equally accurate in both contexts. A correlation between ratings for a particular statement appraised in both contexts would determine the degree to which the statement's meaning is influenced by contextual effects.

High correlations indicate low influence. Low correlations indicate high contextual susceptibility. Statements that are most vulnerable to context influences could be suffering from vague word choices and language. Fuzzy words for probability concepts such as *unlikely, probably, possibly, little chance of, strong possibility*, might not have a shared system of meaning (Reagan, Mostellar, & Youtz, 1989).

Variations on the design are possible, such as substituting clinicians for examinees as raters, or other differences among raters as the between subjects effect (Prince & Guastello, 1990). The name, full Barnum design, was intended as a distinction from the *half Barnum design*, in which the bogus reports are composed by interpreting the opposite of the meaning of the measurement (Guastello, Guastello, & Craft, 1989). In other words, if a variable was measured on a 1–10 scale the 10 would be interpreted as a 1, and 2 as a 9, and so on. This procedure would have the effect of doubling the discriminant validity of the real report when compared against the bogus report.

The reference group from which the bogus reports are derived could have an important impact on discriminant validity. Discriminant validity can be maximized by choosing a reference sample that is as heterogeneous as possible so that the base rates for interpretive categories would be low. On the other hand, the principle of ecological realism would suggest that reference samples should closely resemble the population of test takers that is intended by the measurement. In other words, a clinician who needs to make treatment decisions regarding psychiatric outpatients in an urban area wants the CBTI to tell more about the patient than what is already known by just knowing that the examinee is a psychiatric outpatient in a particular urban area.

DISCUSSION QUESTIONS

1. In what basic ways did AI add new value or new subsystems to the human–computer or person–machine interface?
2. Suppose you were to pick out a good wine to go with dinner. What questions would a wine expert ask the diner? In what order would the questions be asked? If you are not familiar with this knowledge base already, ask someone who knows.
3. How would a food processing company organize an expert system to manage the production of 20 different types of soup?
4. Look at Figure 13.4 again. What value do veterinary records have for tracking terrorists? Where would the information come from?
5. Face recognition software is now available to individuals on their smartphones. Could this technology be used to facilitate any undesirable social behavior?

6. What does Gödel's theorem tell us about stock market crashes and how to prevent them from going out of control?
7. Consider Amershi et al.'s point #15 about marking items as "favorites." What are its effect on artists and producers in the entertainment industries?

Chapter 14

Complex systems

14.1 COMPLEX ADAPTIVE SYSTEMS (CAS)

A system can be complicated but not necessarily complex. Complicated systems have many parts that work together, usually in a mechanistic clockwork fashion in which the parts interconnect as $A \to B \to C \to D$. A *complex system* is characterized as a group of subsystems that are interconnected by information flows; it has multiple connections between parts, with feedback loops that run backward from D to C, or D to A, or in other combinations. Information flows can be steady, cyclic, or chaotic. A simple complex system (!) was shown in Figure 3.7.

The complex *adaptive* system (CAS) concept, which was first introduced by Gell-Mann (1994), can be applied to any living system. Individuals, works teams, and larger sociotechnical systems are all examples of CAS. Dooley (1997) articulated it as the new dominant mental model of work groups and organizations that incorporates nonlinear dynamical systems (NDS) concepts to study patterns of behavior. The perspective frames new questions regarding how the system recognizes signals and events in the environment, harnesses its capabilities to make effective responses, changes its internal configurations as new adaptive responses require and interacts with the external environment.

The CAS has three hallmark features (Dooley, 1997, p. 83): Its internal order is emergent rather than determined by the designers. It is the self-organized result of people, skills, abilities, tools, and demands for output (markets). Successful ones have a good means of gathering information from the external environment and the internal environment. They recognize patterns of events and information, some of which are simple and other nonlinear. The central themes of the CAS are schema and agents, agent interaction, problem-solving and conflict, supervenience of internal order and agent fitness.

DOI: 10.1201/9781003359128-14

14.1.1 Schemata and agents

Group members scan the environment and develop schemata (Dooley, 1997). A *schema* (*pl.* schemata) is essentially the same as a mental model, although its history in psychology is much older and places additional emphasis on the actions that could be taken in response to the mental models of the situation (Newell, 1991). Schemata shape the rules of interaction with other agents that exist within the team or outside the team's boundaries (Dooley, 1997). A group's schemata are often built from existing building blocks, which are brought into the group when members arrive.

The building blocks reflect particular individual differences in job knowledge and other attributes (Burke et al., 2006). Here we see the role of particular individual differences in job knowledge and other attributes. The integration of individual perspectives on a work situation is not always smooth. Although the individual schemata self-organize into one or more supervenient mental models, individual differences can remain that provide enough entropy for further modification of the schemata. The modifications are known as *symmetry-breaking* events (Sulis, 2021): An errant agent enacts a schema that is different from the dominant ones, other agents replicate it and eventually it becomes an alternative schema.

Schemata change via mutation, recombination, and acquisition of new ideas from outside sources. Change occurs in response to both changing environments and changing internal conditions. Indeed there are numerous sources and types of entropy that could arise from changes in client populations, the workforce, demands for products and new markets, hiccups in the supply chain, and governmental regulations (Bigelow, 1982; Guastello, 2002). The processes of mutation and recombination suggest an analogy between the dynamics of genetics and creative problem-solving.

When schemata change, requisite variety, robustness, and reliability are ideally enhanced (Dooley 1997). *Reliability* denotes error-free action in the usual sense. *Robustness* denotes the ability of the system to withstand unpredictable shock from the environment. Resilience and anti-fragility (Chapter 11) are closely related to robustness. Resilience is the ability of the system to bounce back from an environmental shock. Anti-fragility is a state of preparedness that minimizes the impact of the shock. *Requisite variety* refers to Ashby's (1956) Law: For the effective control of a system, the complexity of the controller must be at least equal to the complexity of the system that is being controlled.

Complexity in this context refers to the number of system states, which are typically conceptualized as discrete outcomes. Complexity in sociotechnical systems can be observed in several forms: (a) the number of objects, states, symbols, and properties of the system components; (b) the number of linkages or relationships between them; (c) the breadth and depth of hierarchical layering and clustering present in the linkages, and (d) the degree of branching present in the logic propositions that lead to conditional operations or modes of operation (Endsley et al. 2003, p. 138). They go on to say that unnecessary complexity can be eliminated when designing equipment or end products by stifling "feature creep," which is the tendency for designers to add new features (often called "upgrades") to keep their products looking "new and improved." Designers should set priorities for which features are most likely to be used the most often, and leave behind the ones that the operators would use only rarely.

Hierarchical layering occurs in multimodal displays and controls; the problem of mode error was discussed in Chapter 8. Branching in logic is essentially what is produced by meta-rules in artificial intelligence (AI) programs. The recommendations here would be to keep the use of modes to a minimum. Similarly, the levels of automation versus manual control should be kept to a minimum as well.

The number of system states has a direct impact on the complexity of the mental model needed to manage them, especially when the complexity of the system increases over time. Boisot and McKelvey (2011) identified three strategies for responding to increasing complexity: *routinizing*, even if it means causing errors by over-simplifying; *maintaining a balance between order and chaos*, which is where meta-stability occurs; and *behaving as chaotically as the system*, which they likened to running around like a chicken with its head cut off. The meta-stable state is also known as *optimum variability*, which is an equilibrium between maintaining sufficient complexity and minimizing the energy required to make the necessary adaptive responses (Guastello, 2015; Schuldberg, 2015) as depicted in Figure 14.1. Although Ashby's law drives the system control to greater levels of complexity, the minimum entropy principle also applies at the same time: It is not efficient to engage unnecessary actions or introduce more levels of complexity than what is actually needed. Putting the two principles together,

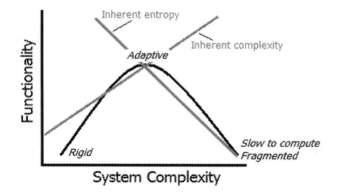

Figure 14.1 The optimum variability principle.

greater complexity of the control system improves system performance up to the point where it meets the improvements associated with minimum entropy. The result is a saddle point between the two directions of system development. Finding schemata that are more complex than the ones in use requires exploratory learning, whereas simplification relies on crystalized past learning (Boisot & McKelvey, 2011).

The concept of optimum variability has been useful for understanding the distinction between healthy and unhealthy organisms, social units and business organizations. It is only a small extrapolation to consider healthy and unhealthy sociotechnical systems (large-scale human–machine systems) in the same fashion.

14.1.2 Agent interaction

As team members interact, there is a flow of information and resources (Dooley 1997). In the early stages of team life, the interaction patterns are often volatile, but they eventually self-organize to enhance collective efficiency, which is in part a reduction in the entropy and uncertainty in how the information flows will occur. The performance of the system can be enhanced by manipulating the levels of decentralization, diversification, and specialization of members' roles. The use of symbolic tags (e.g., job titles) facilitates the formation of subunits or specific functions.

14.1.3 Problem-solving and conflict

The feedback loops associated with team cognition sometimes result in the need to change schemata. The potential for intellectual conflict could involve differences between a current schema and a view of external reality, different views of external reality, different views of current schemata or different alternative schemata. The problem-solving strategies could be dialectic or teleological. The *dialectic* strategy compares contradictory ideas and usually seeks to resolve their conflict. The *teleological* mode relates to the design or purpose, and how a system's purpose would unfold or evolve over time (Dooley 1997; Van de Ven & Poole, 1995). Teleological discussions require the group to recognize that their current situation is not a static one, but a slice of a drama that has been changing over time, perhaps not smoothly. A collective understanding or mental model of the evolving situation would help matters greatly.

14.1.4 Irreversibility and emergent order

If it is indeed a CAS, it remains poised on the edge of chaos ready to reorganize itself in response to new demands. The sequence of states through which it reorganizes is relatively unpredictable, although the understood goal of situation awareness is to envision possible future scenarios and states eventually and predict the outcomes of one's control actions (Endsley 1995, 2015). A team's particular sequence of stages is often subject to initial conditions that contribute to the global unpredictability of the system. The states of team organization are irreversible once they have taken hold and stabilized (Dooley, 1997). Although teams can redeploy old schemata, the effect is not the same because of the history and events that accumulated, both within the team and in the environment, during the time that has elapsed.

The dynamics of agent interaction and problem-solving give rise to the development of new schemata. Once adopted, they are expected to have a supervenient effect on the further actions of the agents. The supervenient effect presents another reason why group events are irreversible: a schema is deployed or changed against a context that could contain little history or precedent, as in a group's early stages of life, but the same schema might have a different impact within a different context where the effect would be introduced.

14.1.5 Agent fitness

A final feature of the CAS involves agent fitness. The notion of fitness arises from genetic algorithm and related studies. A team might generate many work-related ideas or schemata for their internal operation, but some ideas and schemata will be better than others. *Fitness* is the rating of how good they really are. Individual agents have a level of fitness that projects their longevity with the group. Reliability, robustness, resilience, anti-fragility, and optimum variability contribute to fitness at the collective level. Many global and local issues and discrepancies between them could culminate in an agent's level of satisfaction with the team situation. Dissatisfaction with the team's effectiveness, combined with communication and coordination contribute to the overall workload of working with a group that is above and beyond the workload associated with the individual's specific task (Helton, Funke, & Knott, 2014; Table 10.1)

Underlying the three hallmark features are numerous counterpoints between the realities of a CAS and the implicit assumptions of traditionally minded business managers and systems designers. Traditional thinking typically assumes linear relationships among system variables, outcome levels are proportional to input levels, and events can be explained by simple cause-and-effect explanations. In the CAS, small changes can have big consequences, and large interventions could have no effect. Different types of control variables within systems behave differently. Chapter 11 illustrated the limitations of simple single cause models of accidents and the appropriateness of nonlinear and models. Kjellen's (1984b) simulations indicated that small deviations in a procedure or work specification could compound into a large event, resulting in an industrial accident. A CAS would recognize the deviations before they got too far out of hand and correct the situation.

Conventional management thinking tends to ignore what it regards as random blips in system behavior and to maintain equilibrium as its objective of control. A CAS, however, would regard many of the blips as not the result of randomness at all, but possible signs of a larger pattern about to happen. The control objectives of a CAS would focus on identifying the possible patterns that are in motion, and navigating them with a repertoire of non-linear change processes, even when stability remains the desired objective. An intuitive understanding of nonlinear dynamics would be required (Guastello, Futch, Mirabito, Green, & Marsicek, 2021a, 2021b).

Self-organization occurs from the bottom up, whereas hierarchies operate from the top down. This class of phenomena gives rise to new concepts of networks and their relevance to the creation of emergent business activity. The conventional frame of reference might regard hierarchies as inevitable and prefer hierarchical control as a means of dominating the agents at the lower order of the hierarchy. The trend in CAS research, however, is that bottom-up self-organized behavior is more efficient. Hierarchies do emerge nonetheless, often as a means of conserving entropy. They could simultaneously indicate, however, that the system is gravitating toward greater inefficiency, which is a condition that should be deterred. Studies with hierarchical work groups show that the upper levels of the hierarchy can become overwhelmed by the activities from below, although some managers are skilled at maintaining efficiency in the face of acute entropy (Guastello, 2002).

14.1.6 Phase shifts

The evolution of a CAS is characterized by punctuated equilibria—periods of stability that eventually end when new technologies and processes or environmental circumstances are introduced. By the same token the organization's sunk costs in one technical system will inhibit its willingness to change to new systems unless the expected advantages are great and worth the effort of retraining the personnel and changing the equipment; thus there is a type of pragmatic resistance taking place. This sort of resistance has been well-known among consultants engaging in job redesign (Hackman & Oldman, 1976, 1980). In the ergonomics community, this form of resistance has become known as the "active user syndrome."

Phase shifts are essentially cusp catastrophe processes (Chapter 3). Organizational change, or change in a complex human–machine system, can be slow and gradual, and thus *evolutionary*. On other occasions it might be abrupt with widespread implications, and thus *revolutionary*. A cusp catastrophe model for this dynamic appears in Figure 14.2. The two

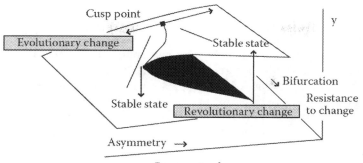

Figure 14.2 Cusp catastrophe model for evolutionary and revolutionary change in organizations. (From *Managing Emergent Phenomena: Nonlinear Dynamics in Work Organizations* (p. 84), by S. J. Guastello, 2002, Mahwah, NJ: Lawrence Erlbaum Associates. Copyright 2002 by Lawrence Erlbaum Associates. Reprinted with permission.)

control parameters are pressure to change and resistance to change. The sources of resistance, such as sunk costs in an existing system, would induce the changes to become more abrupt. Pressure to change could come from numerous sources, as mentioned earlier.

Two tales of technology depict the roles of evolutionary and revolutionary dynamics. Staudenmeyer and Lawless (1999) described a telephone system where the organization's rate of innovation outpaced the decline of older technologies. As new features were added to the system at the users' requests, the technology became "brittle," meaning that a small change required extensive relinking of the many parts of the system so that the system could stay internally coordinated. The technology eventually reached a crisis point where it became more sensible to rewrite the big program from scratch. The demands on the business for doing so were extensive. As the pressure to change the system mounted, the organization's knowledge about how the program was initially assembled was dwindling. Fortunately for the telephone company, there were still a few people left in the organization who still remembered how to manipulate the original 15 million lines of program code. There still remained the problem of how to shut down the existing system while the new system was being installed and tested.

The job of making the transition from the old to new system required not only people programming computers, but a massive reorganization of personnel into work groups that were often very different from their usual work patterns. There was a substantial need for coordination among the teams programming different parts of the system. The dynamics of work group coordination are examined later in this chapter.

The second story of radical technological change involved two competing technologies for a particular type of computer program. Both technologies enjoyed evolutionary growth until one producer got the idea to make his program more compatible with other programs that the users might already be buying and using. The system with the greater number of compatible linkages to other programs suddenly made great market strides, to the detriment of the competitor. The users who had initially selected the dominant technology did not require an adaptive response, except when "upgrades" became available. The other users were required to seriously reconsider their equipment choices and many switched their software (Lange, Oliva, & McDade, 2000).

14.1.7 Complexity catastrophes

The foregoing example of the telephone system concluded successfully. There were some positive contributions from group dynamics and team coordination that helped the process. One should bear in mind, however, that there is no guarantee that self-organization will produce a successful conclusion. Kauffman (1995) introduced the concept of a *complexity catastrophe*, where the system is too complex to reorganize effectively. A real-world example would be an organization that designs and implements one new product within a reasonable amount of time and succeeds. The same organization might be able to introduce two or three such products in the same period of time successfully. But when it tries to introduce, say, eight of them, they all turn into a mess of design errors, system failures, and financial embarrassments.

The introduction of too many new products could also be framed as a group level problem in cognitive workload. Although it would appear that the problem could be alleviated by expanding the workforce to include more designers, engineers, and pertinent others, there still remains the problem of getting all the products out the same corporate door and managing their implementation in the field, especially when the products are interrelated in some way. There is a qualitative difference between eight free-standing products that are equivalent in scope to each other, and one big product that is eight times the size of one of the smaller ones with additional interrelationships among the eight components.

14.2 REAL-WORLD COMPLEXITY

This section expands on the sources of complexity in real-world systems and some of the problems it could present. Problems are considered at the level of the individual operator's experience, the system design perspective, and decisions that fail from a lack of awareness of complex systems principles. Also there are several ways of defining complexity, the Shannon information function (Equation 3.1) is the most applicable for calibrating complexity from the number of states and probabilities (p) of those states.

14.2.1 Simplifying designs

The strategy for simplifying a design and still keeping it effective begins with a search for places where the system demands too much information to operate. The hunt begins by specifying three types of functional requirements: the goals of the user for the design, the intended physical outcomes, and the process that gets from the former to the latter (Suh, 2007). The next step is to identify modules of connectivity that are uncoupled, coupled, and decoupled. In an uncoupled module, the control-result sequence is independent of other control-result sequences; uncoupled modules produce the least information demand, all other things being equal.

Coupled systems often arise when there is a sequence of steps involved. If the operator needed to change one element, other elements would need to be readjusted. A repetition of steps would result. Coupled operations tend to be less resilient against random fluctuations; a perturbation at that affects one subsystem directly affects other subsystems indirectly. One never really knows the total information requirement of a system; one can only work with a particular representation of the system. Nonetheless, the quantity of information in the Shannon sense becomes magnified in a coupled system such that the summation sign become a cross-multiplication.

Coupled systems are typically less transparent than uncoupled systems, which presents another level of information demand that Suh calls "imaginary complexity." Imaginary complexity can start at the design stage where the understanding of the user's task is unclear, and provisions are made for options and features that are not really necessary. From the operator's viewpoint, the control of coupled subsystems is not well understood leading to extraneous operations.

Suh suggested two solutions to the design dilemma once the sources of complexity have been identified. One is to decouple the system as much as possible. Another is to introduce *periodic complexity*, which in principle makes the operations problems go away after a period of time. The illustrative example was the airport and airline confusion that is caused by bad weather conditions – cancelling or rerouting flights and passengers, and take-off delays. Once set in motion the airport confusion will continue until the end of the 24 h cycle, after which time the flight schedule starts over again and the problems of the previous day are no longer in operation. Periodic complexity is very similar to the use of forced periodicity to control chaos.

Helander (2007) recommended reducing the number of feedback loops in the system as a means of decoupling. The rationale is that feedback is valuable if the system requires corrections from the operator. When there is a low need for correction, fewer loops present less complexity. The iterative product design sequence does its best work when couplings are involved.

More couplings would predict that more iterations would be needed to get the system working properly. Fewer couplings would reduce the number of iterations needed. What often happens instead is that software developers release the software before getting all the bugs out because delivery time and revenue override concerns for the integrity of the product. The value judgment is particularly disconcerting when the bugs are security leaks that could expose financial or medical information or impair the operation of a power plant or fuel pipeline (Denning & Denning, 2016).

Endsley et al. (2003) observed that a similar problem arises from the use of automation, which is intended to keep the human input and error rate to a minimum, but often does not. They critiqued the design trend of programming systems to stack jobs sequentially and leave the operator out of the loop once the stack was set in motion. The problem, however, occurs when the automated system fails in some way, the operator is slow to notice and intervene, and concomitant errors accumulate.

14.2.2 Modularity

Hollnagel (2012) introduced a useful method for analyzing complex systems into their component parts and linkages. The Functional Resonance Analysis Method (FRAM) characterizes each component as having six parameters: its inputs, outputs, preconditions for action, resources that the component uses, means of control, and time horizon for operation. The relationship among components is depicted as a flowchart of hexagonal boxes and multiple links from each parameter to parameters of the next component. A small-scale example appears in Figure 14.3, where outputs from two components are inputs to the subsequent components, and a resource provided by the first component is used by the third. All components, parameters, and links should be specified with the understanding that some parameters and links could be more important than others.

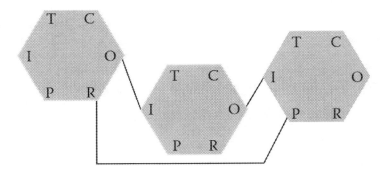

Figure 14.3 Example of FRAM modules.

A component could fail for a number of reasons related to its parameter values: speed of action was too fast or too slow; a distance error where something fell too short or long, a sequence error including repetition, wrong actions were taken on the wrong objects, duration of an act was too long or short, something went in the wrong direction, and timing errors (p. 72). The next step, which makes good use of complex systems thinking, is to analyze the variability in each of the components for sources of error. Here one determines which ranges of tolerance are optimal, and the extent to which variability compounds as one process links to another. The last step is to determine the consequences of errors. A simulation study could be devised around the FRAM configuration to ascertain these critical values and outcomes.

14.2.3 Multiple person–machine systems (PMSs)

Although the concept of the system had been introduced at an earlier time (Meister, 1977); the full implications of the system concept have taken a long time to unravel (DeGreene, 1991). The concepts of self-organization, the CAS, and most other nonlinear dynamics have only become operational for practical applications to human factors and ergonomics since the turn of the century. The early person–machine systems were conceptualized at the interface level where bilateral interactions between one person and one machine occur. Information that was considered relevant was predominately atheoretical in nature, and mostly geared to defining critical numeric values for one function or another.

Newer technologies have enabled a greater number of functions to be controlled by multiple person–machine systems. As depicted in Figure 14.4, information flows between the person and machine pretty much as usual, but there are also flows between machines and between humans. The machines are linked into networks. People can communicate either in real space–time or through networks via machines. Information from one PMS affects another, and delays in communication can result in various forms of uncoordinated action. The information that transfers is often far more complex in nature, and a shared system of meaning for information and symbols becomes all the more essential (Bennett & Flach, 2013; Bennett, Flach, McEwen, & Fox, 2015).

DeGreene (1991) observed that chaotic sequences of events do occur. Sources of chaos can hide in the displays that carry information to each operator and in the output of the PMS that carries information to all other operators. Stimuli often arrive in a flow over time. Humans are highly variable in their ability to interpret patterns arising from chaotic or simpler nonlinear sources (Chapter 7). Thus historical and predictive displays have been growing in their levels of sophistication. Chaotic controllers can take two basic forms. One form might be designed to regularize the flow

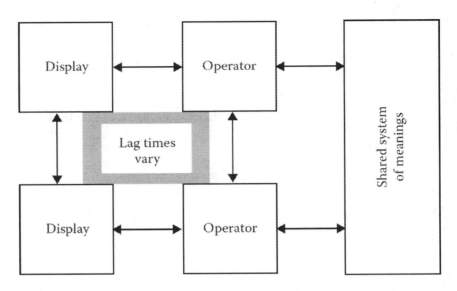

Figure 14.4 Triadic representation of multiple person–machine system interactions.

of information to the operator. The other would recognize and interpret patterns and identify options to the user. The specific means of control were detailed in Chapter 8. Self-organization and synchronization are other outcomes of a chaotic system; the system is not likely to remain in a high entropy for an indefinite period of time.

14.2.4 Revenge effects

The failure to recognize the complex nature of a complex system can produce some untoward consequences. *Revenge effects* occur when our attempts to control a problem by means of a technological intervention actually end up making the problem worse (Tenner, 1996). Here are some examples: An animal species was introduced to control a nuisance insect. The insect population was indeed controlled, but the predator animal overpopulated because its food supply was lush and its natural predators were not available to curb its population. Similar dynamics have occurred with plant species that were transported out of their natural environments to new environments. The eucalyptus tree that was a hardy source of good wood in Australia produced worthless splinters elsewhere in the world while it overtook ecological niches that were occupied by viable crop trees.

Higher quality athletic protection equipment lends itself to a revenge effect when the user feels protected, but takes greater levels of risk that exceed the limits of the protective equipment. In another example, automobile drivers who report using their seat belts more often also drive 8–15 km/h (5–9 mph) over the speed limit, as determined by speed monitors that were installed in their vehicles for experimental purposes (Berlin, Reagan, & Bliss, 2012).

In medicine, the prevention of some common causes of death serve to increase the odds of death by other diseases that are less tractable medically. In recent years, medicine has curbed the odds of death from heart disease, only to meet with an increase in deaths from diabetes, pneumonia, or Alzheimer's disease among the elderly.

In computer technology, the joy that is shared by all workers who have a computer on their desks results in less time spent on expert work, with more time spent on lower level word processing tasks that were once done by non-experts. There are other examples of information technology appearing to save work, but instead only save work for some people while making more work for others. Similarly, the "paperless office" was forecasted since the early 1980s, but a decade later more paper was being consumed not less. Instead of making more information products, more *versions* were produced on the way to getting to the final version.

Sometimes, fortunately, revenge effects happen in reverse. In the case of war, discovering new ways of saving lives carries over to postwar life. Other inventions such as refrigeration had the same fortuitous effect; the original objective was to transport food supplies for the military across the Pacific Ocean without spoiling. Civilians eventually benefited from a radical improvement over the old icebox.

Revenge effects are not the same as side effects or trade-offs. They are the result of the complex nature of the system. When a limited intervention (perhaps one that is based on a simple cause risk model) is taken to change the system in some apparently simple way, another aspect of the system reacts in ways that were not anticipated. Tenner (1996) thus reminded us of the fabled Murphy's Law: Anything that could possibly go wrong will go wrong at the worst possible moment.

Side effects and trade-offs are properties of the intervention technology rather than of the system itself. A medical side effect would occur when a medicine that would relieve one problem could induce another problem, usually in rare circumstances or with prolonged use. A trade-off is similar, but we call it a trade-off if the secondary effect is more probable: The intervention might indeed stop a problem but it gives you a different one instead, and one must choose which negative outcome is better.

Revenge effects should not be confused with any of three other similar nasty habits of complex systems, according to Tenner (1996). A *repeating effect* is where we end up doing the same thing more often rather than gaining time to do other things. In word processing, it appears that people make more versions of a document, but not necessarily any more information products. People who are on the receiving end of these closely similar documents must take care to discard any old versions of the same documents so as not to rely on errors in the earlier versions. Cutting and pasting does indeed save retyping of potentially long passages, however.

A *recomplicating effect* occurs when a technology that was first introduced to simplify some tasks ends up introducing new and different complications. For example, an organization prints and mails a document to a large but defined group of people. Then the organization decides that it wants to restrict distribution of the document to the defined group and not make the document easily available to the world. So it sets up a password system, which takes a lot of work, and then finds out that the password system requires a lot of maintenance because of high turnover in the group membership or because members wanted to change their e-mail addresses, passwords, or Internet service providers frequently.

A *recongesting effect* occurs when an idea that was very attractive at one point in time becomes so popular that the system becomes useless because of the sheer quantity of people trying to use it. Tenner (1996) gave some examples: The electromagnetic spectrum was once sufficient for broadcasting. Now it is overloaded with media and telecommunications channels, such that users were reverting to phone lines for a medium. Meanwhile the telephone companies were running out of phone numbers in some regions of the country. At this time of this writing, landline phone traffic, short e-mails, and a lot of Web browsing are now being offloaded to cell phone bands.

As another example of a recongesting effect, shortcuts that are introduced to an urban highway system to relieve congestion might have the effect of moving greater congestion to another location because everyone is looking for a shortcut. Indeed there are occasions where driving through the city streets is faster than taking the highway route.

14.2.5 Complex systems of systems

Several large-scale systems are currently in development, notably NextGen air traffic control, the smart power grid, multi-robot systems, and new automation capabilities for private automobiles. The former two are described next. Multi-robot operation is considered later in this chapter after introducing a few more principles of human operation. The automobile applications are considered in Chapter 15 in conjunction with other roadway issues.

14.2.5.1 NextGen air traffic control

The workload issues facing air traffic controllers have been well recognized along with growing volumes of air traffic, even though solutions have been slow to materialize. A mega-solution is underway with the slogan "NextGen." It involves new forms of automation for both the aircraft and the control tower.

The take-off and landing portions of the flight put heavy workload demands on both the pilot and the air traffic controllers. The last portion of a landing usually requires the pilot to de-automate and negotiate the landing to the target location based on cues received from direct perception of the interface, possibly some heads-up displays, and the control tower. The new initiatives involve replacing the human vision and control operations with synthetic vision. Synthetic vision involves a system of cameras outside the aircraft that produce a digital video feed to the pilot; at present it is inconclusive as to how much information should be presented to the console and how much to heads-up displays (Ellis, Kramer, Shelton, Arthur, & Prinzel, 2011); human factors issues with heads-up displays were discussed in Chapter 5. In principle, synthetic vision will improve the visual display and connect to the automated landing operation at the tower.

The second portion of the innovation is a digital communication link between the aircraft and the control tower. The hypothetical new control system would land properly equipped aircraft automatically. One lingering challenge is that controllers in different parts of the country work with different interfaces and symbol systems, and a standard would need to be developed based on icons that are not widely shared at present (Billinghurst et al., 2012). A survey of controllers, however, showed a split in opinions regarding the relative merits of NextGen (Cho & Histon, 2012). On the one hand, it would shift workload associated with routine conditions to automation thus freeing up human channel capacity to respond to unusual conditions. On the other hand, it could place more load on the air traffic controller for having to figure out what equipment is available on a particular aircraft and whether it would interface properly with the control tower (Gregg, Martin, Homola, Lee, & Mercer, 2012; Lee & Prevor, 2012).

The phase-in period for even the best system design is not going to be immediate. The lifespan of a commercial aircraft is approximately 30 years, and many are still relatively young or middle aged. The result is a complexity issue that is not completely different from what was observed in the power plants in previous chapters – cascades of technologies from different generations of equipment design all in simultaneous operation. The NextGen systems, once implemented would need to contend with a cascade of technologies and aircraft. The complexity of the technology mix would eventually dissipate – until someone decides that it is time to upgrade all the software.

14.2.5.2 The smart power grid

The electric power grids that operate in North America and the rest of the industrialized work provide a continuous source of power to industries and households through a system of generating facilities and distribution facilities. Irregularities in the cost of production, sources of production, and demand are neatly flattened through the sourcing and delivery systems, which are essentially chaotic controllers. Micro-outages still occur when the volatility of supply and demand exceeds the tolerance of the switching systems, which are mediated by humans to some extent.

The so-called *smart* power grid involves two aspects of technological evolution. One is the source of power. The primary forms of power currently come from the processing of fossil fuels, nuclear power, and water turbines. Solar, wind, and geothermal energy are new sources that have become very attractive because of the rising costs and limited supplies of fossil fuels and the unpopularity of nuclear sources in the wake of the incidents at Three Mile Island and Chernobyl. The challenge for large-scale solar power is that the power needs to be created when and where the sun shines, stored, and transferred to requesting power plants at disparate locations. The challenge for wind power and geothermal energy is similar.

The switching concept behind the smart grid is essentially the same idea behind the relatively local grids that exist now except that the scale of the system increases several orders of magnitude to encompass entire continents (Bushby, 2011; Holmberg, 2011; Ivanoff, Yakimov, & Djaniykov, 2011; Sarfi, Tao, & Gemoets, 2011). The implementation is easier said than done as it requires all participating plants to have state of the art equipment and compatible technologies; many installations are a long way from that standard. The smart grid system runs through the Internet, which of course, offers numerous opportunities for intrusions by unwanted agents.

The compatibility of software and equipment is a recognized problem in grid computing, especially when data mining becomes involved (Dubitzky, 2008). Data mining architectures for "decision making" have been considered, however, along with ways for making the results transparent to the user.

The implications for usability and the new roles for human operators are still in flux. The system is complex with three levels of activity taking place. At the ground level, the power is distributed to through the network to specific users. At the midrange level, the power grid is channeling power from different sources to segments of the distribution system, which are all connected. At the top level, there is a group of perhaps 20 humans watching everything and responding to varying load demands from different segments.

One of the top challenges for power plant operation is to prevent cascading outages that can be instigated by severe weather, excessive loads from air conditioners, or cyberattacks. Cyberattacks can take on many forms, such as "scanning attacks, denial of service, man-in-the-middle attacks, replay attacks, jamming channels, false overloading, altering the grid parameters, manipulation of demand, and time-delay switching" (Shuvro, Das, Jyoti, Abreu, & Hayat, 2022). When a compromising event is instigated, the operators' responses need to be successful in the first hour before the cascade to a widespread outage occurs. Cyberattacks often involve repeated assaults, so the battle can rage on considerably longer.

14.2.5.3 Health care systems

The health care system in any country is also an example of a system of systems that operate at the micro-, meso-, and societal levels. The micro-level involves two interacting systems – patient care and safety and employee safety. Employee safety became a critical problem during the COVID-19 pandemic. Health care employees in the United States and United Kingdom were 13 times more likely to contract infection than the surrounding general community (Rodriguez & Hignett, 2021). At the macro-level, hospitals often reached maximum capacity, particularly the high demand for intensive care units; in some locations, temporary facilities had to be erected to meet the overflow.

At the societal level, complexities already existed among health care providers, pharmaceutical companies, and health insurance. During the pandemic, new complications arose when other types of patients could not receive needed surgeries or other treatments, because too much of a risk had built up in the hospital environment that precluded allowing in other types of susceptible patients, for example those requiring treatments for cancer.

Telemedicine became prominent during the COVID-19 pandemic in an effort to provide care while keeping patients out of high exposure areas or bringing in further virus load to the facility. With advances in software that coalesced at the time, telemedicine became comfortable for many patients and providers, and it seems to have become a norm for many patient–provider interactions. Telemedicine offered particular advantages for connecting patients from remote locations to providers of high-quality care. Safety issues that still proliferate are considered in a later section of this chapter on safety in complex systems.

14.3 COLLECTIVE INTELLIGENCE

The cover image on Kelley's (1994) book *Out of Control* depicted a stylized office building designed as a grid with large windows and huge bees flying in and out. The implicit message was that work processes resemble a swarm of insects more closely than a machine that is designed for exact reproductions of objects. The queen bee does not give orders to the worker bees. The worker bees figure out their own work routines, which we observe as a swarm, based on one-on-one interactions with each other.

The concept of collective intelligence originated with studies of social insects. An ant colony self-organizes into jobs such as foraging for food and nest maintenance. If there is a labor shortage in one of those areas, ant personnel are diverted from another task. Food foraging follows a pattern whereby the foraging patrol heads out in one direction on the first foray, then systematically adjusts its course by a few degrees for the next successive forays. No one ant actually knows the entire plan, but the collective executes it well (Sulis, 1997, 2021).

The concept crossed over to human cognitive phenomena when it became apparent that decentralized networks of people produce ideas, plans, and coordinated actions without being present in the same physical location simultaneously. The interaction among people is greatly facilitated by computer-based systems such standard e-mail, listservers, and Web-based technologies (Bockelman, Morrow, & Fiore, 2013; Guastello & Philippe, 1997). The growth of virtual communities gravitates to an attractor that represents a stable population. The study of collective communication patterns in technology-driven systems, which often facilitate tracking specific statements and their temporal sequencing, has led to a rethinking of human interactions in real time as well (Gureckis & Goldstone, 2006). By the same token, inserting a technology in the human interactions changes the nature of the task and could introduce a new element of workload (Roth, Sushereba, Militello, Diiulio, & Ernst, 2019); typing instead of talking would be one example. In either case, the collective thought process has become known as *distributed cognition*.

The boundaries usually associated with an organization are semipermeable, meaning that a great deal of information flows across organizational boundaries and might not be centralized within the organization. This phenomenon together with analogies to insect colonies was the underlying theme in Kelly (1994). With decentralized or network-based communication and decision patterns, the notion of controlling a human system necessarily changes. Consistent with the idea behind ant colonies, the top-down control that is usually inherent in organizational structures simply does not operate well any longer: Events self-organize from the bottom up.

An ant colony displays some highly organized activities, but each ant does not have a big plan in its little head. Rather, each ant is equipped with elementary schemata that synchronize with those of other ants when larger scale events occur. Sulis (2009, 2021) identified several principles of ant collective intelligence from which it is possible to extrapolate analogous functions in human systems. The first two are *interactive determinism* and *self-organization*: The interaction among individuals gives rise to the final collective result. The final result stabilizes in interesting ways and without external intervention; in other words, the Queen Ant or Bee is not barking (buzzing) orders to the others. The stabilization of a collective action pattern is a *phase transition*.

The distinction between embedded and embodied cognition would be helpful here. The embodied portion operates automatically, assimilating nuances in the environment. The embedded portion is aware of the nuances and perturbations in the environment and permits adaptations or accommodations to be made. Environmental nuances, nonetheless, have an impact on the final collective result; the phenomenon is known as *stochastic determinism*.

When probability structures that underlie collective outcomes remain stable over time, they are regarded as *nondispersive*. They remain as such until a substantial adaptation is made by the collective, producing *broken ergodicity*. Similar disruptions occur at the individual level as an experience of one agent impacts the behavior of another, thereby amplifying the feedback to a third. With enough uncontrolled fluctuation, the ants' foraging path can shift suddenly, producing *broken symmetry*.

Some further principles are likely to take different forms in human in contrast to insect contexts. One is *salience*: The environmental cues that could induce broken symmetry are likely to work to the extent that they are sufficiently visible compared to other cues. In human contexts salience is complicated by *meaning*, which can be operationally defined as the connection to other ideas, personal intentions, and system goals (Kohonen, 1989). Ants do not appear to express much variety in personal intention, but humans do so regularly. Humans and human systems often have several competing goals. That brings us to the last element of Figure 14.4, the shared system of meaning which is essential for communication.

Computational experiments often assume that individual agents are destined to interact. Sometimes that is the case, of course, but people also reserve the choice to interact or not. They can also interact by moving around in the same common space without actually talking to each other. Whatever rules they invoke to decide whether to interact are likely to play out in emergent patterns of social organization (Sulis, 2008, 2009; Trofimova, 2002).

14.3.1 Asynchronous problem-solving in e-communication

The comparative study of real-time and electronic problem-solving media has uncovered four new group phenomena. One is the *unblocking effect*: In real-time brainstorming groups, some participants have difficulty getting a word in edgewise, particularly if the flow of ideas changes direction. Electronic media, in which communication is asynchronous, give individuals the time they need to consolidate and phrase their thoughts and post them to the group. The downside of unblocking is *flaming*: It is just as easy to make snarky remarks and insult others deliberately when an in-person setting would inhibit those behaviors.

Another special effect is the *critical mass*. A review of group productivity studies showed that group brainstorming sessions produce more ideas than are produced by the most competent individuals in about half the occasions where such comparisons were reported. At other times, the group is not more productive than the most productive individual. Some of the disappointing behavior of groups can be explained by whether tasks and rewards were defined for groups or for individuals; social loafing was another explanation (Shepperd, 1993). Dennis and Valacich (1993) found, however, that a critical number of group members is needed to gain an advantage for groups using computer media.

The third effect pertains to channels of communication and filtering. In direct human-to-human interaction there is the task-oriented part of the messaging, but also social content, words spoken conversational structure where context can be assumed, vocal intonations, and facial expressions. The filtering process that occurs when a group's communication and work products occur solely through a computer network has an advantage for keeping everyone focused on the problem-solving task (Walther, 1996). Participants can prepare outgoing messages without the stress of interpersonal interaction (Walther, p. 24), and because of the time asynchrony, task and interpersonal messages can become disentrained. According to Walther, the filtering of social content is sometimes overstated because friendships and romances do form in virtual media. The anticipation of continued interaction predisposes individuals to act in a friendly and cooperative manner. Social remarks also tend to increase in cases where participants are not close to their maximum available communication time (Walther, Anderson, & Park, 1994).

Videoconferencing, which should be able to convey more channels of communication than text, was in limited use prior to the COVID-19 pandemic. Allen, Golden, and Shockley's (2015) review indicated that videoconferencing did not restore channels of communication particularly well because the person would have to look at the camera and the screen at the same time. Normal eye contact was clumsy at best.

Videoconferencing suddenly went into widespread use during when the pandemic started; local and national governments were issuing stay-at-home orders and rules for wearing masks and social distancing. Distance work thus became a necessity for keeping the flow for some types of work. Once the pandemic started, users reported that videoconferences were more efficient uses of time than face-to-face meetings, they were shorter, and produced better outcomes. The benefits were counteracted by unusual fatigue after a videoconference that often extended for several hours (Bennett, Campion, Keeler, & Keener, 2021). The fatigue was specific to videoconferencing, and attributable to higher levels of concentration and focus of attention. Some participants would turn off their cameras and mute their microphones to reduce the stress, but at the same time felt disengaged from the work group.

Shockley et al. (2021) reported similar results. Their theory attributed the fatigue produced by the camera was attributed to uncomfortable levels of self-focused attention. Fatigued individuals were less engaged and less participative in the conversations. The results suggest that greater engagement could be accomplished by with the camera off, but the effects of many cameras off on other coworkers has not been ascertained.

14.3.2 Sensemaking and situation awareness

Sensemaking in emergency response and other contexts has also been studied as *metacognition*, which focuses on the development of mental models by groups and teams. Mental models play a significant role in group coordination, as described later in this chapter, and the experimental paradigms have different with regard to whether the experimenters assumed that a fixed mental model should be shared at the outset of the task sequence, or whether the mental model should be allow to develop on the fly, which it appears to do in many cases in the real world. The question is not particularly helpful, however, because in a complex dynamic environment, events are changing quickly, and new features are likely to emerge. Whereas coordinated groups interact and perform tasks on a rule-based system, groups in a dynamic environment are going through a process of finding those rules for the first time known as *metacognition*. The metacognitive process involves utilizing acquired knowledge and responding to the demands of new situations (Fiore, Rosen, Smith-Jentsch, Salas, & Letsky, 2010).

Sensemaking is closely tied to situational awareness, particularly at level 2 where a mental model of the situation has been formed. A complex system would be made up a team of individuals with a technology that provides multiple sources of information. An individual can only attend to one object at time (attention cycling) and tends to engage some attention tunneling to limit the scope of information search to known items and locations. As a result, the individual is likely to become blind to changes in other parts of the information flow. Sensemaking and situation awareness thus become distributed activities. It also becomes plausible that no one individual really has a complete mental model of the situation (Chiappe, Strybel, & Vu, 2012, 2015; Stanton, Salmon, & Walker, 2015).

Humans in the system rely on their prior knowledge and working memory, but offload the mental computation and visualization to the machine parts of the system due to the usual problems with workload capacity. The technology part of the solution is to develop new software designs with multimedia person–machine interfaces for multiple users have been designed for different contexts. Figure 14.5 is an example of an interface (Riley, Endsley, Boldstad, & Cuevas, 2006) that was designed for multiple military operatives to encode, store, and share information about the movement of troops and equipment. The system also allows the operators to plan tactics from dispersed locations. Agents' real positions are represented by symbols that are neatly drawn by machine. Hypothetical new agent positions that the team is discussing can be entered by using symbols that look as though they were hand drawn with a crayon.

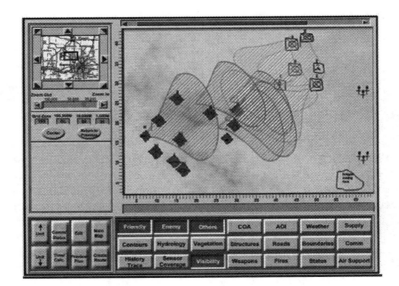

Figure 14.5 Example multimodal interface for situation awareness. (Reprinted from Riley et al. (2006) with permission of Taylor and Francis.)

14.3.3 Networks

Information about a network can be valuable for understanding sources and destinations of messages, influences that produce distortions of the message en route, and the social or work group processes taking place. They might also serve as the basis for engineering telecommunications equipment or the location of transportation hubs (Freeman, 1979; Kantowitz & Sorkin, 1983). They have had some value for uncovering "organized crime" activities, understanding communities, markets, social change (Jacobsen & Guastello, 2011; Wellman & Berkowitz, 1988), job mobility (Levine & Spedaro, 1998), and the transactions within and between discussion groups on the Internet (Wellman, 1997) as well.

The content of communication could be as important for understanding emerging networks as the raw quantity of communication. For instance, Bales and Cohen (1979) classified communications within a 3-D taxonomy of dominant versus submissive, friendly versus unfriendly, and emotionally expressive versus controlled. These parameters of communication in combination could lead to cohesive (and probably cooperative) or polarized (and probably competitive) groups (Axelrod & Hamilton, 1981; Flache & Macy, 1997).

At one level of analysis, communication (in the conventional use of the word) and social exchange are not appreciably different, but the content of exchanges or communications can promote some different dynamics. In a task group, the exchange might consist of approval for approval, or perhaps approval for task compliance (Flache & Macy, 1997). In friendship ties, however, a bit of time is required for a link to fully establish. A link is likely to form if the friendship initiation attempt of one actor is reciprocated by the other (Zeggelink, Stokman, & van de Bunt, 1997).

Smith and Stevens (1997) introduced a motivational component to network formation among individuals. According to the prevailing theory in social psychology, motivation to affiliate takes the form of arousal. When two people affiliate there is an arousal reduction, which is mediated by the endorphin mechanism. Once an affiliation link has been established, a set of four feedback loops form for arousal and arousal reduction within and across the two participants.

14.3.4 Network growth

The fourth special effect of e-communications pertains to the natural formation of large discussion groups on the Internet. The formal analysis of network growth is known as *percolation*. The following phenomena are relatively simplistic but pertinent examples. Networks can consist of interacting individuals, but also of larger social units such as organizations wherein specific people are usually responsible for maintaining the connection between one business organization and another. Thus a node in a network might not be a single agent, but a member of other networks that are implicitly combined with the network under analysis.

In a high-entropy social milieu, the formation of links begins with interactions among people, which are analogous to the interactions among atoms, at least in part (Galam, 1996). In this line of thinking, something akin to temperature occurs in the social environment that causes the people within to bounce around faster as molecules would do within a container. Random contacts occur until a drop in entropy occurs when the right combinations of people or atoms are found, which is when a link occurs. If the process is allowed to continue further, social structures evolve that support the emergence of primary leaders and technical leaders in either real-time or virtual network contexts (Guastello, 2007; Hazy, 2008).

Smith and Stevens observed, furthermore that, unlike the molecule analogy, additional people can join the aggregate until a group is formed. The extent to which the aggregate grows might depend on context variables such as the complexity of the task or activities involved, the type of relationship between individuals in the groups, and whether competitive exclusion is part of the group rules of the relationship. Additionally, individual differences in the desired number of friends can affect the growth of networks in terms of size or density (Zeggelink, Stokman, & van de Bunt, 1997).

Extensive networks of weak linkages are often the best sources of news and technical information (Constant, Sproull, & Kiesler, 1997). They can expand more easily than dense networks because the threshold between being in and out of the network is relatively low. Networks with weak ties have another advantage in that they can provide new information more readily than dense networks; dense networks share information quickly, but the store of content can be expended quickly as well, and needs to be refueled by input from other contacts.

Guastello and Philippe (1997) argued, furthermore, that the dynamics for joining a network could be expressed as a *bandwagon* game. Bandwagon is one of several theoretical coordination scenarios that are considered later in this chapter. As a general rule, a bandwagon occurs when the cost of joining an activity decreases as more participants join the activity. In the case of network growth, the two participants would be a specific person who is contemplating joining the group (or network), and the rest of the group would be conceptualized as the other unitary player. Participation by either the individual or collective may be low, moderate, or intense. The extent to which the information flow is positive for an individual, relative to the cost of involvement, predicts increases in network members. Information flow, assuming that it is relevant to individuals' needs, becomes greater and more varied when networks are larger (Guastello, 2002).

Equation 14.1, describes the dynamics of how some participants could enter a network and grow long-standing links to other network members, while other participants could drop out of the network eventually:

$$z_2 = \theta_1 z_1 \, \exp[\theta_2 z_1] + \theta_3 \qquad (14.1)$$

Equation 13.5 is essentially a May–Oster population model, where θ_1 is comparable to a birth rate for a population of organisms. θ_2 is the group's carrying capacity for members, and reflects the attrition or death rate when the carrying capacity is exceeded. Equation 14.1 was shown to be a close fit for the development of a virtual organization composed of individual members (Guastello & Philippe, 1997). Within this structure it is possible for a network population to reach an asymptotically stable size in which new linkages would enter and form while some would dissipate and die out.

A third model of potential relevance to the growth of networks is the cusp catastrophe model (Guastello, 2002). In the empirical example, the linkages consisted of joint patents among organizations in the telecommunications industries, and the dependent measure was the number of links between an organization and all other organizations in the set. The two stable states would be whether the organization stabilizes within the network or drops out of it. The two control variables were whether the organizations that were contributing the data points came from the telecommunications, semiconductor, or computer-manufacturing industries. Different dynamics were apparently occurring in those industries, which promoted different network growth rates, but further information was not available to reconstruct the differences between industries in more specific detail.

A new phenomenon that has come to the foreground, however, that could affect the broader landscape of person–machine interactions and system design. *Coopertition* is the behavior pattern whereby firms compete in some markets, but cooperate to build hybrid products for another. Pathak, Pokharel, and Mahadevan (2013) found in a simulation study that coopertition is likely to be a stable and productive relationship if the business environment favors building on strength, and unlikely to sustain if the environment favors exploring new ideas and strategies. It would follow, therefore, that some people in the research and development fields would need to adopt different strategies for communicating with the same nodes in their networks. Task-switching phenomena thus move into a different level of scale.

14.3.5 Dynamics of feedback loops

Several researchers have suggested that electronic media are catalyzing the self-organization of a collective intelligence (Guastello, 1998, 2002; Guastello & Philippe, 1997; Mayer-Kress, 1994; Smith, 1994). If that were true, it should be possible to locate chaos dynamics prior to such self-organization. Contractor and Siebold (1993) used simulation techniques to determine "boundary conditions under which groups reproduce, sustain, and change existing structures" (p. 535). They noted further that such structures apparently change in a punctuated equilibrium fashion. Structures would be produced in part by the group's expertise in the use of the decision support system, and the group's perception and awareness of norms. *Structures* implied conversational behaviors of very general varieties. Some common examples of norms were: quality of the thought product, quantity of production or apparent effort, civility of social interaction, and use of gatekeeping initiatives when participants' input strayed beyond a desired norm.

Their simulations were based on the quadratic logistic map function such that change in the time spent contributing to the group (order parameter) was a function of the expertise contributed by the other participants and the group norms. The quadratic function was capable of producing constant, cyclic, and chaotic activity clusters. The simulation results showed different phase diagrams for participation time depending on prior expertise with the decision support system, presence of prior norms, and intensity of training between the sessions of activity.

Contractor and Siebold's (1993) findings were corroborated to a great extent by a separate line of research that showed that the quantity of individuals' output did change over time throughout a discussion according to a modified logistic map function (Guastello, 1995). In the latter example, the dependent measure was the length of a person's response in a conversation on a listserver. There were two active control parameters in the process. One control parameter was cumulative elapsed response, which was the quantity of contribution by other participants in the conversation that occurred in between consecutive responses by a single person. The second control parameter was personal response style, which was essentially a black box containing individual differences in output levels that were not otherwise accounted for. There was, furthermore, a global linear trend whereby the cumulative elapsed response at Time 1 predicted an increased output at Time 2.

In another effort, conversations among three small groups of problem solvers (with real, rather than simulated, problem objectives) were studied over a period of approximately 100 days (Guastello, 1998). The medium was electronic with communication in American Standard Code for Information Interchange (ASCII) text through specialized conferencing software. The amount of total output from each group per day was recorded, and fluctuations over time were analyzed for dynamic content. (Note the shift in perspective here from an independent measure that is based on individual behavior to one that is based on group output.) Density, the dependent measure, was defined as the number of responses generated by the participants in a 4-day time period. Group 1 contained three discussants, Group 2 contained seven, and Group 3 contained eight.

It was determined that the groups' production over time was chaotic, and simultaneously driven by a second periodic dynamic. Equation 14.2 was obtained for density over time:

$$Z_z = e^{(0.25_z)} + 0.43A - 0.26C - 0.34, \tag{14.2}$$

where z_t was the density at two consecutive observations in time, A was the number of active threads during the time interval of z_1 and C represented differences among the conferences. The Lyapunov exponent (regression weight in the exponent to e in Equation 14.2) was positive, indicating chaotic behavior and a dimensionality of 1.28. Conference differences were measured as a three-value effect-coded variable indicating the overall quantity of response from each conference; some discussion forums, which are combinations of topics and people, generate greater quantities of discussion than others.

The second function was the number of discussion threads, A, that were operating on a given day. The greater the number of threads, the greater the total group output was. Equation 14.3 was obtained for the change in the number of subtopics or threads over time:

$$A_2 = 0.75A_1e^{(-0.36A1)} + 0.33. \tag{14.3}$$

The Lyapunov exponent was negative, indicating that the function was not chaotic; the dimensionality of the time series was 1.70, which was actually close to a parabolic function. It was interpreted as periodic because it repeated across conference discussions.

The conclusion from the study was that there were two important dynamics operating. One dynamic was a periodic driver, and the other was a chaotic slave, to use Haken's (1984) nomenclature. The driver was the unpacking of the conversation into separate threads. The chaotic output over time was observed when the discussion volume reached a critical level of intensity. Thus if a discussion seems slow to warm up, breaking the topic into separate threads running simultaneously would facilitate greater input.

14.3.6 Other temporal dynamics

The interest in temporal dynamics of problem-solving interactions resurfaced as a research topic in human factors, both with (Gorman, Cooke, Amazeen, & Fousse, 2012; Russell, Funke, Knott, & Strang, 2012; Strang, Horwood, Best, Funke, & Knott, 2012) and without (Fu & Pirolli, 2013; McComb, Kennedy, Perryman, Warner, & Letsky, 2010) explicit use of nonlinear dynamics. Information search strategies are often shaped by other agents in the network environment (Fu & Pirolli, 2013). Small-world phenomena have some interesting implications for information searches. Ideas spread through networks, but not every idea "goes viral." Most network schemata follow "narrow tree" diffusion structures that end the diffusion much sooner than what would occur through a viral epidemic-type transmission.

The research strategy for the problem-solving process requires recording a conversation, then coding each utterance for features of interest. McComb et al. compared teams' processes in a planning task and how they arrived at their shared mental model. Those who worked face-to-face tended to explore more possibilities and options before arriving at their target model compared to those who worked in distributed fashion through conferencing software. The software-mediated groups took somewhat longer to arrive at their endpoint, but there was no actually difference between the two groups in quality of solution. The explanation for the differences was simply that talking was easier and faster than typing, all other things being equal.

An example of a nonlinear approach (Gorman, Cooke, Amazeen, & Fouse, 2012) involved coding conversations during a drone aircraft mission and analyzing the content for signs of adaptability versus rigidity and exploring conditions. The nonlinear analysis was looking for recurring conversation patterns over time. Pattern formation, entropy levels, and what constitutes a repetition are predicated on the codes that one applies to the data, however (Guastello, 2000); it is possible to code one conversation according to two different protocols and arrive at two different pictures of the nonlinear dynamics involved.

14.3.7 Learning organizations

Allen (2009) reported a simulation study that examined four learning strategies and the relative effectiveness of each: (a) Darwinian learning, where the organizations start with a random strategy, organizations with good strategies survive, and organizations with poor strategies go bankrupt and are replaced by new organizations with random strategies; (b) imitate the winner, which means that organizations copy others in the environment that have apparently functional strategies; (c) trial and error, where organizations explore possible strategies, try some, observe results, reevaluate and perhaps try something else while continuing to consider new options; (d) mixed strategies, where all of the previous three exist in the organizational environment. Results showed that Darwinian learning produced the worst results for the industry as a whole with the largest proportion of bankruptcies. Imitating the winner worked much better, although it was subject to large fluctuations in profitability levels; it involved imitating the winner's limitations too. Overall the greatest success was recorded for the trial and error strategy where agents learn from their mistakes and continually seek out possible improvements by exploring what he characterized as their landscape of opportunities.

Next, consider the situation where the organization acts a whole unit. Work groups within the organization would be coordinated within a larger hierarchy of activities. The notion of a learning organization became a fashionable view of an organization shortly before the notion of the organization as a CAS took hold (Seo, Putnam & Bartunek, 2004). In its earlier manifestations, learning organizations were those that had evolved processes or structures analogous to individual perception, cognition, memory, and adaptation processes. In later studies, learning processes in organizations were seen to promote self-organization of dominant strategies or schemata from the bottom up. Individuals and teams adopt processes that produce ideas, schemata, mental models, and meanings that are eventually shared with other teams. Some schemata become dominant enough in the organization to shape new schemata for newcomers or new responses to new challenges (Van de Ven & Hargrave, 2004).

The perception, situation awareness, or sensemaking processes in organizational contexts require information exchange networks that extend outside the organization to other organizations in the same industry, other organizations in different industries, and of course customers. Van de Ven and Hargrave gave an example of a successful use of bottom-up development of wind turbine technology. Danish industries started with relatively simple technology, and through close interaction with customers and their needs, shaped a premier technology that is financially successful for the organizations involved. Would-be competitors from the United States, however, took an isolationist strategy and attempted to leapfrog the stages of development by developing an advanced technology quickly. They maintained little communication with customers and were generally unsuccessful in their efforts.

14.4 TEAM COORDINATION

Systems composed of multiple person–machine systems do require coordination to function optimally. Although it is tempting to consider the full range of group dynamics phenomena here, this section of the chapter is necessarily limited to principles that have been specifically implicated in person–machine system interaction. The interested reader should see Zander (1994), Hackman (1990), for further principles and examples from social-organizational settings. Coordination is relevant to person–machine systems and has some known implications for CASs.

Coordination occurs when two or more people do the same task or complimentary tasks simultaneously. Numerous examples can be found in manufacturing work teams, hospital emergency rooms, military operations, sports teams, and musical or theatrical performances. Coordination has been measured in empirical studies as the time delay between a group member's action and another member's contingent action (Brannick, Roach, & Salas, 1993), the quality of communication between members of a group (Daily, 1980; Gittell, 2000), or the performance of a group in a task wherein members must take correct actions in a correct sequence (Guastello, 2002; Guastello, Bock, Caldwell, & Bond, 2005; Guastello & Guastello, 1998). There are several theoretical aspects to coordination dynamics that emanate from theories of implicit learning, shared mental models, games, and nonlinear dynamics.

14.4.1 Implicit learning

Groups acquire coordination as an implicit learning process, which is a form of learning that takes place while the group members are explicitly learning to do something else (Frensch & Runger, 2003; Seger, 1994). The group members might be less attentive to the implicit learning process than they are to explicit learning goals. In the case of coordination learning, the implicit learning objective is to learn how to respond correctly to actions and nonverbal signals given by other group members. A group performance experiment demonstrated, however, that coordination that was acquired while learning one task transferred to subsequent tasks where the performance goal that required coordination was different (Guastello & Guastello, 1998).

Although implicit learning is usually studied as an individual process, it can take the form of a group process also. The group process involves co-adaptation and mutual entrainment among the participants, and thus

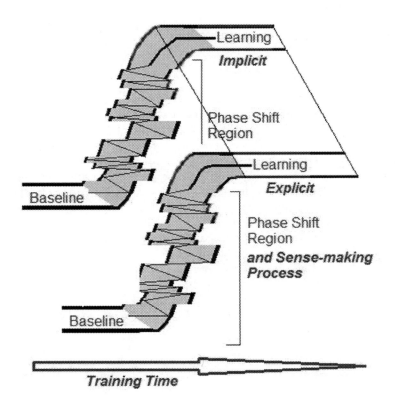

Figure 14.6 Twin learning curves for an explicit task and implicit coordination.

qualifies as a self-organizing process. There is now sufficient evidence from neurological, individual behavioral, and group behavior levels of analysis that learning processes are self-organizing processes. The initial phases of behavior are chaotic in performance or output, with asymptotic learning curves taking over as the process becomes complete (Guastello, Bock, Caldwell, & Bond, 2005). As a result, there are often two learning curves in progress, one for the explicit task and one for the implicit coordination task. Figure 14.6 depicts the two connect learning curves as jagged trajectories, which is more realistic than the smooth curves that are usually shown in textbooks on learning theory.

14.4.2 Shared mental models

Studies indicate that the coordination among work group members is greatly enhanced if the members have a shared mental model of their tasks, procedures, and group processes (Banks & Millward, 2000;

Cannon-Bowers, Salas, & Converse, 1990, 1993; Druskat & Pescosolido, 2002; Stout, Cannon-Bowers, Salas, & Milanovich, 1999). Shared mental models may be induced by cross-training the group members in each others' roles, or by discussions and presentations of groups' task models (Marks, Sabella, Burke, & Zaccaro, 2002; Matthieu, Heffner, Goodwin, Salas, & Cannon-Bowers, 2000). The sharing of mental models also requires consistent organizational support if the group is not a freestanding group (Druskat & Pescosolido, 2002).

In spite of the advantages of shared mental models, however, NDS studies on coordination indicated strong analogies to psychomotor coordination and animal models, which of course did not require discussion groups, cross-training, verbal communication, or leadership for coordination to occur. This does not mean that something akin to a shared mental model did not emerge in human contexts; rather, group members had to arrive at the model through their own individual observations. Game theory would suggest that any mental models that are relevant to coordinated action can be reduced to rules, options, utilities, and sometimes trust in the rationality of the other players.

14.4.3 Role of verbalization

Clear, precise, and prompt verbal communication produces a positive impact on coordination. In one example situation where the task participants were military aircrews working on simulated flights, a panel of raters evaluated the crewmembers on communication and other variables (Brannick, Prince, Prince, & Salas, 1995). Ratings of overall task expertise were correlated with the ratings of communication effectiveness.

Xiao, Patey, and Mackenzie (1995) found that breakdowns in coordination among hospital emergency room staff could be traced to two groups of faults. One group of explanations consisted of interruptions of task flow resulting from conflicting plans, inadequate delegation or diffusion of responsibility, or inadequate support. The other group of explanations was comprised of verbal communication faults. The contribution of verbal communication motivated more comprehensive investigations on the interrelationships among emergency events, communication faults and disruptions, and neurodynamic disruptions within the team as measured by team members' EEGs (Gorman, Grimm, Stevens, Galloway, & Willemsen-Dunlap, 2020. It also triggered the question as to whether verbal communication is necessary or sufficient to produce good team coordination.

14.4.4 Game theory

The central premise of game theory (von Neumann & Morgenstern, 1953) is that people make (economic) decisions that will optimize their own outcomes. The utility associated with an option usually depends on which options are selected by one or more of the other economic agents. Games can be classified as strictly competitive, strictly cooperative, and mixed competitive and cooperative. Prisoners' Dilemma is perhaps the best known game because of its mixed competitive and cooperative properties. Prisoners' Dilemma, Intersection, and Stag Hunt are all classifiable as coordination games because optimal solutions require that participants select the same options as the other participants at the same time. Intersection and Stag Hunt are two strictly cooperative games that have found relevance in complex work group dynamics. Bandwagon is another strictly cooperative game, and it was used as part of an explanation for the development of virtual communities. An important point here is that not all coordination dynamics in work units share the same underlying utility structure.

14.4.5 Intersection games

The Intersection game is analogous to the decision process that occurs at a four-way stop intersection. Each car and driver must figure out which rule is in play for determining which car takes a turn proceeding through the intersection at a given time. The rules in the state's driver's manual might be in effect, or a different variation might be occurring. The drivers, nonetheless, must figure out the rule and apply it to their own behavior. Each driver must decide correctly or else – bam! – negative utility occurs. Car wrecks are too expensive and dangerous for experimental purposes, so the basic principles of Intersection were captured in a card game (Guastello & Guastello, 1998; Guastello et al., 2005). The card game required four participants to play a sequence of cards in a particular order in order for the group to garner game points. Eight hands of cards constituted a round of the game. Different rounds of the game involved rules of equal or greater difficulty.

Game theory investigations often manipulate, or at least make use of, the utilities associated with the game's strategies. The utilities for the Intersection card game experiments remained constant as shown in Table 14.1. If a game

Table 14.1 Matrix of utilities for the Intersection coordination game

		Group of three players		
		All coordinated	Partially coordinated	Not coordinated
Target	Coordinated	4,4	1,1	0,0
Player 4	Not	1,1	0,0	0,0

Note: Reprinted from Guastello and Guastello (1998, p. 435) with permission the American Psychological Association.

is played in an iterative tournament, a strategy that eventually dominates, known as an evolutionarily stable state (ESS), will be close to the Nash equilibrium that corresponds to a simple one-shot exchange (Maynard Smith, 1982). If, however, we add some complexity to the possible game options and utilities and give up the assumption of strict competitiveness, the ESS cannot be guessed from any knowledge of the Nash equilibrium. Rather, the ESS becomes highly dependent on initial conditions as they affect options and utilities available to the participants (Samuelson, 1997). Thus ESS experiments are needed to identify long-run behavior patterns from games that emulate real-world problems and decisions.

The card game experiments (Guastello & Guastello, 1998; Guastello et al., 2005; Guastello & Bond, 2007) showed the following results. Coordination is a nonlinear dynamical process. Coordinated behavior can be observed as chaos in its early stages of formation, and followed soon afterward by a process of self-organization. Coordination tasks of greater difficulty might reflect only the chaos portion of the dynamic for some groups, but groups that are more adept may reflect both dynamics. The transition of players' game strategy choices from the first exchange to the ESS is a learning phenomenon. Learning phenomena themselves are nonlinear and dynamical, as previously mentioned. The outcome of a learning set is typically observed as a fixed-point attractor in the learned behavior.

14.4.5.1 Nonverbal communication

Human coordination, like animal coordination (Reynolds, 1987; Semovski, 2001), is fundamentally a nonverbal process. The nonverbal nature of coordination has been observed in the mainstay of game theory studies in economics, in which the experimenters are concerned that discussion will alter the perceived utilities of strategic choices. Coordination without verbalization was observed for Intersection card games as well. Verbalization can augment human coordination, however. This is not a trivial point; verbalization is neither necessary nor sufficient to produce coordination. In Intersection games it was shown that verbalization promoted sharper learning curves and better total performance. Verbalization did not compensate for the effects of changing personnel within a coordinated group, however (Guastello et al., 2005). The newcomer still needed to develop an entrainment with the rest of the group at a nonverbal level.

Figure 14.7 depicts the performance or learning curves for four rounds of an Intersection game: starting rule, switch to a rule of equal difficulty, switch to a more difficult rule, and switch back to the starting rule (Guastello & Bond, 2007). The learning curves were based on the means of 13 groups who acted nonverbally, and 13 that were allowed to talk during the activity. The trend after the first rule switch looks a continuation of the curve from the first round of the game. By the end of the fourth round,

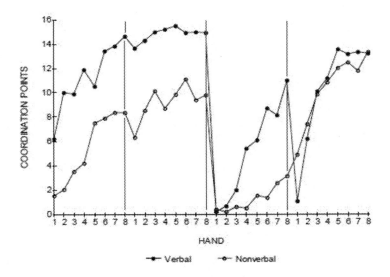

Figure 14.7 Coordination acquisition curves for groups acting verbally and nonverbally. (From Guastello and Bond, 2007, p. 104. Reprinted with permission of the Society for Chaos Theory in Psychology and Life Sciences.)

verbal and nonverbal groups had reached comparable levels of coordination performance.

Shah and Brazael (2010) examined nonverbal communication further with the objective of isolating key behaviors that could be transferred to human–robot interaction; robots should interact with humans in a manner that humans feel comfortable. They investigated team performance that resulted from a mixture of verbal and nonverbal (usually gestural) behaviors, implicit and explicit commands or requests, and conditions of uncertainty, ambiguity, and time pressure, all of which were known to contribute to stress. Overall 90% of the teams' communications were coordination messages; this point did not vary with the stress conditions. The teams produced *more* coordination messages, however, under time stress.

14.4.5.2 Minimum entropy principle

Not all teams experience a drop in performance or coordination when they have to work under greater time stress or uncertainty (Serfaty, Entin, & Deckert, 1993). Shah and Brazael (2010) remarked that, "It appeared during these experiments that teams were changing their coordination and information-seeking strategies" (p. 236). This observation is consistent with the minimum entropy principle discussed previously in conjunction with individual workload.

Coordination itself reflects a reduction in degrees of freedom in group mental workload compared to relatively random parallel action. Experiments from other areas of the human resources literature show that greater task complexity (load) produces greater levels of coordination, and coordination efforts play a larger role in overall performance (Cheng, 1983; Shah & Brazael, 2010; Stevens & Campion, 1994).

14.4.5.3 Changes in team membership

Group coordination, when sufficiently established, can withstand small changes in group membership without affecting the level of coordination in the remaining group. Studies in laboratory cultures illustrated how a group's activity pattern can persist even when every member of the group of ten members was eventually changed in a gradual sequence (Weick & Gilfillan, 1971). For groups of four players in an Intersection game, however, a sharp decline in coordination was observed when three players were replaced, leaving one original player. If only one or two players were changed, however, the group's performance was unaffected (Guastello et al., 2005).

Gorman, Amazeen, and Cooke (2010) found a different result when comparing groups that worked together in intact groups or those that involved swapping personnel from other groups that were already trained to do the same task. Their groups were three-person military reconnaissance teams. Groups that exchanged personnel actually performed better in terms of analyzing information about targets that was produced by unmanned aerial vehicles and finding targets. The margin of improvement for the mixed groups was traced to the stability of their coordination dynamics. Intact teams tended to restrict their interpretations of information or search strategies, using a narrower repertoire of sensemaking strategies. They tended to react to unpredictable events as corrections to their schemata already in motion. Mixed groups, on the other hand, mutually adjusted their cognitive schemata and searched more thoroughly.

14.4.6 Group size

Group size acts as a control parameter in ant collective intelligence (Sulis, 2009). A critical mass of ants is required to produce a sufficient momentum of interactions to instigate a nest-building project. The same principle appears to resonate in research with human creative problem-solving groups. Groups outperform the best-qualified individuals if the groups are large enough to produce a critical mass of ideas (Dennis & Valacich, 1993). Groups also have the potential for outperforming individuals because they can review the available information more thoroughly, retrieve errors more reliably, and rely on the skills and knowledge bases of more people to formulate a solution to a problem (Laughlin, 1996); here the size of the critical mass of people would depend on the complexity of the information processing task.

Campion, Papper, and Medsker (1996) observed that groups need to be large enough to contain enough resources to get their work accomplished, but not be so large as to induce coordination difficulties. Social loafing is more likely in larger groups, however. Loafers, or free riders, would benefit from joining the group with the expectation that someone in the group would get the job done, and all members would benefit (Latané, Williams, & Harkins, 1979). Hierarchical group structures can introduce more opportunities for inefficiency (Guastello, 2002).

By using a research hypothesis concerning group size one can assess the potential trade-off between critical mass and deficits in coordination (Guastello, 2010). If there is a group emergence effect at all, there would be an optimal group size associated with the teams' performance. If larger groups perform better, a group dynamic is operating that would be consistent with the critical mass principle. If mid-size groups perform better, critical mass would be associated with the mid-size groups and coordination loss with the larger groups. If smaller groups perform better, the group dynamics would suggest widespread loafing. If there were no effect for group size, then the teams' task was carried out by the most competent individuals; it would then be debatable whether the others were loafing or just not competent at the task.

14.4.7 Stag hunt and emergency response

Response teams for emergencies, such as earthquakes, require coordinated and self-organized efforts for rescue, medical services, damage containment, and evacuation (Comfort, 1999a; Guastello, 2002; Koehler, 1999). The unfolding of events depends on initial conditions of specific time, location, and preparedness of the response teams. In a real emergency groups of rescuers, health service providers, and food and shelter providers seem to materialize out of nowhere, according to the classic stories. Official response teams are typically overtaxed by the demand to fight fires, carry the injured to medical facilities, manage evacuation in an orderly fashion, work around self-organized evacuation attempts, stifle criminal behavior, and contain damage. Failure to make the right response at the right time could compromise any of the objectives.

Comfort (1999b) analyzed the details of the emergency response efforts in 11 then-recent earthquake disasters. Up to 400 organizations could be involved between the efforts of international government, national government, local government, and civilian nongovernmental organizations. Coordination of efforts could be assessed in terms of information search strategies, information exchange, and organizational learning during the emergency response process, and the effectiveness of actions taken. It was noted that organizational learning enhances lower level action, thus making reliance on hierarchies less critical.

The relative success of emergency response efforts depended on the articulation of common meaning, trust between leaders and citizens, resonance between the responding systems and the physical environment, and the sufficiency of resources. The least effective adaptive responses were severely hampered by communication outages with no backup systems such as satellite-based technical systems. Other critical sources of dysfunction were traced to political problems such as an ongoing civil war, which placed political objectives at crosscurrents with the common good.

Comfort (1999b) characterized any form of adaptive response as an edge-of-chaos phenomenon. There are rapid swings between periods of instability and stability, even in the best of conditions. Most adaptive final outcomes depended greatly on good sociotechnical infrastructures. Preparedness to reduce risk before the disaster allowed the community to minimize damages and bounce back effectively after the disaster.

The coordination requirements for emergency response strongly resemble the structure of a Stag Hunt game (Guastello & Bond, 2004). In the classic Stag Hunt story, all hunters within a group must decide whether they want to join the group to hunt a stag or run off to hunt rabbits alone. If enough people join the collective and exert enough effort, the stag will be caught in theory. The rewards are then distributed among the stag-hunting agents. The decision to join depends on the individual agent's self-efficacy as a hunter and the agent's estimation of the efficacy of the prospective team.

In natural disasters such as earthquakes, some people evacuate immediately (hunt rabbits) and take care of their own families and property first (hunt rabbits). Others join collectives to locate and help survivors (hunt stag). Official emergency response teams are usually hunting stag. Although there are many unpredictable events associated with forest fires, earthquakes, and floods, such as wind shifts, aftershocks, and the specific geography of the affected area, natural disasters are predictably indifferent to the humans involved. The situation self-organizes differently when the source of the disaster is a sentient being—a terrorist of sorts—whose goal is to wreak as much havoc and fear as possible and elude captors.

Military and governmental agencies responding to an attack may pursue an orchestrated agenda (hunt stag) or follow disjointed and uncoordinated plans (hunt rabbits). Table 14.2 lists some differences, however, between a theoretical game of Stag Hunt, a real stag hunt, emergency response to a natural disaster, and emergency response to a sentient attacker. In a theoretical version of the game, the two basic options are to hunt stag or rabbits; these options are the same for the hunters in real stag hunts, natural disaster management, and attack management as well.

Some differences among the situations can be observed when we consider the role of the stag. In a theoretical game, the stag does not contribute to the

Table 14.2 Comparison between the Stag Hunt game and its real-world applications

	Stag Hunt game	Real stag hunt	Natural disaster	Sentient attacker
Hunters' options	Stag vs. Rabbits	Stag vs. Rabbits	Stag vs. Rabbits	Stag vs. Rabbits
Stag's options	Not part of process	Evasive actions	Trivial strategy	Evade or attack
Utility structure	Enough effort leads to reward	Enough effort leads to reward unless stag is smarter	Enough effort leads to reward unless disaster is stronger	Enough effort may lead to reward but attacker often competes effectively
Evolutionary properties	Subgame perfect, or manipulated by experimenter	Stags get smarter over time, lowering hunters' odds	Hunters' odds may decay over time; hunters' coordination deteriorates if uncertainty increases	Attacker rewrites the game and affects odds; hunters' coordination deteriorates if uncertainty increases

Note: Reprinted from Guastello and Bond (2004, p. 350) with permission the Society for Chaos Theory in Psychology and Life Sciences.

options or outcomes for the hunter. Although a real stag does not attack, it can take evasive action or stand and fight. If it survives repeated exchanges with the hunters, it might get smarter, thus decreasing the odds of success for subsequent hunters who do not also get smarter. Herd animals learn to perceive threats and learn evasive maneuvers after their herdmates have fallen prey to a new predator.

A natural disaster plays an active role according to what is known in game theory as a *trivial strategy*: The disaster does exactly what it was going to do—in accordance with the pertinent laws of physics and geography—without any regard for the actions of the humans. For the hunters, success is a matter of degree and a matter of their strength and cooperation against the strength of natural forces. A trivial game is not necessarily trivial in terms of the importance of outcomes to the gaming parties. Importantly, the presence of a trivial strategy characterizes the game as competitive, with respect to the interactions between the humans and the disaster, even though the interactions among the humans may be strictly cooperative.

In a confrontation between the humans and a sentient attacker, the attacker by definition is utilizing offensive tactics in addition to evasion and defense. The competitive interaction is clearly nontrivial. Unlike a conventional stag hunt, this type of stag occasionally eats the hunters. As part of their overall strategies, the competing parties try to expand their own options, and limit the opponent's options and utilities. The situation is not *subgame perfect*, meaning that it is not possible to predict the outcome of an iterated game from the outcome of a single exchange; that would only be possible if the options and utilities facing the players were the same on each exchange. In a disaster situation, however, a turn of events on the part of the natural phenomenon or the attacker may shift the utilities associated with the maintenance of a group effort versus self-interested responses. *Oligarchic reaction functions* (Friedman & Samuelson, 1994), such as regulatory agencies or the United Nations in the case of international conflicts, can drastically alter the utilities and options for the competitors.

Some interesting evolutionary effects have been recorded for iterative Stag Hunt games. Players often gravitate to suboptimal choices (Carlsson & van Damme, 1993; Rankin, Van Huyck, & Battalio, 2000) that are usually tied to the perceived uncertainty of attaining the win-win outcome. Much of the uncertainty arises from a player's estimate of whether the other players will decide in favor of a particular option. In the case of a team trying to coordinate against a hostile competitive agent, the full participation of the group can be seen to fluctuate over time, with downturns occurring when the competitor gains some advantage against the group (Guastello, 2010; Guastello & Bond, 2004; Guastello et al., 2017). Verbal communication does help the coordination efforts, but a trained group can anticipate the proper responses of teammates when a communication outage has occurred. The experiments showed that teams were more readily influenced by the attackers' performance than vice versa. The teams with of 9 or 12 participants were more likely to prevail against the attacker compared to teams of 4 or 6 participants; only teams of 12 people were effective at dampening the adaptive responses of the attacker, however. Adaptability was measured as the size of the Lyapunov exponent (a measure of turbulence in a time series) that was calculated from performance trends. In all cases, when the attackers scored points, the teams' performance on the next move declined. The attackers' performance patterns showed greater levels of turbulence (adaptability) than the teams' overall.

14.4.8 Human–robot interaction

A robot is a semi-autonomous machine that is capable of perception–action sequences and a modicum of autonomy of movement. Unlike its disembodied cousin, the autonomous agent, the robots under consideration here are capable of moving around the physical environment and taking action on it. Although humanoid embodiments are being developed gradually, other design paradigms might emulate the physical mobility of dogs, fish, or insects. Unmanned aircraft are another class of designs.

Three types of configurations for human–robot system are shown in Figure 14.8. The simplest system contains a single person, a single robot, and a machine controller in the middle, with control and display links at both

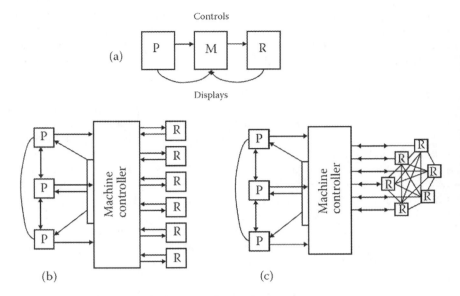

Figure 14.8 Configurations for human–robot interaction.

connections in the process. It is essentially the plan behind the operation of industrial robots, although large industrial systems can have many manu-facturing robots connected to the machine controller. Importantly, there have been several reports of robot arms that fatally crushed workers against building structures when they swung into place suddenly (Chapter 11). The general solution was to build safe operation areas for the humans who could then control the robots from a suitably safe distance. Although some accidents could be eliminated in this fashion, the remote control solution does not address the potential for injury to other humans that might be in the work area and not visible to the controller (Gethman, 2001). Thus an evaluation and control of the workspace around the robot is essential for safe operation, and should be implied by the boxes in Figure 14.8 that represents the robots.

The connection between multiple robots to an operation controlled by one human is a system feature known as *fan-out*, which is similar to the idea of *span of control* in other theories of human-only organizational behavior. The ideal level of fan-out is a standard question in any system where auto-mation is involved (Goodrich, 2013). With contemporary experimental systems, the fan-out ratio is more favorable for drone aircraft, where one person might control two or three drones, than it is for land-roving robots, where the fan-out ratio tends to run negative requiring two or more people to service one robot.

Multi-robot systems require attention to each robot, so some attention switching is involved. As the number of robots increases management strategy shifts from management by consensus to management by exception. Instead of the human determining whether they agree with each move made the robot, they shift attention toward simply fixing what is wrong (Goodrich, 2013). Finding the right level of autonomy is a design question; giving the robots higher levels of autonomy (more automatic processes) does not always make the job easier. Robots get into trouble, get stuck in loops, and need corrective interventions from the humans in order to continue operation. The performance of ground-based robots does not deteriorate gradually; problems tend to arise suddenly. Other robots should be able to function autonomously, or "tread water" until the human can finish one task before servicing another.

As a means of responding to the shocks to the human workload, the configuration in Figure 14.8b might be adopted. Several humans would observe the full display, and when problems arise the human operators would coordinate their actions to provide a complex response to a complex situation. Meanwhile, intelligent display designs continue to be developed. Figure 14.9 shows a processed video feed from a drone aircraft (Abedin, Lewis, Brooks, Owens, & Scerri 2011, p. 92). The drone was equipped with synthetic vision, which it coupled with a map and

Figure 14.9 Software-enhanced video feed from drone aircraft. (Reprinted from Abedin et al. (2011); in the public domain.)

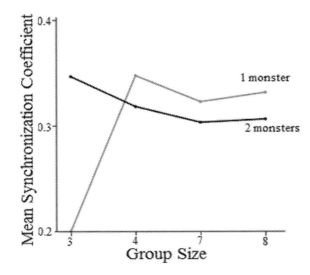

Figure 14.10 Size by attacker (monster) interaction. (Reprinted from Guastello et al. (2019, p. 361) with permission of the Society for Chaos Theory in Psychology and Life Sciences.)

geographic information system program. The human operator searched the display for targets. The artificially intelligent version of the display contained a preprocessor that automatically identified the targets sought by the humans and marked them accordingly. The evaluation results showed that the humans missed fewer targets, which was the goal, but at the expense of situation awareness. The additional time required to process the image disrupted the sense of time between the collection of the image and the current location of the drone, which in turn compromised the operator's ability to act on the targets.

The humans depicted associated with the drone operation in Figure 14.11 are not always the only stakeholders in the video feed. The other system users are the dismounted ground troops. For instance, ground troops in the Israeli army regularly conduct operations searching for ground targets with the aid of a feed from a drone aircraft (Ophir-Arbelle, Oron-Gilad, Borowsky, & Parmet, 2013). The open questions were whether the search would be enhanced by adding an unmanned ground vehicle that provides another video feed from another angle, and how much of an increment of improvement can be expected when the search targets are people versus vehicles, or open-air environments around the target versus dense urban environments. The additional workload associated with the additional feed would be substantial, so the size of the payoff is important. One visitor to

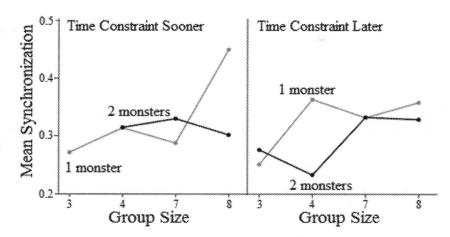

Figure 14.11 Three-way interaction for time condition, number of attackers (monsters), and group size. Data points for time constraint sooner, two monsters, groups of three, were missing. (Reprinted from Guastello et al., (2019, p. 365) with permission of the Society for Chaos Theory in Psychology and Life Sciences.)

a technology conference sponsored by the U.S. Department of Defense was quoted as saying:

> [E]veryone tacitly acknowledge[s] that sensemaking is a problem – cameras are sprouting like mushrooms from every conceivable mounting space in the hopes of giving the operator that magic missing viewpoint. However, it is all ad hoc; there is no understanding of what viewpoints are needed for which tasks or how to transition between them. So-called perceptual interfaces remain Neolithic.
>
> (Ophir-Arbelle, Oron-Gilad, Borowsky, &
> Parmet, 2013, p. 27)

A third type of configuration is the *swarm*, which is modeled after the swarm of insects, flock of birds, or school of fish. In this configuration (Figure 14.9c) each robot is an autonomous agent, which is itself complex with respect to cognitive and psychomotor components (Trianni, 2008). The self-organizing properties of swarms require information loops between the units, and the current challenge is to develop sensors and response structures that keep the cluster of robots functioning even if one of them should become impaired. A human controller is still involved, especially where the goals of the system's actions need to be defined for a given purposes, but the design questions are: How much control could be, or should be, allocated to the human "executive"? How much of the behavior of the system is going to be self-organized? How much broken symmetry can be tolerated? Several science fiction movies have centered on this theme, with nothing good happening to the humans.

Counterintuitively relative to humans, it is probably better not to have all information available to all robots. Instead, the system should be balkanized so that information is filtered into the sub-networks on a need-to-know basis. The alternative outcome would be one in which excessive information induces random reactions, or lock up, similar to other situations involving small world networks. "Structural holes" or elements of disconnectedness could be an asset in this context (Burt, 2004; Fu & Pirolli, 2013; Goodrich, 2013; Hazy, 2008).

Current design issues are concerned with types of interactions among swarming agents and their effect on supervision and control by the humans. At the most complex system level are distinctions between homogeneous agents, which execute all the same rules with the same collective goal, and heterogeneous agents, such as combinations of aerial and ground-level robots, which do not. There are numerous design issues that need to be reconciled for the simpler case of homogeneous agents, which are: (a) trade-offs between the assets and limitations of centralized and decentralized swarms; (b) supervision at the global level of abstraction inasmuch as self-organization is a key feature of swarm systems; (c) extent to which agents are working with a map of a physical terrain versus concentrating on ground-level issues that affect their mobility, (d) support for a large number of agents to function as desired, (e) maintaining energy supply and efficiency so that depleted agents do not shut down the system, (f) algorithms for path planning and obstacle avoidance, and (g) algorithms for transporting objects that could require two or more agents (Barca & Sekercioglu, 2013, p. 353).

Centralization in this context refers to designs that rely on leadership roles among the agents; controlling one leader or hub in the network could have a broader impact on the behavior of the swarm than control operations directed at minor agents. Decentralized designs would not contain leaders; the swarm would operate more like a colony with bottom-up patterns developing but little top-down influence on emergent configurations and activity (Roundtree, Goodrich, & Adams, 2019). Another perspective on centralization is focused less on leadership in the social sense, and focused instead on planning agents in the human–machine system in which planners have a capacity to delegate tasks to other agents (Heilemann, Lindner, & Schulte, 2021); planning agents appear to have a strong positive impact on unmanned aerial vehicle (UAV) operators, which might not be using swarming configurations necessarily.

The studies that contributed to Barca and Sekercioglu's (2013) review usually considered only one or two design issues at a time; there is much to be determined regarding how designs that address all the foregoing issues would evolve. The top control question for any design is to decide how much of the swarm's operation should remain self-organized and how much of it should be determined by rigid programming. The control question is rooted in a search for optimum variability and adaptability.

Transparency in AI-infused systems plays an important role in the development of trust in automation to the extent that the human understands the machine's "thought process." It is an important step toward robots and humans adapting to each other (Amirshirzad, Kumru, & Oztop, 2019). In another extensive review of research findings, Roundtree, Goodrich, and Adams (2019) examined the role of transparency in swarm operation from the perspective of humans' situational awareness. Swarms by definition operate without input from external agents, and they self-organize through their myriad interactions. The humans need to know at level 1, the swarm's status and what it is currently doing. Transparency plays a larger role at level 2, in which the human acquires a mental model of the operation in progress. The mental model is necessary but not always sufficient for level 3, predicting the future state of the swarm and the impact of any corrective action on the part of the human. Transparency can be enhanced through the right choice and quantity of information displays, but often at the expense of greater workload demands; a swarm of 50 agents is likely to push the human well beyond the limits of channel capacity.

14.4.9 Human–autonomy teaming

A possible means for improving human–automation interaction is to reframe the relationship from thinking of the human controller and the machine as a tool to thinking of the automation as a team mate. The interface design should start with a CWA that will inform allocation of function between person and machine (IJtsma, Ma, Pritchett, & Feigh, 2019). The traditional perspective would separate "the tasks people are better at doing" from "the tasks machines are better at doing." The recommended shift is to use the CWA to identify interactive tasks, how many different types of them exist, and how often are they performed.

Roth, Sushereba, Militello, and Diiulio (2019) proposed building abstraction hierarchies starting with mission goals, followed by priorities and values, groups of functions, and specific functions. An in aircraft landing, for example, the groups of tasks would include managing, aircraft control, the landing trajectory, aircraft systems, flight regulations, and communication. For each of the specific tasks, the CWA should identify the humans' task and the information required and how the automation can help. Ashoori, Burns, d'Entremont, and Momtahan (2014) proposed a similar strategy for all-human teams in health care settings.

The orientation of treating the automation as a teammate rather than tool was shown to positively affect team satisfaction and other affective responses (Walliser, deVisser, Wiese, & Shaw, 2019). Anthropomorphizing the automation appears to help the interactions. Human–autonomy teams that underwent formal team-building activities, such as those typically done with all-human teams showed improved performance along with improved affective responses.

Coordination between the humans and the automation have also been investigated (Demir, Likens, Cooke, & Amazeen, 2019). The medium was the UAV reconnaissance task used by Gorman, Amazeen, and Cooke (2010), with performance measured as communication flow, the number of photos taken, waiting time during interactions (shorter is better). Teams were assigned to three groups: traditional control relationships with the automation, synthetic or teammate relationship, and an expert human. A Wizard of Oz methodology was used (and also in Walliser, 2019) was to produce responses from different types of automation or human: All automation responses were actually produced by humans working from a script, but here hidden from view of the human research participants.

Demir, Likens, Cooke, and Amazeen (2019) performed some NDS analyses on the time series data – phase-space diagrams, Lyapunov exponents, and recurrence quantification analysis. The Lyapunov exponents were used as indices of adaptability (similar to Guastello, 2010), with larger positive values indicated greater adaptivity, and lower or negative values indicating a gravitation toward a rigid pattern. The human expert condition was the most adaptive, followed by the control condition and then the synthetic condition. The recurrence analysis showed greatest determinism (rigid patterning), followed by the experimenter, and then the control condition. The results indicated that the synthetic teams had gravitated to a routine interaction pattern too quickly and would do well to become more flexible, and the human–human teams should stabilize their patterns more than they did. The recommendation from both directions was to reach a point of optimum variability or optimum adaptability.

14.4.10 Group cognitive workload

In the extant literature on group workload (GWL), researchers have attempted to integrate behavioral, subjective, and biometric data from individuals to ascertain a group-level measure of workload (Funke, Knott, Salas, Pavlas, & Strang, 2012). Theoretical clarity or generalizability of results were slow to arrive, although it appeared reasonable that measures of GWL should include principles of individual cognitive workload, coordination issues, and task switching between team members (Cooke, Amazeen, Gorman, Guastello, & Likens., 2012). Helton et al. (2014) soon developed set of rating scales that paralleled the NASA–TLX at the group level: coordination, communication, time sharing demand, team efficacy, team support and team dissatisfaction (Table 10.1). The GWL rating scales were responsive to experimental manipulations of team workload (Guastello & Marra, 2018).

For group performance criteria, the cusp catastrophe models for cognitive workload and fatigue (Chapter 10) were found to be equally applicable to team performance (Guastello, Correro, & Marra, 2019), although there were some nuances regarding the control variables. The task involved an emergency response simulation that was repeated six times in two experimental sessions that were one week apart. The elasticity–rigidity variables, TLX ratings and GWL ratings were the averages of the responses from members of each experimental group. In the workload model that resulted from the analysis, the effective asymmetry variable was a combination of three TLX ratings: mental, physical, and temporal demand. The effective bifurcation variable was a combination of the other three TLX ratings: performance, effort, and frustration. The GWL ratings were combined into one variable as a possible bifurcation variable; they did not make a contribution to the model, but they were strongly correlated with both trios of TLX ratings. The elasticity–rigidity variables corresponding to fluid intelligence and emotional intelligence that contributed to bifurcation parameter in the individual models were correlated with both sets of TLX ratings. The implication was that the experience of workload could arise from different directions within a group, but the overall group averages had the direct relationship to team performance.

For the fatigue model, greater variability (higher bifurcation) was associated with smaller groups (in a range from 3 to 8 members), lower task difficulty, and greater experience with the game (second session). Fluid intelligence variables were the compensatory abilities (asymmetry) in the model.

14.5 TEAM SYNCHRONIZATION

Synchrony in group dynamics can be observed as a close mimicry in the behavior of human individuals, such as their body movements, autonomic arousal, and EEG activity. The actions or physiological responses of one person in the group to another are usually measured on a time-scale of 1 s, with a range between 0.7 s and 5 s, and occasionally up to 20 s (Guastello, Futch, & Mirabito, 2020; Kazi, Khaleghzadegan, Dinh, Shelhamer, & Sapirstein, 2021). Synchrony of physical movements such as postural sway, hand and facial expressions, or speech patterns can occur spontaneously when mutual mimicry of behavior is not a performance objective, or when two people are engaged in a conversation. Synchrony in autonomic arousal levels among team members occurs when two or more people engage a joint activity or conversation. The total amount of synchrony in a group is modulated by the empathy levels of the individuals, common focus of attention, (Guastello et al., 2020; Palumbo, Marraccini, Weyandt, Wilder-Smith, & McGee, 2017; Stevens, Galloway,

& Lamb, 2014), and workload (Guastello et al., 2019). Synchrony can also be induced by temporal regularities in the activity or environment that are induced by repetitive actions, turn-taking among the team members (Guastello, Marra, Castro, Equi, & Peressini, 2017; Henning, Boucsein, & Gil, 2001), or by external timing cues (Torrents, Balagué, & Hristovski, 2016).

Synchrony is not the same construct as coordination, based on the game theory definition give previously. There is a window of overlap, however, when close mimicry is the performance goal; in those instances behavioral and physiological synchrony would be expected in high-performing groups. It is possible, however, to have physiological synchrony without coordination as when a group shares a common focus of attention but no immediate interacting task. It is also possible to have coordination without behavioral mimicry when a group members are taking different but compatible actions toward a group goal.

Synchrony is a special case of self-organizing phenomena. It starts with local interactions among team members and patterns emerge eventually. The majority of studies on synchrony in human relationships has been based on dyads rather than larger groups (Kazi et al., 2021). Group studies that involve human performance issues computed indices of team synchrony based on a variety of methods, such as phase clustering (Elkins, Muth, Hoover, Walker, & Carpenter, 2009), patterns of recurrences of values of the physiological or behavioral metrics (Mønster, Håkonsson, Eskildsen, & Wallot, 2016), symbolic dynamics (Stevens & Galloway, 2016), and the driver-empath model, which utilizes statistical time series analyses that culminate in a single number that quantifies the total amount of synchrony within a group (Guastello & Peressini, 2017). The driver-empath model, on which many of the performance and workload effects described next are based, begins with a matrix indicating the influence of each group member on each other group member. The diagonal elements are autocorrelations, which represent the change in one person's behavior over time that is independent of the effects of other group members. The off-diagonal elements are the unique influences of other group members, which can be asymmetrical. The group member that has the largest effect on the group is the *driver*, and the member who is most influenced by other members of the group is the *empath*. The final calculation of the synchronization coefficient (S_E, or *synchronization with the empath*) is the largest effect that group members have on anyone else in the group.

14.5.1 Autonomic arousal

Workload plays the role of the control variable that speeds up the interaction process prior to when synchrony and phase-lock occur. As discussed previously, cognitive workload is cross product of the quantity of information that needs to be processed and length of time available to process it. In the emergency response studies (Guastello et al., 2019) synchrony varied with workload manipulations in the experiment – one or two adversaries to conquer and time constraints that were imposed sooner or later in the game (simulation) tournament. The effect of the two workload variables interacted with group size, as shown in Figures 14.10 and 14.11. There was a marked schism between the outcomes for group sizes of three versus four or more participants that requires further research. It is also well-known that three-way statistical interactions (Figure 14.11) of any type are also prone to situationally specific influences, and might not replicate in the same pattern in new situations, so some caution should be taken when interpreting them.

Of further interest, the level of synchrony was higher when coordination demands were perceived to be low and communication demands were high in the early phase of team interaction and when team dissatisfaction was high (indicating higher workload; Helton et al., 2014) in the later phases of group activity. Similar connections were reported in a later study in which the participants had emerged from a hypothetical plane crash and had to make a plan for their survival until they were rescued (Guastello, Mirabito, & Peressini, 2020). Thus it was possible to draw connections between manipulated workload effects, subjective workload, autonomic arousal, and synchrony.

The first studies of autonomic synchrony and performance were based on the partial relationship between synchrony and coordination: If synchrony resembles coordination in some circumstances and coordination is correlated with performance on coordination-intensive tasks, then autonomic synchrony should be positively correlated with performance (Elkins et al., 2009; Guastello, 2016; Henning, Boucsein, & Gill, 2001); the hypothesis was supported in those studies. The relationship between autonomic synchrony and performance in the emergency response simulation was conditional, however. In the short term – on a scale of about 30 minutes for an emergency response simulation within an experimental session – performance success led to greater synchrony and team participation in decisions (Guastello et al., 2018). In a longer time frame – one week between experimental sessions – synchrony during the first session predicted performance on the second session a week later (Guastello, Witty, Johnson, & Peressini, 2020).

14.5.2 EEG and team neurodynamics

The EEG studies for *team neurodynamics* targeted alpha and beta wave activity in brain regions associated with mirror neurons and social

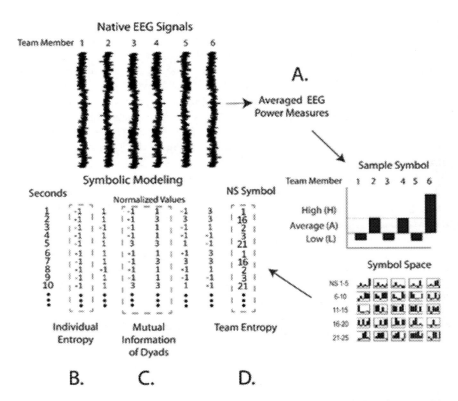

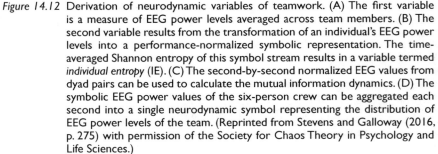

Figure 14.12 Derivation of neurodynamic variables of teamwork. (A) The first variable is a measure of EEG power levels averaged across team members. (B) The second variable results from the transformation of an individual's EEG power levels into a performance-normalized symbolic representation. The time-averaged Shannon entropy of this symbol stream results in a variable termed *individual entropy* (IE). (C) The second-by-second normalized EEG values from dyad pairs can be used to calculate the mutual information dynamics. (D) The symbolic EEG power values of the six-person crew can be aggregated each second into a single neurodynamic symbol representing the distribution of EEG power levels of the team. (Reprinted from Stevens and Galloway (2016, p. 275) with permission of the Society for Chaos Theory in Psychology and Life Sciences.)

interaction (Stevens & Galloway, 2016). For a team of six submariners in a navigation scenario, the high, medium, and low activity levels for each team member form a configuration for the group that was rendered as a symbol. The symbols changed over one-second time intervals and reflected both synchronous and asynchronous states. The variability in neurodynamic symbols as they changed over time was measured as

Shannon entropy. An overview of their database schematic appears in Figure 14.12.

Bursts of entropy were expected, and indeed found, when teams needed to make an adaptive response during the scenario. Changes in entropy levels were also observed across segments of the activity such as the scenario, a deliberate pause, and a debriefing session where the group's attention was train on an instructor or a team member answer a question from the instructor. Better performing groups were those they exhibited lower entropy and relative synchrony during the debriefing phase and greater entropy during the scenario where individual roles were involved (Stevens et al., 2014).

Stevens and Galloway (2019) have since developed a computer program that detects entropy changes while the group activity is in progress. This information format could be useful in team training when paired with other events during the activity. Subsequent research has been directed toward making connections between neurodynamic symbol patterns, adaptive or maladaptive responses to unplanned disruptions of an activity, and team cognition as expressed through problem-solving discussions (Gorman et al., 2020).

14.6 SAFETY IN COMPLEX SYSTEMS

As technologies evolve, they involve more complex human interactions and higher degrees of potential entropy. The complexity of human interaction engenders multimodal controls and mode errors (Chapter 8) and the numerous dynamics discussed in this chapter thus far. According to Gethmann (2001), the proper approach for accident analysis and prevention is to ask the question, "How did it happen?" rather than "How could you?" The human may have made an error that appeared to be the cause of the accident, but combinations of system design and management policies made the error possible. Although human factors psychology has always played an important role in system safety, it must now play a more aggressive role at the stage of system design.

Although complex technical systems may perform many functions automatically, the humans, must still keep track of the state of the system and its recent history (situation awareness again); some control actions by the humans are permissible during some states of operation, but not in others (Hudson, 2001). Situational awareness is often operating at the collective level. For instance, investigations into airline errors have examined the extent to which errors made by air traffic controllers, flight captains, or crew are detected and corrected by one another (Casner, 1994; Fischer, McDonnell, Davison, Haars, & Villeda, 2001). Some problems in transportation, information technology, and medicine are described next.

14.6.1 Transportation

Kirwan (2001) observed that the landscape of safety and risks in the transportation industry was changing due to increased volume, increased competition, and changing relationships between various forms of air, marine, rail, and highway transportation systems. Engineers must imagine more complex accident scenarios that transcend the more traditional localized boundaries among transportation systems and subsystems. For instance, the safety analysis of a flight should begin with flight planning, and not simply when the airplane pushes off from the flight gate. Delays in one part of the system, especially following an accident, may produce unplanned difficulties for passengers, pilots, and air traffic controllers:

> A fire at an airport causing a shutdown at the airport will divert many aircraft and cause chaos in the skies. The effects of a major airport closure will be felt all over Europe, and the ATM system will take hours to stabilize ... The whole system will come under severe strain. Furthermore, many passengers will have to find alternative transportation..., placing strain on other transportation sections and increasing risk on those sectors. This can be called *risk export*, and will probably become more important as a concept as transportation systems become more interconnected and integrated.
>
> (Kirwan, 2001, p. 329, footnotes deleted)

Another class of problems is related to the privatization of transportation organizations in Europe. Large transportation units have been split up among several private owners, each with their own myopic safety concerns. In the case of marine transportation, Hormann (2001) remarked, "Procedures relevant to safety are not limited to the onboard activities. The performance of the operator's shore base and possibly his offices in outports is likewise decisive" (p. 364). Additionally, it has become increasingly difficult to attract young people to seafaring occupations. The pool of available employees thus consists of increasing numbers of marginalized workers. There is an increased need for education, training, and the definition of new careers in marine transportation.

14.6.2 Information technology

Computer users have experienced increasing numbers of attacks by waves of viruses, worms, and Trojan horses in the wake of the attack on the World Trade Center in New York on September 11, 2001. Analyses of virus wave patterns indicate a power law distribution, according to Liebovitch (2004). The power law distribution offers some clues to the origin and operation of networks that are producing the attacks.

Brunnstein (2001) explained a fundamental problem: The Internet, IP protocol, and hypertext markup language "were developed for free exchange of information without *any* reference to requirements of security (confidentiality, integrity, availability) and safety (functional behavior, timeliness, persistency)" (p. 585). As a result, the contemporary users for e-commerce and other applications have had to plug numerous and potentially severe leaks.

Furthermore, what you see is not what you get. The fundamental sources of dysfunction originate and persist because they are not visible to the user. Wilpert (2001) reported several growing problems: (a) Fatal crashes of planes and automobiles can be perpetrated with automatic navigation systems that produce incorrect interpretations of data. (b) Autonomous agents in packages for financial markets might be infiltrated and mimicked by criminal agents. (c) Similarly, criminal agents have developed mobile code that can crack the security of checking accounts and other banking activities. Firewalls might not provide intended security. Espionageurs have been occasionally successful in reversing critical elements of firewall systems. (d) Limitations might be placed on the use of cookies, which are bits of code inserted to a user's system when the user visits certain commercial websites. Cookies provide information to the system about the user's past visits, and may be used to tailor subsequent transactions with that user.

Although most cookies are restricted to benign interactions between one human user and a website, the basic idea of a cookie has morphed into a beacon. This variation sits in a computer and records financial and other transactions by the human with most any merchant or bank, and then reports back to a business operation that collects and sells the information. Computer security systems can detect the presence of beacons on the end-user's system and remove them, but the cat-and-mouse game of computer security continues.

One might question how computer viruses came into being. According to Levy (1992), they were an unfortunate by-product of the artificial life vision. Well-meaning computer scientists wanted to fabricate a silicon life form, and the logical place to start was with the simplest life form, the virus. The first report of a malicious virus was traced to a computer dealer in Pakistan during the early 1980s. The dealer apparently had local customers as well as an enclave of customers from the United States. The dealer loaded the virus into computers that were intended for the U.S. customers, but sold clean computers to the local customers.

Another unfortunate by-product of artificial life was the rise of ubiquitous computing: Since all life can be explained by the flow of information, any time there is a flow of information between two agents there is an opportunity to insert a new piece of software. Whereas well-intentioned software can facilitate desired results or turn out to be an overblown waste of time, cyberterrorist can find links that can be used to extract data or insert bad information that will disrupt the system, or take it hostage for an extended period of time while demanding random.

Gödel's theorem indicates that there is a limit to how much cybersecurity can be accomplished by software solutions alone because there is always a back door. Thus the human – who supplies the logical statement that falls outside the internal logic of the program – must remain in the loop. According to Ben-Asher and Meyer (2018), better security software requires a better understanding of the multiple tasks performed by the human. For each type of task, there are three aspects of human input that are relevant to security: exposure to risks, use of the security features, and response to indications of a security issue generated by the system. A person could be risky or cautious in any of these areas. Their experiments showed that there is an inverse relationship between the use of protective actions and productivity. The experience of damage from an attack affected the level of security chosen by the individual but protective action costs did not do so. The research participants modified their exposure to risk of the cost of protective action was too high. They varied in their level of *compliance*, which is responding to an alert as determined by the security program, and their level of *reliance*, which is to take action only if an alert was received. Either orientation was related to the reliability of the security system.

14.6.3 Medicine

At one point in time, biohazard safety programs were largely borrowed from other safety areas, and thus needed to be redeveloped for biotechnology applications. For instance, according to a 1997 study cited by Zinn (2001), 3.5% of patients in German hospitals contracted infections that were traceable to the overprescription of antibiotics. Then-current prescription practices have led to several strains of resistant bacteria, and patients can contract more than one pathogen in this fashion. The estimated health care costs required to remedy the secondary illnesses was roughly $9 billion per year, and the problem was thought to be growing.

Electronic storage of medical documents requires a complex process of handling, labeling, and encoding. The benefit of electronic patient records is that the original documents are not subject to loss in transit between one physician and another, and it is possible for multiple diagnoses to be made at a distance, especially where specialized expertise is thinly distributed. On the other hand, a study of incorrect handling and labeling of data records showed that 90% of records contained errors (Zinn, 2001). Perhaps one of the most tragic errors was the report of an amputation performed to the wrong leg of the patient.

Genetic products can become biotechnological hazards. Even before we consider whether the production of particular genetic products will have positive or deleterious effects on the ecosystem, there are potential problems in the laboratories and production plants themselves with the handling of materials (Czermak, 2001). New materials may produce specialized risks, and regulations regarding the award for patents on genetic materials are not likely to interface with safety standards (Baram, 2001). The same issues persist with the management and production of nanotechnologies for health maintenance (Calderón-Jiménez, Johnson, Montoro Bustos, Murphy, & Winchester, 2017).

As control is regained over some the foregoing problems, others keep popping up. Some topics: the process of making diagnoses and intervention with attention to the specialists, the equipment that could be involved, and timeliness and coordination (Yousef, Sutcliffe, McDonald, & Newman-Toker, 2022); communication among emergency response teams and physicians, particularly when high-risk situations emerge (Catchpole, Privette, Roberts, Alfred, & Carter, 2022; Joseph, Madathil, Jafarifiroozabadi, Rogers, & Mihandoust, 2022); teamwork, leadership, situation awareness, and cognitive workload (Cha & Yu, 2022, pp. 42–43); the particular role of emergency departments (Austin, Blakely, Salmon, Braithwaite & Clay-Williams, 2022; Perry, Jaffe, & Bitan, 2022); factors affecting clinical performance (Hysong, SoRelle, & Hughes, 2022); versatility and adaptability of elder-care practitioners (Naweed, Stahlut, & O'Keefe, 2022); system monitoring and response to auditory alarms at a distance, for example, nursing stations (Patterson, Rayo, Edworthy, & Moffatt-Bruce, 2022); emotional demands and other stress on health care providers (Ahmadi et al., 2022); experts' and novices' decision-making processes with regard to procuring optimal prosthetics for particular patients' needs (Saravanan & Menold, 2022); and the particular needs of patients with post-traumatic stress disorder (Sadeghi, Sasangohar, McDonald, & Hegde, 2022). Some of the emerging demands include the growing number of older adults, further uses of telemedicine, patients' satisfaction with their health care experiences, and overall management of medical errors (Keebler, Rosen, Sittig, & Salas, 2022).

In the bigger picture of system safety, variables and situations affecting safety culture remain a priority (Kilcullen et al., 2022). The goal is to build systems that contain sufficient back-up processes in the sense of the Swiss cheese model (Reason, 1997). Patient and employee safety need to be reviewed in concert with changing service demands such as total capacity, broader ranges of health care services, services to underserved populations, consistency of internal checks and balances, and overall resilience when unanticipated events occur, such as pandemics and natural disasters (Rodríguez & Hignett, 2021).

14.6.4 Butterfly effects

Can a shortage of rainfall in the central U.S. produce a political revolution in Syria? This is essentially the question posed by the "butterfly effect," more formally known as sensitivity to initial conditions (Chapter 3), which has become an iconic representation of chaotic processes (Dooley, 2009). Karwowski (2012) observed that sensitivity to initial conditions is a serious concern with complex systems that has yet to be reconciled.

When the complexities of the medical system are combined with the complexities of information technologies, one obtains more complexity. Freeny (2007) reported a situation in which a major hospital in California received an e-mail from a woman in Pakistan saying that if the hospital did not pay her "her $500" she would distribute the attached medical files all over the Internet. What happened? Come to find out, hospitals that were attempting to reorganize their medical data into digitized files in anticipation of the new national health care system were resorting to offshore contractors to convert the original data into digital files. Somebody forgot to pay the contractor. Fortunately the problem was caught and fixed soon enough that time, but the story plants one of many seeds of doubt regarding where the next surprise from complexity might arise.

A clinical psychologist at a different site in California recorded its therapy sessions into digital video files and stores them on a cloud system, which was management's idea, not the clinician's (anonymous communication, 2013). Access to the cloud service was interrupted for three weeks because of an exceptional storm that was centered over the Northeastern U.S. on the other side of the continent. Eventually service was restored, but the control over the privacy of the files became a serious concern.

The financial crisis of 2009, according to some analysts, was triggered by the collapse of a market for mortgage securities that were being traded for more than they were worth. One could trace the proximal event to a progressive combination of financial deregulations (Dore & Singh, 2009). The problem became worse as the risks traveled through the financial networks very quickly, but any risk management innovations to be had did not travel as fast; the risks had enough of a head start to reach the crisis point before risk management could possibly catch up (Gallegati, Greenwald, Richiadi, & Stiglitz, 2008).

To return to the question about rainfall and revolution, the following story captures what could happen when unrelated systems interact. According to Lagi, Bar-Yam, Bertrand, and Bar-Yam (2011) a shortage of rainfall in the Midwestern U.S. where most of the corn and grain products are grown resulted in a shortage in the harvest in 2009, which resulted in higher prices for corn according to the usual dynamics of supply and demand. Two other factors contributed to the further rise in corn prices. Commodities traders, whose numbers suddenly increased because the mortgage traders had to find a new job, noticed the price increases and were speculating on future prices of grain products. Meanwhile, the use of corn as a source of ethanol for fuel was increasing, adding to the demand. The higher prices for grain products traveled through the market networks until they reached actual food sales in the Middle East. Several countries in early 2010 were experiencing very difficult political times that were apparently getting close to revolutionary proportions. The high food prices added one more stressor – the proverbial straw that broke the camel's back – and set off the events that became known as Arab Spring.

Given that there are other politically precarious countries in the world, Lagi et al. recommended that policies be adopted to limit the use of corn for fuel so as to increase food supplies and keep revolutionary tensions subcritical. Is this a good solution? Consider what could be a better one, in light of two problems with the corn solution. (a) Not all corn, however and wherever it is grown, is fit for human consumption. Lower quality goes to animals, and the leftovers go into fuel; diverting fuel corn into human corn is not going to work well. (b) The speed and connectedness of the networks are at least as responsible for the outcome as the price of corn. The same dynamic could just as readily happen to a completely different commodity of worldwide importance. The problem was that the commodities markets were too interconnected and overpopulated. The solution, in light of what is currently known about small worlds, would be to disrupt the interconnections among the market trading programs. This solution would not guarantee the prevention of political strife, but it would eliminate spikes in food prices as a contributing variable.

DISCUSSION QUESTIONS

1. Three mathematical models were introduced in the course of explaining collective intelligence phenomena. What are they, and how are they different? Do they have any common properties?
2. Describe a complex person–machine system where coordination is an important component of the system's performance.
3. How might the Bandwagon game explain trends in technology adoption?
4. Reconsider your answer to Question 3. Can you imagine a scenario where Bandwagon dynamics might override good human factors or ergonomic design?
5. What are some limitations to conducting business through e-mail, compared to face-to-face transaction?
6. Describe some situations where an innovation in information technology appeared to save work, but instead it only saved work for some people and made more work for others.
7. Describe some other types of revenge effects in systems with which you are familiar.
8. Consider the innovations that are inherent in NextGen air traffic control. Will they assist or inhibit situation awareness? Make a case in support of either the pro or con position.

9. Consider the connection between iterations in system design and coupling as expressed by Helander (2007) and the cascades of technologies that are currently expected in airport manufacturing and uptake. How many iterations will it take to bring NextGen close to its ideal goal? What would be required to reach the goal?

10. Consider again Suh's (2007) and Helander's (2007) advisements about coupling. Is a large-scale Internet platform a good idea for the smart power grid?

11. Are there any visible similarities between programs for high-frequency trading of commodities and the brokering of electric power resources?

12. How much additional complexity in NextGen air traffic control is expected to be cause by differences in equipage? Would it offset any advantages associated with the new automation feature? How would you conduct a study to answer these questions?

13. Consider the interface shown in Figure 14.5 for a military situation awareness application. What are the assets and limitations of this design?

14. In spite of any lessons learned from the analysis of small world networks, person–machine systems and information-sharing systems are growing toward greater centralization. Who benefits from all the centralization? At what costs to whom?

15. Electric cars have become more available as alternative designs that should reduce air pollution and emissions that are responsible for global warming. Where does the electric power that recharges the batteries come from? What problems are likely to arise in the near future, and how would they be resolved (in broad strokes, of course).

Chapter 15

Environmental design

15.1 MICROENVIRONMENTS

A microenvironment is larger than an immediate workstation, and closer in scale to a room or an office. A private home might also be considered a microenvironment because the occupants usually have easy access to all parts of the home (Kantowitz & Sorkin, 1983). A macroenvironment is a much larger facility, and closer in scale to a hospital, a shopping mall, an airport, an apartment complex, or an industrial site.

15.1.1 Offices

One logical point that proceeds from known human factors principles is to organize the office by function. Communications equipment should be aggregated into one location, large displays that are meant to be shared by several occupants to another, and information storage to another (Kantowitz & Sorkin, 1983). The small office shown in Figure 15.1 is actually organized by function. It has other problems, such as a very small space for so much activity, and overall clutter.

Offices should be constructed around the user patterns. For instance, psychotherapists tend to organize their office space and furniture around a more businesslike theme when counseling individuals, but they build an atmosphere more similar to a living room when counseling couples or families (Anthony & Watkins, 2002). An easily visible clock is important in this type of office as it is in classrooms and other microenvironments where appointments need to begin and end on time. The need for extra seating and good visibility among people talking to each other probably influences choices of furniture and décor in addition to comfort and familiarity. Discrete entrances and exits might also be desirable in this type of environment.

Office configurations should reflect the autonomy and interactions among the occupants (Duffy, 1997). A hive would involve little of either; people

DOI: 10.1201/9781003359128-15

Figure 15.1 Two views of an office organized by function.

work in small cubicles doing prescribed tasks. A cell would accommodate a talented worker who generally works alone and who benefits from being left alone (high autonomy, low interaction). A den accommodates teams of people who produce group products (low autonomy, high interaction). A club atmosphere accommodates a work group where individual-intensive work and group work occur at changing intervals; this group would rate highly on both autonomy and interaction, and their workspace would include private and common areas.

Windows help. Office occupants who have views of attractive scenery outdoors report less job stress than those who have views of parking lots and other buildings or no windows at all (McCoy, 2002). The impact of natural surroundings is considered later in this chapter.

15.1.2 Homes

The allocation of space to rooms in the private home has changed in the last centuries. Homes that were built in the United States during the 17th century were designed around a large fireplace, which served as the central heating system and kitchen for cooking and baking. The fireplace was built so that its floor was flush to the floor of the room rather than raised; thus it was relatively easy to walk in, hang cooking pots, and manage the fuel. The family spent the majority of its time around the fireplace, more so in the winter; it is fairly obvious how the *living room* got its name. Bedrooms were relatively small and were often appointed with a wash-up stand.

The 19th century brought central heating, flush toilets, and electricity, the benefits of which were obvious. The distribution of heat allowed the dwellers to allocate more time to other rooms. The bedrooms grew in size in the late 20th century. The children and adolescents arrange their personal worlds in their rooms, including their televisions, stereos, and computers. The master bedroom provides some opportunity to withdraw from the other household members, and many more functions of life take place in the bedroom that used to take place in the common rooms. Some modern homes are built with a study off the master bedroom, although many households prefer to utilize a room that is shaped like a separate bedroom or perhaps a "den" for this purpose. The television room, or family room, occupies the space that was previously known as the (formal) living room. Sometimes the amount of space allocated to these two types of rooms is merged into a "great room." Overall people exercise some flexibility with the uses of various spaces within a house.

The trending idea is that the house should be built, or rebuilt, according to the way people expect to use the rooms. Obviously households will differ in their habits, and houses will vary accordingly. At one point in history, (Ahrentzen, Levine, & Michelson, 1989), women spent relatively more time in the locations that facilitated household chores, whereas the men spent more time in the spaces devoted to passive leisure.

The age of the occupants also matters. The most critical issues are captured by the statistics on accidental deaths in home environments. Poisoning and drug overdoses, specifically opioids, are the leading cause of preventable accidental death (National Safety Council, 2022). Here most victims are between ages 25 and 44. The warning on medicine contained to "keep this and all other medications out of reach of children" is a good one. Accidental misuse of medications, for instance picking up the wrong prescription bottle (when they all look alike) without reading (or fully processing) the label first, could be a source of mismedication. The odds of this type of casualty increase with age, as older people use more medications; complications among the elderly could be classified as death or illness due to a known preexisting medical condition, however, and not categorized as accidental.

Falls are currently the third most frequent cause of preventable accidental death after opioid poisoning and automobile accidents (National Safety Council, 2022). Most victims of falling deaths are either under 5 years old or over 75 years old (National Safety Council, 1989, 2002). The elderly often prefer fewer stairs. Ladders are another source of falling incidents.

Drowning is the third leading cause of death in the home. The bathroom is dangerous for slipping and falling, and possibly drowning in the bathtub. Common countermeasures include slip-free flooring, tub, and shower surfaces, adequate handrails, and possibly a phone located in the bathroom in case of emergency.

Fires and suffocation rank fourth and fifth in the home. Again most victims are under 5 years old or over 75 years old (National Safety Council, 1989). Age makes them less likely to vacate a dangerous situation unassisted.

15.1.3 Kitchens

The kitchen is a center for work activity, most of which involves food preparation, but some of it does not. Food-preparation tasks are facilitated by a triangular arrangement among the stove, sink, and refrigerator. The work triangle facilitates the largest among of work with the smallest amount of walking between points (Figure 15.2). The less attractive alternative would place the three main devices in a straight line, which often occurs in small apartments.

Storage and countertops are two other critical elements of kitchen design. On the one hand, the quantity of both will depend on the planned size of the dwelling and the costs associated with the size. On the other hand, added

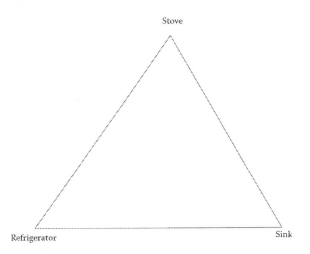

Figure 15.2 The kitchen work triangle.

quantities of storage and countertops can be made possible through the efficient use of space.

Not all countertops are created equal. Some of the space is occupied by decorative items, storage of some other items, food preparation, and messes that are headed for the dishwasher as soon as someone unloads it. One common requirement is a surface on which one may place hot pans and pots. Many decorative countertops become discolored by utensils coming right off the stovetop or from the oven. With a bit of forethought, however, one can install heat-resistant tiles and thus turn the entire countertop area into a hotplate.

If the kitchen is large enough, some of the central floor area can be allocated to a work island, with a work surface on top and additional storage cabinets below. Sometimes glamour conflicts with function. Granite countertops look good on cooking islands, but knives and other cutting tools scratch it easily. Other homeowners prefer the look and utility of a butcher block, which is essentially a large cutting board. The large, heavy, and permanent slab of wood is friendly to knives and cleavers, easy to clean and maintain, but requires some light sanding every five to ten years.

The stove in Figure 15.3 represents a classic dilemma in control panel design, which is where to locate the controls for the stove and how to make

Figure 15.3 Stove with flat even surface; controls located at the rear, behind the burners; countertops left and right are heat-resistant tiles. A portion of a butcher-block top on a work island is seen in the foreground.

the knobs correspond with the burners in the most efficient way. In this example, the user must reach across the pots and pans to use the controls, and a large pot might obscure the controls. The disadvantages of the usability of the control placement could be outweighed by the smooth surface of the stovetop. Here the burners are located below the surface rather than above it. Burns and spills are easier to clean than the conventional stovetops where the wells around the burners frequently collect caked-on crud.

15.1.4 Stairs

Stairs should be built to accommodate normal patterns of walking. Figure 15.4 is a diagram of stairway features from the U.S. Department of Defense (2012) specifications. Nothing has ever prevented the staircase in a house from being designed differently; the spiral staircase is a case in point. Nonetheless, the DOD specifications are as follows: The tread depth should fall within 11.0 and 12.0 inches (280–300 mm). The riser height should fall between 6.5 and 7.0 inches (165–180 mm). The depth of the nosing should not be more than 1.0 inch (25 mm). The overhead clearance should be a minimum of 78 inches (1,980 mm).

The width between handrails should be 36 inches (910 mm) for a one-way staircase and 51 inches (1,300 mm) for a two-way staircase. The

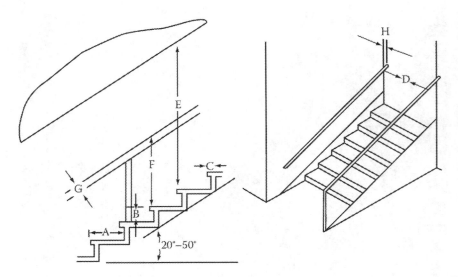

Figure 15.4 U.S. Department of Defense (2012) specifications for stairs. A: Tread depth. B: Riser height. C: Depth of nosing. D: Width between handrails. E: Overhead clearance. F: Height of the handrail from leading edge of tread. G: Handrail diameter. H: Handrail clearance from wall.

recommended handrail diameter is 1.5 inches (36 mm). The clearance between the rail and wall should be 3.0 inches (75 mm).

15.2 MACROENVIRONMENTS

15.2.1 Building and facility complexes

Many types of macroenvironment involve large facilities, which often, in turn, involve multiple buildings. Examples would include hospitals, manufacturing plants, universities, shopping malls, and airports. Obviously there is a large variation in size and intended uses. As a general rule, the building complexes should be organized by function also. Linkages are also important; ideally, facilities users would have to travel only the shortest possible distance between two areas where tasks are performed. Instead of tracking linkages with helmets and eye movements, linkages could be determined by asking people to keep a log of which points they visited in which order during the course of a week. The needs of the employees and facilities managers might be different from those of customers and clients, however; thus the design should balance the needs of the two groups of users.

Clark (1975) catalogued several common faults in facilities design. One group pertains to the interface between the inside and the outside. There should be sufficient space in and around loading docks to allow for the maneuvering of trucks and enough space in the platform area to accommodate the expected flows of materials going in and out. Rain canopies are recommended along with a buffer area between the inside, where climate is controlled for indoor comfort, and the outside, which is not controlled.

The second group of design faults in facilities is related to walkways and production areas. Walkways should be wide enough to accommodate two-way traffic along with handcarts and other haulage devices that have to move throughout the space. Production areas often suffer from lack of sufficient floor space also. There might be enough room for the machines and people, but not always enough room for materials and for storing work in progress. The results are clutter, tripping hazards, and other harmful encounters.

Although work supervisors spend a substantial portion of their time on the shop floor, they also need office space. The usual requirements are desks, relatively private and quiet areas to converse with employees or to use the telephone, and secure storage for records and reference materials. Unlike the standard office environment, the foreperson's office is often an isolated structure that is equipped with large windows for viewing the shop floor. Some manufacturing facilities have found that they had insufficient spaces of this type, especially as the technical organization of work changed significantly.

15.2.2 Facilities management systems

Large buildings often require electronic control systems. These facilities management systems typically contain five integrated systems: (a) heating, ventilation, and air conditioning; (b) water utilization and flow; (c) fire and toxin safety; (d) building security and monitoring; and (e) energy management. Some examples of the control and displays for facilities management appeared in Chapters 5 and 12.

These systems rely on sensor technologies to measure events throughout the building complex, transform the sensations into electronic signals, and display the information in a suitable location along with appropriate controls. Part of the strategy in system design is to aggregate enough sensor information into a small number of centralized control stations so that a small number of vigilant humans can monitor and control large portions of the system. At the same time, the system should be designed with enough entry points so that emergency corrective actions can be taken near the location where the problem is occurring. It is not efficient to require maintenance personnel to walk a great distance to a centralized control station and back again to see if the control action had the desired effect. Alternatively, a route to a centralized control station might be impassable due to fire or other hazards in the area. In essence, the better large systems are hierarchically organized with different points of entry.

Figure 15.5 depicts a sensor display system that detects leaks in refrigeration gas. Leaks beyond a certain intensity are fatal to any human venturing

Figure 15.5 Alarm unit that detects leaks in poisonous gas.

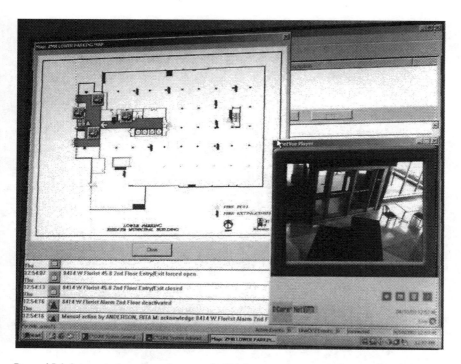

Figure 15.6 Integrated screen from a facilities management program showing open windows for building security camera and floor map.

into the area without a gas mask. When the system detects a critical number of offending parts per million, a warning horn sounds repeatedly, and the red light at the top spins around and flashes. Many types of toxins can be detected and reported in the same manner.

Figure 15.6 shows a computer screen from an integrated security display. The menu that is more or less in the background lists each location in the system (a six-building complex in this case); sensors of different types are located in each of the areas. The operator may click on the screen to activate a pop-up map or perhaps see what a security camera in the area sees.

A sixth component of facilities management systems is communication throughout the building complex. To some extent, communication involves telephones and cell phones that are external to the remaining five areas of facility management systems. Computer technologies, however, allow for intranet communication among operators that are located at computer terminals. In any case, communication systems need to accommodate the volume and flow of communications among facilities management personnel and anticipate the need for communication during emergencies

Figure 15.7 The atrium of Milwaukee City Hall showing six out of seven balconies.

when one medium of communication may be impaired by the nature of the emergency.

System engineers often find a need to anticipate inappropriate actions of facilities users. Figure 15.7 depicts the atrium area of Milwaukee City Hall. Six out of seven balconies are shown. The building was erected in the 1890s. Shortly after the great stock market crash of 1929, people started using the balconies as diving boards. To prevent any further suicides, the City installed cage-like barriers all along each balcony. They were removed after 50 years, when aesthetics eventually won out over those particular security concerns.

15.2.3 Defensible space theory

Some buildings, particularly large residential facilities, are more susceptible to criminal invasions than others. According to the theory of defensible space (Newman, 1972), facilities are less vulnerable if they are designed to allow residents to exert their territorial imperative. Figure 15.8, left, is a high-rise low-cost urban apartment building of early 1960s vintage. The entrance is secluded and not visible from the apartments. Several buildings of this type are lined up in rows to form a housing project. The passageways between the buildings are not easily monitored from the apartments or by police or private security patrols. It is also not possible to monitor what could be going on in the halls from the outside. Hall monitoring is typically not a concern for higher rent buildings, but higher crime levels are anticipated here. In a variation of this design, hallways are arranged on the outer perimeter of the building where balconies are located in Fig. 15.8 (left). They are protected on the outer edge by sturdy steel cages.

The building shown in Fig. 15.8 (right) is an improved design that was introduced in the 1970s. The X-shaped layout allows visibility of the entrance and surrounding grounds from several apartments. The hallways around the central elevator are lit and visible from the outside. Buildings in the complex are staggered to enhance surveillance of one building from another and by the facility's security patrols.

Single-unit housing of the suburban variety is also susceptible to break-ins and theft. According to Newman (1972), houses would be less vulnerable if they were well maintained and appointed with symbols of territoriality such

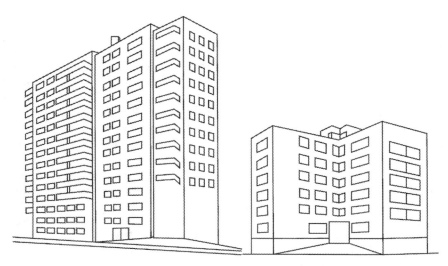

Figure 15.8 Two urban housing designs showing relatively less (left) and relatively greater (right) defensible space.

as low fences. The surrounding land should be well lit and with little dark area for perpetrators to hide in. Poorly maintained properties, according to the theory, would be more vulnerable because they would communicate that the owners were not exerting any control over the property.

MacDonald and Gifford (1989) tested the propositions of defensible space theory by preparing a set of photographs of homes that varied in defensible space characteristics. Photographs were shown to a sample of incarcerated home invaders who rated the homes for vulnerability as theft targets. The surveillance principle was supported; thieves avoided homes where three fourths of the surroundings were visible from the street or adjacent homes by virtue of lighting and the absence of landscape barriers. Contrary to the theory, however, homes with poor upkeep were less likely to be targeted because they communicated that their owners did not have anything that they were concerned about protecting. The symbolic barriers were actually more inviting because they signified that there was something worth protecting inside and sometimes because the barriers themselves actually obscured visibility from the street.

15.2.4 Navigation through facilities

Large and complex facilities often present a challenge to the users to find the place to which they are trying to go. In spite of claims by architects and designers about the navigability of their creations, there is little behavioral research on record to support most claims (Carpman & Grant, 2002). We know from common experience, however, that people differ in their sense of direction; some people always get to where they are going, whereas others find themselves turning right when they should be turning left and vice versa. It is probable that any subjective ratings as to how well a person can navigate a facility will depend on this ability, whatever it is. People who tend to get lost often will have one opinion about the navigability of a building, whereas people with good directional sense would be less affected by unintuitive designs.

Tolman's (1932) principle of expectancy (Chapter 7) simultaneously gave rise to the concept of a *cognitive map*. The experiments with rats used radial mazes (Figure 15.9) designed so that the rat would be allowed to enter the maze from any point, but the (maximum) cheese would always be in the same target location. The rat would need to use different patterns of left and right turns to reach the target. When rats were successful in doing so, the conclusion was that they had attained a cognitive map of the maze. This conclusion contradicted the strict behavioral interpretation, which was that the cheese reinforced a pattern of left and right turns. This is not to say that a sequence of turns could not be learned, however.

Humans navigate through buildings and spaces using three modalities—route following, which is a sequence of turns, following a cognitive map of

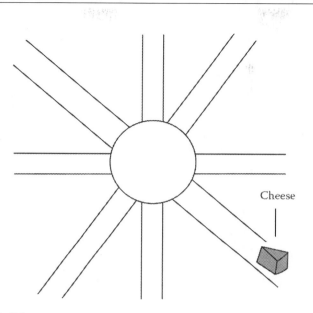

Figure 15.9 Radial maze.

the building or space that is used as a reference, or movement relative to a landmark. A complex route within a building can disrupt the cognitive map (Carlson, Hölscher, Shipley, & Dalton, 2010). The loss of orientation would happen with fewer or more turns depending on the completeness of the individuals' cognitive map and the complexity of the building's internal layout. Figure 15.10 depicts layouts of two hospitals showing primary, secondary, and tertiary routes. The layout on the left, which should *look* more compact, was the easier of the two for building users to navigate (Haq & Zimring, 2003).

Building designs that conflict with users' expectations impede orientation. For instance, users expect the floor plans of multistory buildings to be the same on each floor and tend to become lost if not (Carlson et al., 2010). The Seattle Public Library is an extreme example; it has several floors that are designed as a book spiral. Although award-winning for its striking aesthetics, it still confuses long-term regular visitors.

Signage helps with navigation. The best design advice at present is to position direction information at critical choice points in the path of a user traversing the facility. Important choice points occur when hallways separate into two or more paths and where hallways intersect. Sometimes a list of locations with directional arrows is preferable to a kiosk that shows "you are here." Sometimes the kiosk conveys better information about the overall facility layout. It is important, however, that the signs be salient

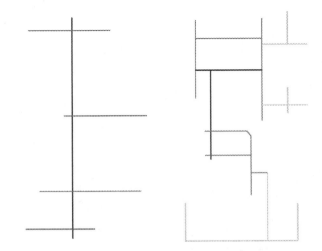

Figure 15.10 Two building layouts showing primary (black), secondary (thinner gray), and tertiary (thin light gray) corridors.

but not aesthetically annoying. The content of the signs should be complete enough to cover all the possible locations that users might like to visit; users might not be aware of all the possible places that they could visit. It also appears that people are less likely to become lost or misdirected when hallways intersect at right angles instead of narrower angles (Carpman & Grant, 2002). Landmarks are helpful for outdoor navigation but it is not currently known what makes an effective landmark for navigation within a building (Carlson et al., 2010).

15.2.5 Special populations

Residential facilities for the elderly with dementia present some special challenges. One's personal residence may look like all the other doors on the hallway with a number on it. One technique that seems to work is to place a small display outside the door to the personal room that contains a few of the resident's personal objects; those would be immediately recognized as signifying one's proper location in the facility (Day & Calkins, 2002).

The residents' need for some fresh air should be balanced with preventing the residents getting lost or escaping. One technique that appears to be successful is not to place exits at the end of the hallway where the residents can see the door from the opposite end. The vision of the door seems to present the wrong idea at the wrong time. Rather, a door to the outside should be located just to the side of the end of the hall; it will be where it is needed for emergency purposes, or loading and unloading, but it is less likely to trigger unscheduled exits (Day & Calkins, 2002).

15.2.6 Emergency exits

OSHA has a set of standards for emergency exits for workplaces. Although they are written with employees in mind, they should fill the needs for public safety in facilities where the general public may be present.

One set of regulations applies to exit routes. An exit route must be a permanent part of the facility construction and maintained during construction, alterations, and renovations. It should be made of fire-resistant materials. It should serve specific work locations only, so as not to create a bottleneck of traffic heading out one emergency door. Thus multiple exits are probably necessary for larger facilities. Emergency doors should be unlocked from the inside so that employees will not have to find keys or look up special instructions to use the exit. The emergency doors must be hinged from the side, swing out, and support the expected occupant load. Once outside, the escaping personnel should not be confronted with dead ends; rather the walkways should lead to open areas to allow the employees to get a safe distance away from the building (U.S. Department of Labor, 2004).

Exit signs should say "EXIT" and be at least 6 inches (15.2 cm) high with principle strokes on the lettering of at least 0.75 inches (1.9 cm). Fire-retardant materials in the exit route should be good for one hour in buildings of three or fewer floors, and for four to five hours in larger buildings. Fire-retardant paints and chemicals must be refreshed periodically. Emergency alarms need to be in place and operating correctly. Employees must be informed and trained in emergency evacuation procedures (U.S. Department of Labor, 2004).

15.2.7 Sick building syndrome

The Center for Disease Control fielded numerous reports of sick building syndrome during the 1970s and 1980s. The buildings were typically newer office buildings with little outside air circulation. The workers in those buildings reported unusually high frequencies of symptoms such as skin irritations, watery eyes, runny noses, abnormal taste or smell sensations, tiredness and fatigue, or reduced mental concentration (Hedge, Erickson, & Rubin, 1992; Rothman & Weintraub, 1995). The first response was to take assessments of airborne chemical toxins, but the consistent finding was that chemical toxins were not present in levels above OSHA standards.

Workers reporting symptoms were more often female and incumbents of lower level jobs. Studies of their work contexts revealed that they were experiencing high amounts of job stress and lower job satisfaction, both of which were more extreme for workers who used computer terminals full time (Hedge et al., 1992). Hedge et al. did consider the possibility that benign levels of airborne particles and insufficient air exchange could be aggravating factors that brought about the particular symptoms. Job stress continues to be the primary explanation for the sick building phenomenon (McCoy, 2002; Sulsky & Smith, 2005). The large numbers of workers in a building, or a floor of a building, reporting symptoms reflects a form of contagion that is otherwise known as mass hysteria (Rothman & Weintraub, 1995).

15.3 THE GREAT OUTDOORS

15.3.1 Aesthetics and stress

There is some sense in the health psychology literature that pleasant outdoor environments reduce stress and enhance the feeling of well-being. Thus offices are sometimes built around reflection ponds or in attractive semiwooded environments; most office dwellers do not have windows that overlook such landscapes, however. Apartment dwellers that have a choice probably prefer to overlook a lake or a park instead of a brick wall or a neighbor's apartment window, all other aspects of the apartment being equal.

Herzog and Bosley (1992) investigated the features that make one outdoor environment more attractive than another. It appears the participants in their study responded to five contributing factors when identifying their favorites: tranquility, mystery, coherence, spaciousness, and focus. *Tranquility* was the sense of peace that one might get from being in the environment. *Mystery* was the impression that more could be seen if one walked into it further. *Coherence* was the sense of being able to structure the environment visually rather than perceive it as an endless jumble of rocks or trees. *Spaciousness* meant that there was plenty of room to wander around. *Focus* referred to the presence of a focal point that commanded one's attention; Figure 15.11 is an example of a focal point within a desert scene.

Herzog and Bosley's analysis of ratings for 66 pictures showed that there were seven categories of image that were organized along four bipolar dimensions. They were: mountains versus fields and streams; desert versus large bodies of water; rushing water (e.g., waterfalls) versus gardens; and misty mountains versus anything that was not a misty mountain.

15.3.2 Navigation

Once again, people find their way by means of route knowledge, cognitive maps, and reference to landmarks. Lapeyre, Hourlier, Servantie, N'Kaoua, and Sauzéon (2011) assessed the extent to which synthetic vision systems

Figure 15.11 Window Rock in Arizona is an example of a focal point in a desert scene used as an experimental stimulus by Herzog and Bosley (1992).

improved wayfinding of pilots using them. The non-pilot participants in the study operated a flight simulator and had to rely either on the standard display or the display that was graphically enhanced. The standard display superimposed direction metrics over a two-color background with blue for sky and brown for land. The synthetic vision display superimposed the same metrics over a more detailed picture of the terrain. Other experimental conditions varied the outdoor visibility conditions that accompanied the directional display.

The participants made a flight to eight waypoints, then after a rest period were asked to recall landmarks, routes and directions using pictures taken from the simulated flight. In the scene recognition task they had to pick 8 out of 13 pictures that represented correct approaches to the waypoints. In the scene recognition task, the eight waypoints had to be put into the correct order. In the sketch map task, they were handed a sheet of graph paper with the departure point marked and were asked to draw the pattern of directional changes that took play in their trip with landmarks as a reference. In order to determine their wayfinding capability they had to replicate the route on the simulator without the synthetic vision system in both high and low visibility conditions. The synthetic vision system made a positive impact on all three levels of spatial knowledge.

Other types of navigation tasks involve two or more people trying to coordinate their locations and directions. When people are approaching a target location from different directions, "turning left" has different meanings. One person working alone with a map might do better with north pointing up or the directional heading pointing up, depending on the task. When two people are working together, however, they should have their maps oriented in the same direction for best performance (van der Kleij & te Brake, 2010). Thus, in the case of multiuser electronic displays, the system should not permit different operators from using different orientations of the map.

Walker and Lindsay (2006) investigated the use of three types of non-verbal audio cues in a navigation aid intended for visually impaired people. Participants followed the path designated by three different maps. Audio cues informed them of when to turn and in which direction. Three cues were generated with a 3-D audio system: sonar "ping," a pure tone at 1,000 Hz, and pink noise. Sonar ping had a history of being successful as a direction locator in other applications. Pink noise was intended as a broadband sound, but less harsh on the ears than white noise. Researchers also varied the *capture radius*, which was the proximity to the turn point when the signal was emitted. Dependent measures were efficiency of staying on the path, time to complete the map, and learning effects across the three maps. All signals produced what was regarded as "good results," although the best combination was the broadband paired with a capture radius of 1.5 m.

15.4 PLAYING IN TRAFFIC

Some of the human factors issues related to automobile operation were covered in Chapter 2 concerning automation, Chapter 5 concerning visual displays, Chapter 8 concerning controls, Chapter 7 concerning cognitive workload, and Chapter 9 concerning anthropometry. This section is concerned with the issues connected to the roadway and its users.

According to one undocumented story, there were four automobiles registered in the city of St. Louis in 1894. Two of them managed to crash at an intersection. The first fatal crash was documented in Britain in 1899 when a single vehicle lost a wheel and crashed. Witnesses reported that the vehicle was traveling unusually fast at the time of the accident (Evans,

2002). Debates to separate fault drivers and vehicles took shape immediately. Inasmuch as the majority of road accidents today are associated with driver variables, driver variables are considered first here.

15.4.1 Exposure

The rates of motor vehicle fatalities are somewhat related to road miles of exposure, but there are other trends to indicate that other large effects are involved. It is not possible to obtain drivers' mileage of exposure from most countries for comparison, thus the number of fatalities per 1,000 vehicles is a more manageable index. Moreover, data from single-vehicle crashes is preferred for many purposes of traffic accident analysis because they circumvent the problem of figuring out which driver was at fault.

Curiously, the nations with the lowest number of vehicles per 1,000 capital units have reported the highest fatality rates per 1,000 vehicles, whereas the most heavily motorized countries reported the lowest rates (Evans, 2002). The nexus of sparse roads, poor-quality roads, and low utilization by motor vehicles is associated with poor regulation, poorly developed driving habits, and uninformed risk perception. The situation is akin to a revenge effect: The absence of obvious dangers makes the situation all the more dangerous (Tenner, 1996).

Accident rates per 1,000 vehicles have declined in all nations since 1980, although the reduction rates vary substantially. The nations with the highest rates in 1980 showed the sharpest drop (Evans, 2002). When the COVID pandemic started in 2020, however, "compared to 2019, miles traveled decreased 11%, the number of registered vehicles slipped 0.2%, and the population grew 0.4%. As a result, the mileage death rate increased 21.7%, the vehicle death rate increased 8.5%, and the population death date jumped 7.9%" (National Safety Council, 2022).

15.4.2 Driver's age

It is a robust finding that drivers aged 18 to 25 years are twice as likely to suffer fatal crashes as drivers aged 40 to 50 years. The fatality rates for male drivers are three times that for females throughout the age span. Given that a one-car crash occurred and someone in the car died, however, the rate of female driver deaths is 50% greater than the rate for male drivers (Bedard, Guyatt, Stones, & Hirdes, 2002). In other words, women do not appear to survive the crash as well as men.

The popular opinion that elderly drivers are more dangerous than younger or middle-aged drivers may be ill founded for similar reasons. Evans (2002) reported a database in which the rate of single-vehicle crashes per million population were plotted against drivers' age; there was no increase in accident rate for drivers aged 70 to 80. Bedard et al. (2002) found, however, that given that a crash occurred with fatality resulting, drivers aged 80 and above are five times more likely to die than younger drivers. In other words, the elderly drivers do not survive the experience well. Similar results were reported by the National Safety Council (2002) for fatalities that could have been either drivers passengers; the highest morbidity rate was associated with people aged 75 and older. Thus the evidence favors the interpretation that the elderly are less likely to survive a crash than younger people, but not as likely to be responsible for one.

There is still some concern among some policymakers about elderly drivers in light of their increasing numbers. De Raedt and Ponjaert-Kristoffersen (2001) studied 84 licensed drivers aged 65 to 94 who participated in a research study at the behest of their insurance companies because of their accident involvement. In-depth interviews with the participants indicated that 63% of them were at fault in at least one of their accidents. De Raedt and Ponjaert-Kristoffersen administered a set of neuropsychological tests, a driving simulation, and the Belgian standard driving test. They found that it was possible to predict accident involvement from those tests if each type of accident—for example, intersections while making left turns, intersections while making right turns, front-rear collisions, side-swipes, and parking—were considered separately. Separate tests predicted different types of accidents because of the specific visual, memory, and motor functions involved in each type of maneuver. The prediction rates were about 19% better than chance (50%). The accidents could thus be explained by declines in perceptual and motor skills. These prediction levels are probably inflated relative to the general population of drivers in that age bracket, however, because of the way the participants were referred to the study. Statistical prediction levels reach maximum accuracy when the chance rate is 50%. Real-world base rates are very different from 50%, however.

15.4.3 Blood alcohol concentration

Alcohol intoxication is thought to be responsible for approximately 55% of driving fatalities (National Safety Council, 1989). It is not certain, however, to what extent fault is assigned to a driver in the databases solely because that driver was intoxicated rather than because that driver actually did something to cause the accident.

As of 2006, the legal limit for alcohol in the bloodstream (BAC) while driving was 0.10% in 36 of the U.S. states. This level is the midpoint between 0.00% and 0.20%, which is the average level of intoxication for drivers involved in fatal crashes, if they were under the influence at all; this average value continues to be reported (Foss, Stewart, & Reinfurt, 2001). Small deficits in drivers' judgment actually set in at 0.07%. Thus most states have now lowered the legal BAC limit to 0.08%. Evans (2002) noted that when the U.S. states raised the legal age for the purchase and consumption of alcohol from 18 to 21 years, there was an 18% drop in traffic fatalities. Foss et al. (2001) reported that no effect could be detected in traffic fatality data after North Carolina lowered its legal BAC limit to 0.08%, and the results from other states were too murky to draw a conclusion about the effectiveness of lowering the limit. A later review (Fell & Voas, 2014) of alcohol-related traffic fatalities was just as murky on the reduction of the BAC level, although they did cite one study showing an 18% reduction in fatal crashes associated with drivers whose BAC was >0.15%. Fell and Voas were advocating lowering the limit further to 0.05% in the United States, noting that most industrialized countries have already adopted that level, with the United States, Canada, and Britain being exceptions.

15.4.4 Seat belts

Seat belts were required by law as standard equipment in automobiles sold in the United States beginning with the 1963 model year. The federal legislature was strongly influenced by the testimonies of John Paul Stapp and Edward Murphy, Jr. Stapp was an Air Force test pilot for experimental aircraft. He claimed to have crashed so many times he would never have survived without his seat belt. Murphy was his chief engineer who is credited with the famous *Murphy's law*: If something can possibly go wrong, it will, at the worst possible time (Tenner, 1996).

State laws requiring that seat belts actually be used did not appear until 1984 (Houston & Richardson, 2002). It is now estimated that the use of seat belts reduces a driver's rate of fatality by 42% to 54% (Bedard et al., 2002; Evans, 2002). By 1996, however, it was estimated that only 68% of drivers actually used their seat belts. Part of the low utilization was attributed to the states' rules for enforcement of their seat belt laws. Most states only permitted ticketing of drivers for failing to use seat belts if they were stopped for another traffic offense. The latest legislative trend, however, is toward "primary enforcement," whereby police may stop and ticket a driver for seat belt failure alone. In the state of California, which switched to primary enforcement, the new law resulted in a 4.9% reduction in traffic injuries (Houston & Richardson, 2002). Michigan reported a 13% increase in seat belt use (Eby, Vivoda, & Fordyce, 2002), but no injury data were reported in that study.

In spite of laws and occasional improvements in technologies, driver behavior often attenuates the impact of the innovation. This principle is known as *selective recruitment*: Careful drivers are likely to respond to a new law or safety idea, but risky drivers remain risky.

15.4.5 Driving speed

According to Bedard et al. (2002), drivers crashing at speeds greater than 69 mph (111 kph) were 2.65 times more likely to die in the crash than drivers crashing at 35 mph (56 kph). In two-car head-on crashes each vehicle only needs to be traveling at half those speeds to attain the same effect.

Speed limits for roadways are usually set in response to several variables: road capacity, traffic density, sharpness of curves, and population or pedestrian density. The latter includes the number of driveways or intersections opening onto a particular road. Prior to 1974, the speed limits for U.S. Interstate highways would range from 50 to 70 mph (80–112 kph). The OPEC Oil Embargo in 1973 produced a petroleum supply shortage in the United States, which quickly resulted in automotive design changes and legislation. One important item was the drop in the national highway speed limit to 55 mph. This limit was more related to the fact that automobile engines at the time were more efficient at 55 mph than at higher speeds than it was to control accidents, although the public was more likely to hear about the projections for highway safety. The predictions were correct, nonetheless: The average travel speed dropped from 63.4 mph (101.4 kph) to 57.6 mph (92.2 kph), with a 34% drop in highway fatalities (Evans, 2002).

The interstate speed limit in the United States was raised to 65 (104 kph) mph for rural areas in the late 1980s in response to the increased use of overland trucking and the commercial value of increasing speed limits. It has since been raised to 70 mph (112 kph) and higher in some areas of the country. The political pressure on petroleum supplies had dissipated during that epoch. Reports on the impact of the change in the national speed limit are not available, although the hypothesis would likely be that highway fatalities in rural areas would have increased.

Excessive speed at lower scales, for instance in urban areas, can be problematic as well. According to one now-aging report (Thompson, Fraser, & Howarth, 1985), 36% of drivers exceeded the speed limit in school zones when children were present. The presence of students in groups of 10 or more promoted a reduction in speed of only 1 mph. Drivers are thus not well prepared to respond to children darting into the street. Thompson et al. opined that, all other factors being equal, children are at greater risk of injury in a car–pedestrian accident because they are short and thus not as readily visible as adults: "A child involved in an accident is frequently assumed responsible for the accident's causation. Accident-involved drivers often say, and probably believe, that the sudden appearance of the child they hit 'never gave them a chance'" (Thompson et al., p. 1473).

15.4.6 Risk homeostasis theory

Driving speed and the other antics of drivers are the end result of personal risk assessments, according to risk homeostasis theory (Wilde, 1988).

Drivers evaluate the relative rewards and costs of the safe options and the risky options and choose a level of risk that is best suited to their proclivities. Getting to the office sooner might outweigh the odds of a speeding ticket. Municipalities vary as to whether the traffic police will stop a car at 5, 10, or 15 mph above the limit. Speeding in excess of 20 mph over the limit could result in more than detainment and fines; it could result in a mandatory court appearance, which is an inconvenience that is often far greater than arriving at most destinations sooner.

The odds of an accident are another deterrent, and drivers decide for themselves whether the odds of an accident outweigh the merits of arriving at a destination sooner or the sheer fun of opening up the engine. Evans (2002) noted that highly skilled drivers are not necessarily safer drivers. Drivers with higher skill levels could perceive their odds of a wreck being lower than average, and thus they would convince themselves to take higher risks than average. Similarly, straight, open roads may appear easier to drive than other roads, but that would only induce drivers to drive much faster than they ordinarily would (Tenner, 1996).

Shortly after the introduction of risk homeostasis theory, there were several studies that qualified the basic idea. Individuals' risk assessments are not exact, but rather loose and not always well informed (Lonero et al., 1994). Salient positive incentives should contribute to safe roadways within the rationale of risk homeostasis theory. Risk homeostasis theory also appears to explain, however, why some people are willing to work in hazardous occupations whereas others would not consider those jobs (Halgrave & Colvin, 1998).

Risk perception may be a complicating variable. For instance, many drivers imagine that speed and traffic concentration are complimentary safety variables. They see speed as the more important risk factor when traveling at high speeds, but concentration as the more important risk factor at lower speeds. If the drivers were more objective, they would see both factors as contributing to their safety under all conditions (Jorgensen & Pedersen, 2002).

Recent efforts to automate the automobile and move more forms of control away from the driver to the car produced the idea of putting speed limit warnings in the car's operating system to alert drivers when they exceeded the speed limit. Berlin, Reagan, and Bliss (2012) outfitted government-owned cars with such a warning and tested the effect on experimental participants who drove them in a 30-mile radius in Michigan. There was some positive effect on reducing the time spent in excess of speed. Perhaps a more interesting result, however, was that drivers who regularly wore their seatbelts exceeded the speed limit by 6 to 10 mph more often those who did not wear their seatbelts.

The speeding alert system requires that the car be equipped with a GPS, an internet connection, and a database of speed limits. Thus, the viability of rigging a car to give out warnings might be limited outside the driver's usual haunts, although the granularity of traffic speed databases has been increasing.

15.4.7 Driver response times

State driving manuals typically provide charts and graphs showing the stopping distance for cars at varying speeds. The response time that is required to gauge a situation and step on the brake pedal is usually around 0.75 s, which is figured into the projections of stopping distances given particular speeds. A driving simulator study indicated that the driver response time to hazards approaching from peripheral directions, ranges from 0.65 to 0.85 s (Lovsund, Hedin, & Tornros, 1991).

It is important to recognize that simulation drivers are primed to make optimal responses. Their alertness levels are high. The typical driver, on the other hand, is operating at suboptimal conditions with multiple distractions in place, and a probable experience of nothing bad ever happening while driving. If a driver is operating under lax conditions and suddenly encounters a surprise on the roadway, the response time could be considerably greater (Eubanks, 1993). Nonmoving objects in the road (other than stopped vehicles with brake lights on) may take additional time to notice and process compared to other types of hazards. Pedestrians entering the street or car doors opening may be more challenging to detect for drivers entering the segment of road from around a blind corner rather than from the head-on direction. Situations with these perceptual challenges combined with suboptimal driver readiness could produce driver response times as high as 3 s.

15.4.8 Roadway configurations

Figure 15.12 (from Evans, 2002) characterizes roadway fatalities for interstate, noninterstate, rural, and urban locations with different configurations. Interstate highways are less often fatal than other roadways, and rural environments are more often fatal than urban environments. Within the urban and rural environments, arterial, collector, and local routes have different orderings of fatality rates. Of additional interest, death rates at night are approximately triple those for daytime accidents in both urban and rural settings (National Safety Council, 1989).

The three-lane road design predisposed drivers to head-on collisions. The three lanes were North (or East), South (or West), and a third lane that was shared by the two directions for passing. The road markings would indicate when it was each lane's turn to use the center lane for passing (overtaking). The markings told drivers when their turn to use the lane would start, but

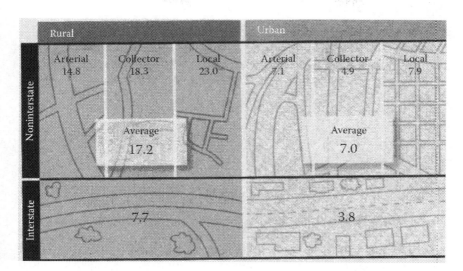

Figure 15.12 Roadway configurations and fatality rates per billion kilometers traveled. (From Evans, 2002, p. 248, illustration by Barbara Aulicino. Reprinted with permission. Copyright 2002 by Sigma Xi, the Scientific Research Society.)

gave them no warning as to when their turn would end, hence the inevitable head-on crash. As the collision rates increased on three-lane roads, often as a result of increasing traffic, they were slowly replaced by wider four-lane configurations, but not entirely so. High-speed two-lane highways are probably more frequent; they tend to be straight, clear roads in rural areas where accident rates are comparatively high.

In spite of the drama associated with head-on collisions, they represent only about 3% of all two-car collisions in the United States (National Safety Council, 1989). Front-to-rear collisions and angle collisions are each about eight times more frequent. Retting, Weinstein, Williams, and Preusser (2001) reported that 90% of urban arterial crashes in the Washington, DC area could be accounted for by seven different accident types: (a) making a left turn and colliding with an oncoming vehicle that has the superior right of way; (b) being rear-ended by another vehicle while waiting to make a left turn; (c) running red lights; (d) vehicle is stopped or just starting in a travel lane and is rear-ended by another vehicle; (e) changing lanes; (f) being run off the road, often into a parked vehicle; and (g) hitting a pedestrian (p. 725). Retting et al. offered several options for design correction: constructing left-turn lanes, widening the roads, increasing signal duration, restricting left turns by constructing median barriers, and restricting driveways onto the thoroughfare.

15.4.9 Lighting and signals

Rural roads are often poorly lit, which may contribute to their crash rates, especially at night. When lighting is installed, accident rates on those roads do decrease, but a generalizable rate is not forthcoming because of all the other roadway factors that could be involved. Not making things any clearer is the finding that once the lighting has been installed, drivers tend to increase their driving speeds by 5% on straight roads and 1% on curved roads (Jorgensen & Pedersen, 2002).

The popular perception is that municipalities do not install traffic signals at intersections unless the crash or death rate is high enough. This perception is probably not entirely wrong, although congestion may be another contributing factor in the decision to install traffic light systems. On the other hand, the increased use of an intersection, perhaps due to population mobility in a suburban area, increases the number of opportunities for accidents.

The unregulated intersection, particularly the four-way stop intersection, poses a challenge to drivers, especially in times of high traffic density. Most state driving manuals specify the driving rules in these situations, but only up to a point. Rule 1 in Wisconsin is that the car that stops at the intersection first proceeds first. Rule 2 is when in doubt about which vehicle arrived first or when two vehicles arrive simultaneously, the one on the right proceeds first. Right here we have the memory challenge: Which car arrived first—I thought it was mine. Given three or four cars, which car on the right of whom goes first? A popular variation occurs when drivers proceeding in east-west directions alternate with those proceeding north-south. The variation works until a car wants to make a left turn. In any case, the drivers must figure out which rule is in play, and when it is their turn. The cognitive processes suggest coordination dilemmas that could generalize to other types of situations (Guastello & Guastello, 1998).

Regulated intersections could pose different opportunities for human factors issues. A cluttered signal configuration in a city could confuse newcomers as to which signal is theirs (Chen, 1999). How often do drivers observe the lights when there are no problems with signal clarity? Red-light running accounted for 22% of urban crashes in the early 1990s (Retting, Williams, Preusser, & Weinstein, 1995), and 3% of fatalities during the 1992–1996 period (Porter & Berry, 2001; Retting, Ulmer, & Williams, 1998). Porter and Berry conducted a telephone survey of a stratified sample of U.S. drivers, of whom 19.4% reported running a red light in the last 10 intersections they used. Drivers who failed to observe lights reported that they were in a hurry, had no passengers, and figured that the odds of being stopped by police were very low. The typical scenario was trying to squeeze through an intersection as the light was changing from yellow to red. The survey results also indicated that drivers who run lights are likely to be younger (18–35 years), less likely to wear seat belts, and also predisposed to weaving through traffic, speeding, tailgating, and gesturing to other drivers. Contrary to their hypothesis, however, Porter and Berry found that running lights was not associated with reported frustration; weaving, tailgating, and gesturing were more closely correlated to reports of frustration.

15.4.10 Driver distractions

Driving is a preattentive process, meaning that driving requires attention to visual stimuli on a continuous basis, some well-honed emergency responses, but not a lot of memory or executive decision-making activity. In other words there is a substantial amount of excess channel capacity that could be devoted to talking, listening to the radio, and thinking about job-related problems. The list of alternative activities now includes operation of a cell phone, onboard computer, and GIS. One automobile manufacturer advertised DVD systems that were meant to entertain the passengers—and the driver as well when the car is parked while waiting for something or someone. The video system at the driver's screen automatically turned off when the car was put in gear. Before the advent of the newest gadgets, bored drivers with questionable situation awareness had been observed spreading business papers across the dashboard and examining them while driving.

Driver distractions often have low-tech origins—use of conventional maps, talking to passengers, tuning radios, operating cassette players, eating, and drinking. Driver distractions were recorded in 3% of police reports of fatal accidents in England and Wales during the 1986–1995 period (Stevens & Minton, 2001), and in 11% of accidents in the United States with any level of severity reportable to police (Wang, Knipling, & Goodman, 1996). Both reports came from a time period shortly before the widespread boom in the use of cell phones. By some estimates, driver distractions could be responsible for up to 30% of contemporary automobile crashes (Shelton, 2001). These crashes could be the result of drivers not attending to road stimuli in favor of something else or attending to the correct stimuli but not devoting enough channel capacity to driving (Kantowitz, 2001) and thus not taking appropriate action soon enough when needed.

The risks associated with some driving distractions are now available (Dingus et al., 2011), as reported in Chapter 6. It is not usually possible, however, to reconstruct what the driver was doing just prior to the crash, or to what extent drivers were doing the same things when they were not crashing. Legislatures have been responding by applying legal limits on driver distractions, for example, the driver must keep one hand on the steering wheel while using a phone. Manufacturers are introducing safety features such as voice-activated controls or a control system that prevents the driver from programming a GIS unless the vehicle is stopped.

By other standards of imagination, however, the drivers who might have had some personal privacy in their automobiles should be captive customers for advertising, interacting with products on their touchscreen panels (where the radio and ventilation system controls used to be). As of early, the National Highway and Traffic Safety Association (NHTSA), which facilitates dialog between manufacturers and the U. S. Department of Transportation, had set a standard in 2012 of not allowing more than 30 characters of text on a display screen for a non-driving task. Manufacturers wanted more text. One scenario was to introduce advertising through the vehicles GPS system: While driving past a shopping mall, the driver can find out all the special deals going on, and hopefully pull off the road to go buy something. Another would facilitate the instant purchase of downloaded music (Wiese & Lee, 2007).

While all the distractions are in various stages of development, so are means for counteracting the distractions by automating the vehicle (Merat & Lee, 2012). Possibilities include total automatic driving, or limited automation with synthetic vision systems that are linked to a heads-up display on the windshield that will alert the driver to road hazards and options for avoiding them. Other ideas for semiautomatic driving involve sensors that alert drivers when they are following the car ahead of them too closely, or veering laterally out of the driving lane, and automatic braking mechanisms.

15.4.11 Automated vehicles

Vehicle automation levels can be put on a scale from 0 to 5 (Hancock et al., 2020). At levels 0–1, the driver is in full control, and automation features serve to assist the driver. Levels 2–3 are human–autonomy partnerships where the driver is prepared to take over full control. At levels 5–6 the vehicle is in full control.

Fully automated vehicles could be attractive to the elderly or other physically challenged people who are no longer able to drive. Elderly drivers currently tend to limit their driving radius and road usage to their own comfort and security levels (Hancock et al., 2020). It is an open question, however, whether higher levels of automation will encourage more extensive driving on their part with an increase in associated risk.

Another category of potential users are parents who spend a lot of time driving their children from one place to another. They would only need to put the children in the vehicle, set the navigation controls, and press a remote-control start-go button. A survey of parents who like this idea indicated that parents would need to keep close cell phone contact with children during the ride, the drop-off, and return trip (Tremoulet et al., 2020). The need for almost constant contact could defeat the purpose of handing over the responsibility for the ride to automation.

Although it might appear at first blush that automation could compensate for distraction and fatigue, drivers who engage the available automation only become more bored and fatigued (Mkrtchyan, Macbeth, Solovey, Ryan, & Cummings, 2012). Their control over the level of automation does not lessen the additional fatigue or boredom, and could actually create an additional distraction when the driver's trust in the automation becomes challenged (Neubauer, Matthews, Langheim, & Saxby, 2012). Some drivers respond to automation by engaging in more non-driving tasks (Carsten, Lai, Barnard, Jamson, & Merat, 2012).

Participants in a driving simulator study drove manually during the first phase of the experiment, followed by an automation phase and a resumption of manual control (Kaduk, Roberts, & Stanton, 2021). Fatigue was significantly greater and performance poorer after the automation phase. This result is consistent with previous research indicating that taking the human out of the loop when deploying automation slows response time when the driver needs to resume control.

Enough accidents with fully automated vehicles have accumulated (167 in a California database from 2014 to 2019) to spot some trends. Rear-end collisions are incurred by automated vehicles at 1.5 times the rate for normal vehicles (Biever, Angell, & Seaman, 2020). Technically the automated vehicles were not at fault because they were hit by other cars and did not rear-end another car themselves. The results suggest, nonetheless, that the very slow or odd motion of the automated vehicles was misinterpreted by the drivers of the other cars.

A few collisions occurred when the automated vehicle behaved erratically prompting the human in charge to resume control (Biever et al., 2020). It is unknown whether the accident would have occurred without human intervention, but the blame for the accident was placed on the driver because the human did intervene. Other notable collisions occurred when the automated vehicle did not recognize a large truck, emergency vehicle, bus, a pedestrian (fatally), or that it was driving off a cliff (Biever et al., 2020; Gawron, 2020; Hancock et al., 2020).

15.5 OUTER SPACE

The developments in manned spaceflight during the second half of the 20th century were outgrowths of engineering and design ideas, including basic terrestrial flight, from the first half of the century (Caprara, 2000; Dewaard & Dewaard, 1984; Kiselev, Medvedev, & Menshikov, 2003). Out of necessity, this final section of the chapter sticks close to human factors issues as they have emerged.

15.5.1 Brief history

Space missions fall into three basic categories. Sortie missions are flights that go out to do a job and then return. Temporarily manned space stations are intended for habitation for six months or less. Permanently manned stations are intended for habitation for durations in excess of six months. Personnel in the latter group are expected to rotate periodically, but the operation of the station is meant to be continuous.

The first manned spaceflights (Project Mercury in the United States) were single-person vehicles that were launched into space at the tip of a three-stage rocket (Figure 15.13). The capsule landed in the ocean with a parachute for a braking device when it returned to earth, and it had to be retrieved by an aircraft carrier. These capsules were not reusable. Reusable space vehicles that landed on solid ground were not available until the series of space shuttles started in 1980. There were parallel missions conducted by the Soviet Union, which is actually credited with the first manned spaceflight, the first space walk, and the record-setting spaceflight of 438 days.

The second series of spaceflights involved two astronauts (Project Gemini in the United States). Here we had the opportunity to see how well two people could get along when stuck in a tin can together (Figure 15.14). One technical goal of the Gemini series was for an astronaut to exit the space capsule and walk in space while tied to the craft with a cord. This goal was fundamental to eventual work in space and for making repairs.

The third series of spaceflights involved three astronauts (Project Apollo in the United States) and culminated with a landing on the moon in 1969 (Figure 15.15). The landing was accomplished by launching a smaller two-person craft from the main capsule (Figure 15.16), which continued to orbit the moon under the auspices of the third astronaut. This landing unit was capable of lifting off from the moon and returning to the Apollo.

Figure 15.13 Mercury capsule accommodating one astronaut.

Figure 15.14 Gemini capsule accommodating two astronauts.

Figure 15.15 Apollo capsule accommodating three astronauts.

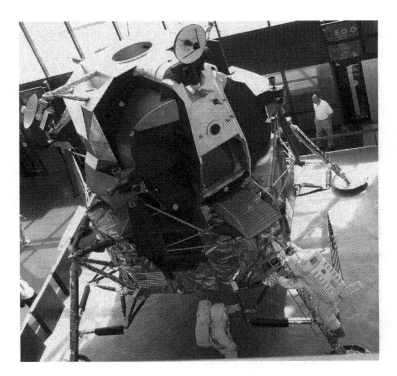

Figure 15.16 Eagle moon Lander.

The launch of Skylab (Figure 15.17), a temporarily manned space station, similar Russian projects, and drone spaceflights to other planets dominated space adventuring during the 1970s. During this time there was much to be learned to prepare the space shuttles for the next decade.

Space shuttles that were launched in the 1980s and 1990s typically involved seven-person crews. For the first time women and civilians were included among the voyagers. Both the United States and Russian space programs during this period included personnel from other interested space-faring countries that did not have an operational space program of their own, short of launching some telecommunications satellites. Space shuttle flights have both scientific and commercial objectives, such as conducting biological and chemical experiments where microgravity is a variable and repairing or reorienting a telecommunications satellite. Figure 15.18 depicts astronauts working outside a space shuttle using jetpacks for mobility.

The 1990s and the 21st century saw the advent of permanently manned space stations, beginning with the Russian project Mir. An international space station has been operational since 1998 (Figure 15.19). Personnel

Figure 15.17 Skylab, exterior view.

Figure 15.18 Astronauts working outside a space shuttle. NASA photograph in the public domain.

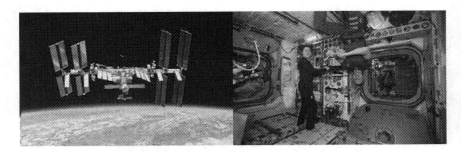

Figure 15.19 International Space Station. NASA photograph in the public domain.

rotations to the space station expanded to one-year intervals in 2015. The space shuttles that transported personnel and supplies have been replaced by newer models operated by a private corporation. Meanwhile, the United States has been sending robotic missions to Mars, and manned stations on the moon and Mars are now in preparation.

A comprehensive systems view of space travel is essential for any further development of the space program. The *human factors* in this context include everything that is not hardware engineering: medical health and wellness, psychological or emotional health, work behavior, limits of perceptual and motor performance, work–rest and sleep cycles, spacecraft habitability, sociology of space communities, culture and living patterns, hazards and emergency management, finances, earthbound politics, space law, specialized education and training, management of space work, and communication systems development (Christensen & Talbot, 1986; Kring, 2001; Harris, 1989, 1992; Suedfeld, 2003).

15.5.2 Personnel selection

When the space program first started in the United States in 1958, NASA considered the possibility that the ideal astronauts would be drawn from the small population of Air Force experimental aircraft pilots. Those individuals had already traveled beyond the speed of sound and into the stratosphere. As it turned out, the experimental aircraft pilots who were available at the time did not want the job (Dewaard & Dewaard, 1984). The first astronauts were selected from a broader range of Air Force personnel who met the highest possible standards for physical and psychological capabilities as they were known at the time. The thinking was that if the top seven could not cope with a few days in space, no one could.

By the late 1970s enough information had accumulated from Antarctic expeditions to allow for some plausible generalizations to outer space personnel. Antarctic conditions are not the same as those found in outer space, but they are perhaps the most extreme terrestrial environments, featuring thin air, social isolation, and severe diurnal disruption. At the time of this writing, there are as many as 5,000 people living in Antarctica during its six months of summer, and as few as 1,000 in its winter. The personality characteristics of the successful Antarctican include independence, task orientation, and achievement motivation. One group of successful personnel, mostly research scientists, is gung ho about the importance of their work, whereas another successful group manages to think of their work as just another 9-to-5 job (Natani, 1980).

Antarcticans who winter over are prone to *dysadaptation syndrome*. U.S. personnel who have succumbed to this malady show all symptoms comparable to clinical depression, as measured by standardized personality inventories. The Russian interpretation of their dysadaptation casualties was "cardiovascular insufficiency" (Boyd, 2001; Natani, 1980), possibly because successful soldiers in severe combat, underwater divers, and bomb-disposal operators displayed lower levels of physiological activation when confronted with stressful situations than the less successful personnel. Thus psychophysiological testing is recommended to screen personnel for work in extreme conditions (Halgrave & Colvin, 1998).

There are personnel selection issues that reach beyond adjustment to the physical environment. A space crew is cooped up with each other in a small living space for an extended period of time. A manned mission to Mars is expected to take close to three years with virtually no help from mission control except for verbal communications that require 45 minutes to transmit. In addition to technical skills, teamwork, communication skills, and adaptability are critical as well as the ability to work autonomously. Crews, which have now expanded to teams of 14 individuals, are now selected as a group with attention to mutual compatibility, rather than individually (Kanas et al., 2002; Rosen, 2021). Because of the small number of people in a Mars mission, an astronaut must be prepared to assume a leadership role or follow a leader interchangeably as situations arise (Holland, Vessey, & Barrett, 2014; Kring, 2001; Rosen, 2021). Facility in a common language is also highly desirable; Russian is strongly recommended for current U. S. Personnel (Alyson, 2018).

15.5.3 Gravitational forces

Although outer space travel occurs in microgravity, there are excessive gravitational forces on the operators during acceleration, especially at the levels required to break free of the earth's orbit. Large gravitational forces are sometimes found in other unusual environments as well. Extreme forces can be harmful and painful, thus anti-G suits were developed to give operators reasonable protection. Even with anti-G production human tolerance is only a matter of minutes and depends on the direction of the thrust. For 6 Gs pushing at an operator's chest, the average endurance time is 550 s; endurance drops to 250 s for a thrust of 10 Gs. For 6 Gs pushing at an operator's back, the average endurance is 350 s, and endurance drops to 100 s for a thrust of 10 Gs. For a downward thrust of 6 Gs, the average endurance is 150 s and drops to 0 s for a thrust of 9 Gs (Chambers, 1963).

The gravitational forces that impinge on an astronaut during takeoff have important effects that start under less extreme conditions than the brief exposure limits just described. As acceleration starts, blood begins to pool and drain from the upper head region to the lower body extremities. The result is a near blackout when acceleration reaches about 4 Gs. Operators do not lose consciousness during this time, but perception and motor coordination are impaired under these high-G conditions: "Linear and angular accelerations themselves produce vestibular, kinesthetic, tactual, and visual cues that, when compared against normal 1 G experiences, appear to be unusual unique, and misleading" (Chambers, 1963, p. 218). The anti-G suit worn by the astronauts is a technical response to this phenomenon. The suits counteract the changes in blood flow by applying pressure to the lower extremities and preventing blood from flowing too far downward. Another technical response is to rely on automatic controllers during this critical phase of takeoff when the astronauts are experiencing psychomotor deficits from the G-forces.

15.5.4 Allocation of function

The initial plan for the Mercury spacecraft allocated functions between mission control personnel and the equipment that they operated. The astronauts were expected to sit in their chairs and not do much of anything. The Mercury prototype did not even have a window. The astronaut team in training strongly objected, insisting that they wanted a set of controls in case something happened. They also figured if they were going to be shot into space, they wanted to look out the window and take a few pictures.

The spaceflight system quickly adopted a more complex view of the person–machine system whereby ground control would work its equipment and remotely control the spacecraft where possible, and the astronauts would control the spacecraft in all other circumstances. As it turned out, the astronauts had the right idea; ground control lost communication contact with John Glenn's Mercury capsule during the third spaceflight, and Glenn had to find his way back to earth on his own.

The marked allocation of functions between the space crew and the ground control has its assets and limitations, especially in emergency situations. Apollo 13, which was meant to be a third trip to the moon, returned to earth prematurely because of life-threatening technological failures. Engineers in ground control were able to work out solutions to some of the problems and report back to the astronauts who were working on other aspects. Another astronaut worked in a spaceflight simulator in the ground station to figure out the successful control sequence that was needed to return the Apollo to earth.

Space station personnel also had some dubious experiences. The first Skylab mission met with some difficulty when ground control asked the astronauts to carry out some unplanned work activities. The astronauts experienced considerable difficulty convincing ground control of the adverse conditions under which they were working that made the requests impossible. The result was that Skylab refused to talk to ground control for an entire day (Bechtel, 2003).

Cosmonauts on Mir reported that their mission control often blamed them unfairly for things that went wrong, such as a fire and an atmosphere leakage. To compound the problem, Mir was operated by a private enterprise that provided bonuses for the cosmonauts for projects that went right, and pay reductions when something went wrong. Cosmonauts thus adopted the strategy of minimizing negative reports (Bechtel, 2003).

15.5.5 Anthropometry

Anthropometric issues showed up early. The first spacecraft were designed for a specially targeted group of seven astronaut trainees. Although a relatively personalized fit could be maintained for the operator's workstation and environment suit, designers did not realize until after the first spaceflight that the human body grows about 2 inches in zero gravity because the pressure on the spinal column is gone. Thus design allowances were needed for subsequent flights (Louviere & Jackson, 1980).

Once in flight and allowed to move around the spacecraft, the floating human body takes on a semifetal position known as the *neutral body position*, as shown in Figure 15.20. This position results in a decline in the line of sight from –5° to –35°. Control panels need to be oriented accordingly (Louviere & Jackson, 1980). Figure 15.21 shows two views of a space shuttle control pod that is operated from a seated position.

The lack of gravitational forces in space results in loss of muscle strength. The impact is most clear after four weeks in space and continues through the first three months. Different parts of the body are affected to different extents, with strength losses of up to 20% in the legs and 10% in the biceps (Fujii & Patten, 1992). Astronauts thus spend part of their day on an exercise treadmill to build back their strength. To use the treadmill, the operator must be anchored with cords to prevent floating away (Joels, Kennedy, &

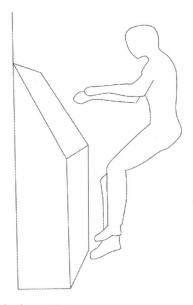

Figure 15.20 The neutral body position.

Figure 15.21 Space shuttle control and display panels.

Larkin, 1982). Cosmonauts reportedly once spent 2.5 hours per day exercising (Fujii & Patten, 1992). Comparable exercise rates from the United States are not available, except to say that some do engage in regular exercise and some do not (Bacal, Billica, & Bishop, 2003); the duration of the mission is a factor that affects the astronauts' daily schedules for self-maintenance.

15.5.6 Vision

Since the earliest spaceflights, astronauts have reported flashes of light that were not really there. The visual flashes are caused by radiation that depolarizes retinal neurons (Fujii & Patten, 1992).

The perception of *up* and *down* is significantly impaired during spaceflight. Visual cues that denote the relationship of objects to other objects are disrupted because the usual reference points, which are gravitationally centered, are not operating. Astronauts compensate for the disruption in visual information by relying on the long axis of their own bodies as reference points (Fujii & Patten, 1992).

15.5.7 Vestibular sense and motor control

The disorganization of visual cues disrupts the coordination of the cerebellum and inner ear, which are responsible for body balance and motor coordination (*otolithic system*). The most common physical complaints reported by astronauts after they return to earth are clumsiness in movements, difficulty walking in a straight line, vertigo while walking or standing, nausea, and difficulty concentrating (Bacal et al., 2003; Hlavacka, Dzurková, & Polónyová, 2000). Some of these effects were ameliorated by in-flight exercise.

Disruption of the otolithic system is the primary explanation for *space adaptation syndrome*. Its symptoms include headache, nausea, vomiting, and sensitivity to head and body motion (Fujii & Patten, 1992, p. 1004). It is reported by 67% of astronauts shortly after takeoff and tapers off after four days in space. Its symptoms often return after they return to earth as the body must readjust itself once again. Space adaptation syndrome can be countered with medication that is taken minutes before blastoff. Seasoned space travelers are less likely to experience its effects, but they are not immune to them. The same disruption is also known to occur as a result of using simulators for spaceflight or virtual reality, where it is known as *simulator sickness*.

Microgravity produces some sensorimotor impairments in pointing and tracking tasks. Pointing movements are slowed in tasks that involve a range of whole arm and finger movements (Bock, Fowler, & Comfort, 2001). The primary culprit is thought to be the disruption of visual feedback, particularly when one movement ends and the next is meant to begin.

Tracking tasks performed with a computer screen and joystick show increased errors during spaceflight. The impairment shows up around the 4th day and extends to the 16th day then appears to subside. Impairments return after the cosmonaut has returned to earth for about another ten days (Manzey, Lorenz, Heuer, & Sangals, 2000). The tracking effect is explained in part by disrupted circadian rhythms, disrupted visual cues, and microgravity itself, resulting in a miscalculation of one's muscular forces while working the controls (Heuer, Manzey, Lorenz, & Sangals, 2003).

15.5.8 Sleep

The loss of the 24-hour day-night cues on earth promotes disruption of sleep onset and duration. Sleep reduces to six hours per night during the first 30 days of spaceflight, then extends to seven hours per night afterward. Six-hour sleeps do not promote performance problems, but performance problems tend to occur after one or two nights of five hours sleep or less (Gundel, Drescher, & Polyakov, 2001). Overall, astronauts experience less time in deep sleep and more frequent sleep interruptions compared to their terrestrial conditions and greater fatigue during work hours (Whitmore et al., 2009).

15.5.9 Space habitats

The first capsules were barely large enough to hold a human, and were not what anyone would call a *habitat*. The two-person Gemini capsules were similarly proportioned, but physical crowdedness now became social crowdedness as well. Analysis of recordings of Gemini astronauts' conversations indicated that they oscillated between times of talking socially and quietude, apparently for privacy (Sieber, 1980). It is interesting that these two tendencies were managed by both parties psychologically; it was not possible to say "good night" and go to one's own room.

The living environment in space offers only a limited amount of space per person, and barely enough room to accomplish all the daily personal tasks during extended missions. It is a closed system rather than an open system. Not only is the environment computer controlled, the computers monitor the humans unusually closely. In Chapter 9 on workspace design, the concern was to have enough space to carry out all the intended work movements, and social space was not an issue. The early space travelers trained together and otherwise knew each other fairly well before their mutual space missions. The space personnel of the future are likely to be more diverse with respect to national origin and cultural habits. Sufficient space needs to be organized to accommodate differences in food preparation, social habits, recreation, religious observances, and habitat aesthetics (Kring, 2001).

The living quarters for the three astronauts of Skylab are shown in Figure 15.22. The space was equivalent to a 90-m² (about 825-ft²) apartment (Caprara, 2000). The four areas are for eating and sleeping, toilet facility,

Figure 15.22 Living areas in Skylab.

food preparation, and physical activity and medical needs. Visitors to the
National Aeronautics and Space Museum in Washington, DC can walk
through Skylab. These minimal facilities accommodated the astronauts for
a maximum of 84 days. The International Space Station has more extensive
layouts for larger crews and longer term living.

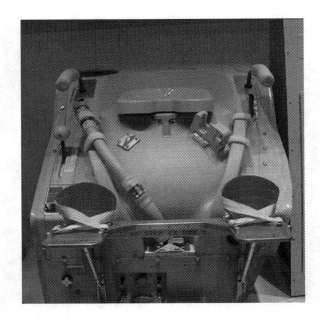

Figure 15.23 Space toilet.

Space toilets (Figure 15.23) operate on a vacuum system. Each droplet and particle must be properly accounted for and not allowed to float around. The equipment compels space travelers to relearn body functions that they thought they had mastered at age two.

Showers must be taken in closed bags. At the time of Skylab, the process of setting up the shower and dissipating the used water required about 2.5 hours (Caprara, 2000). Residents of Mir resorted to using large antiseptic wipes instead of soap and water for stretches of up to five months (Bechtel, 2003).

Space stations can receive supplies of food and other necessities from shuttles, but longer term intergalactic flights present a problem with food (assuming one knows where one is going). Some thought has been given to developing a closed ecosystem wherein the astronauts raise food crops, eat, and recycle waste. Neither the biochemistry nor the agricultural requirements have been fully worked out, and there is a concern that traces of toxins will build up to critical levels after too many cycles have taken place (Bechtel, 2003; Modell & Spurlock, 1980; Rambaut, 1980).

Between the two extremes we have the exploration and possible colonization of Mars. Because of the varying distance between the two planets as they orbit the sun, it is only possible to send flights from earth to Mars every 26 months. Each 26-month period is somewhat different in the distance

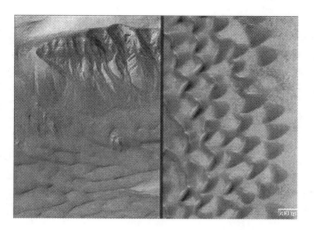

Figure 15.24 Martian landscapes. NASA photograph in the public domain.

calculations, which translate into timing calculations for different parts of the mission. Missions are expected to run upward of 900 days, which would be at least twice the length of the longest mission in space to date. Time and distance create a challenge for the transportation of building supplies, food, and water, and for communication. It is likely that a space station located on Earth's Moon will play a critical role in the development and maintenance of Martian operations. Figure 15.24 shows two views of Martian landscapes taken from drone spacecraft

DISCUSSION QUESTIONS

1. Where else on the stove could you locate the burners? Describe the assets and limitations of the possible configurations.
2. How can the design of a facility such as a factory be a source of stress in the workplace? How about a high-rise apartment building?
3. Compare and contrast the problem of navigation through a large building complex and navigation through a website.
4. What are some of the current traffic laws that pertain to the use of onboard computers, telephones, or GIS? How about texting (on a cellphone) while driving?
5. Motor vehicles have been equipped with a middle rear brake light that is usually mounted in the rear windshield since 1985. Why was this light thought to be necessary? What is the effectiveness of this device?
6. Consider this piece of accident litigation: A car is proceeding down a city street in the left lane of two lanes. The posted speed limit is 35 mph. A 10-year-old boy enters the street from the sidewalk on the right in an

attempt to cross the street and is struck by the car. The car drags the boy 41 feet before stopping. The driver, who is in an emotional state at this time, puts the car in reverse and runs over him again. The boy has substantial medical expenses associated with this accident, and is suing the driver's insurance company. The insurance company's position is that they are not obligated to pay because the accident was the boy's fault. Analyze the human factors involved. What pieces of technical information are required to determine whether substantial fault resides with the driver?

7. If parents were to offload the task of driving their children to various places to an automated vehicle, what logistics would be needed in the AI system to complete the trip without any problems? What about parking the vehicle?

8. In one vision of the future of highways, the cars are all automated, and their movement patterns are coordinated by a central processing unit, probably through the internet or some similar means. Describe some assets and limitations of such a design. Is such a plan realistic? Why?

9. How would teamwork in outer space be similar to or different from the known forms of teamwork on Earth?

References

Abbott, A., Button, C., Pepping, G.-J., & Collins, D. (2005). Unnatural selection: Talent identification and development in sport. *Nonlinear Dynamics, Psychology, and Life Sciences, 9,* 61–88.

Abedin, S., Lewis, M., Brooks, N., Owens, S., Scerri, P., & Sycara, K. (2011). SUAVE: Integrating UAV using a 3D model. *Proceedings of the Human Factors Society, 55,* 91–95.

Ackerman, P. (1987). Individual differences in skill learning: An integration of psychometric and information processing perspectives. *Psychological Bulletin, 102,* 3–27.

Ackerman, P. L. (Ed.). (2011). *Cognitive fatigue.* Washington, DC: American Psychological Association.

Adams, A. E., Rogers, W. A., & Fisk, A. D. (2012). Choosing the right task analysis tool. *Ergonomics in Design, 20*(1), 4–10.

Adams, J. A. (1989). *Human factors engineering.* New York, NY: Macmillan.

Adams, J. S., & Trucks, L. (1976). A procedure for evaluating auditory warning signals. *Proceedings of the Human Factors Society, 20,* 166–172.

Adams, M. A., & Dolan, P. (2005). Spine biomechanics. *Journal of Biomechanics, 38,* 1972–1983.

Agervold, M. (1976). Shiftwork: A critical review. *Scandinavian Journal of Psychology, 17,* 181–189.

Aggarwal, C. C., & Yu, P. S. (Eds.). (2008). *Privacy-preserving data mining: Models and algorithms.* New York, NY: Springer.

Ahmadi, N., Sasangohar, F., Nisar, T., Danesh, V., Larsen, E., Sultana, I., & Bosetti, R. (2022). Quantifying occupational stress in intensive care unit nurses: An applied natural study of correlations among stress, heart rate, electrodermal activity and skin temperature. *Human Factors, 64,* 159–172.

Ahrentzen, S., Levine, D. W., & Michelson, W. (1989). Space, time, and activity in the home: A gender analysis. *Journal of Environmental Psychology, 9,* 89–101.

Aiello, J. R., & Kolb, K. J. (1995). Electronic performance monitoring: A risk factor for workplace stress. In S. L. Sauter & L. R. Murphy (Eds.), *Organizational risk factors for stress* (pp. 163–180). Washington, DC: American Psychological Association.

Akerstedt, T. (1977). Inversion of sleep-wakefulness pattern: Effects on circadian variations in psychophysical activation. *Ergonomics, 10,* 459–474.

Alemi, M. M., Madinei, S., Kim, S., Srinivan, D., & Nussbaum, M. A. (2020). Effects of two passive back-support exoskeletons on muscle activity, energy expenditure and subjective assessments using repetitive lifting. *Human Factors, 62*, 458–474.

Allen, P. A. (2009). Complexity, evolution, and organizational behavior. In S. J. Guastello, M. Koopmans, & D. Pincus (Eds.), *Chaos and complexity in psychology: Theory of nonlinear dynamical systems* (pp. 452–474). New York, NY: Cambridge University Press.

Allen, T. D., Golden, T. D., & Shockley, K. M. (2015). How effective is telecommuting? Assessing the status of our scientific findings. *Psychological Science in the Public Interest, 16*, 40–68.

Alves, E. E., & Kelsey, C. M. (2010). Combating vigilance decrement in a single-operator radar plat-form. *Ergonomics in Design, 18*(2), 6–9.

Alyan, E., Saad, N. M., & Kamel, N. (2021). Effects of workstation type on mental stress: FNIRS study. *Human Factors, 63*, 1230–1255.

Alyson, J. (2018). *What are the special qualities of an astronaut?* Retrieved May 23, 2022, from https://work.chron.com/special-qualities-astronaut-10446.html.

Alzheimer's Association. (2010). *Alzheimer's disease facts and figures*. Retrieved from www.alz.org/alzheimers_disease_facts_and_figures.asp.

Amalberti, R., Carbonell, N., & Falzon, P. (1993). User representations of computer systems in human-computer speech interaction. *International Journal of Man-Machine Studies, 38*, 547–566.

American Civil Liberties Union. (1999). *Drug testing: A bad investment*. New York, NY: Author.

Amershi, S., Weld, D., Vorvoreanu, M., Fourney, A.,Nushi, B., Collisson, P., … Horvitz, E. (2019). *Guidelines for human-AI interaction*. Paper presented to CHI 2019, Glasgow, Scotland, UK. Retrieved from https://doi.org/10.1145/3290 605.3300233

Amirabdollahian, F., Loureiro, R., & Harwin, W. (2002). Minimum jerk trajectory control for rehabilitation and haptic applications. In *Proceedings of the 2002 IEEE-RAS Conference, Washington, DC* (pp. 11–15). Los Alamitos, CA: IEEE.

Amirshirzad, N., Kumru, A., & Oztop, E. (2019). Human adaptation to human-robot shared control. *IEEE Transactions on Human-Machine Systems, 49*, 126–129.

Anderson, P., Rothbaum, B. O., & Hodges, L. F. (2003). Virtual reality exposure in the treatment of social anxiety. *Cognitive and Behavioral Practice, 10*, 240–247.

Andreadis, N., & Quinlan, P. T. (2010). Task switching under predictable and unpredictable circumstances. *Attention, Perception, and Psychophysics, 72*, 1776–1790.

Andreassi, J. L. (1980). *Psychophysiology*. New York, NY: Oxford Press.

Angell, L. S., & Lee, J. D. (2011). Science and technology of driver distraction: Introduction to the special issue. *Ergonomics in Design, 19*(4), 3.

Anthony, K. H., & Watkins, N. J. (2002). Exploring pathology: Relationships between clinical and environmental psychology. In R. B. Bechtel & A. Churchman (Eds.), *Handbook of environmental psychology* (pp. 129–146). New York, NY: Wiley.

Ao, S.-I. (2008). *Data mining and applications in genomics*. Dordrecht, the Netherlands: Springer.

Arecchi, F. T. (2011). Phenomenology of consciousness: From apprehension to judgment. *Nonlinear Dynamics, Psychology, and Life Sciences, 15,* 359–376.

Ariga, A., & Lleras, A. (2011). Brief and rare mental "breaks" keep you focused: Deactivation and reactivation of task goals preempt vigilance decrements. *Cognition, 118,* 439–443.

Ármannsdótir, A. L., Beckerle, P., Moreno, J. C., can Asseldonk, E. H. F., Manrique-Sancho, M-T., del-Ama, A. J., … Briem, K. (2022). Assessing the involvement of users during development of lower limp wearable robotic exoskeletons: A survey study. *Human Factors, 62,* 351–365.

Arthur, W., Jr., Barrett, G. V., & Alexander, R. A. (1991). Prediction of vehicular accident involvement: A meta-analysis. *Human Performance, 4,* 89–105.

Arthur, W., Jr., Barrett, G. V., & Doverspike, D. (1990). Validation of an information-processing-based test battery for the prediction of handling accidents among petroleum-product transport drivers. *Journal of Applied Psychology, 75,* 621–628.

Arthur, W., Jr., & Doverspike, D. (1992). Locus of control and auditory selective attention as predictors of driving accident involvement: A comparative longitudinal investigation. *Journal of Safety Research, 23,* 73–80.

Ash, J. E. (1914). Fatigue and its effects upon control. *Archives of Psychology, 31,* 1–61.

Ashby, W. R. (1956). *Introduction to cybernetics.* New York, NY: Wiley.

Ashoori, M., Burns, C. M., d'Entremont, B., & Momtahan, K. (2019). Using team cognitive work analysis to reveal healthcare team interactions in a birthing unit. *Ergonomics, 57,* 973–986.

Atkins, W. B., Wood, R. E., & Rutgers, P. J. (2002). The effects of feedback format on dynamic decision making. *Organizational Behavior and Human Decision Processes, 88,* 587–604.

Austin, E., Blakely, B., Salmon, P., Braithwaite, J., & Clay-Williams, R. (2022). Identifying constraints on everyday clinical practice: Applying work domain analysis to emergency department care. *Human Factors, 64,* 74–98.

Axelrod, R., & Hamilton, W. D. (1981). The evolution of cooperation. *Science, 211,* 1390–1396.

Baber, C. (2008). Evaluating mobile human-computer interaction. In J. Lumsden (Ed.), *Handbook of research on user interface design and evaluation for mobile technology* (pp. 731–743). Hershey, PA: Information Science Reference.

Baber, C., & McMaster, R. (2016). *Grasping the moment: Sensemaking in response to routine incidents and major emergencies.* Boca Raton, FL: CRC Press.

Baber, C., & Noyes, J. M. (2001). Speech control. In K. Baumann & B. Thomas (Eds.), *User interface design for electronic appliances* (pp. 190–208). New York, NY: Taylor & Francis.

Bacal, K., Billica, R., & Bishop, S. (2003). Neurovestibular symptoms following space flight. *Journal of Vestibular Research, 13,* 93–102.

Bach, D. R., Flandin, G., Friston, K. J., & Dolan, R. J. (2010). Modeling event-related skin conductance responses. *International Journal of Psychophysiology, 75,* 349–356.

Baddeley, A. (2003). Working memory: Looking back and looking forward. *Nature Reviews Neuroscience, 4,* 829–839.

Bahr, G. S., & Ford, R. A. (2011). How and why pop-ups don't work: Pop-up prompted eye movements, user affect and decision making. *Computers in Human Behavior, 27*, 776–763.

Bailey, N. R., Scerbo, M. W., Freeman, F. G., Mikulka, P. J., & Scott, L. A. (2006). Comparison of a brain-based adaptive system and a manual adaptable system for invoking automation. *Human Factors, 48*, 693–709.

Baillargeon, J., & Davis, C. (1984). Barnum meets the computer: A critical test. *Journal of Personality Assessment, 48*, 418–419.

Bak, P. (1996). *How nature works*. New York, NY: Springer.

Balagué, N., Hristovski, R., Agagonés, D., & Tenenbaum, G. (2012). Nonlinear model of attention focus during accumulated effort. *Psychology of Sport and Exercise, 13*, 591–597.

Balci, F., Simen, P., Niyogi, R., Saxe, A., Hughes, J. A., Holmes, P., & Cohen, J. D. (2011). Acquisition of decision making criteria: Reward rate ultimately beats accuracy. *Perception & Psychophysics, 73*, 640–657.

Baldwin, C. L. (2009). Individual differences in navigational strategy: Implications for display design. *Theoretical Issues in Ergonomics Science, 10*, 443–458.

Baldwin, C. L., & Struckman-Johnson, D. (2002). Impact of speech presentation level on cognitive task performance: Implications for auditory display design. *Ergonomics, 45*, 41–74.

Bales, R. F., & Cohen, S. P. (1979). *SYMLOG: A system for the multiple level observation of groups*. New York, NY: Free Press.

Balkin, T. J., Horrey, W. J., Graeber, R. C., Czeisler, C. A., & Dinges, D. F. (2011). The challenges and opportunities of technological approaches to fatigue management. *Accident Analysis and Prevention, 43*, 565–572.

Bankes, S., & Lempert, R. (2004). Robust reasoning with agent-based modeling. *Nonlinear Dynamics, Psychology, and Life Sciences, 8*, 231–258.

Banks, A. P., & Millward, L. J. (2000). Running shared mental models as a distributed cognitive process. *British Journal of Psychology, 91*, 513–531.

Barabási, A.-L. (2002). *Linked: The new science of networks*. Cambridge, MA: Perseus.

Baram, M. (2001). Legal borders necessary for a safe genetic technology. In J. Aithoff (Ed.), *Safety of modern technical systems* (pp. 645–652). Saarbrücken, Germany: TÜV-Saarland Foundation.

Barber, D. J., Reinerman-Jones, L. E., & Matthews, G. (2015). Toward a tactile language for human-robot interaction: Two studies of tacton learning and performance. *Human Factors, 57*, 471–490.

Barbieri, D. F., Brusaca, L. A., Mathiassen, S. E., & Oliveira, A. B. (2020). Effects of time in sitting and standing on pleasantness, acceptability, fatigue, and pain when using a sit–stand desk: An experiment on overweight and normal-weight subjects. *Journal of Physical Activity and Health, 17*, 1222–1230.

Barca, J. C., & Sekercioglu, Y. A. (2013). Swarm robotics reviewed. *Robotica, 31*, 345–359.

Barnes, C. M., Hollenbeck, J. R., Wagner, D. T., DeRue, D. S., Nahgang, J. D., & Schwind, K. M. (2008). Harmful help: The costs of backing-up behavior in teams. *Journal of Applied Psychology, 90*, 529–539.

Barofsky, G. L., & Smith, M. (1993). Reductions in turnover, accidents, and absenteeism: The contribution of a pre-employment screening inventory. *Journal of Clinical Psychology, 49*, 109–116.

Barron, F. (1955). The disposition toward originality. *Journal of Abnormal and Social Psychology, 51*, 478–485.

Barrouillet, P., Portrat, S., & Camos, V. (2011). On the law relating processing to storage in working memory. *Psychological Review, 118*, 175–192.

Bar-Yam, Y. (1997). *Dynamics of complex systems.* Reading, MA: Addison-Wesley.

Baumann, M. R., Sniezek, J. A., & Buerkle, C. A. (2001). Self-evaluation, stress, and performance: A model of decision making under acute stress. In E. Salas & G. Klein (Eds.), *Linking expertise and naturalistic decision making* (pp. 139–158). Mahwah, NJ: Lawrence Erlbaum Associates.

Bechtel, R. B. (2003). On to Mars! In R. B. Bechtel & A. Churchman (Eds.), *Handbook of environmental psychology* (pp. 676–684). New York, NY: Wiley

Bedard, M., Guyatt, G. H., Stones, M. J., & Hirdes, J. P. (2002). The independent contribution of driver, crash, and vehicle characteristics to driver fatalities. *Accident Analysis and Prevention, 34*, 717–728.

Ben-Asher, N., & Meyer, J. (2018). The triad of risk-related behaviors (TriRB): A three-dimensional model of cyber risk taking. *Human Factors, 60*, 1163–1178.

Ben-Naim, A. (2008). *A farewell to entropy: Statistical thermodynamics based on information.* Singapore: World Scientific.

Benner, L., Jr. (1975). Accident investigation: Multinational events sequencing methods. *Journal of Safety Research, 7*, 67–73.

Bennett, A. A., Campion, E. D., Keeler, K. R., & Keener, S. K. (2021). Videoconference fatigue? Exploring changes in fatigue after videoconference meetings during COVID-19. *Journal of Applied Psychology, 106*, 330–344.

Bennett, K. B., & Flach, J. M. (2011). *Display and interface design: Subtle science, exact art.* Boca Raton, FL: CRC Press/Taylor & Francis.

Bennett, K. B., & Flach, J. M. (2013). Configural and pictorial displays. In J. D. Lee & A. Kirlik (Eds.), *The Oxford handbook of cognitive engineering* (pp. 517–533). New York, NY: Oxford University Press.

Bennett, K. B., & Flach, J. M. (2019). Ecological interface design: Thirty-plus years of refinement, progress, and potential. *Human Factors, 61*, 513–525.

Bennett, K. B., Flach, J. M., McEwen, T. R., & Fox, O. (2015). Enhancing creative problem solving through visual display design. In D. A. Boehm-Davis, F. T. Durso, & J. D. Lee (Eds.). *APA handbook of human systems integration* (pp. 419–433). Washington, DC: American Psychological Association.

Bequette, B., Norton, A., Jones, E., & Stirling, L. (2020). Physical and cognitive load effects due to a powered lower-body exoskeleton. *Human Factors, 62*, 411–423.

Bergamasco, M., Bardy, B., & Gopher, D. (2013). *Skill training in multimodal virtual environments.* Boca Raton, FL: CRC Press/Taylor & Francis.

Berlin, S., Reagan, I. J., & Bliss, J. P. (2012). Participant characteristics and speeding behavior during an advisory warning and cash incentive intervention. *Proceedings of the Human Factors and Ergonomics Society, 56*, 1044–1048.

Bernstein, N. (1967). *The coordination and regulation of movements.* Oxford, UK: Pergamon.

Berringer, D. B. (1978). Collision avoidance response stereotypes in pilots and non-pilots. *Human Factors, 20,* 529–536.

Bertin, J. (1981). *Graphics and graphic information processing.* Berlin, Germany: de Gruyter.

Betrancourt, M., & Tversky, B. (2000). Effect of computer animation on users' performance: A review. *Le Travail Humain, 63,* 311–329.

Bettini, C., Jajodia, S., & Wang, S. X. (2000). *Time granularities in databases, data mining, and temporal reasoning.* Berlin, Germany: Springer.

Beveles, A. (1948). A mathematical model for group structures. *Human Organization, 7,* 16–30.

Biever, W., Angell, L., & Seaman, S. (2020). Automated driving system collisions: Early lessons. *Human Factors, 62,* 249–259.

Bigelow, J. (1982). A catastrophe model of organizational change. *Behavioral Science, 27,* 26–42.

Billinghurst, S. S., Morgan, C., Rorie, R. C., Kiken, A., Bacon, L. P., Vu, K.-P. L., Strybel, T. Z., & Battiste, V. (2012). Should students learn general sir traffic control management skills before NextGen tools. *Proceedings of the Human Factors and Ergonomics Society, 56,* 16–20.

Bills, A. G. (1931). Blocking: A new principle in mental fatigue. *American Journal of Psychology, 67,* 230–245.

Binik, Y. M., Servan-Schreiber, D., Freiwald, S., & Hall, K. S. (1988). Intelligent computer-based assessment and psychotherapy: An expert system for sexual dysfunction. *Journal of Nervous and Mental Disease, 176,* 387–400.

Bitire, A. A., & Chuma, L. L. (2022). Effects of occupational health and safety strategies on the organizational performance: A case study on electrical power corporation in Wolaita Sodo District, Ethiopia. *Journal of Legal, Ethical, and Regulatory Issues, 25,* 1–11.

Blankenburger, S., & Hahn, K. (1991). Effects of icon design on human-computer interaction. *International Journal of Man-Machine Studies, 35,* 363–377.

Blattner, M. M., Sumikawa, D. A., & Greenberg, R. M. (1989). Earcons and icons: Their structure and common design principles. *Human-Computer Interaction, 4,* 11–44.

Boccaletti, S., Latora, V., Moreno, Y., Chavez, M., & Hwang, D. U. (2006). Complex networks: Structure and dynamics. *Physics Reports, 424,* 175–178.

Bock, O., Fowler, B., & Comfort, D. (2001). Human sensorimotor coordination during spaceflight: An analysis of pointing and tracking responses during the "Neurolab" space shuttle mission. *Aviation, Space, and Environmental Medicine, 72,* 877–883.

Boer, L., van der Schaaf, M. F., Vincken, K. L., Mol, C. P., Stuijfzand, B. G., & van der Gijp, A. (2018). Volumetric image interpretation in radiology: Scroll behavior and cognitive processes. *Advances in Health Sciences Education, 23,* 783–802.

Boesch, G. (2022). AI emotion and sentiment analysis with computer vision in 2022. Retrieved May 11, 2022, from https://viso.ai/deep-learning/visual-emotion-ai-recognition/

Boisot, M., & McKelvey, B. (2011). Complexity in organization-environment relations: Revisiting Ashby's law of requisite variety. In P. Allen, S. Maguire, & B. McKelvey (Eds.), *The Sage handbook of complexity and management* (pp. 279–298). Thousand Oaks, CA: Sage.

Boles, D. B., Bursk, J. H., Phillips, J. B., & Perdelwitz, J. R. (2007). Predicting dual-task performance with the Multiple Resources Questionnaire (MRQ). *Human Factors, 49*, 32–45.

Boles, D. B., & Phillips, J. B. (2007). A reply to the methodological and theoretical concerns of Vidulich and Tsang. *Human Factors, 49*, 50–52.

Bond, C. F., Jr., & Titus, L. J. (1983). Social facilitation: A meta-analysis of 241 studies. *Psychological Bulletin, 94*, 265–292.

Bond, S., & Cooper, S. (2006). Modeling emergency decisions: Recognition-primed decision making. The literature in relation to an ophthalmic critical incident. *Journal of Clinical Nursing, 15*, 1023–1032.

Botella, J., Barriopedro, M. L., & Suero, M. (2001). A model of the formation of illusory conjunctions in the time domain. *Journal of Experimental Psychology: Human Perception and Performance, 27*, 1452–1467.

Bott, L., & Heit, E. (2004). Nonmonotonic extrapolation in function learning. *Journal of Experimental Psychology: Learning, Memory, and Cognition, 30*, 38–50.

Bousfield, W. A. (1953). The recurrence of clustering in the recall of randomly arranged associates. *Journal of General Psychology, 49*, 229–240.

Boyd, K. (2001). Psychological, emotional studies of Mir Space Station missions show Russians fared better than Americans. *Journal of Human Performance in Extreme Environments, 5*, 96–97.

Braarud, P. Ø. (2001). Subjective task complexity and subjective workload: Criterion validity for complex team tasks. *International Journal of Cognitive Ergonomics, 5*, 261–273.

Braarud, P. Ø., Bodal, T., Helsund, J. E., Louka, M. N., Nihlwing, C., Nystad, E., … Wingstedt, E. (2020). An investigation of speech features, plant system alarms, and operator–system interaction for the classification of operator cognitive workload during dynamic work. *Human Factors, 63*, 736–756.

Brachman, R. (1985). "I lied about the trees," or, defaults and definitions in knowledge representation. *A. I. Magazine, 6*, 80–93.

Branche, C. M., Sniezek, J. E., Sattin, R. W., & Mirkin, I. R. (1991). Water recreation-related spinal injuries: Risk factors in natural bodies of water. *Accident Analysis and Prevention, 23*, 13–18.

Brannick, M. T., Prince, A., Prince, P., & Salas, E. (1995). The measurement of team process. *Human Factors, 37*, 641–651.

Brannick, M. T., Roach, R. M., & Salas, E. (1993). Understanding team performance: A multimethod study. *Human Performance, 6*, 287–308.

Braun, T. W. (2008). Prevention through design (PtD) from the insurance perspective. *Journal of Safety Research, 39*, 137–139.

Brave, S. B. (2003). Agents that care: Investigating the effects of orientation of emotion exhibited by an embodied computer agent. *Dissertation Abstracts International, 64*(5-B), 2437.

Breeden, J. L., Dinkelacker, F., & Hubler, A. (1990). Noise in the modeling and control of dynamical systems. *Physics Review A, 42*, 5827–5836.

Brehmer, B. (1974). Hypotheses about relations between scaled variables in the learning of probabilistic inference tasks. *Organizational Behavior and Human Performance, 11*, 1–27.

Brehmer, B. (1987). Development of mental models for decision in technological systems. In J. Rasmussen, K. Duncan, & J. Leplat (Eds.), *New technology and human error* (pp. 111–120). New York, NY: Wiley.

Brehmer, B. (2005). Micro-worlds and the circular relation between people and their environment. *Theoretical Issues in Ergonomic Science, 6*, 73–94.

Brewer, C. A. (2006). Basic mapping principles for visualizing cancer data using geographic information systems (GIS). *American Journal of Preventive Medicine, 30*, 525–536.

Brewer, C. A., Hatchard, G. W., & Harrower, M. A. (2003). ColorBrewer in print: A catalog of color schemes for maps. *Cartography and Geographic Information Systems, 30*, 5–32.

Brewster, S. A., & Brown, L. M. (2004). Tactons: Structured tactile messages for non-visual information display. In *Proceedings of the 5th Australian User Interface Conference* (pp. 15–23). Sydney, Australia: Australian Computer Society.

Broadbent, D. E. (1958). *Perception and communication.* Elmsford, UK: Pergamon Press.

Bronkhorst, A. W., Veltman, J. A., & van Breda, L. (1996). Application of a three-dimensional auditory display in a flight task. *Human Factors, 38*, 23–33.

Bruce, C. J., & Atkins, F. J. (1993). Efficiency effects of premium-setting regimes under Workers' Compensation: Canada and the United States. *Journal of Labor Economics, 77*(Suppl.), S38–S69.

Brunnstein, K. (2001). About actual threats from the Internet: About viruses, worms, Trojan horses and other computer beasts. In J. Aithoff (Ed.), *Safety of modern technical systems* (pp. 585–586). Saarbrücken, Germany: TÜV-Saarland Foundation.

Brunswick, E. (1955). Representative design and probabilistic theory in a functional psychology. *Psychological Review, 62*, 193–217.

Bryan, W. L., & Harter, N. (1897). Studies in the physiology and psychology of telegraphic language. *Psychological Review, 4*, 27–53.

Bryan, W. L., & Harter, N. (1899). Studies on the telegraphic language: The acquisition of a hierarchy of habits. *Psychological Review, 6*, 345–375.

Buchholz, R. A. (1977). The belief structure of managers relative to work concepts measured by a factor analytic model. *Personnel Psychology, 30*, 567–587.

Buck, J. R., & Kantowitz, B. H. (1983). Decision making with applications to inspection and maintainability. In B. H. Kantowitz & R. L. Sorkin (Eds.), *Human factors: Understanding people-system relationships* (pp. 428–453). New York, NY: Wiley.

Bunn III, W. B., Pikelny, D. B., Slavin, T. J., & Paralkar, S. (2001). Health, safety, and productivity in a manufacturing environment. *Journal of Occupational and Environmental Medicine, 43*, 47–55.

Buntine, W. (1987). Induction of Horn clauses: Methods and the plausible generalization algorithm. *International Journal of Man-Machine Studies, 26*, 499–519.

Buntine, W., & Stirling, D. (1988). *Induction as interaction* (Tech. Rep.). NSW, Australia: University of New South Wales.

Burdea, G. C., & Coiffet, P. (2003). *Virtual reality technology* (2nd ed.). New York, NY: Wiley Interscience.

Burke, C. S., Stagl, K. C., Salas, E., Pierce, L. & Kendall, D. (2006). Understanding team adaptation: A conceptual analysis and model. *Journal of Applied Psychology, 91,* 1189–1207.

Burt, R. (2004). Structural holes and good ideas. *American Journal of Sociology, 110,* 349–399.

Bush, T. R., & Hubbard, R. P. (2008). A comparison of four office chairs using biomechanical measures. *Human Factors, 50,* 629–642.

Bushby, S. T. (2011, November). Information model standard for integrating facilities with smart grid. *BACnet Today & the Smart Grid, ASHRAE Journal,* nv, B18–B22.

Butler, R. J. (1994). *Safety incentives in Workers' Compensation: A review of the evidence and some new findings* (Tech. Rep.). Minneapolis, MN: City University of Minnesota, Industrial Relations Center.

Calderón-Jiménez, B., Johnson, M. E., Montoro Bustos, A. R., Murphy, K. E., Winchester, M. R., & Vega Baudrit, J. R. (2017). Silver nanoparticles: Technological advances, societal impacts, and metrological challenges. *Frontiers in Chemistry, 5*(6). doi: 10.3389/fchem.2017.00006

Campion, M. A. (1983). Personnel selection for physically and demanding jobs: Review and recommendations. *Personnel Psychology, 36,* 527–550.

Campion, M. A., Papper, E. M., & Medsker, G. J. (1996). Relations between work team characteristics and effectiveness: A replication and extension. *Personnel Psychology, 49,* 429–452.

Cannon-Bowers, J., Salas, E., & Converse, S. (1990). Cognitive psychology and team training: Training shared mental models of complex systems. *Human Factors Bulletin, 33,* 1–4.

Cannon-Bowers, J. Salas, E., & Converse, S. (1993). Shared mental models in expert team decision making. In N. J. Castellan (Ed.), *Individual and group decision making: Current issues* (pp. 221–246). Hillsdale, NJ: Lawrence Erlbaum Associates.

Cantwell, R. H., & Moore, P. J. (1996). The development of measures of individual differences in self-regulatory control and their relationship to academic performance. *Contemporary Educational Psychology, 21,* 500–517.

Cao, L. (Ed.). (2009). *Data mining and multi-agent integration.* New York, NY: Springer.

Cao, L., Ni, J., Wang, J., & Zhang, C. (2004, September). Agent services-driven plug and play in the F-TRADE. In *17th Australian Joint Conference on Artificial Intelligence* (pp. 917–922). New York, NY: Springer.

Caplin, S. (2001). *Icon design.* New York, NY: Watson-Guptill.

Caprara, G. (2000). *Living in space: From science fiction to the International Space Station.* Buffalo, NY: Firefly Books.

Card, S. K., English, W. K., & Burr, B. J. (1978). Evaluation of mouse, rate-controlled isometric joystick, step keys, and text keys for text selection on a CRT. *Ergonomics, 21,* 601–613.

Card, S., Moran, T. P., & Newell, A. (1983). *The psychology of human-computer interaction.* Hillsdale, NJ: Lawrence Erlbaum Associates.

Carlson, L. A., Hölscher, C., Shipley, T. F., & Dalton, R. C. (2010). Getting lost in buildings. *Current Directions in Psychological Science, 19,* 284–289.

Carlsson, H., & van Damme, E. (1993). Equilibrium selection in stag hunt games. In K. Binmore, A. Kirman, & P. Tani (Eds.), *Frontiers of game theory* (pp. 237–253). Cambridge, MA: MIT Press.

Carsten, O., Lai, F. C. H., Barnard, Y., Jamson, A. H., & Merat, N. (2012). Control task substitution in semi-automated driving: Does it matter what aspects are automated? *Human Factors, 54,* 747–761.

Carswell, C. M., & Seidelman, W. (2015). Visuospatial displays: Design problems and principles. In D. A. Boehm-Davis, F. T. Durso, & J. D. Lee (Eds.), *APA handbook of human systems integration* (pp. 401–417). Washington, DC: American Psychological Association.

Carter, N., & Menckel, E. (1985). Near-accident reporting: A review of Swedish research. *Journal of Occupational Accidents, 7,* 61–64.

Carter, R. C., Jr. (1978). Knobology underwater. *Human Factors, 20,* 641–647.

Caroux, L., Le Bigot, L., & Vibert, N. (2013). Impact of the motion and visual complexity of the background on players' performance in video game-like displays. *Ergonomics, 56,* 1863–1876.

Carpman, J. R., & Grant, M. A. (2002). Wayfinding: A broad view. In R. B. Bechtel & A. Churchman (Eds.), *Handbook of environmental psychology* (pp. 427–442). New York, NY: Wiley.

Carroll, B. (1987). Expert systems for clinical diagnosis: Are they worth the effort? *Behavioral Science, 32,* 274–292.

Carroll, J. M., & Rosson, M. B. (1987). Paradox of the active user. In J. M. Carroll (Ed.), *Interfacing thought: Cognitive aspects of human-computer interaction* (pp. 80–111). Cambridge, MA: MIT Press.

Casner, S. M. (1994). Understanding the determinants of problem-solving behavior in a complex environment. *Human Factors, 36,* 580–596.

Catchpole, K., Privette, A., Roberts, L., Alfred, M., Carter, B., Woltz, E., … Crookes, B. (2022). A smartphone application for teamwork and communication in trauma: Pilot evaluation "in the wild." *Human Factors, 64,* 143–158.

Catrambone, R., & Seay, A. F. (2002). Using animation to help students learn computer algorithms. *Human Factors, 44,* 495–511.

Cattell, H. E. P. (1994). Development of the 16PF fifth edition. In S. R. Conn & M. L. Rieke (Eds.), *Technical manual, 16PF fifth ed.* (pp. 1–20). Champaign, IL: Institute for Personality and Ability Testing.

Cattell, J. M. (1886). The time it takes to see and name objects. *Mind, 11,* 63–65.

Cattell, R. B. (1971). *Abilities: Their structure, growth and action.* Boston, MA: Houghton Mifflin.

Cattell, R. B., Eber, H. W., & Tatsuoka, M. M. (1970). *Handbook for the sixteen personality factor Questionnaire.* Champaign, IL: Institute for Personality and Ability Testing.

Cave, J., & Eyes, B. (2019). Combining composition and sonic information design in a new electroacoustic work. *Ergonomics in Design, 27*(1), 20–23.

Centers for Disease Control and Prevention. (2007). *Number of deaths for leading causes of death.* Retrieved from www.cdc.gov/nchs/fastats/lcod.html

Centers for Disease Control and Prevention. (2016). *Confined spaces.* Retrieved from www.cdc.gov/niosh/topics/confinedspace

Centers for Disease Control and Prevention. (2022). *Leading causes of death.* Retrieved from www.cdc.gov/nchs/fastats/leading-causes-of-death.htm

Ceschi, A., & Fioretti, G. (2021). From bounded rationality to collective behavior. *Nonlinear Dynamics, Psychology, and Life Sciences, 25*, 385–394.

Ceschi, A., Sartori, R., & Guastello, S. J. (2018). Advanced modeling methods for studying individual differences and dynamics in organizations: Introduction to the special issue. *Nonlinear Dynamics, Psychology, and Life Sciences, 22*, 15–34.

Cha, J. S., & Yu, D. (2022). Objective measures of surgeon non-technical skills in surgery: A scoping review. *Human Factors, 64*, 42–73.

Chaffin, D. B., Herrin, G. D., Keyserling, W. M., & Foulke, J. A. (1977). *Pre-employment strength testing in selecting workers for materials handling jobs* (Report No. CDC 99-74-62). Cincinnati, OH: National Institute for Occupational Safety and Health, Physiology and Ergonomics Branch.

Chambers, A. J., Robertson, M. M., & Baker, N. A. (2019). The effect of sit-stand desks on office workers behavioral and health outcomes: A scoping review. *Applied Ergonomics, 78*, 37–53.

Chambers, R. M. (1963). Operator performance in acceleration environments. In N. M. Burns, R. M. Chambers, & E. Hendler (Eds.), *Unusual environments and human behavior* (pp. 193–320). New York, NY: Free Press.

Chandrasekaran, B., & Goel, A. (1988). From numbers to symbols to knowledge structures: Artificial intelligence perspectives on the classification task. *IEEE Transactions on Systems, Man, and Cybernetics, 18*, 415–424.

Chapanis, A. (1975). Engineering psychology. In M. Dunnette (Ed.), *Handbook of industrial-organizational psychology* (pp. 697–744). Chicago, IL: Rand McNally.

Chapman, L. J., & Chapman, J. P. (1967). Genesis of popular but erroneous psychodiagnostic observations. *Journal of Abnormal Psychology, 72*, 271–280.

Chapman, L. J., & Chapman, J. P. (1969). Illusory correlation as an obstacle to the use of valid psychodiagnostic signs. *Journal of Abnormal Psychology, 74*, 271–280.

Chatterji, G. B., & Sridhar, B. (2001). *Measures for air traffic controller workload prediction.* Paper presented to the 1st AIAA Aircraft, Technology, and Operations Forum, October 2001. Los Angeles, CA: American Institute of Aeronautics & Astronautics.

Chen, B. (1999). *Needle in a haystack.* Retrieved from www.baddesigns.com/many lts.html

Cherry, E. C. (1953). Some experiments on the recognition of speech with one and two ears. *Journal of the Acoustical Society of America, 25*, 975–979.

Chiappe, D., Strybel, T. Z., & Vu, K.-P. L. (2012). Mechanisms for the acquisition of situation awareness in situated agents. *Theoretical Issues in Ergonomics Science, 13*, 625–647.

Chiappe, D., Strybel, T. Z., & Vu, K.-P. L. (2015). A situated approach to the understanding of dynamic systems. *Journal of Cognitive Engineering and Decision Making, 9*, 33–43.

Chien, J. M., & Schneider, W. (2012). The brain's learning and control architecture. *Current Directions in Psychological Science, 21*, 78–84.

Cho, A., & Histon, J. (2012). Factors affecting the learning of a new air traffic control sector for experienced air traffic controllers. *Proceedings of the Human Factors and Ergonomics Society, 56*, 90–94.

Christensen, J. M., & Talbot, J. M. (1986). Psychological aspects of space flight. *Aviation, Space, and Environmental Medicine, 57*, 203–212.

Chueng, J. (1983). Interdependence and coordination in organizations. A rule-system analysis. *Academy of Management Journal, 26,* 156–162.

Cigler, B. A. (2007). The "big questions" of Katrina and the 2005 great flood of New Orleans. *Public Administration Review, 67,* 64–76.

Clark, T. B. (1975, May). Common shortcomings in facilities design. *Industrial Management, X,* 1–4.

Clarke, S. (2006). The relationship between safety climate and safety performance: A meta-analytic review. *Journal of Occupational Health Psychology, 11,* 315–327.

Clayton, K., & Frey, B. B. (1997). Studies of mental "noise." *Nonlinear Dynamics, Psychology, and Life Sciences, 1,* 173–180.

Cloninger, C. (2002). *Fresh styles for web designers: Eye candy from the underground.* Indianapolis, IN: New Riders.

Cohen, P. R., & Howe, A. E. (1989). Toward AI research methodology: Three case studies in evaluation. *IEEE Transactions on Systems, Man, and Cybernetics, 19,* 634–646.

Cohen, S. (1980). Aftereffects of stress on human performance and social behavior: A review of research and theory. *Psychological Bulletin, 88,* 82–108.

Collins, J. (2008). *Nonlinear dynamics and enhanced sensorimotor function.* Retrieved from www.labome.org/grant/r01/hd/nonlinear/dynamics/nonlinear-dynamics-and-enhanced-sensorimotor-function

Colquhoun, W. P. (1971). *Biological rhythms and human performance.* London: Academic Press.

Colquhoun, W. P. (1975). Evaluation of auditory, visual, and dual-mode displays for prolonged sonar monitoring in repeated sessions. *Human Factors, 17,* 425–437.

Comer, D. R. (1995). A model of social loafing in real work groups. *Human Relations, 48,* 647–661.

Comfort, L. (1996). Self-organization in disaster response: Global strategies to support local action. In G. Koehler (Ed.), *What disaster response management can learn from chaos theory* (pp. 94–112). Sacramento, CA: California Research Bureau, California State Library.

Comfort, L. (1999a). Nonlinear dynamics in disaster response: The Northridge, California earthquake, January 17, 1994. In L. D. Kiel & E. Elliott (Eds.), *Nonlinear dynamics, complexity, and public policy* (pp. 139–152). Commack, NY: Nova Science.

Comfort, L. (1999b). *Shared risk: Complex systems in seismic response.* Kidlington, UK: Pergamon/Elsevier.

Conrad, R. (1951). Speed and load stress in a sensorimotor skill. *British Journal of Industrial Medicine, 8,* 1–7.

Constant, D., Sproull, L., & Kiesler, S. (1997). The kindness of strangers: On the usefulness of electronic weak ties for technical advice. In S. Kiesler (Ed.), *Culture of the Internet* (pp. 303–322). Mahwah, NJ: Lawrence Erlbaum Associates.

Contractor, N. S., & Siebold, D. R. (1993). Theoretical framework for the study of structuring processes in group decision support systems. *Human Communication Research, 23,* 528–563.

Conway, A. R. A., Kane, M. J., Bunting, M. F., Hambrick, D. Z., & Engle, R. W. (2005). Working memory span tasks: A methodological review and user's guide. *Psychonomic Bulletin & Review, 12,* 769–786.

Cooke, N. J., Amazeen, P. G., Gorman, J. C., Guastello, S. J., Likens, A., & Stevens, R. (2012). Modeling the complex dynamics of teamwork from team cognition to neurophysiology. *Proceedings of the Human Factors and Ergonomics Society, 56,* 183–187.

Cooke, N. J., & Gorman, J. C. (2013). Microworld experimentation with teams. In J. D. Lee, & A. Kirlik (Eds.), *The Oxford handbook of cognitive engineering* (pp. 327–335). New York, NY: Oxford University Press.

Cooke, N. J., & Shope, S. M. (2004). Designing a synthetic task environment. In S. G. Schiflett, L. R. Elliott, E. Salas, & M. D. Coovert (Eds.), *Scaled worlds: Development, validation, and applications* (pp. 263–296). Burlington, VT: Ashgate.

Corker, K. M., Cramer, M. L., Henry, E. H., & Smoot, D. E. (1992). *Methodology for evaluation of automation impacts on tactical command and control (C-2) systems: Final report* (U.S. AFHRL Tech. Rep. No. 90-91). Washington, DC: AL/HRPP.

Cornum, R., Matthews, M. D., & Seligman, M. E. P. (2011). Comprehensive soldier fitness: Building resilience in a challenging institutional context. *American Psychologist, 66,* 4–9.

Costa, G. (2003). Factors influencing health of workers and tolerance to shift work. *Theoretical Issues in Ergonomic Science, 4,* 263–288.

Costa, P. T., & McCrae, R. R. (1992). Four ways five factors are basic. *Personality and Individual Differences, 13,* 653–665.

Côté, D., Gravel, S., Gladu, S., Bakhiyi, B., & Gravel, S. (2021). Worker health in formal electronic waste recycling plants. *International Journal of Workplace Health Management, 14,* 292–309.

Côté, J. N., Mathieu, P. A., Levin, M. F., & Feldman, A. G. (2002). Movement reorganization to compensate for fatigue during sawing. *Experimental Brain Research, 146,* 394–398.

Cox-Fuenzalida, L.-E., Swickert, R., & Hittner, J. B. (2004). Effects of neuroticism and workload history on performance. *Personality and Individual Differences, 36,* 447–456.

Craig, A., & Tran, Y. (2012). The influence of fatigue on brain activity. In G. Matthews, P. A. Desmond, C. Neubauer, & P. A. Hancock (Eds.), *The handbook of operator fatigue* (pp. 185–196). Burlington, VT: Ashgate.

Craig, S. D., Gholson, B., & Driscoll, D. M. (2002). Animated pedagogical agents in multimedia educational environments: Effect of agent properties, picture features, and redundancy. *Journal of Educational Psychology, 94,* 428–434.

Crawley, S. L. (1926). An experimental investigation of recovery from work. *Archives of Psychology, 85.*

Culbert, C., Riley, G., & Savely, R. T. (1989). An expert system development methodology that supports verification and validation. *ISA Transactions, 28,* 15–18.

Culley, K. E., & Madhavan, P. (2012). Affect and time pressure in a simulated luggage screening paradigm: A fuzzy signal detection theory analysis. *Proceedings of the Human Factors and Ergonomics Society, 56,* 338–342.

Czermak, P. (2001). Safety measures in laboratories and production plants using genetic technology. In J. Althoff (Ed.), *Safety of modern technical systems* (pp. 617–618). Saarbrücken, Germany: TÜV-Saarland Foundation.

Czerwinski, M. P., & Larson, K. (2003). Cognition and the web: Moving from theory to web design. In J. Ratner (Ed.), *Human factors and web development* (2nd ed., pp. 147–165). Mahwah, NJ: Lawrence Erlbaum Associates.

Daily, R. C. (1980). A path analysis of R & D team coordination and performance. *Decision Sciences, 11,* 357–369.

d'Amato, M. R. (1973). *Experimental psychology: Methodology, psychophysics and learning.* New York, NY: McGraw-Hill.

Dang, N. T., Tavanti, M., Rankin, I., & Cooper, M. (2009). A comparison of different input devices for a 3D environment. *International Journal of Industrial Ergonomics, 39,* 554–563.

Darnell, M. J. (1997). *Yellow street lights.* Retrieved from www.baddesigns.com/strlt.htm

Darnell, M. J. (1999). *Don't go to the right?* Retrieved from www.baddesigns.com/dontgo.html

Davis, K. C., Reid, C. R., Rempel, D. D., & Treaster, D. (2020). Introduction to the Human Factors special issue on user centered design for exoskeleton. *Human Factors, 62,* 333–336.

Dawson, D. A. (1994). Heavy drinking and the risk of occupational injury. *Accident Analysis and Prevention, 26,* 655–665.

Dawson, D., Noy, Y. I., Härmä, M., Akerstedt, T., & Belensky, G. (2011). Modelling fatigue and the use of fatigue models in work settings. *Accident Analysis and Prevention, 43,* 549–564.

Day, K., & Calkins, M. P. (2002). Design and dementia. In R. B. Bechtel & A. Churchman (Eds.), *Handbook of environmental psychology* (pp. 374–393). New York, NY: Wiley.

Deatherage, B. (1972). Auditory and other sensory forms of information presentation. In H. Van Cott & R. Kinkade (Eds.), *Human engineering guide to equipment design* (pp. 123–160). Washington, DC: U.S. Government Printing Office.

DeGreene, K. B. (1991). Emergent complexity and person-machine systems. *International Journal of Man-Machine Studies, 35,* 219–234.

Dehn, D. M., & van Mulken, S. (2000). The impact of animated interface agents: A review of empirical research. *International Journal of Human-Computer Studies, 52,* 1–22.

Deininger, R. L. (1960). Human factors engineering studies of the design and use of pushbutton telephone sets. *Bell System Technical Journal, 39,* 986–1012.

DeLosh, E. L., Busemeyer, J. R., & McDaniel, M. A. (1997). Extrapolation: The sine qua non for abstraction in function learning. *Journal of Experimental Psychology: Learning, Memory, and Cognition, 23,* 968–986.

Demir, M., Likens, A. D., Cooke, N. J., & Amazeen, P. G. (2019). Team coordination and effectiveness in human-autonomy teaming. *IEEE Transactions on Human-Machine Systems, 49,* 150–159.

Deng, C.-L., Geng, P., Hu, Y-F., & Kuai, S-G. (2019). Beyond Fitts' law: A three-phase model predicts movement time to position an object in an immersive 3D virtual environment. *Human Factors, 61,* 879–894.

Deng, S., Nan, J., Chang, J., Guo, S., & Zhang, J. J. (2017). Understanding the impact of multimodal interaction using gaze informed mid-air gesture control in 3D virtual objects manipulation. *International Journal of Human-Computer Studies, 105,* 68–80.

Denning, P. J. (1986). The science of computing: Expert systems. *American Scientist, 74*, 18–20.

Denning, P. J., & Denning, D. E. (2016). Cybersecurity is harder than building bridges. *American Scientist, 104*, 154–157.

Dennis, A. R., & Valacich, J. S. (1993). Computer brainstorms: More heads are better than one. *Journal of Applied Psychology, 78*, 531–537.

Denny, V. E., Gilbert, K. J., Erdmann, R. C., & Rumble, E. T. (1978). Risk assessment methodologies: An application to underground mine systems. *Journal of Safety Research, 10*, 24–34.

de Pontbriand, R. J., Allender, L. E., & Doyle, F. J. III. (2008). Temporal regulation and temporal cognition: Biological and psychological aspects of time stress. In P. A. Hancock & J. L. Szalma (Eds.), *Performance under stress* (pp. 143–162). Abingdon, Oxon, UK: Ashgate.

Depue, R. A., & Monroe, S. M. (1986). Conceptualization and measurement of human disorder in life stress research: The problem of chronic disturbance. *Psychological Bulletin, 99*, 36–51.

De Raedt, R., & Ponjaert-Kristoffersen, I. (2001). Predicting at-fault car accidents of older drivers. *Accident Analysis and Prevention, 33*, 809–820.

de Rosis, F., Pelachaud, C., Poggi, I., Carofiglio, V., & De Carolis, B. (2003). From Greta's mind to her face: Modeling the dynamics of affective states in a conversational embodied agent. *International Journal of Human-Computer Studies, 59*, 81–118.

Derr, J., Forst, L., Chen, H. Y., & Conroy, L. (2001). Fatal falls in the U.S. construction industry, 1990-1999. *Journal of Occupational and Environmental Medicine, 43*, 921–931.

Derthick, M. (2007). Where federalism didn't fail. *Public Administration Review, 67*, 36–47.

Desari, D., Crowe, C., Ling, C., Zhu, M., & Ding, L. (2010). EEG pattern analysis for physiological indicators of mental fatigue in simulated air traffic control tasks. *Proceedings of the Human Factors and Ergonomics Society, 54*, 205–209.

Deutsch, C. P. (1964). Auditory discrimination and learning: Social factors. *Merrill-Palmer Quarterly of Behavior and Development, 10*, 277–296.

Deutsch, D. (1992). Hie tritone paradox: Implications for the representation and communication of pit structures. In M. R. Jones & S. Holleran (Eds.), *Cognitive bases of musical communication* (pp. 115–138). Washington, DC: American Psychological Association.

Dewaard, J. E., & Dewaard, N. (1984). *History of NASA: America's voyage to the stars*. New York, NY: Bison/Exter/Simon & Schuster.

Dey, A., & Mann, D. D. (2010). Sensitivity and diagnosticity of NASA-TLX and simplified SWAT to assess the mental workload associated with an agricultural sprayer. *Ergonomics, 53*, 848–857.

Dillard, M. B., Warm., J. S., Funke, G. J., Funke, M. E., Finomore, V. S., Matthews, G., Shaw, T. H., & Parasuraman, R. (2014). The sustained attention to response task (SART) does not promote mindlessness during vigilance performance. *Human Factors, 56*, 1364–1379.

Dillon, R. L., & Tinsley, C. H. (2007). How near-misses influence decision making under risk: A missed opportunity for learning. *Management Science, 54*, 1425–1440.

Dingus, T. A., Hanowski, R. J., & Klauer, S. G. (2011). Estimating crash risk. *Ergonomics in Design, 19*(4), 8–12.

DiNocera, F., Ferlazzo, F., & Gentilomo, A. (1996). [Inter-individual stability in the P300 effect of event-related brain potentials under double-task conditions.] *Psychologie Francaise, 41,* 365–374.

DiPietro, A. (1976). *An analysis of the OSHA inspection program in manufacturing 1972-73.* Washington, DC: U.S. Department of Labor.

Doll, T. J., Folds, D. J., & Leiker, B. S. (1984). *Auditory information systems in military aircraft: Current configurations versus the state of the art* (Rep. No. USAFSAM-TR-84-15). Brooks Air Force Base, TX: U.S. Air Force School of Aerospace Medicine.

Doll, T. J., Hanna, T. E., & Russotti, J. S. (1992). Masking in three-dimensional auditory displays. *Human Factors, 34,* 255–265.

Dooley, K. J. (1997). A complex adaptive systems model of organization change. *Nonlinear Dynamics, Psychology, and Life Sciences, 1,* 69–97.

Dooley, K. J. (2004). Complexity science models of organizational change and innovation. In M. S. Poole & A. H. Van de Ven (Eds.), *Handbook of organizational change and innovation* (pp. 354–373). New York, NY: Oxford University Press.

Dooley, K. J. (2009). The butterfly effect of the "butterfly effect." *Nonlinear Dynamics, Psychology, and Life Sciences, 13,* 279–288.

Doos, M., Backstrom, T., & Samuelson, S. (1994). Evaluation of strategy: Preventing accidents with automated machinery through targeted and comprehensive investigations conducted by safety engineers. *Safety Science, 17,* 187–206.

Dore, M. H. I. (2009). The impact of Edward Lorenz: An introductory overview. *Nonlinear Dynamics, Psychology, and Life Sciences, 13,* 241–242.

Dore, M. H. I., & Singh, R. G. (2009). Turning points in nonlinear business cycle theories, financial crisis, and the global 2007-2009 downturn. *Nonlinear Dynamics, Psychology, and Life Sciences, 13,* 423–444.

Douglas, M., & Wildavsky, A. (1982). *Risk and culture: An essay on the selection of environmental dangers.* Berkeley, CA: University of California Press.

Drag, L. L., & Bieliauskas, L. A. (2010). Contemporary review 2009: Cognitive aging. *Journal of Geriatric Psychiatry, 23,* 75–93.

Drury, C. G. (2015). Sustained attention in operational settings. In J. Szalma, M. Scerbo, P. Hancock, R. Parasuraman, & R. Hoffman (Eds.), *Cambridge handbook of applied perception research* (Vol. 2, pp. 769–792). Cambridge, UK: Cambridge University Press.

Drury, C. G., Holness, K., Ghysin, K. M., & Green, B. D. (2009). Using individual differences to build a common core dataset for aviation security studies. *Theoretical Issues in Ergonomics Science, 10,* 459–480.

Druskat, V. U., & Pescosolido, A. T. (2002). The content of effective teamwork mental models in self-managing teams: Ownership, learning and heedful interrelating. *Human Relations, 55,* 283–314.

Duan, Q. (2013). Service-oriented network virtualization for convergence of networking and cloud computing in next-generation networks. In A.- S. K. Pathen, M. M. Monowar, & Z. M. Fadlullah (Eds.), *Building next-generation converged networks: Theory and practice* (pp. 393–416). Boca Raton, FL: CRC/ Taylor & Francis.

Dubitzky, W. (Ed.). (2008). *Data mining techniques in grid computing environments*. Chichester, UK: Wiley-Blackwell.

Duffy, F. (1997). *The new office*. London: Conran Octopus.

DUIAttorneyTab. (2013). Blood alcohol levels by state. Retrieved from www.duiatto rneytab.com/blood-alcohol-limits-by-state.html

Dunbar, E. 1993. The role of psychological stress and prior experience in the use of personal protective equipment. *Journal of Safety Research, 24,* 181–187.

Dureman, E. I., & Boden, C. (1972). Fatigue in simulated car-driving. *Ergonomics, 15,* 299–305.

Durso, F. T., & Dattel, A. (2004). SPAM: The real-time assessment of SA. In S. Banbury & S. Trembley (Eds.), *A cognitive approach to situation awareness: Theory, measures, and application* (pp. 137–154). New York, NY: Aldershot.

Durso, F. T., & Sethumadhavan, A. (2008). Situation awareness: Understanding dynamic environments. *Human Factors, 50,* 442–448.

Dux, P. E., & Marois, R. (2009). The attentional blink: A review of data and theory. *Attention, Perception, & Psychophysics, 71,* 1683–1700.

Eastman, M. C., & Kamon, E. (1976). Posture and subjective evaluation at flat and slanted desks. *Human Factors, 18,* 15–26.

Eberts, R., & Salvendy, G. (1986). The contributions of cognitive engineering to the safe design and operation of CAM and robotics. *Journal of Occupational Accidents, 8,* 49–67.

Eby, D. W., Vivoda, J. M., & Fordyce, T. A. (2002). The effects of standard enforcement on Michigan safety belt use. *Accident Analysis and Prevention, 34,* 815–824.

Edworthy, J. (1994). The design and implementation of non-verbal auditory warnings. *Applied Ergonomics, 25,* 202–210.

Edworthy, J. R., McNeer, R. R., Bennett, C. L., Dudaryk, R., McDougall, S. J. P., Schlesinger, J. J., ... Osborn, D. (2018). Getting better hospital alarm sounds into a global standard. *Ergonomics in Design, 26*(4), 4–13.

Edworthy, J. R., Reid, S., McDougall, S., Edworthy, J., Hall, S., Bennett, D., ... Pye, E. (2017). The recognizability and localizability of auditory alarms: Setting global medical device standards. *Human Factors, 59,* 1108–1127.

Ein-Dor, T., Mikulincer, M., Doron, G., & Shaver, P. R. (2010). The attachment paradox: How can so many of us (the insecure ones) have no adaptive advantages? *Perspectives on Psychological Science, 5,* 123–141.

Ekman, P. (1993). Facial expression and emotion. *American Psychologist, 48,* 384–392.

Elg, F. (2005). Leveraging intelligence for high performance in complex dynamic systems requires balanced goals. *Theoretical Issues in Ergonomic Science, 6,* 63–72.

Elkins, A. N., Muth, E. R., Hoover, A. W., Walker, A. D., Carpenter, T. L., & Switzer, F. S. (2009). Physiological compliance and team performance. *Applied Ergonomics, 40,* 997–1003.

Elliott, E., & Kiel, L. D. (Eds.). (2004). Agent-based modeling [Special issue]. *Nonlinear Dynamics, Psychology, and Life Sciences, 8,* 121–302.

Ellis, K. K. E., Kramer, L. J., Shelton, K. J., Arthur, J. J. III, & Prinzel, L. J. III. (2011). Transition of attention in terminal area NextGen operations using synthetic vision systems. *Proceedings of the Human Factors and Ergonomics Society, 55,* 46–60.

Elprama, S. A., Vannieuwenhuyze, J. T. A., De Bock, S., Vanderborght, B., De Pauw, K., Meeusen, R., & Jacobs, A. (2020). Social processes: What determines industrial workers' intention to use exoskeletons? *Human Factors, 62,* 337–350.

Enander, A. (1989). Effects of thermal stress on human performance. *Scandinavian Journal of Work and Environmental Health, 15*(1), 27–33.

Endres, L. S., Guastello, S. J., & Rieke, M. L. (1992). Meta-interpretive reliability of computer-based test interpretations: The Karson Clinical Report. *Journal of Personality Assessment, 59,* 448–467.

Endsley, M. R. (1995). Toward a theory of situation awareness in dynamic systems. *Human Factors, 37,* 32–64.

Endsley, M. R. (2015). Situation awareness misconceptions and misunderstandings. *Journal of Cognitive Engineering and Decision Making, 9,* 4–32.

Endsley, M. R. (2021). An objective review and meta-analysis of direct objective measures of situation awareness: A comparison of SAGAT and SPAM. *Human Factors, 63,* 124–150.

Endsley, M. R., Bolté, B., & Jones, D. G. (2003). *Designing for situation awareness: An approach to user-centered design.* Boca Raton, FL: CRC Press.

Epstein, I. (2010). *Clinical data-mining: Integrating practice and research.* New York, NY: Oxford University Press.

Epstein, J. M., & Axtell, R. (1996). *Growing artificial societies.* Cambridge, MA: MIT Press.

Epstein, Y., & Karlin, R. A. (1975). Effects of acute experimental crowding. *Journal of Applied Social Psychology, 5,* 34–53.

Ericsson, K. A., & Simon, H. A. (1980). Verbal reports as data. *Psychological Review, 87,* 215–251.

Estévez-Mujica, C. P., & García-Díaz, C. (2021). Computational modeling approaches to organizational change. *Nonlinear Dynamics, Psychology, and Life Sciences, 25,* 467–506.

Eubanks, J. (1993). *Pedestrian accident prevention.* Tucson, AZ: Lawyers and Judges.

Eubanks, J. L., & Killeen, P. R. (1983). An application of signal detection theory to air combat training. *Human Factors, 25,* 449–456.

Evans, L. (2002). Traffic crashes. *American Scientist, 90,* 244–253.

Evans, M., Hastings, N., & Peacock, B. (1993). *Statistical distributions* (2nd ed.). New York, NY: Wiley.

Evans, R. E., & Kortum, P. (2010). The impact of voice characteristics on user response in an interactive voice response system. *Interacting with Computers, 22,* 606–614.

Eysenck, M. W. (1997). *Anxiety and cognition: A unified theory.* East Essex, UK: Psychology Press/Taylor & Francis.

Fadden, S., Ververs, P. M., & Wickens, C. D. (2001). Pathway HUDs: Are they viable? *Human Factors, 43,* 173–193.

Fan, J., & Smith, A. P. (2017). The impact of workload and fatigue on performance. In L. Longo & M. C. Leva (Eds.), *Mental workload: Models and applications* (pp. 90–105). Cham, Switzerland: Springer.

Fans, S. M. (1997). *Evolution of 3-D technology.* Retrieved from http://vrex.com

Farazmand, A. (2007). Learning from the Katrina crisis: A global and international perspective with implications for future crisis management. *Public Administration Review, 67,* 149–159.

Faris, S. M. (1997). *Evolution of 3-D technology*. Retrieved March 9, 1997, from http://vrex.com

Fechner, G. T. (1964). *Elemente der psychophysik* [Elements of psychophysics]. Amsterdam, the Netherlands: E. J. Bonset.

Feinauer, D. M., & Havlovic, S. J. (1993). Drug testing as a strategy to reduce occupational accidents: A longitudinal analysis. *Journal of Safety Research, 24*, 1–7.

Fell, J. C., & Voas, R. B. (2014). The effectiveness of a 0.05 blood alcohol concentration (BAC) limit for driving in the United States. *Addiction, 109*, 869–874.

Fewster, K. M., Riddell, M. F., Kadam, S., & Callaghan, J. P. (2019). The need to accommodate monitor height changes between sitting and standing. *Ergonomics, 62*, 1515–1523.

Fiedler, F. E., Bell, C. P. L., Chemers, M. M., & Patrick, D. (1984). Increasing mine productivity and safety through management training and organizational development: A comparative study. *Basic and Applied Social Psychology, 5*, 1–18.

Fine, P. E. M. (1979). John Brownlee and the measurement of infectiousness: An historical study in epidemic theory. *Journal of the Royal Statistical Society A, 142*, 347–362.

Fine, S. A., & Getkate, M. (1995). *Benchmark tasks for job analysis: A guide for functional job analysis (FJA) scales*. Mahwah, NJ: Lawrence Erlbaum Associates.

Fiore, S. M., Rosen, M. A., Smith-Jentsch, K. A., Salas, E., Letsky, M., & Warner, N. (2010). Toward an understanding of macrocognition in teams: Predicting processes in complex collaborative contexts. *Human Factors, 52*, 203–224.

Fischer, U., McDonnell, L. K., Davison, J., Haars, K., Villeda, E., & Van Aken, C. (2001). *Detecting errors inflight*. Retrieved from http://olias.nasa.gov/reports/detecting.html

Fitts, P. M. (1954). The information capacity of the human motor system in controlling the amplitude of movement. *Journal of Experimental Psychology, 47*, 381–391.

Fitts, P. M., & Seeger, C. M. (1953). S-R compatibility: Spatial characteristics of stimulus and response codes. *Journal of Experimental Psychology, 46*, 199–210.

Flach, J. M., Guisinger, M. A., & Robinson, A. G. (1996). Fitts' law: Nonlinear dynamics and positive entropy. *Ecological Psychology, 8*, 281–325.

Flache, A., & Macy, M. W. (1997). The weakness of strong ties: Collective action failure in a highly cohesive group. In P. Doreian, & F. B. Stokman (Eds.), *Evolution of social networks* (pp. 19–44). Amsterdam, the Netherlands: Gordon & Breach.

Flanagan, J. C. (1954). The critical incident technique. *Psychological Bulletin, 51*, 327–358.

Flanders, V. (2016). Web pages that suck. Retrieved from www.webpagesthatsuck.com

Flash, T., & Hogan, N. (1985). The coordination of arm movements: An experimentally confirmed mathematical model. *Journal of Neuroscience, 5*, 1688–1703.

Fleishman, E. A. (1954). Dimensional analysis of psychomotor abilities. *Journal of Experimental Psychology, 48*, 437–454.

Fleishman, E. A. (1964). *The structure and measurement of physical fitness*. Englewood Cliffs, NJ: Prentice Hall.

Flin, R., Slaven, G., & Stewart, K. (1996). Emergency decision making in the offshore oil and gas industry. *Human Factors, 38*, 262–277.

Flowers, J. H., Buhman, D. C., & Turnage, K. D. (1997). Cross-modal equivalence of visual and auditory scatterplots for exploring bivariate data samples. *Human Factors, 39,* 341–351.

Forrester, J. W. (1961). *Industrial dynamics.* Cambridge, MA: MIT Press.

Foss, R. D., Stewart, J. R., & Reinfurt, D. W. (2001). Evaluation of the effects of North Carolina's 0.08% BAC law. *Accident Analysis and Prevention, 33,* 507–518.

Fowler, B. (1994). P300 as a measure of workload during a simulated aircraft landing task. *Human Factors, 36,* 670–683.

Fox, J. (1984). Formal and knowledge-based methods in decision technology. *Acta Psychologica, 56,* 303–331.

Frantz, T. L., & Carley, K. M. (2009). Agent-based modeling within a dynamic network. In S. J. Guastello, M. Koopmans, & D. Pincus (Eds.), *Chaos and complexity in psychology: The theory of nonlinear dynamical systems* (pp. 475–505). New York, NY: Cambridge University Press.

Freeman, L. C. (1979). Centrality in social networks: Conceptual clarification. *Social Networks, 1,* 215–239.

Freeny, M. (2007). Whatever happened to clinical privacy? *Annals of the American Psycotherapy Association, 10*(2), 13–17.

Frensch, P. A., & Runger, D. (2003). Implicit learning. *Current Directions in Psychological Science, 12,* 13–18.

Frey, P. W., & Sears, R. J. (1978). Model of conditioning incorporating the Rescorla-Wagner associative axiom, a dynamic attention process, and a catastrophe rule. *Psychological Review, 85,* 321–340.

Frick, K., Jensen, P., Quinlan, M., & Wilthagen, T. (Eds.). (2000). *Systematic occupational health and safety management: Perspectives on an international development.* Oxford, UK: Pergamon.

Friedman, J. W., & Samuelson, L. (1994). The "folk theorem" for repeated games and continuous decision rules. In J. W. Friedman (Ed.), *Problems of coordination in economic activity* (pp. 103–128). Boston, MA: Kluwer.

Friston, K. (2010). The free-energy principle: A unified brain theory? *Nature Reviews: Neuroscience, 11*(2), 127–138.

Fröhlich, B., Plate, J., Wind, J., Wesche, G., & Gobel, M. (2000). Cubic mouse-based interaction in virtual environments. *IEEE Computer Graphics and Applications, X,* 12–15.

Fu, W.-T., & Pirolli, P. (2013). Establishing the micro-to-macro link in cognitive engineering: Multilevel models of socio-computer interaction. In J. D. Lee & A. Kirklik (Eds.), *Oxford handbook of cognitive engineering* (pp. 501–513). New York, NY: Oxford University Press.

Fuchs, C. (2012). *Implications of deep packet inspection (DPI) Internet surveillance for society.* Technical Report No. 1. Uppsala, Sweden: Department of Informatics and Media, Uppsala University.

Fujii, M. D., & Patten, B. M. (1992). Neurology of microgravity and space travel. *Neurologic Clinics, 10,* 999–1013.

Fullarton, C., & Stokes, M. (2007). The utility of a workplace injury instrument in prediction of workplace injury. *Accident Analysis and Prevention, 39,* 28–37.

Fulmer, J. (2022). *The value of emotion recognition technology.* Retrieved April 23, 2022, from www.itbusinessedge.com/business-intelligence/value-emotion-reco gnition-technology/

Funke, G. J., Knott, B. A., Salas, E., Pavlas, D., & Strang, A. J. (2012). Conceptualization and measurement of team workload: A critical need. *Human Factors, 54*, 36–51.

Furnham, A., & Schofield, S. (1987). Accepting personality test feedback: A review of the Barnum effect. *Current Psychological Research and Reviews, 6*, 62–178.

Galam, S. (1996). When humans interact like atoms. In E. White, & J. H. Davis (Eds.), *Understanding group behaviour* (Vol. 1, pp. 293–312). Mahwah, NJ: Lawrence Erlbaum Associates.

Galitz, W. O. (1993). *User-interface screen design.* New York, NY: Wiley.

Gallegati, M., Greenwald, B., Richiadi, M. G., & Stiglitz, J. E. (2008). The asymmetric effect of diffusion processes: Risk sharing and contagion. *Global Economy Journal, 8*(3). doi:10.2202/1524-5861.1365

Gander, P., Hartley, L., Powell, D., Cabon, P., Hitchcock, E., Mills, A., & Popkin, S. (2011). Fatigue risk management: Organizational factors at the regulatory and industry/company level. *Accident Analysis and Prevention, 43*, 573–590.

Gandit, M., Kouabeanan, D. R., & Caroly, S. (2009). Road-tunnel fires: Risk perception and management strategies among users. *Safety Science, 47*, 105–114.

Garosi, E., Kalantari, R., Zanjirani, A., Zuaktafi, M., Roknabadi, E. H., & Bakshi, E. (2020). Concern about verbal communication in the operating room: A field study. *Human Factors, 62*, 940–953.

Gaver, W. W. (1986). Auditory icons: Using sound in computer interfaces. *Human-Computer Interaction, 2*, 167–277.

Gaver, W. W. (1994). Using and creating auditory icons. In G. Kramer (Ed.), *Auditory display: Sonification, audification, and auditory interfaces* (pp. 417–446). Reading, MA: Addison-Wesley.

Gawron, V. J. (2019a). *Human performance and situation awareness measures.* Boca Raton, FL: CRC Press.

Gawron, V. J. (2019b). *Workload measures.* Boca Raton, FL: CRC Press.

Gawron, V. J. (2020). Lessons lost: What we learned about automation in aviation can be applied to autonomous vehicles. *Ergonomics in Design, 28*(4), 24–26.

Gell-Mann, M. (1994). *The quark and the jaguar.* New York, NY: W. H. Freeman.

Gerdes, K. E., Leitz, C. A., & Segal, E.A., (2011). Measuring empathy in the 21st century: Development of an empathy index rooted in social cognitive neuroscience and social justice. *Social Work Research, 35*, 83–93.

Gethmann, C. F. (2001). Ethical aspects of technical safety. In J. Althoff (Ed.), *Safety of modern technical systems* (pp. 45–46). Saarbrücken, Germany: TÜV-Saarland Foundation.

Getty, D. J. (1983). Three dimensional displays: Perceptual research and applications. *Psychological Documents, 13*, 178 (Manuscript No. 2570).

Gibson, J. J. (1950). *The perception of the visual world.* Boston, MA: Houghton Mifflin.

Gibson, J. J. (1979). *The ecological approach to visual perception.* Boston, MA: Houghton Mifflin.

Gilhooly, K. J., Fioratou, E., & Henretty, N. (2010). Verbalization and problem solving: Insight and spatial factors. *British Journal of Psychology, 101*, 81–93.

Gilmore, R. (1981). *Catastrophe theory for scientists and engineers.* New York, NY: Wiley.

Gilster, P. (1993). *The Internet navigator.* New York, NY: Wiley.

Gittell, J. H. (2000). Organizing work to support relationship coordination. *International Journal of Human Resource Management, 11*, 517–539.

Glass, D. C., & Singer, J. E. (1972). *Urban stress: Experiments on noise and social stressors*. New York, NY: Academic Press.

Glover, H. (1982). Four syndromes of post-traumatic stress disorder: Stressors and conflicts of the traumatized with special focus on the Vietnam Combat Veteran. *Journal of Traumatic Stress, 1*, 57–78.

Gnoni, M. G., Andriulo, S., Maggio, G., & Nardone, P. (2013). "Lean occupational" safety: An application for a near-miss management system design. *Safety Science, 53*, 96–104.

Gödel, K. (1962). *On formally undecidable propositions*. New York, NY: Basic Books.

Goldberg, L. R. (1968). Simple models or simple process? Some research on clinical judgments. *American Psychologist, 23*, 483–496.

Goldberg, L. R. (1969). Man versus model of man: A rationale, plus some evidence, for a method of improving clinical inferences. *Psychological Bulletin, 6*, 422–432.

Goldberger, A. (1996). Nonlinear dynamics for clinicians: Chaos theory, fractals and complexity at the bedside. *The Lancet, 347*, 1312–1314.

Goldstein, J. (2011). Emergence in complex systems. In P. Allen, S. Maguire, & B. McKelvey (Eds.), *The Sage handbook of complexity and management* (pp. 79–92). Thousand Oaks, CA: Sage.

Gontier, E., LeDantec, C., Leieu, A., Paul, I., Charvin, H., Bernard, C., Lalonde, R., & Rebai, M. (2007). Frontal and parietal ERPs associated with duration discrimination with or without task interference. *Brain Research, 1170*, 79–89.

Gonzalez, C., Vanyukov, P., & Martin, M. K. (2005). The use of microworlds to study dynamics decision making. *Computers in Human Behavior, 21*, 273–286.

Goodrich, M. A. (2013). Multitasking and multi-robot management. In J. D. Lee & A. Kirklik (Eds.), *Oxford handbook of cognitive engineering* (pp. 379–394). New York, NY: Oxford University Press.

Gopher, D. (1993). The skill of attention control: Acquisition and execution of attention strategies. In D. E. Meyer (Ed.), *Attention and performance XIV: Synergies in experimental psychology, artificial intelligence and cognitive neuroscience* (pp. 299–322). Cambridge, MA: MIT Press.

Gordon, J. E. (1978). *Structures, or why things don't fall down*. New York, NY: Penguin Books.

Gorman, J. C., Amazeen, P. G., & Cooke, N. J. (2010). Team coordination dynamics. *Nonlinear Dynamics, Psychology, and Life Sciences, 14*, 265–290.

Gorman, J. C., Cooke, N. J., & Amazeen, P. G., & Fouse, S. (2012). Measuring patterns in team interaction sequences using a discrete recurrence approach. *Human Factors, 54*, 503–517.

Gorman, J. C., Dunbar, T. A., Grimm, D., & Gipson, C. L. (2017). Understanding and modeling teams as dynamical systems. *Frontiers in Psychology, 8*, 1053. https://doi.org/10.3389/psyc.2017.01053.

Gorman, J. C., Grimm, D. A., Stevens, R. H., Galloway, T., Willemsen-Dunlap, A. M., & Halpin, D. J. (2020). Measuring real-time cognition during team training. *Human Factors, 62*, 825–860.

Governors Highway Safety Association. (2013). *Seat belt laws*. Retrieved from www.ghsa.org/html/stateinfo/laws/seatbelt_laws.html

Governors Highway Safety Association. (2021). *Distracted driving law chart – April 21*. Retrieved July 29, 2021, from https://ghsa.org/state-laws/issues/distracted%20driving

Grant, A. M., & Parker, S. K. (2009). Redesigning work design theories: The rise of relational and proactive perspectives. *Academy of Management Annals, 3,* 317–375.

Greenleaf, W. (1996). VR solutions for rehabilitation, ergonomics, and disability. *Virtual Reality, 3*(1), 13–17.

Greenlee, E. T., Funke, G. J., Warm, J. S., Finomore, V., Patterson, R., Barnes, L. E., Vidulich, M. A., & Funke, M. E. (2015). The effects of stereoscopic depth on vigilance performance and cerebral hemodynamics. *Human Factors, 57,* 1063–1075.

Greenwood, D. (1991). An overview of neural networks. *Behavioral Science, 36,* 1–33.

Greer, J. M., Burdick, K. J., Chowdhury, A. R., & Schlesinger, J. J. (2018). Dynamic alarm systems for hospitals (D. A. S. H.). *Ergonomics in Design, 26*(4), 14–19.

Gregerson, N. P., Brehmer, B., & Moren, B. (1996). Road safety improvement in large companies: An experimental comparison of different measures. *Accident Analysis and Prevention, 28,* 297–306.

Gregg, S., Martin, L., Homola, J., Lee, P. U., Mercer, J., Brasil, C., Cabrall, C., & Lee, H. (2012). Shifts in air traffic controllers' situation awareness during high-altitude mixed equipage operations. *Proceedings of the Human Factors and Ergonomics Society, 56,* 95–99.

Gregson, R. A. M. (1988). *Nonlinear psychophysical dynamics.* Hillsdale, NJ: Lawrence Erlbaum Associates.

Gregson, R. A. M. (1992). *n-Dimensional nonlinear psychophysics.* Hillsdale, NJ: Lawrence Erlbaum Associates.

Gregson, R. A. M. (1998). Herbart and "Psychological Research on the Strength of a Given Impression." *Nonlinear Dynamics, Psychology, and Life Sciences, 2,* 157–167.

Gregson, R. A. M. (2009). Psychophysics. In S. J. Guastello, M. Koopmans, & D. Pincus (Eds.), *Chaos and complexity in psychology: The theory of nonlinear dynamical systems* (pp. 108–131). New York, NY: Cambridge University Press.

Gregson, R. A. M. (2013). Symmetry-breaking, grouped images and multistability with transient unconsciousness. *Nonlinear Dynamics, Psychology, and Life Sciences, 17,* 325–344.

Grieg, G. L. (1990). On the shape of energy-detection ROC curves. *Perception and Psychophysics, 48,* 77–81.

Gruber, T. R., & Cohen, P. R. (1987). Design for acquisition: Principles of knowledge-system design to facilitate knowledge acquisition. *International Journal of Man-Machine Studies, 26,* 143–150.

Guastello, D. D., & Guastello, S. J. (1987a). A climate for safety in hazardous environments: A psychosocial approach (Report No. 2389). *Social and Behavioral Sciences Documents, 17,* 67.

Guastello, D. D., & Guastello, S. J. (1987b). The relationship between work group size and occupational accidents. *Journal of Occupational Accidents, 9,* 1–9.

Guastello, S. J. (1981). Catastrophe modeling of equity in organizations. *Behavioral Science, 26,* 63–74.

Guastello, S. J. (1982). Color matching and shift work: An industrial application of the cusp-difference equation. *Behavioral Science, 27,* 131–139.

Guastello, S. J. (1985). Euler buckling in a wheelbarrow obstacle course: A catastrophe with complex lag. *Behavioral Science, 30,* 204–212.

Guastello, S. J. (1987). A butterfly catastrophe model of motivation in organizations: Academic performance. *Journal of Applied Psychology, 72,* 165–182.

Guastello, S. J. (1988). Catastrophe modeling of the accident process: Organizational subunit size. *Psychological Bulletin, 103,* 246–255.

Guastello, S. J. (1989). Catastrophe modeling of the accident process: Evaluation of an accident reduction program using the Occupational Hazards Survey. *Accident Analysis and Prevention, 21,* 61–77.

Guastello, S. J. (1991). Psychosocial variables related to transit accidents: A catastrophe model. *Work and Stress, 5,* 17–28.

Guastello, S. J. (1992). Accidents and stress-related health disorders: Forecasting with catastrophe theory. In J. C. Quick, J. J. Hurrell, & L. M. Murphy (Eds.), *Work and well-being: Assessments and interventions for occupational mental health* (pp. 262–269). Washington, DC: American Psychological Association.

Guastello, S. J. (1993). Do we really know how well our occupational accident prevention programs work? *Safety Science, 16,* 445–463.

Guastello, S. J. (1995). *Chaos, catastrophe, and human affairs: Applications of nonlinear dynamics to work, organizations, and social evolution.* Hillsdale, NJ: Lawrence Erlbaum Associates.

Guastello, S. J. (1998). Creative problem solving groups at the edge of chaos. *Journal of Creative Behavior, 32,* 38–57.

Guastello, S. J. (2000). Symbolic dynamic patterns of written exchange: Hierarchical structures in an electronic problem solving group. *Nonlinear Dynamics, Psychology, and Life Sciences, 4,* 169–188.

Guastello, S. J. (2001). Nonlinear dynamics and occupational accidents. In J. Althoff (Ed.), *Safety of modem technical systems* (pp. 315–324). Saarbrücken, Germany: TÜV-Saarland Foundation.

Guastello, S. J. (2002). *Managing emergent phenomena: Nonlinear dynamics in work organizations.* Mahwah, NJ: Lawrence Erlbaum Associates.

Guastello, S. J. (2003). Nonlinear dynamics, complex systems, and occupational accidents. *Human Factors and Ergonomics in Manufacturing, 13,* 293–304.

Guastello, S. J. (2005). Statistical distributions in self-organizing phenomena: What conclusions should be drawn? *Nonlinear Dynamics, Psychology, and Life Sciences, 9,* 463–478.

Guastello, S. J. (2006). Comment: Motor control research requires nonlinear dynamics. *American Psychologist, 61,* 77–78.

Guastello, S. J. (2007). Nonlinear dynamics and leadership emergence. *Leadership Quarterly, 18,* 357–369.

Guastello, S. J. (2009). Chaos as a psychological construct: Historical roots, principal findings, and current growth directions. *Nonlinear Dynamics, Psychology, and Life Sciences, 13,* 289–310.

Guastello, S. J. (2010). Nonlinear dynamics of team performance and adaptability in emergency response. *Human Factors, 52,* 162–172.

Guastello, S. J. (2011). Discontinuities: SETAR and catastrophe models with polynomial regression. In S. J. Guastello & R. A. M. Gregson (Eds.), *Nonlinear dynamical systems analysis for the behavioral sciences using real data* (pp. 251–280). Boca Raton, FL: CRC Press/Taylor & Francis.

Guastello, S. J. (2013). Modeling illness and recovery with nonlinear dynamics. In J. P. Sturmberg & Carmel M. Martin (Eds.), *Handbook on complexity in health* (pp. 147–156). New York, NY: Springer.

Guastello, S. J. (2015). The complexity of the psychological self and the principle of optimum variability. *Nonlinear Dynamics, Psychology, and Life Sciences, 19*, 511–528.

Guastello, S. J. (Ed.). (2016). *Cognitive workload and fatigue in financial decision making*. Tokyo: Springer.

Guastello, S. J., Bock, B. R., Caldwell, P., & Bond, R. W., Jr. (2005). Origins of group coordination: Nonlinear dynamics and the role of verbalization. *Nonlinear Dynamics, Psychology, and Life Sciences, 9*, 297–333.

Guastello, S. J., Boeh, H., Gorin, H., Huschen, S., Peters, N. E., Fabisch, M., & Poston, K. (2013). Cusp catastrophe models for cognitive workload and fatigue: A comparison of seven task types. *Nonlinear Dynamics, Psychology, and Life Science, 17*, 23–47.

Guastello, S. J., Boeh, H., Schimmels, M., Gorin, H., Huschen, S., Davis, E., Peters, N. E., Fabisch, M., & Poston, K. (2012). Cusp catastrophe models for cognitive workload and fatigue in a verbally cued pictorial memory task. *Human Factors, 54*, 811–825.

Guastello, S. J., Boeh, H., Shumaker, C., & Shimmels, M. (2012). Catastrophe models for cognitive workload and fatigue. *Theoretical Issues in Ergonomics Science, 13*, 586–602.

Guastello, S. J., & Bond, R. W., Jr. (2004). Coordination learning in Stag Hunt games with application to emergency management. *Nonlinear Dynamics, Psychology, and Life Sciences, 8*, 345–374.

Guastello, S. J., & Bond, R. W., Jr. (2007). The emergence of leadership in coordination-intensive groups. *Nonlinear Dynamics, Psychology, and Life Sciences, 11*, 91–117.

Guastello, S. J., Correro, A. N. II, & Marra, D. E. (2018). Do emergent leaders experience greater workload? The swallowtail catastrophe model and changes in leadership in an emergency response simulation. *Group Dynamics: Theory, Research, and Practice, 22*, 200–222.

Guastello, S. J., Correro, A. N. II, & Marra, D. E. (2019). Cusp catastrophe models for cognitive workload and fatigue in teams. *Applied Ergonomics, 79*, 152–168.

Guastello, S. J., Correro, A. N. II, Marra, D. E., & Peressini, A. F. (2019). Physiological synchronization and subjective workload in a competitive emergency response task. *Nonlinear Dynamics, Psychology, and Life Sciences, 23*, 347–376.

Guastello, S. J., Futch, W., & Mirabito, L. (2020). Cognitive workload and fatigue dynamics in a chaotic forecasting task. *Nonlinear Dynamics, Psychology, and Life Sciences, 24*, 179–213.

Guastello, S. J., Futch, W., Mirabito, L., Green, D., Marsicek, L. & Witty, B. (2021a). Heuristics associated with forecasting chaotic events: A rare cognitive ability. *Theoretical Issues in Ergonomics Science, 22*, 457–487.

Guastello, S. J., Futch, W., Mirabito, L., Green, D., Marsicek, L. & Witty, B. (2021b). Forecasting chaotic events and the prediction of a rare cognitive ability. *Personality and Individual Differences, 170*, 110430. https://doi.org/10.1016/j.paid.2020.11043

Guastello, S. J., Gershon, R. M., & Murphy, L. R. (1999). Catastrophe model for the exposure to blood-borne pathogens and other accidents in health care settings. *Accident Analysis and Prevention, 31*, 739–750.

Guastello, S. J., Gorin, H., Huschen, S., Peters, N. E., Fabisch, M., & Poston, K. (2012). New paradigm for task switching strategies while performing multiple tasks: Entropy and symbolic dynamics analysis of voluntary patterns. *Nonlinear Dynamics, Psychology, and Life Sciences, 16*, 471–497.

Guastello, S. J., Gorin, H., Huschen, S., Peters, N. E., Fabisch, M., & Poston, K., & Weinberger, K. (2013). The minimum entropy principle and task performance. *Nonlinear Dynamics, Psychology, and Life Sciences, 17*, 405–423.

Guastello, S. J., & Gregson, R. A. M. (Eds.). (2011). *Nonlinear dynamical systems analysis for the behavioral sciences using real data.* Boca Raton, FL: CRC Press/Taylor & Francis.

Guastello, S. J., & Guastello, D. D. (1986). The relation between the locus of control construct and involvement in traffic accidents. *Journal of Psychology: Interdisciplinary and Applied, 120*, 293–298.

Guastello, S. J., & Guastello, D. D. (1988). *The occupational hazards survey: Manual and case report.* Milwaukee, WI: Authors.

Guastello, S. J., & Guastello, D. D. (1998). Origins of coordination and team effectiveness: A perspective from game theory and nonlinear dynamics. *Journal of Applied Psychology, 83*, 423–437.

Guastello, S. J., Guastello, D. D., & Craft, L. L. (1989). Assessment of the Barnum effect in computer-based test interpretations. *Journal of Psychology: Interdisciplinary and Applied, 123*, 477–484.

Guastello, S. J., & Johnson, E. A. (1999). The effect of downsizing on hierarchical work flow dynamics in organizations. *Nonlinear Dynamics, Psychology, and Life Sciences, 3*, 347–378.

Guastello, S. J., Koehler, G., Koch, B., Koyen, J., Lilly, A., Stake, C., & Wozniczka, J. (2008). Risk perception when the tsunami arrived. *Theoretical Issues in Ergonomic Science, 9*, 95–114.

Guastello, S. J., Korienek, G., & Traut, M. (1989). Verbal versus pictorial representation of objects in a human-computer interface. *International Journal of Man-Machine Studies, 31*, 99–120.

Guastello, S. J., & Liebovitch, L. S. (2009). Introduction to nonlinear dynamics and complexity. In S. J. Guastello, M. Koopmans, & D. Pincus. (Eds.), *Chaos and complexity in psychology: Theory of nonlinear dynamical systems* (pp. 1–40). New York, NY: Cambridge University Press.

Guastello, S. J., & Lynn, M. (2014). Catastrophe model of the accident process, safety climate, and anxiety. *Nonlinear Dynamics, Psychology, and Life Sciences, 18*, 177–198.

Guastello, S. J., Malon, M., Timm, P., Weinberger, K., Gorin, H., Fabisch, M., & Poston, K. (2014). Catastrophe models for cognitive workload and fatigue in a vigilance dual-task. *Human Factors, 56*, 737–751.

Guastello, S. J., & Marra, D. E. (2018). External validity and factor structure of individual and group workload ratings. *Theoretical Issues in Ergonomics Science, 19,* 229–253.

Guastello, S. J., Marra, D. E., Castro, J., Equi, M., & Peressini, A. F. (2017). Turn taking, team synchronization, and non-stationarity in physiological time series. *Nonlinear Dynamics, Psychology, and Life Sciences, 21,* 319–334.

Guastello, S. J., Marra, D. E., Correro, A. N. II, Michels, M., & Schimmel, H. (2017). Elasticity and rigidity constructs and ratings of subjective workload for individuals and groups. In L. Longo & M. C. Leva (Eds.), *Human mental workload: Models and applications* (pp. 51–76). Cham, Switzerland: Springer.

Guastello, S. J., Marra, D. E., Peressini, A. F., Castro, J., & Gomez, M. (2018). Autonomic synchronization, team coordination, participation, and performance. *Nonlinear Dynamics, Psychology, and Life Sciences, 22,* 359–394.

Guastello, S. J., Marra, D. E., Perna, C., Castro, J., Gomez, M., & Peressini, A. F. (2016). Physiological synchronization in emergency response teams: Subjective workload, drivers and empaths. *Nonlinear Dynamics, Psychology, and Life Sciences, 20,* 223–270.

Guastello, S. J., & McGee, D. W. (1987). Mathematical modeling of fatigue in physically demanding jobs. *Journal of Mathematical Psychology, 31,* 248–269.

Guastello, S. J., & Mirabito, L. (2018). Time granularity, lag length, and down-sampling rates for neurocognitive data. *Nonlinear Dynamics, Psychology, and Life Sciences, 22,* 457–483.

Guastello, S. J., Mirabito, L., & Peressini, A. F. (2020). Autonomic synchronization under three task conditions and its impact on team performance. *Nonlinear Dynamics, Psychology, and Life Sciences, 24,* 79–104.

Guastello, S. J., Nathan, D. E., & Johnson, M. J. (2009). Attractor and Lyapunov models for reach and grasp movements with application to robot-assisted therapy. *Nonlinear Dynamics, Psychology, and Life Sciences, 13,* 99–121.

Guastello, S. J., Nielson, K. A., & Ross, T. J. (2002). Temporal dynamics of brain activity in human memory processes. *Nonlinear Dynamics, Psychology, and Life Sciences, 6,* 323–334.

Guastello, S. J., & Peressini, A. F. (2017). Development of a synchronization coefficient for biosocial interactions in groups and teams. *Small Group Research, 48,* 3–33.

Guastello, S. J., & Philippe, P. (1997). Dynamics in the development of large information exchange groups and virtual communities. *Nonlinear Dynamics, Psychology, and Life Sciences, 1,* 123–149.

Guastello, S. J., Pincus, D., & Gunderson, P. R. (2006). Electrodermal arousal between participants in a conversation: Nonlinear dynamics for linkage effects. *Nonlinear Dynamics, Psychology, and Life Sciences, 10,* 365–399.

Guastello, S. J., Reiter, K., & Malon, M. (2016). Cognitive workload and fatigue in a vigilance dual task: Miss errors, false alarms, and the impact of wearing biometric sensors while working. *Nonlinear Dynamics, Psychology, and Life Sciences, 20,* 509–535.

Guastello, S. J., Reiter, K., Malon, M., & Shircel, A. (2015). When auditory and visual signal processing conflict: Cross-modal interference in extended work periods. *Theoretical Issues in Ergonomics Science, 16,* 232–254.

Guastello, S. J., Reiter, K., Malon, M., Timm, P., Shircel, A., & Shaline, J. (2015). Catastrophe models for cognitive workload in N-back tasks. *Nonlinear Dynamics, Psychology, and Life Sciences, 19*, 173–200.

Guastello, S. J., & Rieke, M. L. (1990). The Barnum effect and validation for computer-based test interpretations: The Human Resource Development report. *Psychological Assessment: Journal of Consulting and Clinical Psychology, 2*, 186–190.

Guastello, S. J., & Rieke, M. L. (1994). Computer-based test interpretations as expert systems: Validity and viewpoints from artificial intelligence theory. *Computers in Human Behavior, 4*, 435–495.

Guastello, S. J., Shircel, A., Malon, M., & Timm, P. (2015). Individual differences in the experience of cognitive workload. *Theoretical Issues in Ergonomics Science, 16*, 20–52.

Guastello, S. J., Shircel A., Malon, M., Timm, P., Gonring, K., & Reiter, K. E. (2016). Experimental analysis of cusp models. In S. J. Guastello (Ed.), *Cognitive workload and fatigue in financial decision making* (pp. 37–68). Tokyo: Springer.

Guastello, S. J., Witty, B., Johnson, C., & Peressini, A. F. (2020). Autonomic synchronization, leadership emergence, and the roles of drivers and empaths. *Nonlinear Dynamics, Psychology, and Life Sciences, 24*, 451–473.

Guilford, J. P. (1967). *The structure of intellect.* New York, NY: McGraw-Hill.

Guion, R. M. (1998). *Assessment, measurement and prediction for personnel decisions.* Mahwah, NJ: Lawrence Erlbaum Associates.

Gun, R. T. (1993). The role of regulations in the prevention of occupational injury. *Safety Science, 16*, 47–66.

Gundel, A., Drescher, J., & Polyakov, V. V. (2001). Quantity and quality of sleep during the record manned space flight of 438 days. *Human Factors and Aerospace Safety, 1*, 87–98.

Gunnar Optics. (2022). *Doctor recommended lens technology.* Retrieved from https://gunnar.com/pages/technology

Gureckis, T. M., & Goldstone, R. L. (2006). Thinking in groups. *Pragmatics & Cognition, 14*, 293–311.

Hackman, J. R. (Ed.). (1990). *Groups that work (and those that don't).* San Francisco, CA: Jossey-Bass.

Hackman, J. R., & Oldham, G. R. (1976). Motivation through the design of work: Test of a theory. *Organizational Behavior and Human Performance, 15*, 250–279.

Hackman, J. R., & Oldham, G. R. (1980). *Work redesign.* Reading, MA: Addison-Wesley.

Haigh, R. D. (1977). Mechanical safety and electrical control gear. In W. Handley (Ed.), *Industrial safety handbook* (2nd ed., pp. 251–259). New York, NY: McGraw-Hill.

Haken, H. (1984). *The science of structure: Synergetics.* New York, NY: Van Nostrand Reinhold.

Haken, H. (1988). *Information and self-organization: A macroscopic approach to complex systems.* New York, NY: Springer.

Haken, H., Kelso, J. A. S., & Bunz, H. (1985). A theoretical model of phase transition in human hand movements. *Biological Cybernetics, 51*, 347–356.

Hakstian, A. R., & Cattell, R. B. (1978). Higher-stratum ability structures on a basis of twenty primary abilities. *Journal of Educational Psychology, 70,* 657–669.

Halgrave, R. J., & Colvin, C. (1998). Selection of personnel for stressful occupations: The potential utility of psychophysiological measures as selection tools. *Human Performance in Extreme Environments, 9,* 121–139.

Halpern, C. A., & Dawson, K. D. 1997. Design and implementation of a participatory ergonomics program for machine sewing tasks. *International Journal of Industrial Ergonomics, 20,* 429–440.

Hammer, J. M. (1999). Human factors of functionality and intelligent avionics. In D. Garland, J. Wise, & V. D. Hopkins (Eds.), *Handbook of human factors in aviation* (pp. 549–565). Mahwah, NJ: Lawrence Erlbaum Associates.

Hammer, M., & Champy, J. (1993). *Reengineering the corporation: A manifesto for business revolution.* New York, NY: HarperCollins.

Hammer, M., & Champy, J. (1996). Reengineering the corporation: The enabling role of information technology. In J. M. Shafritz & J. S. Ott (Eds.), *Classics of organizational theory* (pp. 561–577). Belmont, CA: Wadsworth.

Hancock, M. F., Jr. (2012). *Practical data mining.* Boca Raton, FL: CRC Press.

Hancock, P. A. (2007). On the process of automation transition in multitask human-machine systems. *IEEE Transactions on Systems, Man, and Cybernetics—Part A: Systems and Humans, 37,* 586–598.

Hancock, P. A. (2013). In search of vigilance: The problem of iatrogenically created psychological phenomena. *American Psychologist, 68,* 97–109.

Hancock, P. A. (2014). Automation: How much is too much? *Ergonomics, 57,* 449–454.

Hancock, P. A. (2017). Whither workload: Mapping a path for its future development. In L. Longo & M. C. Leva (Eds.), *Human mental workload: Models and applications* (pp. 3–17). Cham, Switzerland: Springer.

Hancock, P. A., & Desmond, P. A. (Eds.). (2001). *Stress, workload, and fatigue.* Mahwah, NJ: Lawrence Erlbaum Associates.

Hancock, P. A., Hancock, G. M., & Warm, J. S. (2009). Individuation: The N = 1 revolution. *Theoretical Issues in Ergonomics Science, 10,* 481–488.

Hancock, P. A., Kajaks, T., Caird, J. K., Chignell, M. H., Burns, P. C., Feng, J., … Vrkljan, B. H. (2020). Challenges to human drivers in increasingly automated vehicles. *Human Factors, 62,* 310–328.

Hancock, P. A., Ross, J. M., & Szalma, J. L. (2007). A meta-analysis of performance response under thermal stressors. *Human Factors, 49,* 851–877.

Hancock, P. A., & Warm, J. S. (1989). A dynamic model of stress and sustained attention. *Human Factors, 31,* 519–537.

Hancock, P. A., Williams, G., & Manning, C. M. (1995). Influence of task demand characteristics on work and performance. *International Journal of Aviation Psychology, 5,* 63–86.

Haq, S., & Zimring, C. (2003). Just down the road a piece: The development of topological knowledge of building layouts. *Environment and Behavior, 35,* 132–160.

Harary, F., Norman, R. Z., & Cartright, D. (1965). *Structural models: An introduction to the theory of directed graphs.* New York, NY: Wiley.

Hardy, L. (1996a). Testing the predictions of the cusp catastrophe model of anxiety and performance. *Sport Psychologist, 10,* 140–156.

Hardy, L. (1996b). A test of catastrophe models of anxiety and sports performance against multidimensional anxiety theory models using the method of dynamic differences. *Anxiety, Stress, and Coping, 9,* 69–86.

Hardy, L., & Parfitt, C. G. (1991). A catastrophe model of anxiety and performance. *British Journal of Psychology, 82,* 163–172.

Harms-Ringdahl, L. (1987). Safety analysis in design: Evaluation of a case study. *Accident Analysis and Prevention, 19,* 305–317.

Haroush, K., Deouell, L. Y., & Hochstein, S. (2011). Hearing while blinking: Multisensory attentional blink revisited. *Journal of Neuroscience, 31,* 922–927.

Harris, P. R. (1989). Behavioral science space contributions. *Behavioral Science, 34,* 207–227.

Harris, P. R. (1992). *Living and working in space: Human behavior, culture, and organization.* New York, NY: Ellis Horwood/Simon & Schuster.

Harris, W. C., Hancock, P. A., & Harris, S. C. (2005). Information processing changes following extended stress. *Military Psychology, 17,* 115–128.

Hart, S. G., & Staveland, L. E. (1988). Development of the NASA task load index (TLX): Results of experimental and theoretical research. In P. A. Hancock & N. Meshkati (Eds.), *Human workload* (pp. 138–183). Amsterdam, the Netherlands: North-Holland.

Haubner, P. (1992). Design of dialogues for human-computer interaction. In M. Galer, S. Harker, & J. Ziegler (Eds.), *Methods and tools in user-centered design for information technology* (pp. 203–236). Amsterdam, the Netherlands: North-Holland.

Hausdorff, J. M., Ashkenazy, Y., Peng, C.-K., Ivanov, P. C., Stanley, H. E., & Goldberger, A. I. (2001). When human walking becomes random walking: Fractal analysis and modeling of gait rhythm fluctuations. *Physica A, 302,* 138–147.

Havlovic, S. J., & Feinauer, D. (1990). Worker participation groups and safety committees: Is there an interaction effect on OSHA recordable accidents? *Proceedings of the Industrial Relations Research Association, 42,* 514.

Hawkins, H. L., Hillyard, S. A., Luck, S. J., Mouloua, M., Downing, C. J., & Woodward, D. P. (1990). Visual attention modulates signal detectability. *Journal of Experimental Psychology: Human Perception and Performance, 16,* 802–811.

Hazy, J. K. (2008). Toward a theory of leadership in complex systems: Computational model explorations. *Nonlinear Dynamics, Psychology, and Life Sciences, 12,* 281–310.

Hazy, J. K., Goldstein, J. A., & Lichtenstein, B. (Eds.). (2007). *Complex systems leadership theory.* Litchfield Park, AZ: ISCE.

Head, J., Helton, W., Russell, P., & Neumann, E. (2012). Text-speak processing impairs tactile location. *Acta Pscyhologica, 141,* 48–53.

Head, J., Tenan, M. S., Tweedell, A. J., Wilson, K. M., & Helton, W. S. (2020). Response complexity reduces errors on a response inhibition task. *Human Factors, 62,* 787–799.

Heath, R. A. (2002). Can people predict chaotic sequences? *Nonlinear Dynamics, Psychology, and Life Sciences, 6,* 37–54.

Hedge, A., Erickson, W. A., & Rubin, G. (1992). The effects of personal and occupational factors on sick building syndrome reports in air-conditioned offices. In J. M. Quick & J. J. Hurrell, Jr. (Eds.), *Stress and well-being at work* (pp. 286–298). Washington, DC: American Psychological Association.

Hedge, A., & James, T. (2012). Gender effects on musculoskeletal symptoms among physician computer users in outpatient diagnostic clinics. *Proceedings of the Human Factors and Ergonomics Society, 56,* 887–891.

Heilemann, F., Lindner, S., & Schulte, A. (2021). Experimental evaluation of tasking and teaming design patterns for human delegation of unmanned vehicles. *Human-Intelligent Systems Integration, 3,* 223–240.

Heinrich, H. W. (1931). *Industrial accident prevention.* New York, NY: McGraw-Hill.

Helander, M. G. (2007). Using design equations to identify sources of complexity in human-machine interaction. *Theoretical Issues in Ergonomics Science, 8,* 123–146.

Helander, M. G., & Burri, G. J. (1995). Cost effectiveness of ergonomics and quality improvements in electronics manufacturing. *International Journal of Industrial Ergonomics, 15,* 137–151.

Hellesoy, O., Gronhaug, K., & Kvitastein, O. 1998. Profiling the high hazards perceivers: An exploratory study. *Risk Analysis, 18,* 253–259.

Helton, W. S., Funke, G. J., & Knott, B. A. (2014). Measuring workload in collaborative contexts: Trait versus state perspectives. *Human Factors, 56,* 322–332.

Helton, W. S., & Kemp, S. (2011). What basic-applied issue? *Theoretical Issues in Ergonomics Science, 12,* 397–407.

Helton, W. S., & Russell, P. N. (2011). Working memory load and the vigilance decrement. *Experimental Brain Research, 212,* 429–437.

Helton, W. S., & Russell, P. N. (2013). Visuospatial and verbal working memory load: Effects on visuospatial vigilance. *Experimental Brain Research, 224,* 429–436.

Helton, W. S., & Russell, P. N. (2015). Rest is best: The role of rest and task interruptions on vigilance. *Cognition, 134,* 165–173.

Helton, W. S., Weil, L., Middlemiss, A., & Sawers, A. (2010). Global interference and spatial uncertainty in the Sustained Attention to Response Task (SART). *Consciousness and Cognition, 19,* 77–85.

Henao, O. (2002). The ability of competent and poor readers to remember information from hypermedia and printed texts [Capacidad de lectores competentes 7 lectores poco habiles para recordard informacion de un texto hipermedial e impresso]. *Infacia y Aprendizaje, 25,* 315–328.

Henning, R. A., Boucsein, W., & Gil, M. C. (2001). Social-physiological compliance as a determinant of team performance. *International Journal of Psychophysiology, 40,* 221–232.

Heppenheimer, T. A. (1990). How Von Neumann showed the way. *American Heritage of Innovation and Technology, 6*(2), 8–16.

Hertel, G., Kerr, N. L., & Messé, L. A. (2000). Motivation gains in performance groups: Paradigmatic and theoretical developments on the Köhler effect. *Journal of Personality and Social Psychology, 79,* 580–601.

Hertzum, M., & Holmegaard, K. D. (2015). Thinking aloud influences perceived time. *Human Factors, 57,* 101–109.

Herzberg, F., Mausner, B., & Snyderman, D. (1959). *The motivation to work.* New York, NY: Wiley

Herzog, T. R., & Bosley, P. J. (1992). Tranquility and preference as affective qualities of national environments. *Journal of Environmental Psychology, 12,* 115–128.

Hess, R. F. (1990). The Edridge-Green lecture vision at low light levels: Role of spatial, temporal and contrast filters. *Ophthalmic and Physiological Optics, 10,* 351–359.

Heuer, H., Manzey, D., Lorenz, B., & Sangals, J. (2003). Impairments of manual tracking performance during spaceflight are associated with specific effects of microgravity on visuomotor transformations. *Ergonomics, 46,* 920–934.

Hielemann, F., Lindner, S., & Schulte, A. (2021). Experimental evaluation of tasking and teaming design patterns for human delegation of unmanned vehicles. *Human-Intelligent System Integration, 3,* 223–240.

Hlavacka, F., Dzurkova, O., & Pólonyová, A. (2000). CNS adaptation to microgravity and its influence on human balance after spaceflight. *Homeostasis, 40,* 132–133.

Hockey, G. R. J. (1997). Compensatory control in the regulation of human performance under stress and high workload: A cognitive-energetical framework. *Biological Psychology, 45,* 73–93.

Hoffman, R. R., LaDue, D. S., Mogil, H. M., Roebber, P. J., & Trafton, J. G. (2017). *Minding the weather: How expert forecasters think.* Cambridge, MA: MIT Press.

Holbrook, R. I., & Burt de Perera, T. (2013). Three-dimensional spatial cognition: Freely swimming fish accurately learn and remember metric information in a volume. *Animal Behavior, 86,* 1077–1083.

Holcom, M. L., Lehman, W. E. K., & Simpson, D. D. (1993). Employee accidents: Influences of personal characteristics, job characteristics, and substance use in jobs different in accident potential. *Journal of Safety Research, 24,* 205–221.

Holding, D. H. (1983). Fatigue. In G. R. J. Hockey (Ed.), *Stress and fatigue in human performance* (pp. 145–167). New York, NY: Wiley.

Holland, A. W., Vessey, W. B., & Marrett, J. D. (2014, May). *Conducting an exploration focused astronaut job analysis.* Paper presented to the Society for Industrial and Organizational Psychology, Philadelphia, PA.

Holland, J. H. (1995). *Hidden order: How adaptation builds complexity.* Cambridge, MA: Perseus.

Hollis, G., Kloos, H., & Van Orden, G. C. (2009). Origins of order in cognitive activity. In S. J. Guastello, M. Koopmans, & D. Pincus (Eds.), *Chaos and complexity in psychology: The theory of nonlinear dynamical systems* (pp. 206–242). New York, NY: Cambridge University Press.

Hollnagel, E. (2011). The score of resilience engineering. In E. Hollnagel, J. Periès, D. D. Woods, & J. Wreathall (Eds.), *Resilience engineering in practice: A guidebook* (pp. xxix–xxxix). Burlington, VT: Ashgate.

Hollnagel, E. (2012). *FRAM: The functional resonance analysis method.* Burlington, VT: Ashgate.

Hollnagel, E., Woods, D. D., & Leveson, N. (Eds.). (2006). *Resilience engineering.* Burlington, VT: Ashgate.

Holmberg, D. (2011). Demand response and standards: New role for buildings in the smart grid. *BACnet Today & the Smart Grid, ASHRAE Journal, November,* nv, B23–B28.

Holmes, T. H., & Rahe, R. H. (1967). The Social Readjustment Rating Scale. *Journal of Psychosomatic Research, 11,* 213–218.

Hommes, C. H., & Rosser, J. B., Jr. (2001). Consistent expectations equilibria and complex dynamics in renewable resource markets. *Macroeconomic Dynamics, 5,* 180–203.

Hong, S. L. (2010). The entropy conservation principle: Applications in ergonomics and human factors. *Nonlinear Dynamics, Psychology, and Life Sciences, 14,* 291–315.

Hong, W., Thong, J. M., & Tarn, K. Y. (2004). Does animation attract online users' attention? The effects of flash on information search performance and perceptions. *Information Systems Research, 15,* 60–86.

Hormann, H. (2001). Safety at sea. In J. Althoff (Ed.), *Safety of modern technical systems* (pp. 361–369). Saarbrücken, Germany: TÜV-Saarland Foundation.

Horrey, W. J. (2011). Assessing the effects of in-vehicle tasks on driving performance. *Ergonomics in Design, 19*(4), 4–7.

Houston, D. J., & Richardson, L. E., Jr. (2002). Traffic safety and the switch to a primary seat belt law: The California experience. *Accident Analysis and Prevention, 34,* 743–807.

Howell, D. C. (1992). *Statistical methods for psychology* (3rd ed.). Boston, MA: PWS-Kent.

Hoyert, M. S. (1992). Order and chaos in fixed-interval schedules of reinforcement. *Journal of the Experimental Analysis of Behavior, 57,* 339–363.

Hudlick, E., & McNeese, M. D. (2002). Assessment of user affective and belief states for interface adaptation: Application to an Air Force pilot task. *User Modeling and User-Adapted Interaction, 12,* 1–47.

Hudson, P. T. W. (2001). Automation: A problematic solution. In J. Alfthoff (Ed.), *Safety of modern technical systems* (pp. 393–402). Saarbrücken, Germany: TÜV-Saarland Foundation.

Huffington Post. (2012, October 8). Texas school district reportedly threatening students who refuse tracking ID, can't vote for homecoming. Retrieved from www.huffingtonpost.com/2012/10/08/texas-school-district-rep_n_1949415.html

Hurrell, J. J., Jr., & Murphy, L. R. (1996). Occupational stress intervention. *American Journal of Industrial Medicine, 29,* 338–341.

Hurst, H. E. (1951). Long-term storage capacity of reservoirs. *Transactions of the American Society of Civil Engineers, 116,* 770–808.

Husemann, B., Von Mach, C. Y., Borsotto, D., Zepf, K. I., & Scharnbacher, J. (2009). Comparison of musculoskeletal complaints and data entry between a sitting and sit-stand workstation paradigm. *Human Factors, 51,* 310–320.

Hysong, S. J., SoRelle, R., & Hughes, A. M. (2022). Prevalence of effective audit-and-feedback practices in primary care settings: A qualitative examination within veterans health administration. *Human Factors, 64,* 99–108.

IJtsma, M., Ma, L. M., Pritchett, A. R., & Feigh, K. M. (2019). Computational methodology for the allocation of work and interaction in human-robot teams. *Journal of Cognitive Engineering and Decision Making, 13,* 221–241.

Ijspeert, A., Nakanishi, J., & Schaal, S. (2003). *Learning attractor landscapes for motor primitives.* Cambridge, MA: MIT Press.

Ilkowska, M., & Engle, R. W. (2012). Trait and state differences in working memory capacity. In A. Gruszka, G. Matthews, & B. Szymura (Eds.), *Handbook of individual differences in cognition: Attention, memory, and executive control* (pp. 295–320). New York, NY: Springer Science-Business Media.

Illuminating Engineering Society. (1981). *IES lighting handbook*. New York, NY: Author.

International Electrotechnical Commission. (2005). *IEC 60601-1-8: Medical electrical equipment—General requirements, tests, and guidance for alarm systems in medical electrical equipment and medical electrical systems*. Geneva, Switzerland: Author.

International Standards Organization. (1986). *Mechanical vibration: Guidelines for the measurement and assessment of human exposure to hand-transmitted vibration* (Standard No. S3.34). Geneva, Switzerland: Author.

Ioteyko, J. (1920). *La fatigue* [Fatigue] (2nd ed.). Paris: Flammarion.

Ivanoff, R. V., Yakimov, P. I., & Djamiykov, T. S. (2011). Implementation of smart metering as an essential part of advanced metering infrastructure by end-customers. *Annual Journal of Electronics, nv*, 17–20.

Jacko, J. A., & Sears, A. (2003). *The human-computer interaction handbook*. Mahwah, NJ: Lawrence Erlbaum Associates.

Jackson, E. A. (1991a). Control of dynamic flows and attractors. *Physical Review A, 44*, 4839–4853.

Jackson, E. A. (1991b). On the control of complex dynamic systems. *Physica D, 50*, 341–366.

Jacobs, R. R., Conte, J. M., Day, D. V., Silva, J. M., & Harris, R. (1996). Selecting bus drivers: Multiple predictors, multiple perspectives on validity, and multiple estimates of utility. *Human Performance, 9*, 199–217.

Jacobsen, J. J., & Guastello, S. J. (2011). Diffusion models for innovation: S-curves, networks, power laws, catastrophes, and entropy. *Nonlinear Dynamics, Psychology, and Life Sciences, 15*, 307–333.

Jagacinski, R. J., & Flach, J. M. (2003). *Control theory for humans*. Mahwah, NJ: Lawrence Erlbaum Associates.

Janata, P., & Edwards, W. H. (2013). A novel sonification strategy for auditory display of heart rate and oxygen saturation changes in clinical settings. *Human Factors, 55*, 356–372.

Jankowski, J., & Grabowski, A. (2015). Usability evaluation of VR interface for mobile robot teleoperation. *International Journal of Human-Computer Interaction, 31*, 882–889.

Jansen, R. J., Sawyer, B. D., van Egmond, R., de Ridder, H., & Hancock, P. A. (2016). Hysteresis in mental workload and task performance: The influence of demand transitions and task prioritization. *Human Factors, 58*, 1143–1157.

Jarrett, R. F. (1948). Percent increases in output of selected personnel as an index of test efficiency. *Journal of Applied Psychology, 32*, 135–145.

Jensen, P. L. (1997). Can participatory ergonomics become "the way we do things in this firm"?: The Scandinavian approach to participatory ergonomics. *Ergonomics, 40*, 1078–1087.

Ji, Y. G. (Ed.). (2013). *Advances in affective and pleasurable design*. Boca Raton, FL: CRC Press/Taylor & Francis.

Jirsa, V. K., & Kelso, J. A. S. (Eds.). (2004a). *Coordination dynamics: Issues and trends*. New York, NY: Springer.

Jirsa, V. K., & Kelso, J. A. S. (2004b). Integration and segregation of perceptual and motor behavior. In V. K. Jirsa & J. A. S. Kelso (Eds.), *Coordination dynamics: Issues and trends* (pp. 243–260). New York, NY: Springer.

Jobidon, M.-E., Rousseau, R., & Breton, R. (2005). The effect of variability in temporal information on the control of a dynamic task. *Theoretical Issues in Ergonomic Science, 6,* 49–62.

Joels, K. M., Kennedy, G. P., & Larkin, D. (1982). *The space shuttle operator's manual.* New York, NY: Ballantine.

Johannsdottir, K. R., Magnusdottir, E. H., Sigurjonsdottir, S., & Gudnason, J. (2018). The role of working memory capacity in cardiovascular monitoring of cognitive workload. *Biological Psychology, 132,* 154–163.

Johnson, C. (2002). *Novel visualization for the presentation of accident reports.* Retrieved from www.dcs.gla.uk/~johnson/papers

Johnson, S. E. (2007). The predictive validity of safety climate. *Journal of Safety Research, 38,* 511–521.

Jones, L. (2001). *Human factors and haptic interfaces.* Retrieved from www.ims.umn.edu/talks/workshops/6-14-15.2001/jones/jones.pdf

Jones, L. A., & Sarter, N. B. (2008). Tactile displays: Guidance for their design and application. *Human Factors, 50,* 90–111.

Jorgensen, F., & Pedersen, P. A. (2002). Drivers' response to the installation of road lighting: An economic interpretation. *Accident Analysis and Prevention, 34,* 601–608.

Joseph, A., Madathil, K. C., Jafarifiroozabadi, R., Rogers, H., Mihandoust, S., Khasawneh, A., ... McElligott, J. T. (2022). Communication and teamwork during telemedicine-enabled stroke care in an ambulance. *Human Factors, 64,* 21–41.

Juarrero, A. (1999). *Dynamics in action.* Cambridge, MA: MIT Press.

Kaber, D. B., & Endsley, M. R. (2004). The effects of level of automation and adaptive automation on human performance, situation awareness and workload in a dynamic control task. *Theoretical Issues in Ergonomics Science, 5,* 113–153.

Kaduk, S. I., Roberts, A. P. J., & Stanton, N. A. (2021). Driving performance, sleepiness, fatigue, and mental workload throughout the time course of semi-automated driving: Experimental data from the driving simulator. *Human Factors in Manufacturing and Service Industries, 31,* 143–154.

Kahn, R. L., Wolfe, D. M., Quinn, R. P., Snoek, J. D., & Rosenthal, R. A. (1964). *Organizational stress: Studies in role conflict and ambiguity.* New York, NY: Wiley.

Kahneman, D., & Tversky, A. (1979). Prospect theory: An analysis of decision under risk. *Econometrica, 47,* 253–291.

Kahrimanovic, M., Tiest, W. M. B., & Kappers, A. M. L. (2011). Discrimination thresholds for haptic perception of volume, surface area, and weight. *Attention, Perception, and Psychophysics, 73,* 649–2656.

Kanas, N., Salnitskiy, V., Grand, E. M., Gushin, V., Weiss, D. S., Kozerenko, O., Sled, A., Marmar, C. R. (2002). Lessons learned from shuttle/Mir: Psychosocial countermeasures. *Aviation, Space, and Environmental Medicine, 73,* 607–611.

Kane, M. J., & Engle, R. W. (2002). The role of prefrontal cortex in working-memory capacity, executive attention, and general fluid intelligence: An individual-differences perspective. *Psychonomic Bulletin & Review, 9,* 617–671.

Kane, M. J., Hambrick, D. Z., & Conway, A. R. A. (2005). Working memory capacity and fluid intelligence are strongly related constructs: Comment on Ackerman, Beier, and Boyle (2005). *Psychological Bulletin, 131,* 66–71.

Kanfer, R. (2011). Determinant and consequences of subjective cognitive fatigue. In P. Ackerman (Ed.), *Cognitive fatigue* (pp. 189–208). Washington, DC: American Psychological Association.

Kantowitz, B. H. (1985). Channels and stages in human information processing: A limited analysis of theory and methodology. *Journal of Mathematical Psychology, 29*, 135–174.

Kantowitz, B. H. (2001). *Reducing driver distractions in the age of overload.* Retrieved from www.umich.edu/~newsinfo/Releases/2001/Mar01/r030501c.html

Kantowitz, B. H., & Sorkin, D. (1983). *Human factors: Understanding people-system relationships.* New York, NY: Wiley.

Kaplan, A. D., Cruit, J., Edsley, M., Beers, S. M., Sawyer, B. D., & Hancock, P. A. (2021). The effects of virtual reality, augmented reality, and mixed reality as training enhancement methods: A meta-analysis. *Human Factors, 63*, 706–726.

Kaplan, D., & Glass, L. (1995). *Understanding nonlinear dynamics.* New York, NY: Springer.

Karalaia, N., & Hogarth, R. M. (2008). Determinants of linear judgment: A meta-analysis of lens model studies. *Psychological Bulletin, 134*, 404–426.

Karayanni, D. A., & Baltas, G. A. (2003). Web site characteristics and business performance: Some evidence from international business-to-business organizations. *Marketing Intelligence and Planning, 21*, 105–114.

Karwowski, W. (2000). Purely editorial. *Theoretical Issues in Ergonomic Science, 1*, 1–2.

Karwowski, W. (Ed.). (2006). *Handbook of standards and guidelines in ergonomics and human factors.* Mahwah, NJ: Lawrence Erlbaum Associates.

Karwowski, W. (2012). A review of human factors challenges of complex adaptive systems: Discovering and understanding chaos in human performance. *Human Factors, 54*, 983–995.

Karwowski, W., Rahmi, M., & Mihalyi, T. (1988). Effects of computerized automation and robotic safety performance of a manufacturing plant. *Journal of Occupational Accidents, 10*, 217–233.

Kashima, Y., Kirley, M., Strivada, A., & Robins, G. (2016). Modeling cultural dynamics. In R. R. Vallacher, S. J. Read, & A. Nowak (Ed.), *Computational social psychology* (pp. 281–307). New York, NY: Routledge.

Kato, T. (2012). Development of the coping flexibility scale: Evidence for the coping flexibility hypothesis. *Journal of Counseling Psychology, 59*, 262–273.

Katsavelis, D., Muhkerjee, M., Decker, L., & Stergiou, N. (2010a). Variability of lower extremity joint kinematics during backward walking in a virtual environment. *Nonlinear Dynamics, Psychology, and Life Sciences, 14*, 165–178.

Katsavelis, D., Muhkerjee, M., Decker, L., & Stergiou, N. (2010b). The effect of virtual reality on gait variability. *Nonlinear Dynamics, Psychology, and Life Sciences, 14*, 239–256.

Kauffman, S. A. (1993). *The origins of orders: Self-organization and selection in evolution.* New York, NY: Oxford University Press.

Kaufman, L. (1974). *Sight and mind: An introduction to visual perception.* New York, NY: Oxford University Press.

Kazi, S., Khaleghzadegan, S., Dinh, J. V., Shelhamer, M. J., Sapirstein, A., Goeddel, L. A., … Rosen, M. A. (2021). Team physiological dynamics: A critical review. *Human Factors, 63*, 32–65.

Keebler, J. R., Rosen, M. A., Sittig, D. F., Thomas, E., & Salas, E. (2022). Human factors and ergonomics in healthcare: Industry demands and a path forward. *Human Factors, 64*, 250–258.

Kelly, A. J., & Dux, P. E. (2011). Different attentional blink tasks reflect distinct information processing limitations: An individual differences approach. *Journal of Experimental Psychology: Human Perception and Performance, 37*, 1867–1873.

Kelly, K. (1994). *Out of control: The new biology of machines, social systems and the economic world*. Reading, MA: Addison-Wesley.

Kelly, R. V., Jr. (1995). Abundant worlds and virtual personas. *Virtual Reality, 2*(5), 35–43.

Kemeny, J. (1979). *The need for change: The legacy of TMI. Report of the Presidential Commission on Three Mile Island*. New York, NY: Pergamon Press.

Kennedy-Metz, L. R., Dias, R. D., Srey, R., Rance, G. C. Conboy, H. M., Haime, M. E., ... Zenato, M. A. (2021). Analysis of dynamic changes in cognitive workload during cardiac surgery: Perfusionists' interactions with the cardiopulmonary bypass pump. *Human Factors, 63*, 757–771.

Kern, D., Karwowski, W., Franco, E. G., & Murata, A. (2018). Evidence of chaos in a routine watchstanding task. *Nonlinear Dynamics, Psychology, and Life Sciences, 22*, 153–171.

Kerr, R., McHugh, M., & McCrory, M. 2009. HSE management standards and stress-related outcomes. *Occupational Medicine, 59*, 584–479.

Khalid, H. M. (Ed.). (2004). Conceptualizing effective human factors design [Special issue]. *Theoretical Issues in Ergonomics Science, 5*(1), 1–91.

Kiefer, A. W., & Myer, G. D. (2015). Training the antifragile athlete: A preliminary analysis of neuromuscular training effects on muscle activation dynamics. *Nonlinear Dynamics, Psychology, and Life Sciences, 19*, 489–510.

Kiel, L. D., McCaskill, J., Harrington, J., & Elliott, L. D. (2021). Agent-based modeling: Challenges and prospects. In E. Elliott & L. D. Kiel (Eds.), *Complex systems in the social and behavioral sciences: Theory, method and application* (pp. 183–203). Ann Arbor, MI: University of Michigan Press.

Kieras, D. (1988). Toward a practical GOMS model methodology for user interface design. In M. Helander (Ed.), *Handbook of human-computer interaction* (pp. 135–157). Amsterdam, the Netherlands: Elsevier/North-Holland.

Kieras, D. (2004). GOMS models for task analysis. In D. Diaper & N. A. Stanton (Eds.), *The handbook of task analysis for human-computer interaction* (pp. 83–116). Mahwah, NJ: Lawrence Erlbaum Associates.

Kim, S-H., & Kaber, D. B. (2014). Examining the effects of conformal terrain features in advanced heads-up displays on flight performance and pilot situation awareness. *Human Factors in Manufacturing and Service Industries, 24*, 386–402.

Kim, S.-H., Kaber, D. B., Alexander, A. L., Stelzer, E. M., Kaufman, K., & Veil, T. (2011). Multidimensional measure of display clutter and pilot performance for advanced head-up display. *Aviation, Space, and Environmental Medicine, 82*, 1013–1022.

Kirlik, A. (2006). *Adaptive perspectives on human-technology interaction: Methods and models for cognitive engineering and human-computer interaction*. New York, NY: Oxford University Press.

Kirwan, B. (2001). Transportation safety: A changing landscape. In J. Althoff (Ed.), *Safety of modern technical systems* (pp. 325–334). Saarbrücken, Germany: TÜV-Saarland Foundation.

Kiselev, A., Medvedev, A. A., & Menshikov, V. A. (2003). *Astronautics: Summary and prospects*. Vienna, Austria: Springer.

Kjellen, U. (1984a). The deviation concept in occupational accident control—1: Definition and classification. *Accident Analysis and Prevention, 16,* 289–306.

Kjellen, U. (1984b). The deviation concept in occupational accident control—2: Data collection and assessment of significance. *Accident Analysis and Prevention, 16,* 307–323.

Kjellen, U. (1990). Safety control in design: Experiences of an offshore project. *Journal of Occupational Accidents, 12,* 49–61.

Kjellén, U., & Hovden, J. (1993). Reducing risks by deviation control—A retrospection into a research strategy. *Safety Science, 16,* 417–438.

Kjellen, U., & Larsson, T. J. (1981). Investigating accidents and reducing risks: A dynamic approach. *Journal of Occupational Accidents, 3,* 129–140.

Klahr, P., & Waterman, D. A. (1986). *Artificial intelligence: A Rand perspective* (Tech. Rep. No. P-7172). Santa Monica, CA: Rand.

Klein, G. (1989). Recognition-primed decisions. *Advances in Man-Machine Systems Research, 5,* 47–92.

Klein, G., Snowden, D., & Pin, C. L. (2011). Anticipatory thinking. In K. L. Mosier & U. M. Fisher (Eds.), *Informed by knowledge: Expert performance in complex situations* (pp. 235–246). New York, NY: Psychology Press.

Klein, J., Moon, Y., & Picard, R. W. (2002). The computer responds to user frustration: Theory, design, and results. *Interacting with Computers, 14,* 119–140.

Klen, T., & Vayrynen, S. T. (1983). The protective influence of logger's safety equipment. *Scandinavian Journal of Work and Environmental Health, 9,* 201–203.

Koehler, G. (1995). Fractals and path-dependent processes: A theoretical approach for characterizing emergency medical responses to major disasters. In R. Robertson & A. Combs (Eds.), *Chaos theory in psychology and the life sciences* (pp. 199–216). Hillsdale, NJ: Lawrence Erlbaum Associates.

Koehler, G. (1996). What disaster response management can learn from chaos theory. In G. Koehler (Ed.), *What disaster response management can learn from chaos theory* (pp. 2–41). Sacramento, CA: California Research Bureau, California State Library.

Koehler, G. (1999). The time compacted globe and the high-tech primitive at the millennium. In E. Elliott & L. D. Kiel (Eds.), *Nonlinear dynamics, complexity, and public policy* (pp. 153–174). Commack, NY: Nova Science.

Koehler, G. (2011). Q-consciousness: Where is the flow? *Nonlinear Dynamics, Psychology, and Life Sciences, 15,* 335–358.

Koelega, H. S., Brinkman, J.-A., Hendriks, L., & Verbaten, M. N. (1989). Processing demands, effort, and individual differences in four different vigilance tasks. *Human Factors, 31,* 45–62.

Koesling, H., Zoellner, M., Sichelschmidt, L., & Ritter, H. (2009). With a flick of the eye: Assessing gaze-controlled human-computer interaction. In H. Ritter, G. Sagerer, R. Dillmann, & M. Buss (Eds.), *Human centered robot systems: Cognition, interaction, technology* (pp. 111–120). New York, NY: Springer.

Kohonen, T. (1989). *Self-organization and associative memory*. New York, NY: Springer.

Koller, F. (1992). Multimedia interfaces. In M. Galer, S. Harker, & J. Ziegler (Eds.), *Methods and tools in user-centered design for information technology* (pp. 299–315). Amsterdam, the Netherlands: North-Holland.

Koppen, M., & Doignan, J.-P. (1990). How to build a knowledge space by querying an expert? *Journal of Mathematical Psychology, 34*, 311–331.

Kortschot, S. W., & Jamieson, G. A. (2020). Classification of attentional tunneling through behavioral indices. *Human Factors, 62*, 973–986.

Kovalerchuk, B., & Vityaev, E. (2000). *Data mining in finance: Advances in relational and hybrid methods*. Boston, MA: Kluwer.

Kramer, G. (1994). An introduction to auditory display. In G. Kramer (Ed.), *Auditory display: Sonification, audification, and auditory interfaces* (pp. 1–78). New York, NY: Oxford University Press.

Kremen, M. A. (1966). The effectiveness of tracking behavior as a function of the mode of the input signal. *Voprosy Psychologii, 5*, 83–92.

Kring, J. P. (2001). Multicultural factors for international spaceflight. *Journal of Human Performance in Extreme Environments, 5*, 11–32.

Kroll, W. (1981). The C. H. McCoy research lecture: Analysis of local muscular fatigue patterns. *Research Quarterly for Exercise and Sport, 52*, 523–539.

Kumar, S. (2002). Trunk rotation: Ergonomic and evolutionary perspectives. *Theoretical Issues in Ergonomic Science, 3*, 235–256.

Kundi, M. (2003). Ergonomic criteria for the evaluation of work schedules. *Theoretical Issues in Ergonomic Science, 4*, 302–318.

Kurz, M. J., Markopoulou, K., & Stergiou, N. (2010). Attractor divergence as a metric for assessing walking balance. *Nonlinear Dynamics, Psychology, and Life Sciences, 14*, 151–164.

Lacherez, P., Seah, E. L., & Sanderson, P. (2007). Overlapping melodic alarms are almost indiscriminable. *Human Factors, 49*, 637–645.

Lacker, S. J., Salcedo, J. N., Szalma, J. L., & Hancock, P. A. (2016). The stress and workload of virtual reality training: The effects of presence, immersion and flow. *Ergonomics, 59*, 1060–1072.

LaFond, D., Roberge-Vallières, V., & Tremblay, S. (2017). Judgment analysis in a dynamic multitask environment: Capturing nonlinear policies using decision trees. *Journal of Cognitive Engineering and Decision Making, 11*, 122–135.

Lagi, M., Bar-Yam, Y., Bertrand, K. Z., & Bar-Yam, Y. (2011). The food crises: A quantitative model of food prices including speculators and ethanol conversion. Retrieved from http://arxiv.org/abs/1109.4859

Landauer, T. K. (1987). Relations between cognitive psychology and computer system design. In J. M. Carroll (Ed.), *Interfacing thought: Cognitive aspects of human-computer interaction* (pp. 1–25). Cambridge, MA: MIT Press.

Landauer, T. K., Galotti, K. M., & Hartwell, S. (1983). Natural command names and initial learning: A study of text-editing terms. *Communications of the CAM, 26*, 495–503.

Landsbergis, P. A., Schnall, P. L., Schwartz, J. E., Warren, K., & Pickering, T. G. (1995). Job strain, hypertension, and cardiovascular disease: Empirical evidence, methodological issues, and recommendations for further research. In S. K. Sauter

& L. R. Murphy (Eds.), *Organizational risk factors for job stress* (pp. 97–112). Washington, DC: American Psychological Association.

Laner, S., & Sell, R. G. (1960). An experiment on the effect of specially designed safety posters. *Occupational Psychology, 34,* 153–169.

Lange, R., Oliva, T., & McDade, S. (2000). An algorithm for estimating multivariate catastrophe theory models: GEMCAT II. *Studies in Nonlinear Dynamics in Economics, 4,* 138–168.

Lanyon, R. I. (1984). Personality assessment. *Annual Review of Psychology, 35,* 667–701.

Lanyon, R. I. (1987). The validity of computer-based personality assessment products: Recommendations for the future. *Computers in Human Behavior, 3,* 225–238.

Lapeyre, B., Hourlier, S., Servantie, X., N'Kaoua, B., & Sauzéon, H. (2011). Using the landmark-route-survey framework to evaluate spatial knowledge obtained from synthetic vision systems. *Human Factors, 53,* 647–661.

Lardent, C. L., Jr. (1991). Pilots who crash: Personality constructs underlying accident prone behavior of fighter pilots. *Multivariate Experimental Clinical Research, 10,* 1–25.

Large, D. R., Clark, L., Quandt, A., Burnett, G., & Skrypchuk, L. (2017). Steering the conversation: A linguistic exploration of natural language interactions with a digital assistant during simulated driving. *Applied Ergonomics, 63,* 53–61.

Larsson, T. J. (1993). Investigating accidents and reducing risks: A dynamic approach (Kjellen & Larsson, 1981); its relevance for injury prevention. *Safety Science, 16,* 439–444.

Latané, B., Williams, K., & Harkins, S. (1979). Many hands make light the work: The cases and consequences of social loafing. *Journal of Personality and Social Psychology, 37,* 822–832.

Laughlin, P. R. (1996). Group decision making and collective induction. In E. Witte & J. H. Davis (Eds.), *Understanding group behavior: Consensual action by small groups* (pp. 61–80). Mahwah, NJ: Lawrence Erlbaum Associates.

Leary, M. R. (1990). Anxiety, cognition, and behavior: In search of a broader perspective. *Journal of Social Behavior and Personality, 5,* 39–44.

Leary, M. R., & Kowalski, R. M. (1995). *Social anxiety.* New York, NY: Guilford Press.

LeBlanc, L. A., & Rucks, C. T. (1996). A multiple discriminant analysis of vessel accidents. *Accident Analysis and Prevention, 28,* 501–510.

Lee, P. U., & Prevor, T. (2012). Prediction of traffic complexity and controller workload in mixed-equipage NextGen environments. *Proceedings of the Human Factors and Ergonomics Society, 56,* 100–104.

Lehner, P. E. (1989). Toward an empirical approach to evaluating the knowledge base of an expert system. *IEEE Transactions on Systems, Man, and Cybernetics, 19,* 658–662.

Lehtola, M. M., van der Molen, H. F., Lappalainen, J., Hoonakker, P. L. T., Hsiao, H., Haslam, R. A., Hale, A. R., & Verbeek, J. H. (2008). The effectiveness of interventions for preventing injuries in the construction industry: A systematic review. *American Journal of Preventive Medicine, 35,* 77–85.

Leonhardt, J., Macchi, L., Hollnagel, E., & Kirwan, B. (2009). A white paper on resilience engineering for ATM. Eurocontrol. Retrieved from www.eurocontrol. int/esp/gallery/content/public/library

Lerch, F. J., & Harter, D. E. (2001). Cognitive support for real-time dynamic decision making. *Information Systems Research, 12*, 63–82.

Leroy, S., Schmidt, A. M., & Madjar, N. (2021). Working from home during COVID-19: A student of the interruption landscape. *Journal of Applied Psychology, 106*, 1448–1465.

Lesgold, A. (2000). On the future of cognitive task analysis. In J. M. Schraagen, S. F. Chipman, & V. L. Shalin (Eds.), *Cognitive task analysis* (pp. 451–465). Mahwah, NJ: Lawrence Erlbaum Associates.

Levi, K. (1989). Expert systems should be more accurate than human experts: Evaluation procedures from human judgment and decision making. *IEEE Transactions on Systems, Man, and Cybernetics, 19*, 658–662.

Levine, J. H., & Spedaro, J. (1988). Occupational mobility: A structural model. In B. Wellman & S. D. Berkowitz (Eds.), *Social structures: A network approach* (pp. 452–475). New York, NY: Cambridge University Press.

Levulis, S. J., DeLucia, P. R., & Kim, S-Y. (2018). Effects of touch, voice, and multi-modal input and task load on multiple UAV monitoring performance during simulated manned-unmanned teaming in a military helicopter. *Human Factors, 60*, 1117–1129.

Levy, S. (1992). *Artificial life: A report from the frontier where computers meet biology.* New York, NY: Random House.

Li, T., Ogihara, M., & Tzanetakis, G. (Eds.). (2012). *Music data mining.* Boca Raton, FL: CRC Press.

Liao, M.-J., & Johnson, W. W. (2004). Characterizing the effects of droplines on target acquisition performance on a 3-D perspective display. *Human Factors, 46*, 476–496.

Liberty Mutual Insurance. (2005). *Liberty Mutual materials handling tables.* Boston, MA: Author. Retrieved from http://libertymmhtables.libertymutual.com/ CM_LMTablesWeb

Liberty Mutual Insurance (2020). *Manual materials handling.* Retrieved May 11, 2022, from https://libertymmhtables.libertymutual.com/

Lidwell, W., Holden, K., & Butler, J. (2011). *Universal principles of design: 125 ways to enhance usability, influence perception, increase appeal, make better design decisions, and teach through design.* Beverly, MA: Rockport.

Liebovitch, L. S. (2004, July). *Arrival time of e-mail and viruses provide information about the structure and dynamics of the Internet.* Paper presented at the 14th annual international conference of the Society for Chaos Theory in Psychology and Life Sciences, Milwaukee, WI.

Lietz, C. A., Gerdes, K. E., Sun, F., Geiger, J. M., Wagaman, M. A., & Segal, E. A. (2011). The Empathy Assessment Index (EAI): A confirmatory factor analysis of a multidimensional model of empathy. *Journal of the Society for Social Work and Research, 2*, 104–124.

Likens, A. D., Amazeen, P. G., Stevens, R., Galloway, T., & Gorman, J. C. (2014). Neural signatures of team coordination are revealed by multifractal analysis. *Social Neuroscience, 9*, 218–234.

Likert, R. (1932). A technique for the measurement of attitudes. *Archives of Psychology, 140,* 1–55.

Likert, R. (1961). *New patterns of management.* New York, NY: McGraw-Hill.

Lin, Y., & Cai, H. (2009). A method for building a real-time cluster-based continuous mental work-load scale. *Theoretical Issues in Ergonomic Science, 10,* 531–544.

Linde, L. K. (1986). What is domain skill? Frames in information seeking. *Behavioral Science, 31,* 89–102.

Lindholm, E., & Cheatham, C. M. (1983). Autonomic activity and workload during learning of a simulated aircraft carrier landing task. *Aviation, Space, and Environmental Medicine, 54,* 435–439.

Lipscomb, H. J., Li, I., & Dement, J. (2003). Work-related falls among union carpenters in Washington State before and after the Vertical Fall Arrest Standard. *American Journal of Industrial Medicine, 44,* 157–161.

Lipscomb, J. S. (1996). *Stereoscopic and head-mounted display on IBM RISC System/6000 workstations, and on IBM PCs.* Retrieved from www.research.ibm.com/people/lipscomb/www-stereo-print-export.html

Lisle, L., Merenda, C., Tanous, K., Kim, H., Gabbard, J. L., & Bowman, D. A. (2019). Effects of volumetric augmented reality displays on human depth judgments: Implications for heads-up displays in transportation. *International Journal of Mobile Human Computer Interaction, 11*(2), 1–18.

Little, K. B., & Schneidman, E. S. (1959). Congruencies among interpretations of psychological test and anamnestic data. *Psychological Monographs, 75*(6, Whole No. 476), 1–42.

Liu, M. (2004). Examining the performance and attitudes of sixth graders during their use of a problem-based hypermedia learning environment. *Computers in Human Behavior, 20,* 357–379.

Loft, S., Sanderson, P., Neal, A., & Mooij, M. (2007). Modeling and predicting mental workload in en route air traffic control: Critical review and broader implications. *Human Factors, 49,* 376–399.

Logie, R. H. (2011). The functional organization and capacity limits of working memory. *Current Directions in Psychological Science, 20,* 240–245.

Lonero, L. P., Clinton, K., Wilde, G. J. S., Roach, K., McKnight, A. J., MacLean, H. (1994). *The roles of legislation, education and reinforcement in changing road user behavior.* Toronto, Canada: Ministry of Transportation, Safety Research Office, Safety Policy Branch.

Longo, L. (2018). Experienced mental workload, perception of usability, their interaction and impact on task performance. *PLoS ONE, 13*(8), e0199661. doi: 10.1371/journal.pone.0199661

López-Sánchez, J. I., & Hancock, P. A. (2019). Diminishing cognitive capabilities in an ever hotter world: Evidence from an applicable power-law distribution. *Human Factors, 61,* 906–919.

Lorenz, E. N. (1963). Deterministic periodic flow. *Journal of the Atmospheric Sciences, 20,* 130–141.

Lorist, M. M., & Faber, L. G. (2011). Consideration of the influence of mental fatigue on controlled and automatic cognitive processes. In P. Ackerman (Ed.), *Cognitive fatigue* (pp. 105–126). Washington, DC: American Psychological Association.

Lorist, M. M., Klein, M., Nieuwenhaus, S., De Jong, R., Mulder, G., & Meijman, T. F. (2000). Mental fatigue and task control: Planning and preparation. *Psychophysiology*, 37, 614–625.

Louviere, A. J., & Jackson, J. T. (1980). Man-machine design for spaceflight. In T. S. Cheston & D. L. Winter (Eds.), *Human factors of outer space production* (pp. 97–112). Boulder, CO: Westview Press and American Association for the Advancement of Science.

Love, L., & Johnson, C. (2002). *Using diagrams to support the analysis of system "failure" and operator "error."* Retrieved from www.dcs.gla.uk/~johnson/papers/aft.html

Lovsund, P., Hedin, A., & Tornros, J. (1991). Effects on driving performance of visual field effects: A driving simulator study. *Accident Analysis and Prevention*, 23, 231–342.

MacCoun, R. J. (2016). Computational models of social influence and behavior. In R. R. Vallacher, S. J. Read, & A. Nowak (Ed.), *Computational social psychology* (pp. 258–280). New York, NY: Routledge.

MacDonald, J. E., & Gifford, R. (1989). Territorial cues and defensible space theory: The burglar's point of view. *Journal of Environmental Psychology*, 9, 193–205.

Macek, A. J. (1987). Successful implementation of an injury prevention program. *Proceedings of the Human Factors Society*, 31, 1138–1140.

Mackworth, N. H. (1948). The breakdown of vigilance during prolonged visual search. *Quarterly Journal of Experimental Psychology*, 1, 6–21.

MacLean, M. H., & Arnell, K. M. (2010). Personality predicts temporal attention costs in the attentional blink paradigm. *Psychonomic Bulletin and Review*, 17, 556–562.

Madhavan, P., & Wiegman, D. A. (2007). Similarities and differences between human-human and human-automation trust: An integrative review. *Theoretical Issues in Ergonomics Science*, 8, 277–302.

Madhyastha, T. M., & Reed, D. A. (1994). A framework for sonification design. In G. Kramer (Ed.), *Auditory display: Sonification, audification, and auditory interfaces* (pp. 267–290). New York, NY: Oxford University Press.

Maikala, R. V., Cirello, V. M., Dempsey, P. G., & O'Brien, N. V. (2009). Psychophysiological response in women during cart pushing on different frictional walkways. *Human Factors*, 51, 681–693.

Malaviya, P., & Ganesh, K. (1976). Shift work and individual differences in the productivity of weavers in an Indian textile mill. *Journal of Applied Psychology*, 61, 714–176.

Malaviya, P., & Ganesh, K. (1977). Individual differences in productivity across type of work shift. *Journal of Applied Psychology*, 62, 527–528.

Mandelbrot, B. B. (1983). *The fractal geometry of nature*. New York, NY: Freeman.

Manzey, D., Lorenz, B., Heuer, H., & Sangals, J. (2000). Impairments of manual tracking performance during spaceflight: More converging evidence from a 20-day space mission. *Ergonomics*, 43, 589–609.

March, J. G. (2002). Forward. In A. Lomi & E. R. Larsen (Eds.), *Dynamics of organizations: Computational modeling and organization theories* (pp. ix-xvii). Cambridge, MA: MIT Press.

Marken, R. S. (1991). Degrees of freedom in behavior. *Psychological Science, 2*, 86–91.

Marklin, R. W., Saginus, K. A., Seeley, P., & Freier, S. H. (2010). Comparison of anthropometry of U.S. electric utility field-workers with North American general populations. *Human Factors, 52*, 463–662.

Marks, E. S. (1935). Individual differences in work curves. *Archives of Psychology, 186*.

Marks, M. A., Sabella, M. J., Burke, C. S., & Zaccaro, S. J. (2002). The impact of cross-training on team effectiveness. *Journal of Applied Psychology, 87*, 3–13.

Marks, M. L., Mirvis, P. H., Hackett, E. J., & Grady, J. F., Jr. (1986). Employee participation in a quality circle program: Impact on quality of work life, productivity, and absenteeism. *Journal of Applied Psychology, 71*, 61–69.

Marti, S., Sigman, M., & Dehaene, S. (2012). A shared cortical bottleneck underlying attentional blink and psychological refractory period. *NeuroImage, 59*, 2883–2898.

Martin, K., McLoed, E., Périard, J., Rattray, B., Keegan, R., & Pyne, D. B. (2019). Impact of environmental stress on cognitive performance: A systematic review. *Human Factors, 61*, 1205–1246.

Martinez, R. M., & Brito, S. V. (1994). *Looking for participation: Cuban experiences applying the feedback model* (Tech. Rep.). Havana, Cuba: Faculty of Industrial Engineering, ISPJAE.

Masalonis, A. J., & Parasuraman, R. (2003). Fuzzy signal detection theory: Analysis of human and machine performance in air traffic control, and analytic considerations. *Ergonomics, 46*, 1045–1074.

Masud, M., Khan, L., & Thuraisingham, B. (2012). *Data mining tools for malware detection.* Boca Raton, FL: CRC Press/Taylor & Francis.

Matsumoto, A., & Szidarovszky, F. (2012, March). *Discrete-time delay dynamics of boundedly rational monopoly.* Paper presented to the 5th International Nonlinear Science Conference, Barcelona, Spain.

Matthews, G., & Campbell, S. E. (2009). Sustained performance under overload: Personality and individual differences in stress and coping. *Theoretical Issues in Ergonomics Science, 10*, 417–443.

Matthews, G., Hancock, P. A., & Desmond, P. A. (2012). Models of individual differences in fatigue for performance research. In G. Matthews, P. A. Desmond, C. Neubauer, & P. A. Hancock (Eds.), *The handbook of operator fatigue* (pp. 155–170). Burlington, VT: Ashgate.

Matthews, G., & Reinerman-Jones, L. E. (2017). *Workload assessment: How to diagnose workload issues and enhance performance.* Santa Monica, CA: Human Factors and Ergonomics Society.

Matthews, G., Reinerman-Jones, L. E., Barber, D. J., & Abich, J. (2015). The psychometrics of mental workload: Multiple measures are sensitive but divergent. *Human Factors, 57*, 125–143.

Matthieu, J. E., Heffher, T. S., Goodwin, G. F., Salas, E., & Cannon-Bowers, J. A. (2000). The influence of shared mental models on team process and performance. *Journal of Applied Psychology, 85*, 273–283.

Mavrommati, I. (2001). Creativity techniques. In K. Baumann & B. Thomas (Eds.), *User interface design for electronic appliances* (pp. 51–76). New York, NY: Taylor & Francis.

Mayer, J. D. (2001). A field guide to emotional intelligence. In J. Ciarrochi, J. P. Forgas, & J. D. Mayer (Eds.), *Emotional intelligence in everyday life: A scientific inquiry* (pp. 3–24). Philadelphia, PA: Taylor & Francis.

Mayer, J. D., & Salovey, P. (1997). What is emotional intelligence? In P. Salovey & D. Sluyter (Eds.), *Emotional development and emotional intelligence: Educational implications* (pp. 3–34). New York, NY: Basic Books.

Mayer, R. E., Dow, G. T., & Mayer, S. (2003). Multimedia learning in an interactive self-explaining environment: What works in the design of agent-based microworlds? *Journal of Educational Psychology, 95,* 806–812.

Mayer, R. E., Heiser, J., & Lonn, S. (2001). Cognitive constraints on multimedia learning: When presenting more material results in less understanding. *Journal of Educational Psychology, 93,* 187–198.

Mayer-Kress, G. (1994, November). *Global brains as paradigm for a complex adaptive world.* Paper presented at the conference on Evolving Complexity: Challenges to Society, Economy and the Individual, University of Texas at Dallas, Richardson.

Mayer-Kress, G., Newell, K. M., & Liu, Y.-T. (2009). Nonlinear dynamics of motor learning. *Nonlinear Dynamics, Psychology, and Life Sciences, 13,* 3–26.

Mayhew, D. J. (2003). Introduction. In J. Ratner (Ed.), *Human factors and web development* (2nd ed., pp. 3–20). Mahwah, NJ: Lawrence Erlbaum Associates.

Maynard Smith, J. (1982). *Evolution and the theory of games.* Cambridge, UK: Cambridge University Press.

McAfee, R. B., & Winn, A. R. (1989). The incentives/feedback to enhance work place safety: A critique of the literature. *Journal of Safety Research, 20,* 7–19.

McAnally, K. I., & Martin, R. L. (2007). Spatial audio displays improve the detection of target messages in a continuous monitoring task. *Human Factors, 49,* 688–695.

McCallum, M., & Richard, C. (2008). *Human factors analysis of pipeline monitoring and control operations: Final technical report.* Retrieved from http://primis.phmsa.dot.gov/crm/FinalTechnicalReportNov2008.pdf

McComb, S., Kennedy, D., Perryman, R., & Warner, N. (2010). Temporal patterns of mental model convergence: Implications for distributed teams interacting in electronic collaboration spaces. *Human Factors, 52,* 264–281.

McCormick, E. J. (1976). *Human factors engineering and design* (4th ed.). New York, NY: McGraw-Hill.

McCoy, J. M. (2002). Work environments. In R. B. Bechtel & A. Churchman (Eds.), *Handbook of environmental psychology* (pp. 443–360). New York, NY: Wiley.

McDaniel, R. R., Jr., & Driebe, D. J. (Eds.). (2005). *Uncertainty and surprise in complex systems.* New York, NY: Springer-Verlag.

McDonald, A. D., Ferris, T. K., & Wiener, T. A. (2020). Classification of driver distraction: A comprehensive analysis of feature generation, machine learning, and input measures. *Human Factors, 62,* 1019–1035.

McEvoy, L. K., Smith, M. E., & Gervins, A. (1998). Dynamic cortical networks of verbal and spatial working memory: Effects of memory load and task practice. *Cerebral Cortex, 8,* 563–574.

McGrath, J. E. (1976). Stress and behavior in organizations. In M. D. Dunnette (Ed.), *Handbook of industrial and organizational psychology* (pp. 1031–1062). Chicago, IL: Rand McNally.

McGregor, D. M. (1960). *The human side of enterprise.* New York, NY: McGraw-Hill.

McKelvey, B., & Lichetenstein, B. B. (2007). Leadership in the four stages of emergence. In J. K. Hazy, B. B. Lichtenstein, & J. Goldstein (Eds.), *Complex systems leadership theory* (pp. 93–107). Litchfield Park, AZ: ISCE.

McKendree, J., & Anderson, J. R. (1987). Effect of practice on knowledge and use of basic Lisp. In J. M. Carroll (Ed.), *Interfacing thought: Cognitive aspects of human-computer interaction* (pp. 236–259). Cambridge, MA: MIT Press.

McVay, J. C., & Kane, M. J. (2009). Conducting the train of thought: Working memory capacity, goal neglect, and mind wandering in an executive-control task. *Journal of Experimental Psychology: Learning, Memory and Cognition, 35,* 196–204.

Meehl, P. E. (1954). *Clinical versus statistical prediction: A theoretical analysis and a review of the evidence.* Minneapolis, MN: University of Minnesota Press.

Meehl, P. E. (1956). Wanted: A good cookbook. *American Psychologist, 11,* 262–272.

Meister, D. (1977). Implications of the system concept for human factors research methodology. *Proceedings of the Human Factors Society, 21,* 453–456.

Menckel, E., & Carter, N. (1985). Development and evaluation of accident prevention routines: A case study. *Journal of Safety Research, 20,* 73–82.

Menckel, E., Carter, N., & Hellbom, M. (1993). The long-term effects of two group routines on accident prevention activities and accident statistics. *International Journal of Industrial Ergonomics, 12,* 301–309.

Mendeloff, J. (1979). *Regulating safety: An economic and political analysis of occupational safety and health policy.* Cambridge, MA: MIT Press.

Merat, N., & Lee, J. D. (2012). Preface to the special section on human factors and automation in vehicles: Designing highly automated vehicles with the driver in mind. *Human Factors, 54,* 681–686.

Meshkati, N. (1989). An etiological investigation of micro- and macro-economic factors in the Bhopal disaster: Lessons for both industrialized and developing countries. *International Journal of Industrial Ergonomics, 4,* 161–175.

Metzger, M. A., & Theisz, M. F. (1994). Forecast: Program to obtain forecasts from subjects for successive values of chaotic time series. *Behavioral Research Methods, Instruments, & Computers, 26,* 387–294.

Meyer, D. E., Smith, J. E., Kornblum, S., Abrams, R. A., & Wright, C. E. (1990). Speed-accuracy tradeoffs in aimed movements: Toward a theory of rapid voluntary action. In M. Jeannerod (Ed.), *Attention and performance* (Vol. 13, pp. 173–226). Hillsdale, NJ: Lawrence Erlbaum Associates.

Michalewicz, Z. (1993). A hierarchy of evolution programs: An experimental study. *Evolutionary Computation, 1,* 51–76.

Milgram, S. (1967). The small world problem. *Psychology Today, 2,* 60–67.

Militello, L. G., & Klein, G. (2013). Decision-centered design. In J. D. Lee & A. Kirlik (Eds.), *The Oxford handbook of cognitive engineering* (pp. 261–271). New York, NY: Oxford University Press.

Miller, G. A. (1956). The magical number seven plus or minus two: Some limits on our capacity for processing information. *Psychological Bulletin, 63,* 81–97.

Minelli, T. A. (2009). Neurodynamics and electrocortical activity. In S. J. Guastello, M. Koopmans, & D. Pincus (Eds.), *Chaos and complexity in psychology: Theory of nonlinear dynamical systems* (pp. 73–104). New York, NY: Cambridge University Press.

Mintz, A., & Blum, M. L. (1949). A re-examination of the accident proneness concept. *Journal of Applied Psychology, 33*, 195–221.

Miyake, A., Friedman, N. P., Emerson, M. J., Witzki, A. H., & Howerter, A. (2000). The unity and diversity of executive functions and their contributions to complex "frontal lobe" tasks: A latent variable analysis. *Cognitive Psychology, 41*, 49–100.

Mkrtchyan, A. A., Macbeth, J. C., Solovey, E. T., Ryan, J. C., & Cummings, M. L. (2012). Using variable-rate alerting to counter boredom in human supervisory control. *Proceedings of the Human Factors and Ergonomics Society, 56*, 1441–1445.

Moacdieh, N., & Sarter, N. (2015). Display clutter: A review of definitions and measurement techniques. *Human Factors, 57*, 61–100.

Modell, M., & Spurlock, J. M. (1980). Rationale for evaluating a closed-loop food chain for space habitats. In T. S. Cheston & D. L. Winter (Eds.), *Human factors of outer space production* (pp. 133–146). Boulder, CO: Westview Press and American Association for the Advancement of Science.

Moemeng, C., Gorosetsky, V., Zuo, Z., Yang, Y., & Zhang, C. (2009). Agent-based distributed data mining: A survey. In L. Cao (Ed.), *Data mining and multi-agent integration* (pp. 49–60). New York, NY: Springer.

Moemeng, P., Cao, L., & Zhang, C. (2008). F-TRADE 3.0: An agent-based integrated framework for data mining experiments. In *Proceedings of IEEE/WIC/ACM International Conference on Web Intelligence and Intelligent Agent Technology* (pp. 612–615). Los Alamitos, CA: IEEE Computer Society.

Mohr, A., & Clemmer, D. L. (1989). Evaluation of an occupational injury intervention in the petroleum drilling industry. *Accident Analysis and Prevention, 21*, 263–271.

Mønster, D., Håkonsson, D. D., Eskildsen, J. K., & Wallot, S. (2016). Physiological evidence of interpersonal dynamics in a cooperative production task. *Physiology and Behavior, 156*(Suppl. C), 24–34.

Moore, J. S., & Garg, A. 1998. The effectiveness of participatory ergonomics in the red meat packing industry: Evaluation of a corporation. *International Journal of Industrial Ergonomics, 21*, 47–58.

Morgan, H. M., Klein, C., Boehm, S. G., Shapiro, K. L., & Linden, D. E. J. (2008). Working memory load for faces modulates P300 N170 N250r. *Journal of Cognitive Neuroscience, 20*, 989–1002.

Morgan, J. F., & Hancock, P. A. (2011). The effect of prior task loading on mental workload: An example of hysteresis in driving. *Human Factors, 43*, 75–86.

Morineau, T., & Flach, J. M. (2019). The heuristic version of cognitive work analysis: A first application to medical emergency situations. *Applied Ergonomics, 79*, 98–106.

Morineau, T., Frénod, E., Blanche, C., & Tobin, L. (2009). Turing machine as an ecological model for task analysis. *Theoretical Issues in Ergonomics Science, 10*, 511–530.

Morris, C. (1906). *The San Francisco calamity, earthquake, and fire.* Np: W. E. Scull.

Morris, J. C., Morris, E. D., & Jones, D. M. (2007). Reaching for the philosopher's stone: Contingent coordination and the military's response to hurricane Katrina. *Public Administration Review, 67*, 94–106.

Morrow, P. M., & Fiore, S. M. (2013). Team cognition: Coordination across individuals and machines. In J. D. Lee & A. Kirklik (Eds.), *Oxford handbook of cognitive engineering* (pp. 200–215). New York, NY: Oxford University Press.

Mosso, A. (1894). *La fatigue intellectuelle et physique* [Mental and physical fatigue]. Paris, France: Alcon.

Mosso, A. (1915). *Fatigue*. New York, NY: Putnam.

Mueller, S. T. (2020). Cognitive anthropomorphism and AI: How humans and computers classify images. *Ergonomics in Design, 28*(3), 12–19.

Murphy, L. R. (1984). Occupational stress management: A review and appraisal. *Journal of Occupational Psychology, 57*, 1–16.

Murphy, L. R., DuBois, D., & Hurrell, J. J. (1986). Accident reduction through stress management. *Journal of Business and Psychology, 1*, 5–18.

Murray, D. J. (1990). Fechner's later psychophysics. *Canadian Psychology, 31*, 54–60.

Mykytyn, P. P., Jr. (1989). Group embedded figures test (GEFT): Individual differences, performance, and learning effects. *Educational and Psychological Measurement, 49*, 951–959.

Naber, A. M., McDonald, J. N., Asenuga, O. A., & Arthur, W., Jr. (2015). Team members' interaction anxiety and team training effectiveness: a catastrophic relationship? *Human Factors, 57*, 163–176.

Nadolski, J. N., & Sandonato, C. E. (1987). Evaluation of an employee assistance program. *Journal of Occupational Medicine, 29*, 32–37.

Nagashima, S., Suwazono, Y., Okubo, Y., Mirei, U., Kobayaski, E., Kido, T., & Nogawa, K. (2007). Working hours and mental and physical fatigue in Japanese workers. *Occupational Medicine, 57*, 449–452.

Naitoh, P., Kelly, T. L., Hunt, P., & Goforth, H. W. (1994). *Sleep during SEAL delivery vehicle (SDV)/dry dock shelter exercises analyzed by a graphic approach* (Tech. Rep. No. 94–30). San Diego, CA: U.S. Naval Health Research Center.

Nash, J. (1951). Non-cooperative games. *Annals of Mathematics, 54*, 286–295.

Natani, K. (1980). Future directions for selecting personnel. In T. S. Cheston & D. L. Winter (Eds.), *Human factors of outer space production* (pp. 25–63). Boulder, CO: Westview Press and American Association for the Advancement of Science.

Nathan, D. E., Guastello, S. J., Prost, R. W., & Jeutter, D. C. (2012). Understanding neuromotor strategy during functional upper extremity tasks using symbolic dynamics. *Nonlinear Dynamics, Psychology, and Life Sciences, 16*, 37–59.

Nathan, D. E., Johnson, M. J., & McGuire, J. R. (2009). Design and validation of low-cost assistive glove for hand assessment and therapy during activity of daily living-focused robotic stroke therapy. *Journal of Rehabilitation Research and Development, 46*, 587–602.

Nathan, D. E., Prost, R. W., Guastello, S. J., & Jeutter, D. C. (2014). Understanding the importance of natural neuromotor strategy in upper extremity neuroprosthetic control. *International Journal of Bioinformatics Research and Applications, 10*, 217–234.

National Aeronautics and Space Administration. (2004). *Lesson 2: Technology interface design*. Retrieved from http://humanfactors.arc.nasa.gov/web/hfl01/technology.html

National Institute for Occupational Safety and Health. (1994). *Applications manual for the revised NIOSH lifting equation*. Washington, DC: Author. Retrieved from www.cdc/gov/niosh/docs/94-110/

National Institute for Occupational Safety and Health. (1997). *Elements of ergonomics programs: A primer based on workplace evaluations of musculoskeletal disorders* (DHHS NIOSH Publication No. 97-117). Cincinnati, OH: Author.

National Institute for Occupational Safety and Health. (2007). *Assistant grain elevator supervisor dies after being engulfed in shelled corn—North Carolina*. Washington, DC: Author. Retrieved from www.cdc.gov/niosh/face/In-house/full9416.html

National Institute for Occupational Safety and Health. (2011). *Confined spaces*. Washington, DC: Author. Retrieved from www.cdc.gov/niosh/topics/confin edspace/

National Motorist Association. (2013). *State speed limit chart*. Retrieved from www.motorists.org/speed-limits/state-chart

National Safety Council. (1989). *Accident facts, 1989 edition*. Chicago, IL: Author.

National Safety Council. (2002). *International accident facts* (3rd ed.). Lyle, IL: Author.

National Safety Council (2022). *NSC injury facts*. Itasca, IL: Author. Retrieved from https://injuryfacts.nsc.org

Naweed, A., Stahlut, J., & O'Keefe, V. (2022). The essence of care: Versatility as an adaptive response to challenges in the delivery of quality aged care by personal care attendants. *Human Factors, 64*, 109–105.

Nayar, S. N., & Anand, V. N. (2007). 3D display using passive optical scatterers. *IEEE Computer Magazine, 40*(7), 54–63.

Nayar, S. N., & Anand, V. N. (2013). Volumetric displays: Passive optical scatters. Retrieved from www.cs.columbia.edu/CAVE/projects/3d_display

Neal, A., & Griffin, M. A. (2006). A study of the lagged relationships among safety climate, safety motivation, safety behavior, and accidents at the individual and group levels. *Journal of Applied Psychology, 91*, 946–953.

Neerincx, M. A., & Griffioen, E. (1996). Cognitive task analysis harmonizing tasks to human capabilities. *Ergonomics, 39*, 543–561.

Neisser, U. (1976). *Cognition and reality: Principles and implications of cognitive psychology*. San Francisco, CA: W. H. Freeman.

Neitzel, R. L., Seixas, N. S., Harris, M. J., & Camp, J. (2008). Exposure to fall hazards and safety climate in the aircraft maintenance industry. *Journal of Safety Research, 39*, 391–402.

NEOS Technologies. (2004). *3-D volumetric display*. Retrieved from http://neost ech.com/neos/3d.htm

Neubauer, C., Matthews, G., Langham, L., & Saxby, D. (2012). Fatigue and voluntary utilization of automation in simulated driving. *Human Factors, 54*, 734–746.

Neuhoff, J. G., Wayand, J., & Kramer, G. (2002). Pitch and loudness interact in auditory displays: Can the data get lost in the map? *Journal of Experimental Psychology: Applied, 8*, 17–25.

Neuringer, A., & Voss, C. (1993). Approximating chaotic behavior. *Psychological Science, 4*, 113–119.

Newell, A., & Card, S. K. (1985). The prospects for psychological science in human-computer interaction. *Human-Computer Interaction, 1*, 209–242.

Newell, K. M. (1991). Motor skill acquisition. *Annual Review of Psychology, 42,* 213–237.

Newhouse, R., Ruelle, D., & Takens, F. (1978). Occurrence of strange attractors: An axiom near quasiperiodic flows on T^m, m≥3. *Communications in Mathematical Physics, 64,* 35–41.

Newman, M., Barabási, A.-L., & Watts, D. J. (Eds.). (2006). *The structure and dynamics of networks.* Princeton, NJ: Princeton University Press.

Newman, O. (1972). *Defensible space.* New York, NY: Macmillan.

Ngo, M. K., Vu, K.-P. L., Thorpe, E., Battiste, V., & Strybel, T. Z. (2012). Intuitiveness of symbol features for air traffic management. *Proceedings of the Human Factors and Ergonomics Society, 56,* 1804–1808.

Nicolis, G., & Prigogine, I. (1989). *Exploring complexity:* New York, NY: Freeman.

Nielsen, K. T. (2000). Organization theories implicit in various approaches to OHS management. In K. Frick, P. Jensen, M. Quinlan, & T. Wilthagen (Eds.), *Systematic occupational health and safety management: Perspectives on an international development* (pp. 99–123). Oxford, UK: Pergamon Press.

Niu, Y-F., Zuo, H-R., Yang, X., Xue, C.-Q., Peng, N-Y., … Jin, T. (2021). Improving accuracy of gaze-control tools: Design recommendations for optimum position, sizes, and spacing of interactive objects. *Human Factors and Ergonomics in Manufacturing and Service Industries, 31,* 249–269.

Nolfi, S., & Floreano, D. (2000). *Evolutionary robotics.* Cambridge, MA: MIT Press.

Norman, D. A. (1988). *The psychology of everyday things.* New York, NY: Basic Books.

Norman, K. L. (2008). Better design of menu selection systems through cognitive psychology and human factors. *Human Factors, 50,* 556–559.

Noyes, J. M., & Bruneau, D. P. J. (2007). A self-analysis of the NASA-TLX workload measure. *Ergonomics, 50,* 514–519.

Noyes, J. M., & Frankish, C. R. (1992). Speech recognition technology for individuals with disabilities. *Augmentative and Alternative Communication, 8,* 297–303.

Nusbaum, E. C., & Silvia, P. J. (2011). Are intelligence and creativity really so different? Fluid intelligence, executive processes, and strategy use in divergent thinking. *Intelligence, 39,* 36–45.

Oberauer, K., & Kleigel, R. (2006). A formal model of capacity limits in working memory. *Journal of Memory and Language, 55,* 601–626.

Occupational Safety and Health Administration. (1996a). *Specifications for accident prevention signs and tags 1910-145.* Retrieved from www.osha.gov/pls/oshaweb/uwadisp.show_document?p_table=standards&p_id-9794

Occupational Safety and Health Administration. (1996b). *Occupational noise exposure compliance standard 1910.95.* Retrieved from www.osha.gov/SLTC/noisehearingconservation/compliance.html

Occupational Safety and Health Administration. (2002). *Effective ergonomics: A four-pronged comprehensive approach. U.S. Department of Labor.* Retrieved from http://osha.gov/ergonomics/ergofact02.html

O'Connor, P., O'Dea, A., & Melton, J. (2007). A methodology for identifying human error in U.S. navy diving accidents. *Human Factors, 49,* 214–226.

Ogawa, S., Lee, T. M., Kay, A. R., & Tank, D. W. (1992). Magnetic resonance imaging with contrast dependent on blood oxygenation. *Proceedings of the National Academy of Science (USA), 87,* 9968–9872.

Ohsawa, Y., & Yada, K. (2009). *Data mining and marketing.* Boca Raton, FL: CRC Press.

Oldham, G. R., & Fried, Y. (1987). Employee reactions to workspace characteristics. *Journal of Applied Psychology, 72,* 75–80.

Oliver, M. N., & Berke, E. (2010, January). *The raw and the cooked: Smoothing spatial data.* Paper presented to the 22nd annual Primary Care Research Methods & Statistics Conference, San Antonio, TX.

Olson, J. R. (1987). Cognitive analysis of people's use of software. In J. M. Carroll (Ed.), *Interfacing thought: Cognitive aspects of human-computer interaction* (pp. 260–293). Cambridge, MA: MIT Press.

Omodei, M. M., McLennan, J., & Wearing, A. J. (2005). How expertise is applied in real-world decision environments: Head-mounted video and cued recall as a methodology for studying routines of decision making. In T. Betsch & S. Haberstroh (Eds.), *The routines of decision making* (pp. 271–288). Mahwah, NJ: Lawrence Erlbaum Associates.

O'Neill, B. (1984). Structures for nonhierarchical organizations. *Behavioral Science, 29,* 61–77.

Open Ergonomics. (2000). *PeopleSize 2000: The shape of customers to come.* Retrieved from www.openerg.com/psz/onestop.htm

Ophir-Arbelle, R., Oron-Gilad, T., Borowsky, A., & Parmet, Y. (2013). Is more information better? How dismounted soldiers use video feed from unmanned vehicles: Attention allocation and information extraction considerations. *Journal of Cognitive Engineering and Decision Making, 7,* 26–48.

Osman, M. (2010). Controlling uncertainty: A review of human behavior in complex dynamic environments. *Psychological Bulletin, 136,* 65–86.

Ott, E., Grebogi, C., & Yorke, J. A. (1990). Controlling chaos. *Physical Review Letters, 64,* 1196–1199.

Ott, E., Sauer, T., & Yorke, J. A. (Eds.). (1994). *Coping with chaos.* New York, NY: Wiley.

Palinkas, L. A. (1987). *Antarctica as a model for the human exploration of Mars* (Tech. Rep. No. 87-16). San Diego, CA: Naval Health Research Center.

Palumbo, R. V., Marraccini, M. E., Weyandt, L. L., Wilder-Smith, O., McGee, H. A., Liu, S., & Goodwin, M. S. (2017). Interpersonal autonomic physiology: A systematic review of the literature. *Personality and Social Psychology Review, 21,* 99–141.

Papantonopoulos, S., & Galvendy, G. (2008). Analytic cognitive task allocation: A decision model for cognitive task allocation. *Theoretical Issues in Ergonomics Science, 9,* 155–185.

Parasuraman, R. (2003). Neuroergonomics: Research and practice. *Theoretical Issues in Ergonomics Science, 4,* 5–20.

Parasuraman, R. (2013). Neuroergonomics: Brain-inspired cognitive engineering. In J. D. Lee & A. Kirlik (Eds.), *The Oxford handbook of cognitive engineering* (pp. 159–177). New York, NY: Oxford University Press.

Parasuraman, R., Masalonis, A. J., & Hancock, P. A. (2000). Fuzzy signal detection theory: Basic postulates and formulas for analyzing human and machine performance. *Human Factors, 42,* 636–659.

Parasuraman, R., Sheridan, T. B., & Wickens, C. D. (2000). A model of types and levels of human interaction with automation. *IEEE Transactions on Systems, Man, and Cybernetics, A: Systems and Humans, 50,* 468–474.

Parasuraman, R., Warm, J. S., & Dember, W. N. (1987). Vigilance: Taxonomy and utility. In L. S. Mark, J. S. Warm, & R. L. Houston (Eds.), *Ergonomics and human factors: Recent research* (pp. 11–32). New York, NY: Springer-Verlag.

Parasuraman, R., & Wickens, C. D. (2008). Humans: Still vital after all these years of automation. *Human Factors, 50,* 511–520.

Park, S. H., & Woldstad, J. C. (2000). Multiple two-dimensional displays as an alternative to three-dimensional displays in telerobotic tasks. *Human Factors, 42,* 592–601.

Parkes, K. R. (1998). Psychosocial aspects of stress, health and safety on North Sea installations. *Scandinavian Journal of Work, Environment & Health, 24,* 321–333.

Pascuale-Leone, J. (1970). A mathematical model for the transition rule in Piaget's developmental stages. *Acta Psychologica, 32,* 301–345.

Pashler, H., & Johnston, J. C. (1998). Attention limitations in dual-task performance. In H. Pashler (Ed.), *Attention* (pp. 155–189). East Sussex, UK: Psychology Press.

Pasqualotto, E., Tamara, M., Ruf, C. A., Bartl, M., Olivetti Belardinelli, M., Birbaumer, N., & Halder, S. (2015). Usability and workload of access technology for people with severe motor impairment: A comparison of brain-computer interfacing and eye tracking. *Neurorehabilitation and Neural Repair, 29,* 950–957.

Pastukov, A., Lisner, A., Füllekrug, J., & Braun, J. (2014). Sensory memory of illusory depth in structure-from-motion. *Attention, Perception, and Psychophysics, 76,* 123–132.

Pathak, S., Pokharel, M. P., & Mahadevan, S. (2013). Hyper-competition, collusion, free-riding or coopetition: Basins of attraction when firms simultaneously complete and cooperate. *Nonlinear Dynamics, Psychology, and Life Sciences, 17,* 133–157.

Patrick, J., & Morgan, P. L. (2010). Approaches to understanding, analyzing and developing situation awareness. *Theoretical Issues in Ergonomics Science, 11,* 41–57.

Patterson, E. S., Rayo, M. F., Jr., Edworthy, J. R., & Moffatt-Bruse, S. D. (2022). Applying human factors engineering to address the telemetry alarm problem in a large medical center. *Human Factors, 64,* 126–142.

Pauchant, T. C., & Mitroff, I. I. (1992). *Transforming the crisis-prone organization.* San Francisco, CA: Jossey-Bass.

Pazzani, M. (1987a). Explanation-based learning for knowledge-based systems. *International Journal of Man-Machine Studies, 26,* 413–433.

Pazzani, M. (1987b). *Inducing causal and social theories: A prerequisite for explanation-based learning* (Tech. Rep.). Los Angeles: University of California Artificial Intelligence Laboratory.

Pederson, L. M., Nielson, K. J., & Kines, P. (2012). Realistic evaluation as a new way to design and evaluate occupational safety interventions. *Safety Science, 50,* 48–54.

Perry, O., Jaffe, E., & Bitan, Y. (2022). Dynamic communication quantification model for measuring information management during mass-casualty incident simulations. *Human Factors, 64,* 228–249.

Peterson, J. L. (1981). *Petri net theory and the modeling of systems.* Englewood Cliffs, NJ: Prentice Hall.

Peterson, L. E., Pepper, L. J., Hamm, P. B., & Gilbert, S. L. (1993). Longitudinal study of astronaut health: Mortality in the years 1959-91. *Radiation Research, 133*, 257–264.

Peterson, W. W., Birdsall, T. G., & Fox, W. C. (1954). The theory of signal detectability. *IRE Professional Group on Information Theory PGIT-4, X*, 171–212.

Pheasant, S. (1996). *Bodyspace: Anthropometry, ergonomics, and the design of work*. Bristol, PA: Taylor & Francis.

Pheasant, S., & Haslegrave, C. (2005). *Bodyspace: Anthropometry, ergonomics and the design of work* (3rd ed.). London: Taylor & Francis.

Pidgeon, N. F. (1991). Safety culture and risk management in organizations. *Journal of Cross-Cultural Psychology, 22*, 129–140.

Pikovsky, A., Rosenblum, M., & Kurths, J. (2001). *Synchronization: A universal concept in nonlinear sciences*. Cambridge, UK: Cambridge University Press.

Pimental, K., & Texiera, K. (1995). *Virtual reality: Through the new looking glass* (2nd ed.). New York, NY: McGraw-Hill.

Pinch, T., & Trocco, F. (2002). *Analog days: The invention and impact of the Moog synthesizer*. Cambridge, MA: Harvard University Press.

Pincus, D., & Metten, A. (2010). Nonlinear dynamics in biopsychosocial resilience. *Nonlinear Dynamics, Psychology, and Life Sciences, 14*, 353–380.

Pincus, S. M. (1991). Approximate entropy as a measure of system complexity. *Proceedings of the National Academic of Sciences USA, 88*, 2297–2301.

Pincus, S. M., & Goldberger, A. L. (1994). Physiological time-series analysis: What does regularity quantify? *American Journal of Physiology, 266*(Pt. 2), H1643–1656.

Pirolli, P. (2007). *Information foraging theory*. New York, NY: Oxford University Press.

Plant, K. L., & Stanton, N. A. (2012). Why did the pilots shut down the wrong engine? Explaining errors in context using schema theory and the perceptual cycle model." *Safety Science, 50*, 300–315.

Plant, K. L., & Stanton, N. A. (2015). The process of processing: Exploring the validity of Neisser's perceptual cycle model with accounts from critical decision-making in the cockpit. *Ergonomics, 58*, 909–923.

Poffenberger, A. T. (1928). The effects of continuous work upon output and feelings. *Journal of Applied Psychology, 12*, 259–467.

Porter, B. E., & Berry, T. D. (2001). A nationwide survey of self-reported red light running: Measuring prevalence, predictors, and perceived consequences. *Accident Analysis and Prevention, 33*, 735–742.

Post, D. L., Geiselman, E. E., & Goodyear, C. D. (1999). Benefits of color coding weapons symbology for an airborne helmet-mounted display. *Human Factors, 41*, 515–523.

Postman, K., Hermens, H. J., De Vries, J., Koopman, H. F. J. M., & Eisma, W. H. (1997). Energy storage and release of prosthetic feet, part 1: Biomechanical analysis related to user benefits. *Prosthetics and Orthotics International, 21*, 17–27.

Potvin, J. R. (2012). Predicting maximum acceptable efforts for repetitive tasks: An equation based on duty cycle. *Human Factors, 54*, 175–188.

Potvin, J. R., Ciriello, V. M., Snook, S. H., Maynard, W. S., & Brogmus, G. E. (2021). The Liberty Mutual manual materials handling (LM-MMH) equations. *Ergonomics, 64*, 955–970.

Pratt, N., Willoughby, A., & Swick, D. (2011). Effects of working memory on visual selective attention: Behavioral and electrophysiological evidence. *Frontiers in Human Neuroscience, 5*, article D57. https://doi.org/10.3389/fnhum.2011.00057

Prince, R. J., & Guastello, S. J. (1990). The Barnum effect in a computerized Rorschach interpretation system. *Journal of Psychology: Interdisciplinary and Applied, 124*, 217–222.

Priplata, A. A. (2006). Noise-enhanced balance control in patients with diabetes and patients with stroke. *Annals of Neurology, 59*, 4–12.

Prytz, E. G., Norén, C., & Jonson, C-O. (2018). Fixation differences in visual search of accident scenes by expert and emergency responders. *Human Factors, 60*, 1219–1227.

Quick, J. C., Murphy, L. R., & Hurrell, J. J., Jr. (Eds.). (1992). *Stress and well-being at work.* Washington, DC: American Psychological Association.

Quick, J. C., & Quick, J. D. (1984). *Organizational stress and preventive management.* New York, NY: McGraw-Hill.

Quinlan, M. (1988). Psychological and sociological approaches to the study of occupational illness: A critical review. *Australia and New Zealand Journal of Sociology, 24*, 189–207.

Ralph, J., Gray, W. D., & Schoelles, M. J. (2010). Squeezing the balloon: Analyzing the unpredictable effects of cognitive workload. *Proceedings of the Human Factors and Ergonomics Society, 54*, 299–303.

Rambaut, P. C. (1980). Nutritional criteria for closed-loop space food systems. In T. S. Cheston & D. L. Winter (Eds.), *Human factors of outer space production* (pp. 113–132). Boulder, CO: Westview Press and American Association for the Advancement of Science.

Rankin, F. W., Van Huyck, J. B., & Battalio, R. C. (2000). Strategic similarity and emergent conventions: Evidence from similar stag hunt games. *Games and Economic Behavior, 32*, 315–337.

Rasmussen, J., Pejtersen, A. M., & Goodstein, L. P. (1994). *Cognitive systems engineering.* New York, NY: Wiley.

Rasmussen, J., & Vicente, K. J. (1989). Coping with human errors through system design: Implications for ecological interface design. *International Journal of Man-Machine Studies, 31*, 517–534.

Raymond, J. E., Shapiro, K. L., & Arnell, K. M. (1992). Temporary suppression of visual processing in an RSVP task: An attentional link? *Journal of Experimental Psychology: Human Perception, and Performance, 18*, 849–860.

Reagan, R. T., Mostellar, F., & Youtz, C. (1989). Quantitative meanings of verbal probability expressions. *Journal of Applied Psychology, 74*, 433–442.

Reason, J. (1997). *Managing the risks of organizational accidents.* Brookfield, VT: Ashgate.

Redick, T. S., Calvo, A., Gay, C. E., & Engle, R. W. (2011). Working memory capacity and go/no-go task performance: Selective effects of updating, maintenance, and inhibition. *Journal of Experimental Psychology: Learning, Memory, and Cognition, 37*, 308–324.

Rees, J. V. (1988). *Reforming the workplace: A study of self-regulation in occupational safety.* Philadelphia, PA: University of Pennsylvania Press.

Rege, A., & Agogino, A. M. (1988). Topological framework for representing and solving probabilistic inference problems in expert systems. *IEEE Transactions on Systems, Man, and Cybernetics, 18*, 402–414.

Reid, G. B., & Nygren, T. E. (1988). The subjective workload assessment technique: A scaling procedure for measuring mental workload. In P. A. Hancock & N. Mehtaki (Eds.), *Human mental workload* (pp. 185–218). Amsterdam, the Netherlands: North Holland.

Reinerman-Jones, L., Taylor, G., Sprouse, K., Barber, D., & Hudson, I. (2011). Adaptive automation as a task switching and task congruence challenge. *Proceedings of the Human Factors and Ergonomics Society, 55,* 197–201.

Reising, J. M., & Curry, D. G. (1987). A comparison of voice and multifunctional controls: Logic design is the key. *Ergonomics, 30,* 1063–1077.

Reiss, E. (2012). *Usable usability: Simple steps for making stuff better.* Indianapolis, IN: Wiley.

Reivich, K. J., Seligman, M. E. P., & McBride, S. (2011). Master resilience training in the U.S. army. *American Psychologist, 66,* 25–34.

Rempel, D. (2008). The split keyboard: An ergonomics success story. *Human Factors, 50,* 385–392.

Rempel, D., Willms, K., Anshel, J., Jaschinski, W., & Sheedy, J. (2007). The effects of visual display distance on eye accommodation, head posture, and vision and neck symptoms. *Human Factors, 49,* 830–838.

Reniers, G. L. L., & Dullaert, W. (2007). DomPrevPlannng©: User-friendly software for planning domino effects prevention. *Safety Science, 45,* 1060–1081.

Renshaw, P. F., & Wiggins, M. W. (2007). A self-report critical incident assessment tool for army night vision goggle helicopter operations. *Human Factors, 49,* 200–213.

Retting, R. A., Ulmer, R. G., & Williams, A. F. (1998). *Prevalence and characteristics of red light running crashes in the United States.* Arlington, VA: Insurance Institute for Highway Safety.

Retting, R. A., Weinstein, H. B., Williams, A. F., & Preusser, D. F. (2001). A simple method for identifying and correcting crash problems on urban arterial streets. *Accident Analysis and Prevention, 33,* 723–734.

Retting, R. A., Williams, A. F., Preusser, D. F., & Weinstein, H. B. (1995). Classifying urban crashes for countermeasure development. *Accident Analysis and Prevention, 27,* 283–294.

Reuter-Lorenz, P. A., & Cappell, K. A. (2008). Neurocognitive aging and compensation hypothesis. *Current Directions in Psychological Science, 17,* 177–182.

Reynolds, C. W. (1987). Flocks, herds, and schools: A distributed behavioral model. *Computer Graphics, 21,* 25–34.

Rice, S., & Geels, K. (2010). Using system-wide trust theory to make predictions about dependence on four diagnostic aids. *Journal of General Psychology, 137,* 362–375.

Riley, J. M., Endsley, M. R., Boldstad, C. A., & Cuevas, H. M. (2006). Collaborative planning and situation awareness in Army command and control. *Ergonomics, 49,* 1139–1153.

Rittle-Johnson, B., & Star, J. R. (2009). Compared with what? The effects of different comparisons on conceptual knowledge and procedural flexibility for equation solving. *Journal of Educational Psychology, 101,* 529–544.

Rittle-Johnson, B., Star, J. R., & Durkin, K. (2009). The importance of prior knowledge when comparing examples: Influences on conceptual and procedural knowledge of equation solving. *Journal of Educational Psychology, 101,* 836–852.

Ritz, H., Nassar, M. R., Frank, M. J., & Shenhav, A. (2018). A control theoretic model of adaptive learning in dynamic environments. *Journal of Cognitive Neuroscience, 30,* 1405–1421.

Riva, G., Davide, F., & Ijsselsteijn, W. A. (Eds.). (2003). *Being there: Concepts effects and measurements of user presence in synthetic environments.* Amsterdam, Netherlands Antilles: IOS Press.

Rizzo, A. A., Neumann, U., Enciso, R., Fidaleo, D., & Noh, J. Y. (2001). Performance-driven facial animation: Basic research on human judgments of emotional state in facial avatars. *Cyberpsychology and Behavior, 4,* 471–487.

Robertson, C. C., & Morgan, B. J. T. (1990). Aspects of the design and analysis of signal detection experiments. *British Journal of Mathematical and Statistical Psychology, 42,* 7–14.

Robertson, I. H., Manly, T., Andrade, J., Baddeley, B. T., & Yiend, J. (1997). "Oops!": Performance correlates of everyday attentional failures in traumatic brain injured and normal subjects. *Neuropsychologia, 35,* 747–758.

Robertson, L. S., & Keeve, J. P. (1983). Worker injuries: The effects of workers' compensation and OSHA inspections. *Journal of Health Policy Law, 8,* 581–597.

Robson, L. S., Clarke, J. A., Cullen, K., Bielecky, A., Severin, C., Bigelow, P. L., Irwin, E., Culyer, A., & Mahood, Q. (2007). The effectiveness of occupational health and safety management system interventions: A systematic review. *Safety Science, 45,* 329–353.

Rodríguez, Y., & Hignett, S. (2021). Integration of human factors/ergonomics in healthcare systems: A giant leap in safety as a key strategy during Covid-19. *Human Factors in Manufacturing and Service Industries, 31,* 570–576.

Roebuck, J. A., Jr. (1995). *Anthropometric methods: Designing to fit the human body.* Santa Monica, CA: Human Factors and Ergonomics Society.

Roebuck, J. A., Jr., Kroemer, K., & Thomson, W. (1975). *Engineering anthropometry methods.* New York, NY: Wiley Interscience.

Roscoe, S. N. (2006). The adolescence of engineering psychology. In J. Stuster (Ed.), *The Human Factors and Ergonomics Society: Stories from the first 50 years* (pp. 12–14). Santa Monica, CA: Human Factors and Ergonomics Society.

Rose, C. L., Murphy, L. B., Byard, L., & Nikzad, K. (2002). The role of the big five personality factors in vigilance performance and workload. *European Journal of Personality, 16,* 185–200.

Rosen, M. (2021, February). *Cross-analog validity evidence for unobtrusive measurement strategies for individual and team competencies.* Paper presented to the Bioastronautics@hopkins Inaugural Event. Online. Baltimore, MD: Johns Hopkins University School of Medicine.

Rosenbaum, D. A. (2005). The Cinderella of psychology: The neglect of motor control in the science of mental life and behavior. *American Psychologist, 60,* 308–317.

Rosenbaum, D. A., Slotta, J. D., Vaughn, J., & Plamondon, R. (1991). Optimal movement selection. *Psychological Science, 2,* 92–101.

Ross, S. N., & Ware, K. (2013). Hypothesizing the body's genius to trigger and self-organize its healing: 25 years using a standardized neurophysics therapy. *Frontiers in Physiology, 4.* doi: 10.3389/fphys.2013.00334

Rosser, J. B., Jr. (1997). Speculations on nonlinear speculative bubbles. *Nonlinear Dynamics, Psychology, and Life Sciences, 1,* 275–300.

Roth, E. M., Sushereba, C., Militello, L. G., & Diiulio, J. (2019). Function allocation considerations in the era of human autonomy teaming. *Journal of Cognitive Engineering and Decision Making, 13,* 199–220.

Rothman, A. L., & Weintraub, M. I. (1995). The sick building syndrome and mass hysteria. *Neurologic Clinics, 13,* 405–412.

Rothstein, M. A. (2007). Health privacy in the electronic age. *Journal of Legal Medicine, 28,* 487–501.

Rotter, J. B. (1966). Generalized expectancies for the internal versus external locus of control of reinforcement. *Psychological Monographs, 80.*

Roundtree, K. A., Goodrich, M. A., & Adams, J. A. (2019). Transparency: Transitioning from human-machine systems to human-swarm systems. *Journal of Cognitive Engineering and Decision Making, 13,* 171–195.

Rowe, W. D. (1977). *An anatomy of a risk.* New York, NY: Wiley.

Rubinov, M., & Sporns, O. (2010). Complex network measures of brain connectivity: Uses and interpretations. *NeuroImage, 52,* 1059–1069.

Rubinstein, J. S., Meyer, D. E., & Evans, J. E. (2001). Executive control of cognitive processes in task switching. *Journal of Experimental Psychology: Human Perception and Performance, 27,* 763–797.

Russell, S. M., Funke, G. J., Knott, B. A., & Strang, A. J. (2012). Recurrence quantification analysis used to assess team communication in simulated air battle management. *Proceedings of the Human Factors and Ergonomics Society, 56,* 468–472.

Saal, F. E., & Knight, P. A. (1988). *Industrial/organizational psychology: Science and practice.* Belmont, CA: Wadsworth.

Saarela, K. L. (1990). An intervention program utilizing small groups: A comparative study. *Journal of Safety Research, 21,* 149–156.

Sadeghi, M., Sasangohar, F., McDonald, A. D., & Hegde, S. (2022). Understanding heart rate reactions to post-traumatic stress disorder (PTSD) among veterans: A naturalistic study. *Human Factors, 64,* 173–187.

Sadeghipour, A., Yaghoubzadeh, R., Rüter, A., & Kopp, S. (2009). Social motorics: Towards an embedded basis of social human-robot interaction. In H. Ritter, G. Sagerer, R. Dillmann, & M. Buss (Eds.), *Human centered robot systems: Cognition, interaction, technology* (pp. 193–203). New York, NY: Springer.

Sakei, K. (2001). *Nonlinear dynamics and chaos in agricultural systems.* Amsterdam, the Netherlands: Elsevier.

Salas, E., Grossman, R., Hughes, A. M., & Coultas, C. W. (2015). Measuring team cohesion: Observations from the science. *Human Factors, 57,* 365–374.

Salminen, S., & Heiskanen, M. (1997). Correlations between traffic, occupational, sports, and home accidents. *Accident Analysis and Prevention, 29,* 33–36.

Salmon, P. M., Stanton, N. A., Walker, G. H., Barber, C., Jenkins, D. P., McMaster, R., & Young, M. S. (2008). What really is going on? Review of situation awareness models for individuals and teams. *Theoretical Issues in Ergonomics Science, 9,* 297–323.

Salovey, P., & Mayer, J. D. (1990). Emotional intelligence. *Imagination, Cognition, and Personality, 9,* 185–211.

Salzer, Y., Oron-Gilad, T., Ronen, A., & Parmet, Y. (2011). Vibrotactile "on-high" alerting system in the cockpit. *Human Factors, 53,* 118–131.

Samuelson, D. A., & Macal, C. M. (2006, August). Agent-based simulation comes of age: Software opens up many new areas of application. *OR/MS Today*. Retrieved November 5, 2017, from www.orms-today.org/orms-8-06/fragent.html

Samuelson, L. (1997). *Evolutionary games and equilibrium selection*. Cambridge, MA: MIT Press.

Sandal, G. M. (2000). Coping in Antarctica: Is it possible to generalize results across settings? *Aviation, Space, and Environmental Medicine, 71*, A37–A43.

Sandal, G. M., Endresen, I. M., Vaernes, R., & Ursin, H. (1999). Personality and coping strategies during submarine missions. *Military Psychology, 11*, 381–404.

Sanders, M. S., & McCormick, E. J. (1993). *Human factors in engineering design*. New York, NY: McGraw-Hill.

Sangali, L. M. (2018). The role of statistics in the era of big data. *Statistics and Probability Letters, 136*, 1–3.

Sarason, I. G., Sarason, B. R., Potter, E. H., & Antoni, M. H. (1985). Life events, social support, and illness. *Psychosomatic Medicine, 47*, 156–163.

Saravanan, P., & Menold, J. (2022). Deriving effective decision-making strategies of prosthetists: Using hidden Markov modeling and qualitative analysis to compare experts and novices. *Human Factors, 64*, 188–206.

Sarfi, R. J., Tao, M. K., & Gemoets, L. (2011). Making the smart grid work for community energy delivery. *Information Polity, 16*, 267–281.

Sarter, N. (2008). Investigating mode errors on automated flight decks: Illustrating the problem-driven, cumulative, and interdisciplinary nature of human factors research. *Human Factors, 50*, 506–510.

Sartory, G., Heine, A., Muleeler, B. W., & Elvermann-Hallner, A. (2002). Event- and motor-related potentials during the continuous performance task in attention-deficit/hyperactivity disorder. *Journal of Psychophysiology, 16*, 97–106.

Sauter, S. L., & Murphy, L. R. (Eds.). (1995). *Organizational risk factors for job stress*. Washington, DC: American Psychological Association.

Sawyer, R. K. (2005). *Social emergence: Societies as complex systems*. New York, NY: Cambridge University Press.

Scerbo, M. W. (2001). Adaptive automation. In W. Karwowski (Ed.), *International encyclopedia of ergonomics and human factors* (pp. 1077–1079). London: Taylor & Francis.

Schels, M., Thiel, C., Schwenker, F., & Palm G. (2009). Classifier fusion applied to facial expression recognition: An experimental comparison. In H. Ritter, G. Sagerer, R. Dillmann, & M. Buss (Eds.), *Human centered robot systems: Cognition, interaction, technology* (pp. 121–130). New York, NY: Springer.

Scherer, S., Schwenker, F., Campbell, N., & Palm, G. (2009). Multimodal laughter detection in natural discourses. In H. Ritter, G. Sagerer, R. Dillmann, & M. Buss (Eds.), *Human centered robot systems: Cognition, interaction, technology* (pp. 111–120). New York, NY: Springer.

Schiflett, S. G., Elliott, L. R., Salas, E., & Coovert, M. D. (Eds.). (2004). *Scaled worlds: Development, validation, and applications*. Burlington, VT: Ashgate.

Schifrman, S. S., Reynolds, M. L., & Young, F. W. (1981). *Introduction to multidimensional scaling: Theory, methods, and applications*. New York, NY: Academic Press.

Schmidt, R. A., Zelaznik, H. N., Hawkins, B., Frank, J. S., & Quinn, J. T., Jr. (1979). Motor output variability: A theory for the accuracy of rapid motor acts. *Psychological Review, 86*, 415–451.

Schmidt, W. C. (2001). Presentation accuracy of web animation methods. *Behavioral Research Methods, Instrumentation, and Computers, 33*, 187–200.

Schmorrow, D. D., & Stanney, K. M. (2008). *Augmented cognition: A practitioner's guide*. Santa Monica, CA: Human Factors and Ergonomics Society.

Schneider, B., & Reichers, A. E. (1983). On the etiology of climates. *Personnel Psychology, 36*, 19–39.

Schneider, M., Rittle-Johnson, B., & Star, J. R. (2011). Relations among conceptual knowledge, procedural knowledge, and procedural flexibility in two sample differing in prior knowledge. *Developmental Psychology, 47*, 1525–1538.

Schneider-Garces, N. J., Gordon, B. A., Brumback-Peltz, C. R., Shin, E., Lee, Y., Sutton, B. P., Maclin, E. L., Gratton, G., & Fabiani, M. (2009). Span, CRUNCH, and beyond: Working memory capacity and the aging brain. *Journal of Cognitive Neuroscience, 22*, 655–669.

Schraagen, J. M., Chipman, S. F., & Shute, V. J. (2000). State-of-the-art review of cognitive task analysis techniques. In J. M. Schraagen, S. F. Chipman, & V. L. Shalin (Eds.), *Cognitive task analysis* (pp. 467–487). Mahwah, NJ: Lawrence Erlbaum Associates.

Schuldberg, D. (2015). What is optimum variability? *Nonlinear Dynamics, Psychology, and Life Sciences, 19*, 553–568.

Schulz, E., Tenenbaum, J. B., Duvenaud, D., Speekenbrink, M., & Gershman, S. J. (2017). Compositional inductive biases in function learning. *Cognitive Psychology, 99*, 44–79.

Schutte, N. S., Malouf, J. M., Hall, L. E., Haggerty, D. J., Cooper, J. T., Golden, C. J., & Dornheirn, L. (1998). Development and validation of a measure of emotional intelligence. *Personality and Individual Differences, 25*, 167–177.

Schutz, M., & Stefanucci, J. K. (2019). Exploring the effects of "sound shape" on consumer preference. *Ergonomics in Design, 27*(1), 16–19.

Secchi, P. (2018). On the role of statistics in the era of big data: A call for a debate. *Statistics and Probability Letters, 136*, 10–14.

See, J. E., Howe, S. R., Warm, J. S., & Dember, W. N. (1995). Meta-analysis of the sensitivity decrement in vigilance. *Psychological Bulletin, 117*, 230–249.

Seger, C. A. (1994). Implicit learning. *Psychological Bulletin, 115*, 163–196.

Seligman, M. P. (1975). *Helplessness: On depression, development and death*. San Francisco, CA: Freeman.

Sellers, J., Helton, W. S., Näswall, K., Funke, G. J., & Knott, B. A. (2014). Development of the Team Workload Questionnaire (TWLQ). *Proceedings of the Human Factors and Ergonomics Society, 58*, 989–993.

Sellnow, T. L., Seeger, M. W., & Ulmer, R. R. (2002). Chaos theory, informational needs, and natural disasters. *Journal of Applied Communications Research, 30*, 269–292.

Selye, H. (1976). *The stress of life* (2nd ed.). New York, NY: McGraw-Hill.

Seminara, J. L., Gonzales, W. R., & Parsons, S. O. (1977). *Human factors review of nuclear power plant control room design* (Tech. Rep. No. EPRINP-309). Sunnyvale, CA: Lockheed Missiles and Space.

Semovski, S. V. (2001). Self-organization in fish school. In W. Sulis & I. Trofimova (Eds.), *Nonlinear dynamics in the life and social sciences* (pp. 398–406). Amsterdam, the Netherlands: IOS Press.

Sengupta, S., & O'Brien, K. J. (2012, 21 September). Facebook can ID faces, but using them grows tricky. *Wall Street Journal*, p. A1. Retrieved from www.nytimes.com/2012/09/22/technology/facebook-backs-down-on-face-recognition

Seo, M.-G., Putnam, L. L., & Bartunek, J. M. (2004). Dualities and tensions of planned organizational change. In M. S. Poole & A. H. Van de Ven (Eds.), *Handbook of organizational change and innovation* (pp. 73–107). New York, NY: Oxford University Press.

Serfaty, D., Entin, E., & Deckert, J. (1993). *Team adaptation to stress in decision making and coordination with implications for CIC team training* (Rep. No. TR-564). Burlington, MA: ALPHATECH.

Sesto, M. E., Irwin, C. B., Chen, K. B., Chourasia, A. O., & Weigman, D. A. (2012). Effect of touch screen button size and spacing on touch characteristics of users with and without disabilities. *Human Factors, 54,* 425–436.

Shah, J., & Breazeal, C. (2010). An empirical analysis of team coordination behaviors and action planning with application to human-robot teaming. *Human Factors, 52,* 234–245.

Shakioye, S. O., & Haight, J. M. (2010). Modeling using dynamics variables: An approach for the design of loss prevention programs. *Safety Science, 48,* 46–53.

Shannon, C. E. (1948). A mathematical theory of communication. *Bell System Technical Journal, 27,* 379–423.

Shannon, C. E., & Weaver, W. (1949). *The mathematical theory of communication.* Urbana, IL: University of Illinois Press.

Shannon, H. S., & Guastello, S. J. (1996, February). *What do we know really works in safety?* Paper presented at the third International Conference in Injury Prevention and Control, Melbourne, Australia.

Shannon, H. S., & Guastello, S. J. (1997, June). *Workplace safety interventions: What can we say about what works?* Paper presented at the International Ergonomics Association Conference, Tempere, Finland.

Shannon, H. S., Robson, L.S., & Guastello, S. J. (1999). Methodological criteria for evaluating occupational safety intervention research. *Safety Science, 31,* 161–179.

Sharma, S., & Ivancevic, V. G. (2010). Nonlinear dynamical characteristics of situation awareness. *Theoretical Issues in Ergonomics Science, 11,* 448–460.

Shelhamer, M. (2007). *Nonlinear dynamics in physiology: A state-space approach.* Singapore: World Scientific.

Shelton, R. L. (2001). *Driver distractions: Electronic devices in the automobile.* Retrieved from www-nrd.nhtsa.dot.gov/PDF/nrd-13/Nov1_ppt.pdf

Shepard, R. N. (1962). The analysis of proximities: Multidimensional scaling with an unknown distance function. II. *Psychometrika, 27,* 219–246.

Shepard, R. N. (1982). Geometrical approximation to the structure of musical pitch. *Psychological Review, 89,* 305–332.

Shepperd, J. A. (1993). Productivity loss in performance groups: A motivation study. *Psychological Bulletin, 113,* 67–81.

Sheridan, T. B. (2002). *Human and automation: System design and research issues.* New York, NY: Wiley.

Sheridan, T. B. (2008). Risk, human error, and system resilience: Fundamental ideas. *Human Factors, 50,* 418–426.

Shockley, K., Richardson, D. C., & Dale, R. (2009). Conversation and coordinative structures. *Topics in Cognitive Science, 1,* 305–319.

Shockley, K. M., Gabriel, A. S., Robertson, D., Rosen, C. C., Chawla, N., Ganster, M. L., & Ezerins, M. E. (2021). The fatiguing effects of camera use in virtual meetings: A within-person field experiment. *Journal of Applied Psychology, 106,* 1137–1155.

Shneiderman, B. (1987). *Designing the user interface: Strategies for effective human-computer interaction.* Reading, MA: Addison-Wesley.

Shneiderman, B. (2000). *The eyes have it: User interfaces for information visualization.* Retrieved from www.cs.umdedu/hcil/pubs/presentations

Shortliffe, E. H. (1976). *Computer-based medical consultations: MYCIN.* New York, NY: Elsevier.

Shuvro, R. A., Das, P., Jyoti, J. S., Abreu, J. M., & Hayat, M. M. (2022). Data-integrity aware stochastic model for cascading failures in power grids. *IEEE Transactions on Power Systems.* doi: 10.1109/TPWRS.2022.3164671

Sieber, J. E. (1980). Well-being and privacy in space: Anticipating conflicts of interest. In T. S. Cheston & D. L. Winter (Eds.), *Human factors of outer space production* (pp. 65–78). Boulder, CO: Westview Press and American Association for the Advancement of Science.

Simon, H. A. (1957). *Administrative behavior* (2nd ed.). Totowa, NJ: Littlefield Adams.

Singh, I. L., Molloy, R., & Parasuraman, R. (1993). Automation induced "complacency": Development of the complacency-potential rating scale. *International Journal of Aviation Psychology, 3,* 111–122.

Sival, M., Olson, P. L., & Pastalan, L. A. (1981). Effect of driver's age on night-time legibility of highway signs. *Human Factors, 23,* 59–64.

Skarda, C. A., & Freeman, W. J. (1987). How brains make chaos in order to make sense of the world. *Behavioral and Brain Science, 10,* 161–195.

Slater, M. (2004). Presence and emotions. *Cyberpsychology and Behavior, 7,* 121.

Sleimen-Malkoun, R., Temprado, J. J., Jirsa, V. K., & Berton, E. (2010). New directions offered by the dynamical systems approach to bimanual coordination for therapeutic intervention and research in stroke. *Nonlinear Dynamics, Psychology, and Life Sciences, 14,* 435–462.

Smallwood, J., Beach, E., Schooler, J. W., & Handy, T. C. (2008). Going AWOL in the brain: Mind wandering reduces cortical analysis of external events. *Journal of Cognitive Neuroscience, 20,* 458–469.

Smith, C. M. (2001). Human factors in haptic interfaces. Retrieved from www.acm.org/crossroads/xrds.3-3/haptic.html#2

Smith, J. B. (1994). *Collective intelligence in computer-based collaboration.* Hillsdale, NJ: Lawrence Erlbaum Associates.

Smith, K., & Hancock, P. A. (1995). Situational awareness is adaptive, externally directed consciousness. *Human Factors, 37,* 137–148.

Smith, L., & Iskra-Golec, I. (2003). Internal locus of control and shiftwork effects. *Theoretical Issues in Ergonomic Science, 4,* 327–339.

Smith, P. C., Kendall, L. M., & Hulin, C. L. (1969). *The measurement of satisfaction in work and retirement: A strategy for the study of attitudes.* Chicago: Rand McNally.

Smith, R. S. (1976). *The Occupational Safety and Health Act: Its goals and its achievements.* Washington, DC: American Enterprise Institute for Public Policy.

Smith, R. S. (1979). The Impact of OSHA inspections on manufacturing injury rates. *Journal of Human Resources, 14,* 145–170.

Smith, S. F. (1977). Running nips. In W. Handley (Ed.), *Industrial safety handbook* (2nd ed., pp. 186–197). New York, NY: McGraw-Hill.

Smith, T. S., & Stevens, G. T. (1997). Biological foundations of social interaction: Computational explorations of nonlinear dynamics. In R. A. Eve, S. Horsfall, & M. E. Lee (Eds.), *Chaos, complexity and sociology: Myths, models, and theories* (pp. 197–214). Thousand Oaks, CA: Sage.

Smith, W., & Dowell, J. (2000). A case study of coordinative decision-making in disaster management. *Ergonomics, 43,* 1153–1166.

Smithson, M. (1997). Judgment under chaos. *Organizational Behavior and Human Decision Processes, 69,* 59–66.

Snowden, D. (2011). Naturalizing sensemaking. In K. L. Mosier & U. M. Fisher (Eds.), *Informed by knowledge: Expert performance in complex situations* (pp. 223–234). New York, NY: Psychology Press.

Snyder, J. S., & Alain, C. (2007). Toward a neurophysiological theory of auditory stream segregation. *Psychological Bulletin, 133,* 780–799.

Soderquist, D. R., & Shilling, R. D. (1990). Loudness and the binaural masking level difference. *Bulletin of the Psychonomic Society, 28,* 553–555.

Song, J., Lim, J. H., & Yun, M. H. (2016). Finding the latent semantics of haptic interaction research: A systematic literature review of haptic interaction using content analysis and network analysis. *Human Factors and Ergonomics in Manufacturing & Service Industries, 26,* 577–594.

Sorger, G. (1998). Imperfect foresight and chaos: An example of a self-fulfilling mistake. *Journal of Economic Behavior and Organization, 33,* 363–383.

Spangenberg, S., Mikkelson, K. L., Kines, P., Dyreborg, J., & Baarts, C. (2002). The construction of the Oresund link between Denmark and Sweden: The effect of a multifaceted safety campaign. *Safety Science, 40,* 457–465.

Spencer, H. (2004). *Pioneers of modern typography* (2nd ed.). Cambridge, MA: MIT Press.

Spivey, M. J., & Dale, R. (2006). Continuous dynamics in real-time cognition. *Current Directions in Psychological Science, 15,* 207–211.

Sprott, J. C. (2003). *Chaos and time-series analysis.* New York, NY: Oxford.

Sprott, J. C. (2013). Is chaos good for learning? *Nonlinear Dynamics, Psychology, and Life Sciences, 17,* 223–233.

Sridharan, N. S. (1985). Evolving systems of knowledge. *A. I. Magazine, 6,* 108–121.

Stachowski, A. A., Kaplan, S. A., & Waller, M. J. (2009). The benefits of flexible team interaction during crises. *Journal of Applied Psychology, 94,* 1536–1543.

Stamovlasis, D. (2006). The nonlinear dynamical hypothesis in science education problem solving: A catastrophe theory approach. *Nonlinear Dynamics, Psychology and Life Science, 10,* 37–70.

Stamovlasis, D. (2011). Nonlinear dynamics and neo-Piagetian theories in problem solving: Perspectives on a new epistemology and theory development. *Nonlinear Dynamics, Psychology and Life Science, 15*, 145–173.

Stamovlasis, D., & Tsaparlis, G. (2012). Applying catastrophe theory to an information-processing model of problem solving in science education. *Science Education, 96*, 392–410.

Stanton, N. A., Chambers, P. R. G., & Piggot, J. (2001). Situational awareness and safety. *Safety Science 39*, 189–204.

Stanton, N. A., Salmon, P. M., & Walker, G. H. (2015). Let the reader decide: A paradigm shift for situation awareness in sociotechnical systems. *Journal of Cognitive Engineering and Decision Making, 9*, 44–50.

Stanton, N. A., Salmon, P. M., Walker, G. H., & Jenkins, D. P. (2010). Is situation awareness all in the mind? *Theoretical Issues in Ergonomics Science, 11*, 29–40.

Starch, D., & Ash, J. E. (1917). The mental curve of work. *Psychological Bulletin, 24*, 391–402.

Staszewski, J. J. (2004). Models of human expertise as blueprints for cognitive engineering: Applications to landmine detection. *Proceedings of the 48th Annual Meeting of the Human Factors and Ergonomics Society, 48*, 458–462.

Staszewski, J. J. (2006). Spatial thinking and the design of landmine detection training. In G. A. Allen (Ed.), *Applied spatial cognition: From research to cognitive technology* (pp. 231–265). Mahwah, NJ: Lawrence Erlbaum Associates.

Staudenmeyer, N., & Lawless, M. W. (1999, November). *Organizational decay: How innovation stresses organizations.* Paper presented at INFORMS '99, Philadelphia.

Stefański, A. (2009). *Determining thresholds of complete synchronization, and application.* Singapore: World Scientific.

Stergiou, B., & Decker, L. M. (2011). Human movement variability, nonlinear dynamics, and pathology: Is there a connection? *Human Movement Science, 30*, 869–888.

Stergiou, N., Harbourne, R. T., & Cavanaugh, P. T. (2006). Optimal movement variability: A new theoretical perspective for neurologic physical therapy. *Journal of Neurologic Physical Therapy, 30*, 120–128.

Sterman, J. (1988). Deterministic models of chaos in human behavior: Methodological issues and experimental results. *System Dynamics Review, 4*, 148–178.

Sterman, J. (1989). Misperceptions of feedback in dynamic decision making. *Organizational Behavior and Human Decision Processes, 43*, 301–335.

Sternberg, R. J. (1995). *Psychology: In search of the human mind.* Orlando, FL: Harcourt.

Sternberg, R. J. (2004). *Psychology* (4th ed.). Belmont, CA: Thompson-Wadsworth.

Stevens, A., & Minton, R. (2001). In-vehicle distraction and fatal accidents in England and Wales. *Accident Analysis and Prevention, 33*, 539–545.

Stevens, M., & Campion, M. (1994). The knowledge, skill, and ability requirements for teamwork: Implications for human resource management. *Journal of Management, 20*, 503–540.

Stevens, R., & Galloway, T. (2016). Tracing neurodynamic information flows during teamwork. *Nonlinear Dynamics, Psychology, and Life Sciences, 20*, 271–292.

Stevens, R. H., & Galloway, T. L. (2019). Teaching machines to recognize neurodynamic correlates of team and team member uncertainty. *Journal of Cognitive Engineering and Decision Making, 13*, 310–327.

Stevens, R. H., Galloway, T. L., & Lamb, C. (2014). Submarine navigation and team resilience: Linking EEG and behavioral models. *Proceedings of the Human Factors and Ergonomics Society, 58*, 245–249.

Stevens, R., Gorman, J. C., Amazeen, P., Likens, A., & Galloway, T. (2013). The organizational neurodynamics of teams. *Nonlinear Dynamics, Psychology, and Life Sciences, 17*, 67–86.

Stevens, S. S. (1951). Mathematics, measurement, and psychophysics. In S. S. Stevens (Ed.), *Handbook of experimental psychology* (pp. 1–50). New York, NY: Wiley.

Stevens, S. S. (1957). On the psychophysical law. *Psychological Review, 64*, 153–181.

Stewart, L., Dominguez, C. O., & Way, L. W. (2011). A data-frame sensemaking analysis of operative reports: Bile duct injuries associated with laparoscopic cholecystectomy. In K. L. Mosier & U. M. Fisher (Eds.), *Informed by knowledge: Expert performance in complex situations* (pp. 223–234). New York, NY: Psychology Press.

Stirling, L., Kelty-Stphen, D., Fineman, R., Jones, M. L. H., Park, B-K. D., Reed, M. P., ... Choi, H. J. (2020). Static, dynamics, and cognitive fit of exosystems for the human operator. *Human Factors, 62*, 424–440.

Stone, M., Remington, R., & Dismukes, K. (2001). *Remembering intentions in dynamic environments*. Retrieved from http://olias.nasa.gov/reports/detecting.html

Stout, R. J., Cannon-Bowers, J., Salas, E., & Milanovich, D. M. (1999). Planning, shared mental models, and coordinated performance: An empirical link is established. *Human Factors, 41*, 61–71.

Strang, A. J., Horwood, S., Best, C., Funke, G. J., Knott, B. A., & Russell, S. M. (2012). Examining temporal regularity in categorical team communication using sample entropy. *Proceedings of the Human Factors and Ergonomics Society, 56*, 473–477.

Straub, R. O. (2001). *Faculty guide for use with Scientific American Frontiers video collection for introductory psychology* (2nd ed.). New York, NY: Worth.

Strayer, D. L., Drews, F. A., & Crouch, D. J. (2006). Comparison of the cell phone driver and the drunk driver. *Human Factors, 48*, 381–391.

Strayer, D. L., Drews, F. A., & Johnston, W. A. (2003). Cell phone induced failures of visual attention during simulated driving. *Journal of Experimental Psychology: Applied, 9*, 23–52.

Strogatz, S. (2003). *Sync: The emerging science of spontaneous order*. New York, NY: Hyperion.

Suedfeld, P. (2003). Canadian space psychology: The future may be almost here. *Canadian Psychology, 44*, 85–92.

Suh, N. P. (2007). Ergonomics, axiomatic design and complexity theory. *Theoretical Issues in Ergonomics Science, 8*, 101–122.

Sukel, K. E., Catrambone, R., Essa, I., & Brostow, G. (2003). Presenting movement in a computer-based dance tutor. *International Journal of Human-Computer Interaction, 15*, 433–452.

Sulis, W. (1997). Fundamental concepts of collective intelligence. *Nonlinear Dynamics, Psychology, and Life Sciences, 1*, 35–54.

Sulis, W. (2008). Stochastic phase decoupling in dynamical networks. *Nonlinear Dynamics, Psychology, and Life Sciences, 12*, 327–358.

Sulis, W. (2009). Collective intelligence: Observations and models. In S. J. Guastello, M. Koopmans, & D. Pincus. (Eds.), *Chaos and complexity in psychology: Theory of nonlinear dynamical systems* (pp. 41–72). New York, NY: Cambridge University Press.

Sulis, W. (2021). Lessons from collective intelligence. In E. Elliott & L. D. Kiel (Eds.), *Complex systems in the social and behavioral sciences: Theory, method and application* (pp. 263–297). Ann Arbor, MI: University of Michigan Press.

Sulsky, L., & Smith, C. (2005). *Work stress*. Belmont, CA: Thompson-Wadsworth.

Surry, J. (1969). *Industrial accident research: A human engineering appraisal*. Toronto, Canada: Labor Safety Council, Ontario Ministry of Labor.

Suruda, A., Whitaker, B., Bloswick, D., Philips, P., & Sasek, R. (2002). Impact of the OSHA trench and excavation standard on fatal injury rates among residential construction workers, 1994-1998. *American Journal of Industrial Medicine, 45*, 210–217.

Swets, J. A., Dawes, R. M., & Monahan, J. (2000). Psychological science can improve diagnostic decisions. *Psychological Science in the Public Interest, 1*, 1–26.

Szalma, J. L. (2009). Individual differences: Incorporating human variation into human factors/ergonomics research and practice. *Theoretical Issues in Ergonomics Science, 10*, 377–380.

Szalma, J. L. (2012). Individual differences in stress, fatigue, and performance. In G. Matthews, P. A. Desmond, C. Neubauer, & P. A. Hancock (Eds.), *The handbook of operator fatigue* (pp. 75–90). Aldershot, UK: Ashgate.

Szalma, J. L., Hancock, P. A., Warm, J. S., Dember, W. N., & Parsons, K. S. (2006). Training for vigilance: Using predictive power to evaluate feedback effectiveness. *Human Factors, 48*, 682–692.

Szalma, J. L., Schmidt, T. N., Teo, G. W. L., & Hancock, P. A. (2014). Vigilance on the move: Video game-based measurement of sustained attention. *Ergonomics, 57*, 1315–1336.

Szalma, J. L., & Taylor, G. S. (2011). Individual differences in response to automation: The five factor model of personality. *Journal of Experimental Psychology: Applied, 17*, 71–96.

Szalma, J. L., & Teo, G. W. L. (2012). Spatial and temporal task characteristics as stress: A test of the dynamic adaptability theory of stress, workload, and performance. *Acta Psychologica, 139*, 471–485.

Szymura, B. (2010). Individual differences in resource allocation policy. In A. Gruszka, G. Matthews, & B. Szymura (Eds.), *Handbook of individual differences in cognition: Attention, memory, and executive control* (pp. 231–246). New York, NY: Springer.

Tabachnick, B. G., & Fidell, L. S. (2007). *Using multivariate statistics* (5th ed.). Boston, MA: Allyn & Bacon.

Taleb, N. (2012). *Antifragile: Things that gain from disorder*. New York, NY: Random House.

Tasto, D. L., Colligan, M. J., Skjel, E. W., & Polly, S. J. (1978). *Health consequences of shift work*. Washington, DC: U.S. Department of Health, Education, and Welfare, and National Institute for Occupational Safety and Health.

Teas, D. C. (1990). *Review of audition literature: Selection of acoustic signals for use in the synthesis of auditory space* (Rep. No. USAFSAM-TR-90-33). Brooks Air Force Base, TX: U.S. Air Force School of Aerospace Medicine.

Tenner, E. (1996). *Why things bite back: Technology and the revenge of unintended consequences.* New York, NY: Alfred A. Knopf.

Thackray, R., & Touchstone, R. M. (1991). Effects of monitoring under high and low taskload on detection of flashing and coloured radar targets. *Ergonomics, 34,* 1065–1081.

Theeuwes, J., van der Burg, E., Olivers, C. N. L., & Bronkhorst, A. (2007). Cross-modal interaction between sensory modalities: Implications for the design of multisensory displays. In A. F. Kramer, D. A. Wiegman, & A. Kirklik (Eds.), *Attention: From theory to practice* (pp. 196–205). New York, NY: Oxford University Press.

Theiler, J., & Eubank, S. (1993). Don't bleach chaotic data. *Chaos, 3,* 771–782.

Thom, R. (1975). *Structural stability and morphegenesis.* New York, NY: Addison-Wesley.

Thomas, J. C., & Gould, J. D. (1975). A psychological study of query by example. *National Computer Conference: AFIPS Conference Proceedings, nv,* 439–445.

Thompson, H. L. (2010). *The stress effect: Why smart leaders make dumb decisions—and what to do about it.* San Francisco, CA: Jossey-Bass.

Thompson, J., & Parasuraman, R. (2012). Attention, biological motion, and action recognition. *Neuroimage, 59,* 4–13.

Thompson, S. J., Fraser, E. J., & Howarth, C. I. (1985). Driver behaviour in the presence of child and adult pedestrians. *Ergonomics, 10,* 1469–1474.

Thomson, D. R., Besnder, D., & Smilek, D. (2015). A resource-control account of sustained attention: Evidence from mind-wandering and vigilance paradigms. *Perspectives on Psychological Science, 10,* 82–96.

Thurstone, L. L. (1927). A law of comparative judgment. *Psychological Review, 33,* 268–278.

Tolman, E. C. (1932). *Purposive behavior in animals and man.* New York, NY: Century.

Torgerson, W. G. (1958). *Theory and methods of scaling.* New York, NY: Wiley.

Torrents, C., Balagué, N., & Hristovski, R. (2016). Interpersonal coordination in competitive and cooperative performance contexts. In P. Passos, K. Davids, & J. Y. Chow (Eds.), *Interpersonal coordination and performance in social systems* (pp. 85–93). New York, NY: Routledge.

Townsend, J. T., & Wenger, M. J. (2004). A theory of interactive parallel processing: New capacity measures and predictions for a response time inequality series. *Psychological Review, 30,* 708–719.

Trainni, V. (2008). *Evolutionary swarm robotics: Evolving self-organizing behaviors in groups of autonomous robots.* Berlin, Germany: Springer.

Travers, J., & Milgram, S. (1969). An experimental study of the small world problem. *Sociometry, 32,* 425–443.

Travis, C. B., McLean, B. E., & Ribar, C. (Eds.). (1989). *Environmental toxins: Psychological, behavioral, and sociocultural aspects, 1973-1989. Bibliographies in Psychology, No. 5.* Washington, DC: American Psychological Association.

Tremoulet, P. D., Seacrist, T., McIntosh, C. W., Loeb, H., DiPietro, A., & Tushak, S. (2020). Transporting children in autonomous vehicles: An exploratory study. *Human Factors, 62,* 278–287.

Trianni, V. (2008). *Evolutionary swarm robotics: Evolving self-organizing behaviors in groups of autonomous robots*. Berlin: Springer.

Trist, E. L., Susman, G. I., & Brown, G. R. (1977). An experiment in autonomous working in an American underground coal mine. *Human Relations, 30*, 201–236.

Trofimova, I. (2002). Sociability, diversity and compatibility in developing systems: EVS approach. In J. Nation, I. Trofimova, J. Rand, & W. Sulis (Eds.), *Formal descriptions of developing systems* (pp. 231–248). Dordrecht, the Netherlands: Kluwer.

Tucker, P., & Folkard, S. (2012). Work scheduling. In G. Matthews, P. A. Desmond, C. Neubauer, & P. A. Hancock (Eds.), *The handbook of operator fatigue* (pp. 457–468). Burlington, VT: Ashgate.

Tufano, D. R. (1997). Automotive HUDs: The overlooked safety issues. *Human Factors, 39*, 303–311.

Tufte, E. R. (1983). *The visual display of quantitative information*. Cheshire, CT: Graphics Press.

Turing, A. M. (1963). Computing machinery and intelligence. In E. A. Feigenbaum & J. Feldman (Eds.), *Computers and thought* (pp. 11–35). New York, NY: McGraw-Hill.

Turkle, S. (1984). *The second self: Computers and the human spirit*. New York, NY: Wiley.

Turpin, J. A. (1981). Effects of blur and noise on digital imagery interpretability. *Proceedings of the Human Factors Society, 25*, 286–289.

Turvey, M. T. (1990). Coordination. *American Psychologist, 45*, 938–953.

Tušl, M., Rainieri, G., Fraboni, F., De Angelis, M., Depolo, M., Pietrantoni, L., & Pingitore, A. (2020). Helicopter pilots' tasks, subjective workload, and the role of external visual cues during shipboard landing. *Journal of Cognitive Engineering and Decision Making, 14*, 242–257.

Ulam, S., & Schrandt, R. (1986). Some elementary attempts at numerical modeling of problems concerning rates of evolutionary processes. *Physica D, 22*, 4–12.

U.S. Department of Defense. (1991). *Anthropometry of U.S. military personnel (metric)*. Washington, DC: Author. Technical report no. DOD-HDBK-743A. Retrieved from www.everyspec.com/DoD/DoD-HDBK/download.php?spec=DOD=HDBK-743A.016856.pdf

U.S. Department of Defense. (2002). *DARPA's total information awareness program homepage*. Retrieved from www.darpa.mil/ao/TIASystems.htm

U.S. Department of Defense. (2012). *Department of Defense design criteria standard: Human Engineering*. Washington, DC: Author. Retrieved from www.assistdocs.com

U.S. Department of Labor. (1977). *Dictionary of occupational titles*. Washington, DC: U.S. Government Printing Office.

U.S. Department of Labor. (1988). *The role of labor-management committees in safe-guarding worker safety and health*. Washington, DC: Author.

U.S. Department of Labor. (2004). *Occupational Safety and Health Administration Standards 1910.36-38, 1926.200 (a, b, c)*. Retrieved from www.osha.gov/pls/oshaweb

U.S. Department of Labor. (2021). O*Net resource center. Retrieved from www.onetcenter.org/overview.html

U.S. Department of Transportation. (2003). *Manual on uniform traffic control devices for streets and highways.* Retrieved from http://mutcd/fhwa.dot.gov

U.S. Department of Transportation, Pipeline and Hazardous Materials Safety Administration. (2012). *Control room management.* Retrieved from http://primis.phmsa.dot.gov/crm/index.htm

Vaillancourt, D. E., & Newell, K. M. (2000). The dynamics of resting and postural tremor in Parkinson's disease. *Clinical Neurophysiology, 111,* 2046–2056.

Vale, C. D., Keller, L. S., & Bentz, V. J. (1986). Development and validation of computerized interpretation system for personnel tests. *Personnel Psychology, 39,* 525–542.

Valenza, G., Lanata, A., Scilingo, E. P., & De Rossi, D. (2010). Towards a smart glove: Arousal recognition based on textile electrodermal response. *Proceedings of the 32nd Annual International Conference of the IEEE EMBS, 32,* 3598–3601.

Van Dam, N. T., Earleywine, M., & Alterriba, J. (2012). Anxiety attenuates awareness of emotional faces during rapid serial visual presentation. *Emotion, 12,* 796–806.

Van den Oever, F., & Schraagen, J. M. (2021). Team communication patterns in critical situations. *Journal of Cognitive Engineering and Decision Making, 15,* 28–51.

van der Kleij, R., & te Brake, G. (2010). Map-mediated dialogues: Effects of map orientation differences and shared reference points on map location-finding speed and accuracy. *Human Factors, 52,* 526–536.

van der Velde, M., & Class, M. D. (1995). The relationship of role conflict and ambiguity to organizational culture. In S. L. Sauter & L. R. Murphy (Eds.), *Organizational risk factors for stress* (pp. 53–60). Washington, DC: American Psychological Association.

Van de Ven, A. H., & Hargrave, T. J. (2004). Social and institutional change: A literature review and synthesis. In M. S. Poole & A. H. Van de Ven (Eds.), *Handbook of organizational change and innovation* (pp. 259–303). New York, NY: Oxford University Press.

Van de Ven, A. H., & Poole, M. S. (1995). Explaining development and change in organizations. *Academy of Management Review, 20,* 510–540.

van Heerden, I. I. I. (2007). The failure of the New Orleans levee system following hurricane Katrina and the pathway forward. *Public Administration Review, 67,* 24–35.

Van Nes, F. (1992). Design and evaluation of applications with speech interfaces: Experimental results and practical guidelines. In M. Galer, S. Harker, & J. Ziegler (Eds.), *Methods and tools in user-centered design for information technology* (pp. 281–298). Amsterdam, the Netherlands: North-Holland.

Van Orden, G. C., Holden, J. G., & Turvey, M. T. (2003). Self-organization of cognitive performance. *Journal of Experimental Psychology: General, 132,* 331–350.

Verbaten, M. N., Huyben, M. A., & Kemner, C. (1997). Processing capacity and the frontal P3. *International Journal of Psychophysiology, 25,* 237–248.

Verhaegen, P. (1993). Absenteeism, accidents and risk-taking: A review ten years later. *Safety Science, 16,* 359–366.

Vicente, K. J. (1999). *Cognitive work analysis: Toward safe, productive, and healthy computer-based work.* Boca Raton, FL: CRC Press.

Vicente, K. J., Mumaw, R. J., & Roth, E. M. (2004). Operator monitoring in a complex dynamic work environment: A qualitative cognitive model based on field observation. *Theoretical Issues in Ergonomics Science, 5*, 359–384.

Vickers, D., & Lee, M. D. (1998). Dynamic models of simple judgments: I. Properties of a self-regulating accumulator module. *Nonlinear Dynamics, Psychology, and Life Sciences, 2*, 169–194.

Vickers, D., & Lee, M. D. (2000). Dynamic models of simple judgments: II. Properties of a selforganizing PAGAN (parallel, adaptive, generalized accumulator network) model for multichoice tasks. *Nonlinear Dynamics, Psychology, and Life Sciences, 4*, 1–31.

Vidulich, M. A., & Tsang, P. S. (2007). Methodological and theoretical concerns in multitask performance: A critique of Boles, Bursk, Phillips, and Perdelwitz. *Human Factors, 49*, 46–49.

Vicente, K. J. (1999). *Cognitive work analysis: Toward safe, productive, and healthy computer-based work*. Boca Raton, FL: CRC Press.

Voiers, W. D., Sharpley, A. D., & Panzer, I. L. (2002). Evaluating the effect of noise on voice communication systems. In G. M. Davis (Ed.), *Noise reduction in speech applications* (pp. 125–152). Boca Raton, FL: CRC Press.

von Helmholtz, H. L. F. (1962). *Treatise on physiological optics* (3rd ed., J. P. C. Southall, Ed. and Trans.). New York, NY: Dover (Original work published 1909).

von Neumann, J., & Morgenstern, O. (1953). *Theory of games and economic behavior*. Princeton, NJ: Princeton University Press.

Vora, P. W. (2003). Designing friction-free experience for e-commerce sites. In J. Ratner (Ed.), *Human factors and web development* (2nd ed., pp. 225–240). Mahwah, NJ: Lawrence Erlbaum Associates.

VRex, Inc. (2004). *About VRex Inc.* Retrieved from http://vrex.com/about/index/html

Vroom, V. H. (1964). *The motivation to work*. New York, NY: Wiley.

Vroom, V. R., & Jago, A. G. (1988). *The new leadership: Managing participation in organizations*. Englewood Cliffs, NJ: Prentice Hall.

Vu, K-M. L., & Chiappe, D. (2015). Situation awareness in human systems integration. In D. A. Boehm-Davis, F. T. Durso, & J. D. Lee (Eds.), *APA handbook of human systems integration* (pp. 293–308). Washington, DC: American Psychological Association.

Vyal, K., Cornwell, B., Arkin, N., & Grillon, C. (2012). Describing the interplay between anxiety and cognition: From impaired performance under low cognitive load to reduced anxiety under high load. *Psychophysiology, 49*, 842–852.

Wade, C., Redfern, M. S., Andres, R. O., & Breloff, S. P. (2010). Joint kinetics and muscle activity while walking on ballast. *Human Factors, 52*, 560–573.

Waldrop, M. M. (1992). *Complexity: The emerging science at the edge of chaos*. New York, NY: Simon & Schuster.

Walker, B. N., & Lindsay, J. (2006). Navigation performance with a virtual auditory display: Effects of beacon sound, capture radius, and practice. *Human Factors, 48*, 265–278.

Walker, B. N., Lindsay, J., Nance, A., Nakano, Y., Dalladino, D. K., Tingler, T., & Jeon, M. (2013). Spearcons (speech-based earcons) improve navigation performance in advanced auditory menus. *Human Factors, 55*, 157–182.

Walker, C. C., & Dooley, K. J. (1999). The stability of self-organized rule-following work teams. *Computational and Mathematical Organization Theory, 5*, 5–30.

Walker, G. H., Stanton, N. A., Jenkins, D. P., Salmon, P. M., & Rafferty, L. (2010). From the 6 Ps of planning to the 4 Ds of digitization: Difficulties, dilemmas, and defective decision making. *International Journal of Human-Computer Interaction, 26*, 173–188.

Walker, J. L., & Cavenar, J. (1982). Vietnam veterans: Their problems continue. *Journal of Nervous Mental Disorder, 170*, 174–179.

Wallace, J. C., Popp, E., & Mondore, S. (2006). Safety climate as a mediator between foundation climates and occupational accidents: A group-level investigation. *Journal of Applied Psychology, 91*, 681–688.

Waller, L. A., & Gotway, C. A. (2004). *Applied spatial statistics for public health data*. Hoboken, NJ: Wiley.

Walliser, J. C., deVisser, E. J., Wiese, E., & Shaw, T. H. (2019). Team structure and team building improve human-machine teaming with autonomous agents. *Journal of Cognitive Engineering and Decision Making, 13*, 258–204.

Walther, J. B. (1996). Computer-mediated communication: Impersonal, interpersonal, and hyperpersonal interaction. *Communication Research, 23*, 3–45.

Walther, J. B., Anderson, J. F., & Park, D. W. (1994). Interpersonal effects in computer-mediated interactions: A meta-analysis of social and antisocial communication. *Communication Research, 21*, 460–487.

Wandell, B. A. (1982). Measurement of small color differences. *Psychological Bulletin, 89*, 281–302.

Wang, J.-S., Knipling, R. R., & Goodman, N. J. (1996, October). *The role of driver inattention increases, new statistics from the 1995 crashworthiness data system*. Paper presented at the 40th American Association of Automobile Manufacturers conference, Vancouver, Canada.

Ward, L. M. (2002). *Dynamical cognitive science*. Cambridge, MA: MIT Press.

Ward, L. M. (2004). Oscillations and synchrony in cognition. In V. K. Jirsa & J. A. S. Kelso (Eds.), *Coordination dynamics: Issues and trends* (pp. 217–242). New York, NY: Springer.

Ward, L. M., & West, R. L. (1998). Modeling human chaotic behavior: Nonlinear forecasting analysis of logistic iteration. *Nonlinear Dynamics, Psychology, and Life Sciences, 2*, 261–282.

Ward, S. J. (2009). User research by designers. In W. Karwowski, M. M. Soares, & N. A. Stanton (Eds.), *Human factors and ergonomics in consumer product design* (pp. 127–141). Boca Raton, FL: CRC Press.

Warm, J. S., Finomore, V. S., Vidulich, M. A., & Funke, M. E. (2015). Vigilance: A perceptual challenge. In R. R. Hoffman, P. A. Hancock, M. W. Scerbo, J. L. Szalma, & R. Parasuraman (Eds.), *The Cambridge handbook of applied perception research* (pp. 241–283). New York, NY: Cambridge University Press.

Warm, J. S., & Jerison, H. J. (1984). The psychophysics of vigilance. In J. S. Warm (Ed.), *Sustained attention in human performance* (pp. 15–57). New York, NY: Wiley.

Ware, K. (2011, August). *Observing the emergence of and the control of chaos in the human nervous system*. Paper presented to the 21st Annual International Conference of the Society for Chaos Theory in Psychology & Life Sciences, Orange, CA.

Wattenberger, B. L. (1980). A look at the VDT issue. *Proceedings of the International Symposium on Human Factors in Telecommunications, 10,* 173–180.

Watts, D. J., & Strogatz, S. H. (1998). Collective dynamics for "small-world" networks. *Nature, 393,* 440–442.

Wears, R. L. (2010). Exploring the dynamics of resilience. *Proceedings of the Human Factors and Ergonomics Society, 54,* 394–398.

Weber, E. H. (1978). *The sense of touch* (D. J. Murray, Trans.). New York, NY: Academic Press (Original work published 1846).

Weick, K. E. (2005). Managing the unexpected: Complexity as distributed sensemaking. In R. R. McDaniel, Jr., & D. J. Driebe (Eds.). *Uncertainty and surprise in complex systems* (pp. 51–65). New York, NY: Springer-Verlag.

Weick, K. E., & Gilfillan, D. P. (1971). Fate of arbitrary traditions in a laboratory microculture. *Journal of Personality and Social Psychology, 17,* 179–191.

Weinland, J. P. (1927). Variability of performance in the curve of work. *Archives of Psychology, 30.*

Welford, A. T. (1980). Relationships between reaction time, fatigue, stress, age, and sex. In A. T. Welford (Ed.), *Reaction time* (pp. 321–354). New York, NY: Academic Press.

Wellman, B. (1997). An electronic group is virtually a social network: In S. Keisler (Ed.), *Culture of the Internet* (pp. 179–205). Mahwah, NJ: Lawrence Erlbaum Associates.

Wellman, B., & Berkowitz, S. D. (Eds.). (1988). *Social structures: A network approach.* New York, NY: Cambridge University Press.

Wenzel, E. M. (1994). Spatial sound and sonification. In G. Kramer (Ed.), *Auditory display: Sonification, audification, and auditory interfaces* (pp. 127–150). New York, NY: Oxford University Press.

Wenzel, E. M., Wightman, F. L., & Foster, S. H. (1988). Development of a three-dimensional auditory display system. *SIGCHI Bulletin, 20*(2), 52–56.

Wesler, M. (2004). *Tactile displays.* Technical Report, Wright State University College of Engineering and Computer Science). Retrieved from www.cs.wright.edu/~jgalli/Tactile.ppt

Wesley, D., & Dau, L. A. (2017). Complacency and automation bias in the Enbridge pipeline disaster. *Ergonomics in Design, 25*(1), 17–22.

Whitmore, A. M., Leveton, L. B., Barger, L., Brainard, G., Dinges, D. F., Klerman, E., & Shea, C. (2009). Risk of performance errors due to sleep loss, circadian desynchronization, fatigue, and work overload. In J. C. McPhee & John B. Charles (Eds.), *Human health and performance risks of space exploration missions* (pp. 85–118). Houston, TX: National Aeronautics and Space Administration.

Williams, D. J., & Noyes, J. M. (2009). Reducing the risk to consumers: Implications for designing safe consumer products. In W. Karwowski, M. M. Soares, & N. A. Stanton (Eds.), *Human factors and ergonomics in consumer product design* (pp. 3–22). Boca Raton, FL: CRC Press.

West, B. J. (2006). *Where medicine went wrong: Rediscovering the path to complexity.* Singapore: World Scientific.

West, B. J., & Scafetta, N. (2003). Nonlinear model of human gait. *Physics Review E, 67,* 051917-1.

West, B. J., & Scafetta, N. (2010). *Disrupted networks: From physics to climate change.* Singapore: World Scientific.

Whitaker, L. A. (1998). Human navigation. In C. Forsythe, E. Grose, & J. Ratner (Eds.), *Human factors and web development* (pp. 63–71). Mahwah, NJ: Lawrence Erlbaum Associates.

White, R., Engelen, G., & Uljee, I. (1997). The use of constrained cellular automata for high-resolution modelling of urban and land-use development. *Environment and Planning B: Planning and Design, 24,* 323–343.

Whittle, P. (2010). *Neural nets and chaotic carriers* (2nd ed.). London: Imperial College Press.

Wickens, C. D. (2002). Multiple resources and performance prediction. *Theoretical Issues in Ergonomics Science, 3,* 159–177.

Wickens, C. D. (2007). How many resources and how to identify them? Commentary on Boles et al. and Vidulich and Tsang. *Human Factors, 49,* 53–56.

Wickens, C. D. (2008a). Multiple resources and mental workload. *Human Factors, 50,* 449–455.

Wickens, C. D. (2008b). Situation awareness: Review of Mica Endsley's 1995 articles on situation awareness theory and measurement. *Human Factors, 50,* 397–403.

Wickens, C. D., & Alexander, A. L. (2009). Attentional tunneling and task management in synthetic vision displays. *International Journal of Aviation Psychology, 19,* 182–199.

Wickizer, T. M., Kopjar, B., Franklin, G., & Joesch, J. (2004). Do drug-free workplace programs precent occupational injuries? Evidence from Washington State. *Health Services Research, 39,* 91–110.

Wiederhold, B. K., & Wiederhold, M. D. (2003). Three-year follow-up for virtual reality exposure for fear of flying. *Cyberpsychology and Behavior, 6,* 441–445.

Wiener, E. L., & Curry, R. E. (1980). Flight-deck automation: Promises and problems. *Ergonomics, 23,* 995–1011.

Wiese, E. E., & Lee, J. D. (2007). Attention grounding: A new approach to in-vehicle information system implementation. *Theoretical Issues in Ergonomics Science, 8,* 255–276.

Wightman, F. L., & Kistler, D. J. (1989). Headphone simulation of free-field listening: II. Psychophysical validation. *Journal of the Acoustical Society of America, 85,* 868–878.

Wijnants, M. L., Bosman, A. M. T., Hasselman, F., Cox, R. A., & Van Orden, G. C. (2009). 1/f scaling in movement time changes with practice in precision aiming. *Nonlinear Dynamics, Psychology, and Life Sciences, 13,* 79–98.

Wilde, G. J. S. (1988). Risk homeostasis theory and traffic accidents: Propositions, deductions and discussion of dissension in recent reactions. *Ergonomics, 31,* 441–468.

Wilkins, I., & Dragos, B. (2013). Destructive destruction? An ecological study of high frequency trading. Retrieved from www.metamute.org/

Williams, A. M., Vickers, J., & Rodrigues, S. (2002). The effects of anxiety on visual search movement kinematics, and performance in table tennis: A test of Eysenck and Calvo's processing efficiency model. *Journal of Sport and Exercise Psychology, 24,* 428–455.

Williams, S. M. (1994). Perceptual principles in sound grouping. In G. Kramer (Ed.), *Auditory display: Sonification, audification, and auditory interfaces* (pp. 95–126). New York, NY: Oxford University Press.

Williamson, A. (2012). Countermeasures for driver fatigue. In G. Matthews, P. A. Desmond, C. Neubauer, & P. A. Hancock (Eds.), *The handbook of operator fatigue* (pp. 441–455). Burlington, VT: Ashgate.

Wilpert, B. (2001, September). *Summary remarks to the congress.* Paper presented at the World Congress on Safety in Modern Technical Systems, Saarbrücken, Germany.

Wilson, G. T., & Davison, G. C. (1971). Processes of fear reduction in systematic desensitization. *Psychological Bulletin, 76*, 1–14.

Winters, R. M., Tomlinson, B. J., Walker, B. N., & Moore, E. B. (2019). Sonic interaction design for science education. *Ergonomics in Design, 27*(1), 5–10.

Wisneski, K. J., & Johnson, M. J. (2007). Quantifying kinematics of purposeful movements to real, imagined, or absent functional objects: Implications for modeling trajectories for robot-mediated ADL Tasks. *Journal for Neuroengineering and Rehabilitation, 7.* doi: 10.1186/1743-0003-4-7

Witkin, H. A., Oltman, P. K., Raskin, E., & Karp, S. A. (1971). *A manual for the Embedded Figures Test.* Palo Alto, CA: Consulting Psychologists Press.

Wixted, J. T., & Mickes, L. (2012). The field of eyewitness memory should abandon probative value and embrace receiver operating characteristic analysis. *Perspectives on Psychological Science, 7*, 275–278.

Wockutch, R. E., & VanSandt, C. V. (2000). OHS management in the United States and Japan: The DuPont and Toyota models. In K. Frick, P. Jensen, M. Quinlan, & T. Wilthagen (Eds.), *Systematic occupational health and safety management: Prospectives on an international development* (pp. 367–390). Oxford, UK: Pergamon Press.

Wojtezak-Jaroszawa, J. (1978). *Physiological and psychological aspects of night and shift work.* Washington, DC: U.S. Department of Health, Education, and Welfare, and National Institute for Occupational Safety and Health.

Wolfram, S. (Ed.). (1986). *Theory and applications of cellular automata.* Singapore: World Scientific.

Wolfram, S. (2002). *A new kind of science.* Champaign, IL: Wolfram Media.

Woodhill, D., Crutchfield, N., & James, F. S. (1987, September). Successful loss control programs require management's time, money. *Occupational Health and Safety, X, nv,* 61–65.

Woods, D. D., & Wreathall, J. (2008). Stress-strain plots as a basis for assessing system resilience. In E. Hollnagel, C. P. Nemeth, & S. W. A. Dekker (Eds.), *Resilience engineering: Remaining sensitive to the possibility of failure* (pp. 143–158). Aldershot, UK: Ashgate.

Woodworth, R. S. (1899). The accuracy of voluntary movement. *Psychological Review Monograph, 3*, 1–14.

Wu, C., Cha, J., Sulek, J., Zhou, T., Sundaram, C. P., Wachs, J., & Yu, D. (2020). Eye-tracking metric predict perceived workload in robotic surgery skills training. *Human Factors, 62*, 1365–1386.

Xiao, Y., Patey, R., & Mackenzie, C. F. (1995). *A study of team coordination and its breakdowns in a real-life stressful environment examined through video analysis* (Tech. Rep.). Baltimore: University of Maryland, School of Medicine, Department of Anesthesiology.

Ye, L., Cardwell, W., & Mark, L. S. (2009). Perceiving multiple affordances for objects. *Ecological Psychology, 21*, 185–217.

Yeh, M., Merlo, J. L., Wickens, C. D., & Brandenburg, D. L. (2003). Head up versus head down: The cost of imprecision, unreliability, and visual clutter on cur effectiveness for display signaling. *Human Factors, 45,* 390–407.

Yerkes, R. M., & Dodson, J. D. (1908). The relationship of strength of stimulus to rapidity of habit formation. *Journal of Comparative Neurology and Psychology, 18,* 459–482.

Yin, S., Wickens, C. D., Helander, M., & Laberge, J. C. (2015). Predictive displays for a process-control schematic interface. *Human Factors, 57,* 110–124.

Young, M. S., & Stanton, N. A. (2002). Attention and automation: New perspectives on mental underload and performance. *Theoretical Issues in Ergonomics Science, 3,* 178–194.

Yousef, E. A., Sutcliffe, K. M., McDonald, K. M., & Newman-Toker, D. E. (2022). Crossing academic boundaries for diagnostic safety: 10 Complex challenges and potential solutions from clinical perspectives and high-reliability organizing principles. *Human Factors, 64,* 6–20.

Yung, M., Manji, R., & Wells, R. P. (2017). Exploring the relationship of task performance and physical and cognitive fatigue during a day-long precision task. *Human Factors, 59,* 1029–1047.

Zadeh, L. A. (1965). Fuzzy sets. *Information and Control, 8,* 338–353.

Zadeh, L. A. (1981). Possibility theory and soft data analysis. In L. Cobb & R. M. Thrall (Eds.), *Mathematical frontiers of the social and policy sciences* (pp. 69–129). Boulder, CO: Westview Press.

Zagare, F. C. (1984). *Game theory: Concepts and applications* (Quantitative Applications in the Social Sciences Paper Series, No. 41). Newbury Park, CA: Sage.

Zaidel, D. M., & Noy, Y. I. (1997). Automatic versus interactive vehicle navigation aids. In Y. I. Noy (Ed.), *Ergonomics and safety of intelligent driver interfaces: Human factors in transportation* (pp. 287–307). Mahwah, NJ: Lawrence Erlbaum Associates.

Zander, A. (1994). *Making groups effective* (2nd ed.). San Francisco, CA: Jossey-Bass.

Zeeman, E. C. (1977). *Catastrophe theory: Selected papers, 1972-1977.* Reading, MA: Addison-Wesley.

Zeggelink, E. P. H., Stokman, F. N., & van de Bunt, G. G. (1997). The emergence of groups in the evolution of friendship networks. In P. Doreian & F. B. Stokman (Eds.), *Emotion of social networks* (pp. 45–71). Amsterdam, the Netherlands: Gordon & Breach.

Zenger, B., & Fahle, M. (1997). Missed targets are more frequent than false alarms: A model for error rates in visual search. *Journal of Experimental Psychology: Human Perception and Performance, 23,* 1783–1791.

Zhang, X., Khalili, M. M., & Liu, M. (2020). Long-term impacts of fair machine learning. *Ergonomics in Design, 28*(3), 7–11.

Zinn, C. (2001). Hospital infection control as part of an integrated safety concept. In J. Althoff (Ed.), *Safety of modern technical systems* (pp. 643–644). Saarbrücken, Germany: TÜV-Saarland Foundation.

Zohar, D. (1980). Safety climate in industrial organizations: Theoretical and applied implications. *Journal of Applied Psychology, 65,* 96–102.

Zohar, D. (2003). Safety climate: Conceptual and measurement issues. In J. C. Quick & L. E. Tetrick (Eds.), *Handbook of occupational health psychology* (pp. 123–142). Washington, DC: American Psychological Association.

Zohar, D., & Luria, G. (2005). A multilevel model of safety climate: Cross-level relationships between organization and group-level climate. *Journal of Applied Psychology, 90,* 616–628.

Zunjic, A. (2009). Consumer product risk assessment. In W. Karwowski, M. M. Soares, & N. A. Stanton (Eds.), *Human factors and ergonomics in consumer product design* (pp. 23–32). Boca Raton, FL: CRC Press.

Zwerling, C., Daltroy, L. H., Fine, L. J., Johnston, J. J., Melius, J., & Silverstein, B. A. (1997). Design and conduct of occupational injury intervention studies: A review of evaluation strategies. *American Journal of Industrial Medicine, 32,* 164–179.

Index

Note: Page numbers followed by f and t indicate figures and tables, respectively.

Absolute threshold, 65–68, 72,
77, 78, 87
Accident, 7, 10–13, 20, 21, 39, 49, 133,
225, 233, 276, 311, 343–394;
see also Emergency response;
Resilience; Safety –al operation,
238, 248, 269, 274
death and injury trends, 343–346
hazard perception, 388–390
prevention programs, 375–388
proneness, 39, 347–348, 359,
379
risk models, 346–348
Swiss cheese model, 371, 372f, 543
ADA, *see* Americans with Disabilities
Act
Adaptable systems, 20
Adaptive systems, 7, 20, 21, 373, 472;
see also Complex Adaptive
Systems
Aesthetics, 101, 130, 421, 437, 556,
559, 560, 562, 586
Agent-based models, 58, 59, 61–63
fitness, 63, 493
AI, *see* Artificial intelligence (AI)
Aiming, 235, 263, 440f
Air traffic controllers, 80, 125, 185,
365, 502, 503, 540
Alcohol intoxication, 193, 314, 348,
350, 376t, 377, 378, 383, 566,
567
Algebra flexibility, 327–329t, 334t
Algorithmic systems, 62, 213, 217, 380,
381, 429, 447, 459–460, 466,
473, 474, 480, 482, 553;
see also Genetic algorithms

Allocation of function, 15, 20, 26, 33,
533, 582
Alpha waves, 197, 198f
Altitude, 24, 112, 297, 298, 300, 301
American National Standards Institute
(ANSI), 131, 133, 246
Americans with Disabilities Act (ADA),
383
Analog displays, 109, 253
vs. digital displays, 119, 120f, 135
Anchoring effect, 166
Animation, 356, 405, 423, 424
ANSI, *see* American National Standards
Institute
Anthropometry, *see also* Workspace
design
body coordination and equilibrium,
280–281
body measurements, 271–275
faces, 275
flexibility (psychomotor), 280, 298
handtools, 6, 216t, 235, 285, 289
lifting, 280
lean body mass, 281, 282
outer space, 583, 584
seating, 285, 288, 289, 293, 547,
588f
stamina, 281; *see also* Fatigue
strength (human physical), 6, 278,
287, 316, 331, 442, 583
Antifragility, 324
Anxiety, 301, 313, 314, 324, 325, 327,
329t, 333, 337, 353, 357, 358,
361, 362t, 367–369, 377, 383,
445, 446, 464
Apollo capsule, 371, 576, 577f, 583

Apparent motion, 104
Apple Corporation, 410
Approximate entropy, 37
Arm-hand steadiness, 235
ARPANET, 425
Artificial intelligence (AI), 7, 21, 275,
    451, 475–486
  algorithmic systems, *see* Algorithmic
    systems
  architecture, 419, 432, 433, 458–469,
    471, 472, 476, 480, 504
  chaining strategies, 461–463, 465
  classification structures, 80, 420,
    447, 461
  expert systems, 7, 32, 60, 455, 459,
    467, 470, 479, 480; *see also*
    Expert *vs.* novices
  frame-based systems, 464–466
  Gödel, *see* Gödel's incompleteness
    theorem
  recursive systems, 466–467
  rule-based systems, 219, 460, 462,
    465, 467, 509
  Simon, *see* Bounded rationality;
    Satisficing
  Turing test, 455, 456
  Von Neumann, *see* Artificial life
Artificial life
  agent-based models, 7, 51, 59, 451,
    452, 471, 541
  autonomous agents, 53, 61, 62, 471,
    474, 475, 527, 531, 541
  cellular automata, 58–61, 471
  genetic algorithms, 58, 59, 63, 64,
    471, 493
  neural networks, 57, 64, 85, 204,
    207, 471–473; *see also* Machine
    learning
  simulations, *see* Simulations
  societies, 61
Ashby's law, 265, 490, 491
Attention, 171, 172
Attentional blink, 189–190
Attention tunneling, 172, 188, 208, 509
Attitude(s), 68, 70–71, 359
Attractor(s)
  chaotic, 41, 44, 46; *see also* Chaos
  fixed-point, 40, 41
  oscillator, 40
Auditory cortex, 137, 145
Auditory displays, *see also* Gestalt laws;
    Hearing

confusability, 145–148
  desensitization, 148, 446
  localization, 146–150
  nonverbal, 142–144
  speech displays, 144, 146, 150, 151,
    153–155
  speech spectrograms, 154f
  3-D auditory displays, 145, 149–150,
    438
Auditory icons, 123, 148, 150–153
  earcons, 151, 152, 422
  spearcons, 152, 162, 422, 423
Augmented cognition, 200, 226, 312
Augmented reality, 445; *see also* Virtual
    reality
  extended reality, 445
Automation, 1, 15–17, 19, 20, 24, 26,
    81, 124, 173, 185, 210, 213,
    258, 312, 365, 394, 445, 456,
    491, 498, 503, 528, 546;
    see also Human-autonomy teams
  trust in, 5, 7, 13, 21–23, 533
  vehicles, 502, 574, 575
Automatization (of cognitive processes),
    194, 201, 318
  controlled processes, 202, 203
  telegraph operation, 201, 202f
Autonomic nervous system, 44, 196,
    198, 199f, 297, 537–538
Autonomous agents, 53, 61, 62, 471,
    474–476, 541

Back door, to program, 455, 542
Bandwagon, 31, 512, 520
Bang-bang theory, 255
Barnum effect, 485–486
Behavior modification programs, 351,
    375, 376t, 381, 386
Benchmark jobs, 217, 300
Beta testing, 29
Beta waves, 197, 537–538
Bifurcation, 40–42
  logistic map, 41–43f, 86, 182, 212,
    268, 513, 514
Binary code, 400, 452
Binocular depth cues, 101, 117, 142
Biohazard safety programs, 542
Blind spot, 92
Blocking effect, 318, 507
Blood-alcohol concentration, *see*
    Alcohol intoxication
Blue-blindness, 96

Body controllers, 442
Body coordination and equilibrium, 280
Bounded rationality, 185, 458
Brain lateralization theory, 124
Brewster stereoscope, *see* Stereoscope
Brightness, 67, 78, 83, 86, 87, 94–97, 421, 433
Buckling stress
  cognitive demands, 50, 314, 322f, 339
  physical demands, 320–322
Building and facility complexes, 239, 241, 336, 505, 548, 553–554
Butterfly effects, 44, 543–544

Calibration, 117, 235, 242f, 247, 372, 373, 380
Candelus, 133
Capture radius, 564
Card reader, 400
Carpal tunnel syndrome, 6, 285, 289, 290, 345, 398
CAS, *see* Complex adaptive system (CAS)
Catastrophes, 42, 48–50, 57, 204, 268, 312, 322, 330, 331, 337, 340, 355f, 368, 496
  cusp model, 49, 50, 156, 232f, 233, 268, 319–322f, 324–326, 328–331, 33, 334t, 336, 339f, 341, 342, 359–262t, 367, 369, 370, 374, 388, 494, 495f, 535
CBTI (computer-based test interpretation), 459, 481, 484–486
Cellular automata, 58–61, 471
Centers for Disease Control and Prevention, 10, 344, 392, 561
Centrality, 55, 56, 432
Chaining strategies, 461–463, 465
Channels (and stages)
  limited capacity theory, 186–187
  variable capacity theory, 188
Chaos (–otic), 41–47, 52, 60, 64, 85–87, 181t, 182f, 199, 201, 207–209, 213, 231, 256, 259, 265–268, 312, 319, 329t, 334t, 364, 391, 393, 477, 489, 491, 492, 499, 500, 504, 513–515, 518, 521, 525, 526, 531, 540, 544

controllers, 46, 265–268
Circadian rhythm, 87, 292, 307–309, 317, 342, 585–586
Civil liability, 9, 10–12
Closure, principle of, 99, 101, 145
Clouds, 104, 413
Cochlea, 137, 138
Channels (cognitive), 161, 169, 172, 185, 186, 189, 191, 196, 202, 206, 212, 222–225, 260, 275, 304, 313, 367, 409, 424, 438, 503, 508, 583, 573
  limited capacity theory, 186, 187
  multitasking, 191, 192, 194
  variable capacity theory, 188, 189
Cognitive
  fatigue, *see* Fatigue
  inventory, 219
  load, 193, 250, 301, 336, 362, 365, 434, 482; *see also* Cognitive; Workload
  map, 177, 558, 559, 562
  task analysis, 26, 218
Coherence, 203, 305, 562
Collective intelligence, 8, 58, 64, 425, 505–507
  in E-communication, 507–509
  learning organizations, 516
  network growth, 511–513
  sensemaking and situation awareness, 509–510f
  temporal dynamics, 513–515
Color(s)
  of display, 60, 61, 68, 94, 109, 111f, 113, 115, 119, 125, 132, 134, 192, 248, 249, 293, 420, 422, 427, 433, 437, 563
  display contrast, 113, 134, 420, 433
  of noise, 155, 156
Color vision, 83–88, 91, 93–95
  abnormalities (colorblindness), 95–96, 340, 341
Commission
  devices, 1, 2, 22, 74, 125, 153, 364, 392, 393, 414, 425, 447, 476, 499, 502, 503, 510, 511r, 515, 525, 547, 555, 556, 578, 580, 582
  errors of, 23, 265
  nonverbal, 427, 521, 522
  models, 25, 36f, 53–55, 196, 484, 506, 507

Compatibility, 31, 148, 228–229, 268, 504, 581
Compiler, 415, 453, 456
Complacency effect, 21, 23
Completeness, of displays, 112
Complex adaptive system (CAS), 7, 489–496, 499, 516, 517
Complexity theory, 58
Component analysis, see Factor analysis
Comprehensive ergonomics programs, 257–258
Computer
    –based test interpretation (CBTI), 459, 481, 484–486
    programming, 2, 3, 7, 15–18, 21, 23, 29, 32, 60–64, 84, 119, 123f, 151, 153, 155, 159, 169, 183, 195, 201, 207, 218, 220, 225, 250, 259, 294, 352, 395, 396, 399–402, 405, 407, 409–416, 419, 424, 425, 434, 438, 439, 441, 442, 446, 447, 530, 533, 539, 542, 545, 546, 555f, 583; see also Artificial intelligence
    punch card, 400f, 409, 453
    workstations, see Computer work station; Workspace design
Cones, 91–93
Confined spaces, 277, 298
Confusability, auditory displays and, 145–148
Conscientiousness, 323, 324, 328, 329t, 336
Consistent expectation equilibrium, 208
Contributory negligence, 11
Control(s)
    feedback and, 253, 254, 265, 266; see also Chaotic controllers
    labels, 248
    motor control, 258, 259
    multidimensional, 242–244
    order of controls, 263, 264
    resistance, 248, 249
    shape, 246
    size, 244–246
    space of, 247
    types of, 237–242, 442
    voice, 251–253, 269
Control panels, 15, 107, 113, 120–122, 128, 239, 248–252f

Control parameter, 41, 33, 45, 48, 49, 52, 86, 232, 324, 337, 340, 359, 360, 470, 495, 514, 523
Control precision, 235
Coordination, see Gross body coordination; Group coordination
Coping strategies, 311, 312f, 327, 358
    flexibility, 15, 20, 200, 272, 280, 298, 323, 325, 327–329t, 334, 335, 373, 374, 453, 461, 549
Cost-benefit analysis, 29–31
Coupled systems, 497
COVID-19, 25, 344–346, 393, 505, 508, 565
Critical mass effect, 268, 311, 431, 507, 523, 524
Critical point, 19, 30, 42, 52, 150, 210, 233, 234, 263, 268, 281, 323, 360
Cross-modal attentional blink, 190f
Crowding, as social stressor, 277, 302, 303; see also Overcrowding
Crystallized intelligence, 175
Cybersecurity (or cyberattack), 81, 504, 505, 542

"The Daily Sucker," 433
Darwinian learning, 516
Data, see Artificial life, Expert systems
Data-ink ratio, 121
Data mining, 53, 62, 477, 479, 480
Data storage capacity, 2, 412–415
Death
    heat, 37
    highway, 112
    statistics, 343–346f, 348, 354, 379, 512, 549, 550, 565, 570
Decisions, see also Expectancy theory; Production paradox; Prospect theory
    binary, 176
    dynamic, 206, 207, 303
    future state prediction, see Forecasting
    incomplete information, 179
    nonoptimizing, 179
    planning, 179
    optimizing, 176–179
    troubleshooting, see Troubleshooting
Defensible space theory, 557, 558
Degree, centrality, 55

Degrees of freedom (in cognitive task
    organization) 188, 194, 199,
    204–206, 232, 234, 263, 293,
    324, 332, 335, 407, 523
Depth perception, 92, 101–103, 106,
    114, 142
  binocular disparity, 101
  monocular cues, 102, 103
Desensitization
  in auditory displays, 148
  systematic, 446
Desktop computer, 2, 31, 143, 162,
    191, 290, 295, 399f, 402,
    403, 409–411, 428; see also
    Keyboard
Determinism, 156, 534
  interactive, 506
  stochastic, 507
Deterministic process, 36, 38, 42, 155,
    156, 180, 181, 199, 208, 213,
    214, 263, 267, 305, 472
Deviations (in accident investigation),
    347, 361, 363, 366, 369, 493
Diathesis stress, 337, 340–342
Dictionary of Occupational Titles, 215,
    216t
Difference threshold, 65, 66, 72, 84
Digital displays, 119
Digital eye fatigue, 119
Digital video disks (DVDs), 413–415,
    573
Directory assistance operators, 155, 291
Direct perception, 103, 503
Discrimination index, 72, 76
Displays, see Auditory displays; Tactile
    displays; Visual displays
Distinguishability of display, 109–111f,
    122
Distracted driving, 20, 193, 473, 570,
    573–575
Distributed cognition, 506
Divergent thinking, 175, 183
DOD, see U.S. Department of Defense
    (DOD)
Domino models, 351, 352, 354, 393
Donders' response time, 227, 228. 235
Dow-Jones index, 475
Driver(s), see also Distracted driving;
    Traffic and roadway
  age of, 565, 566
  response times, 570
  in synchronization, 536

Drowning, 550
Dual-task methodology, 189, 191
Dwell time, 200, 246
Dynamic
  decisions, see Decisions
  flexibility, 280
  isometric strength, 280
  systems, see Chaotic controller;
    Complex adaptive system
Dynamometers, 278
Dysadaptation syndrome, 581
Dysregulation, 309–310

Eagle Moon Lander, 578f
EAP, see Employee assistance program
Earcons, see Auditory icons
Eardrum, 137, 156
Ecological task approach (ETA), 220
E-communication, 507–509, 511
EEGs, see Electroencephalograms
    (EEGs)
Elasticity, 50, 206, 314, 320–324, 327,
    329, 330, 325, 374, 535
Elastic resistance, 249
Electroencephalograms (EEGs), 20, 81,
    82f, 196–201, 226, 308, 316,
    380, 448, 519, 535, 537–538,
    538f
Electronic monitoring, 303, 337
Emergence, 50, 130, 511, 524
Emergency exits, 280, 561
Emergency response (ER), 53, 149,
    188, 329t, 330, 334t, 380, 388,
    392, 393, 414, 509, 524–525,
    535, 537, 538, 543, 573
  hazard perception, 388–390f
  situation awareness and sensemaking,
    392, 393
  time ecologies, 390–392
Emotion, 81, 275, 315, 323, 326t, 421,
    439, 443–447; see also Anxiety
Emotional intelligence, 325, 535
Empath, 536
Empathy, 325, 446, 535
Employee assistance program (EAP),
    383
Endurance limit, 320
Entropy, 35–37, 45–47, 59, 156, 173,
    174, 199, 200, 213, 263, 265,
    332, 353, 369, 391, 490–492,
    494, 500, 511, 515, 522, 538,
    538f

Environments, outdoor, 8, 298, 308, 423, 548, 560, 562, 563; see also Microenvironments; navigation
Episodic memory, see Memory
Equal Employment Opportunity Commission, 282
Equilibrium, 37, 208, 457, 491, 494, 513, 521
  body coordination and, 280, 281
Ergodicity, 51, 507
Ergonomics, 6
Error, 5, 8, 10, 19, 28, 50, 71, 106, 181, 191, 193, 194, 198, 206, 208, 214, 222, 231, 233–235
  of commission, 23, 265
  in communication, 35, 36, 398, 400, 453
  extraneous acts, 23, 497
  minimization of, signal detection theory and, 73, 74, 306
  mode, 24, 491, 539
  of omission, 23
  probability, 24, 25, 39, 78, 79, 85, 199
  sequential, 24, 499
  timing, 24, 185, 499
Error messages, 282–283, 314
ER systems, see Emergency response (ER)
ESS, see Evolutionarily stable state (ESS)
ETA, see Ecological task approach (ETA)
Euler buckling, 320, 321
Event chain, see Domino models
Evolutionarily stable state (ESS), 521
Evolutionary programs, 63
Example space, 466
Executive function, working memory and, 169, 170, 190, 191, 194, 202, 204, 205, 298, 331, 332, 335, 339
Exoskeleton and exosuit, 285
Expectancy
  life, 477
  tables, 377
  theory, 176–177, 430, 558
Expert knowledge space, 481, 482
Experts vs novices, 18–20, 22, 23, 32, 99, 203, 255, 388, 543, 415, 426, 432

Expert systems, 7, 32, 60, 455, 459, 467, 470, 479, 480; see also Artificial intelligence; Expert vs. novices
Explicit learning, 207, 517
Explicit memory, 169
Explicit timing demands, multitasking and, 192
Explosive strength, 280
Exponential distributions, 39, 68, 178
  accidents and, 345, 347, 348f
Extended reality, 445
Extent flexibility, 280
Extreme temperatures, 298
Eye, structure of, 92f; see also Visual displays
Eye-tracking, 122, 128, 200, 380, 408

Face, 127, 275, 300, 476, 486
Facebook, 476
Face-to-face meetings, 2, 508, 515, 545
Facilities management systems, 109, 111f, 554, 555
Factor analysis, 83, 174, 235–237, 481
Factorial models, 347, 349, 351–353f, 393
Fan-out, 54, 528
Fatigue, 7, 23, 24, 82, 88, 106, 169, 170, 281, 285, 287, 288, 293, 314–317, 380, 381, 404, 408, 508, 561, 574, 575, 586
  cognitive, 188, 197, 317–322f, 328–333, 342, 356, 535, 537
  degrees of freedom, 293, 332
  multitasking and task switching, 173, 333–337
  vigilance dual task, 336–337
  work curve, 315, 316, 319, 330, 332
Fault isolation, see Troubleshooting
Fault tree analysis, 353–356f, 393
"Favorite task" strategy, 174
FCC, see Federal Communications Commission
Feature detection, 97, 98
Fechner's law, 67, 68, 78
Federal Aviation Administration (FAA), 345, 384
Federal Communications Commission (FCC), 9, 426, 437
Feedback, 52, 180, 182, 203, 218, 223, 381, 387, 469, 489, 492, 497, 507, 511, 513

auditory, 253
delay, 206, 208
haptic (or tactile), 159, 246, 254, 261, 303, 401, 442–444
motor control, 260, 305, 585
open and closed loops, 254
positive and negative, 46, 47, 264
visual, 253, 261, 440, 585
F-Finance program, 474, 480
Field independence, 100, 327, 378
Field of vision, 93, 107–109, 128, 222, 408, 441
Field-testing, 26, 28, 29
Figure-ground distinction, 99, 100, 130, 144
File transfer protocol (FTP), 425
Finger dexterity, 162, 235
First-order control, 264
Fitts' law, 254–258
Fixed-point attractors, 40, 41, 48, 521
Flaming, 507
Floppy disk, 410, 413
Flow chart, 53, 58, 59, 354
Fluid intelligence, 171f, 174–176, 324, 327, 328, 333, 334t, 535
fMRI, see Functional magnetic resonance image (fMRI)
Forecasting, 78, 116, 180–183, 208, 213, 214, 329t, 334t, 478
Form, perception of, 98, 433; see also Gestalt laws
FORTRAN, 415, 452, 453
Fourth-order controls, 264
Fovea, 92f
Fractal, 46, 57, 266
Fractal dimension, 46, 48, 156, 199, 263, 304, 305
FRAM, see Functional resonance analysis method
Frame-based systems, 464–466
Frustration, 18, 195, 196, 324–326, 329t, 336, 398, 411, 468, 535, 573
FSDT, see Fuzzy signal detection theory
Function, allocation of, see Allocation of function
Functional job analysis, 215–217
Functional magnetic resonance image (fMRI), 117, 200, 316, 420, 421
Functional resonance, 373
Functional resonance analysis method (FRAM), 498–499

Fuzzy signal detection theory (FSDT), 71, 79–81; see also Signal detection theory

Game theory, 457, 458, 519–521, 526, 536
Gaze control, 408
GEFT, see Group Embedded Figures Test
Gemini capsule, 576, 577f, 586
Genetic algorithms, 58, 59, 63, 471, 493
Geographic information systems, 420, 530
Gestalt laws of perception, 51, 98–100f, 122, 127, 130, 144–146, 433, 471
Gestural interfaces, 241, 405–408, 522
Glare, 96
Gloves, 162, 261, 298, 406–407, 442, 443f
Gödel's incompleteness theorem, 455, 487, 542
Good continuation, principle of, 99, 101, 127, 145
GOPHER, 425, 429
Governmental interventions, for accident prevention, 375, 376t, 385, 387, 388
Graphic design, see Aesthetics
Graphic user interfaces (GUIs), 207, 239, 404f, 415, 417f, 420, 422, 476
Grasp(ing), 117, 260, 261, 263, 286, 403, 445
Gravitational forces, 581–582
Gross body coordination, 235, 280, 281f; see also Motor control and coordination
Group (or team) coordination, 8, 24, 53, 58, 64, 306, 517, 522, 523; see also Synchronization
cognitive workload, 185, 196, 325, 326t, 333, 364, 391, 393, 493, 495, 496, 534, 536–537, 543
game theory, 458, 512, 519, 520
group size, 523, 524
human-machine interaction, 444, 477, 528f, 546; see also Human-automation teaming
implicit learning, 517–518f
intersection games, 520, 521, 572

shared mental models, 509, 518, 519
stag hunt games, 524–527
verbalization, role of, 519, 521, 522
Group dynamics, 196, 361, 362t, 363,
    388, 393, 517, 520, 524, 535;
    see also Group coordination;
    Human-automation teaming
and complex technologies, 364, 365,
    496
Group Embedded Figures Test (GEFT),
    100
GUIs, *see* Graphic user interfaces
    (GUIs)

Habitats, outer space, *see*
    Microenvironments
Hair cells, 137
Hammer, 137, 289, 290, 397, 398
Handtools, *see* Anthropometry,
    Handtools
Haptic perception, 159, 160, 259, 439,
    443, 444
Harmonics, 138, 140, 142
Hazard perception, 388–390
Heads-up displays (HUDs), 119–122,
    172, 439, 503, 574
Health, stress impacts on, 35, 277, 288,
    310, 313, 314, 342, 344,
    374–375, 376t, 562, 580
Health care settings, 152, 275, 362t,
    364, 369, 380, 477, 505, 533,
    543, 544
    workstations in, 293–294
Hearing
    binaural, 142
    human ear, 137, 138
    loudness, 138, 139
    pitch, 139f, 140
    streaming, 145, 147, 171
    timbre, 141
Hearing loss, 35, 156–159
Heart rate variability, 198–200, 251
Heat exposure, *see* Extreme
    temperatures
Helmet, VR, 439–441, 446
HEP, *see* Human error probability
    (HEP)
Historical displays, 114, 115, 499
Horn clause, 464
HUDs, *see* Heads-up displays (HUDs)
Human-automation teaming, 533, 534
Human error, *see* Error, Redundancy

Human error probability (HEP), 24, 25
Human memory, *see* Memory
Human-robot interaction, 54, 241, 522,
    527
Hybrid process, 186
Hypermedia, 423, 424
Hypothermia, 298
Hysteresis, 85, 232, 233, 322, 323,
    330, 368

IBM Compositor®, 409
IBM Selectric®, 398, 409
IEC 60602–1–8, 146
IES Lighting Handbook, 134
Illuminance, 133
Illumination, 133, 134
Image clarity, in visibility, 107
Imaginary complexity, 497
Implicit learning, 207, 517, 518f
Implicit memory, *see* Memory
Implicit timing demands, 192
Impulse, 246, 255, 287
Impulsivity, 328, 329t, 348
Individual differences, 17, 71, 76–78,
    100, 147, 174, 189, 190, 219,
    255, 285, 298, 303, 323, 324,
    330, 340, 379, 408, 490, 511
Industry standards, 8, 10, 127, 148,
    250
Inertial resistance, 249
Information
    emotions as, 446, 447
    foraging, 425, 430–432
    incomplete, 15, 171, 179, 224, 469
    overload, 187
    quantifying, 36, 37, 228, 497
    technology, 2, 53, 225, 258, 470,
        501, 540–542, 546
Injury, causes of, 275, 286, 343, 346f,
    348, 352, 354, 365, 366, 369
Inner ear, structure of, *see* Hearing
Intelligence, *see also* Artificial
    intelligence
    collective, *see* Collective intelligence
    crystallized, 175–176
    fluid, 171f, 174–176, 327–328
    hierarchical theory of, 171f, 175f
    quotient (IQ), 174
Intelligibility of displays, 133. 153.
    395
Interactive determinism, 506
Interactive pages, 297, 435–436

International Electrotechnical
    Commission, 146–148, 152
Internet, see also Information; Foraging
    extreme graphics, 437–438
    origins, 424–427
    search engines, 427–430
    site navigation, 432, 433
    web pages, 433–435
    World Wide Web (WWW), 424–425
Interpretability, of displays, 111–112,
    125
Intersection
    coordination game, 520–523
    traffic, 12, 138, 564, 572
Invariance, 148
Inventory, cognitive, 217, 219
IQ, see Intelligence, Quotient
Ishihara test, 96
Isolation, as social stressor, 302, 303,
    581
Iterative
    calculations, 41, 50, 236
    design or laboratory testing, 26, 29,
        273, 496
    games, 521, 527

JDI, see Job Descriptive Index
JND, see Just noticeable difference
    (JND)
Job descriptions, 214–218, 224
Job Descriptive Index (JDI), 291
Just noticeable difference (JND), 65–67

Keeeeeeeeeeeeeeystroke effect, 398
Keyboards, 159, 231, 236, 239–243,
    254, 288–290, 293, 294,
    397–399, 401–404, 406, 408,
    409, 451
    membrane, 401–402f
Keyline paste-up, 410
Keypunch machines, 399, 400, 410
Kitchen, 229, 289, 549, 551f
Knobs, 105, 160, 228, 229f, 241, 247,
    249
Knowledge, 18–19, 22, 32, 82, 141,
    175, 180, 184, 185, 202, 203,
    207, 209, 211, 215, 221, 343,
    388, 393, 490
    base validity, 411, 460, 470, 480,
        481, 483, 484
    expert space, 455, 465, 467, 481,
        509, 523, 562

extraction of, 364, 382, 427, 458,
    482
hierarchy of, 219, 220

Labels
    on controls, 160, 248, 435
    on displays, 121, 437
    on products, 133
Lean body mass, 281
Learning
    dynamic processes, 180–182, 194,
        201–204, 231–234, 263, 265,
        303, 306, 316, 335, 381, 492,
        521
    implicit, 207, 517
    machine, see Machine learning
    and skill acquisition, 106, 230
Learning organizations, 516–518, 524,
    525
Lifting (biomechanical issues), 280,
    283, 285–287, 369
Lighting, see Illumination; Traffic and
    roadway
Likelihood ratio, 73
Likert scale, 70
Limited capacity theory, 186, 187
Linear perspective, in drawings, 102
Line of sight, 107–109, 117, 134, 583
Load stress, see Cognitive; Cognitive
    load; Fitts' Law; Workload
Localization, auditory displays and,
    146–150
Location, of display, see Control
    panels; Field of view
Lockheed Dialog® search system, 427
Logistic map, 31–43f, 86, 182, 212,
    268, 513, 514
Long-term memory, see Memory
Loudness, 65–67, 78, 83, 138, 144,
    145, 148–150, 156, 247
Low-fidelity simulations, 207
Lumens, 133
Lyapunov exponent, 45, 46, 87, 173,
    262, 263, 514, 515, 527, 534

Machine guards, 275, 276, 379, 388
Machine learning, 63, 64, 447, 467,
    471–474
MagCart® typewriter (IBM), 409
Management interventions, for accident
    prevention, 375, 376t, 386, 387
Manhattan Project, 451

Manual dexterity, 235
Martian landscapes, 589f
Maximin principle, 457
May-Oster population model, 512
Mechanical system, concept of, 1
Medical management, for accident
    prevention, 376t, 383
Membrane keyboards, 401, 402
Memory (computer), 409–412, 472,
    473, 482
Memory (human)
    episodic, 169, 306, 329t, 334t
    explicit, 169
    implicit, 169
    interference, 166–168f, 225
    long-term, 124, 166–168f, 225
    pictorial, 329t, 334t
    procedural, 169
    propositional, 169
    semantic, 168
    short-term, 165–167f, 201, 225, 415
    spatial, 169
    trace, 165
    working, 169–171f, 174, 176, 189,
        190–192f, 197, 205, 208, 210,
        298, 327, 331, 332f, 424, 509
    workload, 170
Mercury capsule, 576, 582
Metacognition, 509; see also Cognition
Meta-interpretive reliability, 484, 485
Metaphorical icons, 151
Metarules, 454, 461, 466, 484
Microenvironments, 7, 8, 547
    homes, 8, 93, 237, 547, 558
    kitchens, 229, 289, 549, 551
    offices, 32, 138, 290, 302, 408, 446,
        501, 547, 548f, 553, 562
    space habitats, 586, 587
    stairs, 550–552
Microscribe® Digitizer, 405f
Microsoft Corporation, 410
Microsoft Windows, 454
Midas touch problem, 408
Mindlessness, 82, 213, 305
Minimum entropy principle, 265, 491,
    492, 522
Mobile devices, 407, 408, 480
Mode error, 24, 122, 244, 491, 539
Modular(ity), 55, 181, 184, 220, 498,
    499
Modulus of elasticity, 320, 321
Monocular cues, 102

Moog synthesizer, 406
Morphemes, 153
Motion perception, 104f
Motion parallax, 103
Motor control and coordination, 258,
    259, 305, 582, 585
    aiming, 236, 263, 440f
    in outer space, 585
    reaching and grasping, 204, 260, 261
    walking, 258–260, 267, 304, 305,
        408, 552, 585
Movement time (MT), 227, 254–257
MT, see Movement time (MT)
Multidimensional
    controls, 242–244f
    nonlinear psychophysics, 85, 86
    scaling, 83, 86
    stimuli, 68–78
Multilimb coordination, 235
Multimedia displays, see also Auditory
    icons
    animation and hypermedia, 423, 424
    speech interfaces, 422
Multiple-cause models, of accidents,
    350
Multiple PMSs, 50, 53, 186
Multitasking, 191–193, 195, 225, 329t,
    333, 334t, 340
Murphy's law, 501, 567
Mystery, 562
Mystery Meat Navigation, 433

NASA, see National Aeronautics and
    Space Administration (NASA)
Nash equilibrium, 457, 521
National Aeronautics and Space
    Administration (NASA), 63f,
    119–122, 203, 365, 579, 580f,
    589f
    Task Load Index (TLX), 195, 196,
        325, 326t, 336, 535
National Highway and Traffic Safety
    Association (NHTSA), 574
National Institute of Occupational
    Safety and Health (NIOSH), 10,
    277, 285–287
National Safety Council, 344–346, 377,
    549, 550, 565, 566, 570, 571
National Science Foundation, 426
Navigation, 8, 20, 159, 177, 192, 205,
    221, 246, 251, 538, 541,
    562–564, 574

facilities, 558–560, 589
web sites, 422, 423, 425, 429, 432, 433, 436, 437, 589
Near-infrared spectroscopy, 200, 316
Near-miss
accident reporting program, 345, 363, 376t, 381, 384
signal detection, 81
Near point, 292, 293
Negative afterimages, 94
Negligence, 10–12
Networks, 62, 63f, 357, 425, 426, 431, 441, 501, 506, 508, 510
centrality, 55, 56, 532
growth, 511–513
nonhierarchical structures, 54
small worlds, 56, 58f, 515
social, 53, 55, 511
Neural networks, 471–473; see also Machine learning
Neutral body position, 583, 584f
NextGen air traffic control, 22, 502, 503, 546
NHTSA, see National Highway and Traffic Safety Association
NIOSH, see National Institute of Occupational Safety and Health (NIOSH)
Noise
colors of, 144, 155–156, 373, 564
exposure, hearing loss and, 10, 35, 156–159
neural, 319, 473
as physical stressor, 7, 35, 297, 301, 359, 362t, 367, 388
signal detection and, 35, 36f, 72–77, 80, 93, 143, 144
statistical, 36, 45, 180, 199, 201, 267, 268
web sites, 437
Nomic icons, 151
Nonlinear psychophysics, 85–87
Nonoptimization decisions, 180–182
divergent thinking, 183
future state prediction, 180–182, 533
planning, 179, 180
Nonverbal communication, 447, 521, 522
Nonwork-related social stressors, 301, 302
Novices, see Experts vs. novices
Numeric keypad, 28f, 240f, 401

Occupational health and safety (OHS), 157, 342, 370, 371; see also Diathesis-stress; Occupational Safety and Health Administration; Safety
Occupational Safety and Health Administration (OSHA), 10, 12, 131, 133, 157, 158, 298, 299, 361, 362, 376t, 383, 385, 387, 388, 561
OHS, see Occupational health and safety
Oligarchic reaction functions, 527
O*NET, 217, 218
Omission, errors of, 23
Open-textured rule, 466
Opponent process theory, 93, 94, 96
Optic disk, 92
Optimization, decisions, 175, 176
expectancy theory, 176, 177
incomplete information, 179
prospect theory, 178–170
Optimum variability, 259, 265, 492, 493, 533, 534
Order of controls, 263, 264
Ordering of rules, 463, 464
Ordering of tasks, 333–335
Orientation, on web pages, 434
Oscillators, 40, 41, 45, 52, 87, 181, 267, 268
OSHA, see Occupational Safety and Health Administration (OSHA)
Otolithic system, 259, 585
Outcome feedback, 208
Out of control, 264, 351, 369, 487, 505
Overconfidence bias, 177
Overcrowding, safety, 277, 278f; see also Crowding

Parallax effect, visual displays and, 106, 113
Parallel process (cognitive), 59, 124, 186f, 187, 189, 210, 373, 424
Participatory ergonomics, 366, 385
PDF, see portable document format
Perception, 4, 67, 78, 85, 86, 87, 112, 122, 129, 135, 139, 140, 209, 257, 261, 327, 333, 334t, 392, 470, 503, 513, 516, 582, 585
auditory, 139f, 140–142, 144, 145, 150, 157

of change, 38, 39
color differences, 68, 96
constructivist perspective, 98, 99
depth, 92, 102, 106, 114
feature detectors, 97, 98
form, 99, 100f, 327
figure *vs.* ground, 99, 100, 130, 144;
    see also Field independence
Gestalt laws, 99, 100f, 144, 145, 423
haptic, 159, 259, 388–390f, 565,
    569
hazards (danger or risk), 358,
    388–390f, 565, 569
motion, 103, 104
time, 224
Perception-action, 171, 172, 206, 210,
    219, 262, 527
Perceptual cycling
Percolation, 364
Perfect forecasting equilibrium, 208;
    see also Forecasting
Performance, *see also* Buckling stress;
    Yerkes-Dodson law
    criteria, 8, 9, 101, 111, 153, 288,
        535
    cusp models for, 319–342
    degrees of freedom, 194, 199,
        204–206, 234, 235, 374
    levels of, 232, 315, 338, 340, 373
Periodic complexity, 497
Periodic entrainment, chaotic
    controllers and, 267, 268
Personnel selection, 6, 282–285, 375,
    376t, 377, 378, 383, 580
Petri nets, accident analysis and, 59,
    356, 393
Phase shifts, 51, 56, 58, 157f, 204, 206,
    305
Phase transition, 204, 506
Phoneme, 153, 252
Phonological loop, 170
Photograph(–y), 93, 94, 101, 102, 116,
    253, 392, 416, 434, 438
Photopic functions, 96, 97
Physical abilities
    body coordination and equilibrium,
        280, 281f
    flexibility, 280
    lean body mass, 281
    simulation, 282–285
    stamina, 281
    strength, 278–280

Physical fatigue, 281, 285, 287, 288,
    293, 305, 315f, 316, 317f,
    320–322, 330, 331
Physical loudness, 65, 66, 83, 138, 139
Physical stressors, 297, 359, 362t
    extreme altitudes, 300, 301
    extreme temperatures, 298–300
    noise, 301
    toxins, 297
Physiological indicators (of cognitive
    workload), 196–200, 446
Pictograms, 123, 125
Pictorial memory, *see* Memory, human
Pictorial signs, 130
Pitch, 140, 141, 144–148
    envelope, 141f, 152, 153
    helix, 139f
Planning, *see* Nonoptimization
    decisions
Poisoning/drug overdoses, 549, 550,
    554f; *see also* Physical stressors;
    Toxins
Poisson distributions, 39, 347
Policy-capturing regression, 460
Population stereotype, 18, 229, 230,
    256, 401, 411, 435
Pop-up window, 115, 151, 416, 418,
    419, 421, 437, 448, 469, 555
Portable document format (PDF), 426
Positioning devices, 402, 403
Poster campaigns, accident prevention
    and, 376t, 382
Power law (distributions), 39, 48f, 51,
    56, 57, 61, 72, 78, 231, 300,
    431, 540
Preautomation workload, 19
Predictive displays, 115, 116, 209, 264,
    499
Predictive power, 79
Primacy effect, 116
Procedural memory, *see* Memory,
    human
Process-event sequences, accident
    occurrence and, 353
Process feedback, 208
Production paradox, 183
Programming/programs, 451–455
Propositional memory, *see* Memory,
    human
Prospect theory, 178–179
Proximity, principle of, 101, 100f
Psychological Abstracts, 427, 428

Pulse sequence, for standard fire alarm, 149f
P300 wave, 197, 198, 200, 201

Quality circle, 382
Quetelet, Adolph, 271
QWERTY keyboard, 290, 397, 399, 401, 402

RAM, *see* Random access mode (RAM)
Random access mode (RAM), 169, 412
Random process(es), 39
Rapid eye movements (REM) sleep, 308
Rapid prototyping, 28, 29
Rate control, 235
Reaching and grasping, motor control and, 204, 260–263
Reaction time, 227, 235, 298, 333
Real-world complexity, 208, 496–500f, 521
Receiver operating characteristic (ROC) curve, 74–77f, 79
Recency effect, 166
Recognition-primed decision making (RPD), 203, 204, 373, 470
Recomplicating effect, 502
Recongesting effect, 502
Recursion (–ive) 85, 465, 456, 466, 468
Red-green cone, 93
Redistribution principle, 332
Redundancy, 25, 26, 119, 124, 149, 161, 162, 187, 269, 477
Reliability, 22–26, 32, 71, 119, 161, 162, 183, 187, 195, 206, 269, 451, 452, 484, 490, 493, 542
REM sleep, *see* Rapid eye movements
Repeating effect, 501
Residual variability, 205
Resilience (engineering), 49f, 50, 188, 302, 323, 324, 339, 363, 372–374, 393, 490, 493, 543
Resistance of controls, 248, 249
Resource competition model, 189f, 192, 196
Response criterion, 66, 67
Response orientation, 235
Response set, 67, 71, 75
Response time (RT), 8, 86, 122, 144, 173, 190, 191, 194, 219, 227, 231, 234, 241, 250, 253, 254, 306, 321, 323, 324, 380
automobile driver, 570, 575

Donders' RT, 227, 228
Responsive design, 436
Retrieval cue, 167, 168f
Revenge effects, 181, 500, 501, 545
Risk homeostasis theory, 569
Roadway configurations, 570, 571f; *see also* Traffic and roadway
ROC curve, *see* Receiver operating characteristic (ROC) curve
Rods, 91
RPD, *see* Recognition-primed decision making (RPD)
RT, *see* Response time (RT)
Rule-based systems, 219, 460–462, 465, 467, 509
Rule groups, validity of, 173, 438, 465, 466, 473

Safety, 8, 214, 223, 233, 313, 394, 404, 561; *see also* Occupational Health and Safety Administration
climate, 352, 358, 359–362t, 365–370, 378, 381, 386, 387
committees, 375, 376t, 382
in complex systems, 539
culture, 370–371
factor, 347
information technology, 540
machine guards, 275
medicine, 393, 505, 542–543
product design, 11, 28
transportation, 193, 225, 230, 539, 540, 566–573
workspace design, 10, 276f, 277f, 554
SAGAT, *see* Situation Awareness Global Assessment Technique
Sample entropy, 24
Satisficing in decision making, 185, 458
"Scaled worlds," 207
Scaling
multidimensional, 83, 84f, 481
nonpsychophysical stimuli, 70, 71
psychophysical stimuli, 68–70t
relationship, 46, 300, 345
Scotopic functions, 97
Screen organization, visual displays, 416–419
Seasonal affective disorder, 308
Seat belts, 569
Seating, 272, 273, 285, 288

Second-order control, 264
The Second Self, computer as, 3
Securities and Exchange Commission, 475
Selective recruitment, 567
Self-confidence, 337–339f
Self-organization, 37, 46–48, 51, 59, 203, 204, 207, 231, 263, 364, 373, 494, 496, 499, 500, 506, 513, 516, 521
Semantic memory, *see* Memory, human
Sensemaking, 210, 212, 214, 223, 392, 509, 516, 523, 531; *see also* Situation awareness
Sequential errors, 24
Serial process, 186f
"Set-first" strategy, 174
Shadowing, 225
Shannon entropy, 36, 37, 173, 538–539; *see also* Entropy
Shannon-Weaver communications model, 36f, 484
Shared mental models, 515, 517–519
Shell program, 453, 454
Short-term memory, *see* Memory, human
Sick building syndrome, 561, 562
Signal detection (theory), 68, 71–78, 83, 84, 88, 112, 134, 150, 156, 176, 234, 306, 430, 471, 483, 484; *see also* Fuzzy signal detection theory
Signal-to-noise ratio, 91
Signs, 130–133, 559, 560, 561; *see also* Poster campaigns
Similarity, principle of, 100, 145
Simon, on AI, 458, 482; *see also* Bounded rationality
Simulations, system, 22, 58–64, 207, 208, 211, 212, 282–285, 321, 329t, 334t, 357, 393, 438, 439, 458, 499, 513, 416, 535, 537, 566, 570
Single-cause models, of accidents, 347–349
Single-unit housing, 349
Site navigation, 432, 433
Situation awareness, 9, 19, 24, 38, 58, 122, 209–214, 226, 266, 373, 392, 470, 493, 509, 510f, 516, 530, 539, 543, 545, 573

Situation Awareness Global Assessment Technique (SAGAT), 211
Situation Present Awareness Method (SPAM), 211
Size-distance constancy, 97, 102, 159, 160, 257
Skill acquisition, 202, 230, 231
Skylab, 578, 579f, 583, 586, 587f
Sleep, 82, 197, 307, 308
disturbances, 297, 309, 310, 315, 330, 380
in outer space, 580, 586
Small worlds, 56–58f, 207, 208, 515, 532, 545
Smart display, 12, 111, 112, 415, 469, 470; *see also* Visual displays
Smart phones, 407
Smart power grid, 502, 504, 546
Social networks, 53, 54, 302, 310
Social stressors, 301, 302
crowding and isolation, 302, 303
electronic monitoring, 303
Sociometric analysis, 53
Sociotechnical embeddedness, 32
Sound waves, 138–141; *see also* Hearing
Space (outer) environment
gravitational forces, 581, 582
habitats, 586–589
neutral body position, 584f
Space adaptation syndrome, 584, 585
Space of controls, 247–248
Spaciousness, 562
Spam, 426
SPAM (method), *see* Situation Present Awareness Method
Span of control, 54, 528
Spatial memory, *see* Memory
Spatial separation, 144
Speaker identification, in speech displays, 154
Spearcons, *see* Auditory icons
Special populations, macroenvironments and, 560
Speech
displays, 144, 146, 150, 151, 153–155
interfaces, 422
interpretation area of brain, 137
spectrogram, 154, 155
Speed-accuracy trade-off, 42, 223, 233, 234, 306, 350, 473

Speed limits, 12, 112, 248, 501, 568, 569, 589
Speed of arm movement, 235
Speed stress, 188
    and load stress, 304
Spin-glass phenomenon, 41
Stability, 40, 181
    in auditory perception, 145
    meta- 491
    postural, 159, 288
    of (system) performance, 106, 169, 258, 416, 494, 523, 525
    walking, 259
Stag Hunt game, 520, 524–527
Stamina, 218
Static (isotonic) strength, 278
Static resistance, 249
Statistical analysis programs, 454
Statistical template matching, 184
Stereophonic sound, 142
Stereoscope (–ic), 101, 102f, 116, 438, 442
Stimulus-response compatibility, 228, 229f
Stirrup, 137
Stochastic
    determinism, 507
    process, 38, 180, 208
    resonance, 156, 157f, 267, 473
Stocks
    financial, 63
    and flows, 59
Strain, 287, 293, 307, 310, 314, 320, 344, 372, 540
Streaming, 145; see also Hearing
Strength (human physical), 278–280
Stress, see Physical stressors; Social stressors; Speed and load stress
Strict liability, 10
Stroboscopic motion, 103, 104f
Stylus, 404, 405
Swarm(–ing agents), 117, 181, 475, 505, 532, 533
Swiss cheese model, 371, 372f, 543
Symbolic dynamics, 37, 173, 456, 536
Symbolic icons, 151
Sympathetic nervous system, 297
Synchronicity (synchronization), 81, 51–53, 258, 305, 500, 536
System
    concept of, 1

design, 15, 29, 183, 209, 454, 465, 481, 496, 503
    dynamics, 59
    feedback, 208
    reengineering, 32

Tactile displays, 160–162, 347, 253–254; see also Haptic perception
    gloves, 161, 162, 261, 298, 396, 406, 409, 442, 443f, 448
    knobs, 160
    vibration, 162
Tacton, 162
Tactor, 161
Task analysis, cognitive, 26, 218, 219, 221–224
    cognitive inventory, 219
    ecological task approach, 220
    goals, hierarchy of (GOMS), 220
    rules, skills, and knowledge, hierarchy of, 220
Task-based job analysis, 217
Task difficulty, defined, 187, 329t, 535
"Task-first" strategy, 174
Task switching, 37, 173, 174, 188, 192, 211, 268, 335, 513, 534
Teams, 52, 58, 109, 180, 185, 188, 196, 199, 209, 211, 218, 221, 223, 233, 251, 325, 327, 474, 489, 490, 495, 509, 515–519, 522–527, 534, 543, 548, 581, 582, 590; see also Human-autonomy teaming; Synchronization
Team neurodynamics, 537–538, 538f
Technical manuals, 29, 30
Technology interventions (accident prevention and), 376t, 379, 386
Telecommuting, 508
Telegraph operation, 201–203
Telemedicine, 505, 543
TELNET, 428, 429
Template matching (troubleshooting), 184
Temporal dynamics, 40, 87, 317, 515
Text-to-speech (TTS) conversions, 422
Texture, 159, 433, 434, 439, 443, 444
Texture gradient, 103
Theremin, 405, 406
Thermodynamics, second law of, 37
Think-aloud technique, 224

Third-order controls, 264
3–D
    auditory displays, 142, 145, 149,
        150, 438, 564
    controller, 244, 405, 407, 442–444
    photography, 101–102
    visual displays, 116–118, 257f, 439,
        441
Threshold
    absolute, 65, 66–68f, 72, 77, 78, 87
    difference, 65, 66, 72, 82
    response, 75, 81, 82, 89, 91, 93, 137,
        148, 150
    shift, 157
    synchronization, 305
Timbre, 83, 141, 145, 148, 422
Time horizons (time ecologies), 116,
        194, 340, 390, 391, 498
Timing demands, multitasking and,
        192
Timing errors, 24, 499
Topological entropy, 37, 46, 173, 174
Total Information Awareness system,
        475f, 477
Touchscreen, 241, 246, 253, 403, 404,
        407, 408, 574
Toxin(s), 297, 298, 362t, 380, 391,
        408, 565
Traffic and roadway, 20, 351, 362t,
        380, 391, 408, 565
    lighting and signals, 96, 134, 572
    roadway configurations, 570,
        571–573
    speed limits, 568, 569, 589
Training programs, 17, 30, 203, 204,
        222, 393
Training systems, VR, 18, 444, 445
Tranquillity, 562
Transparency, 468, 533
Transportation, 310, 344, 362t, 364,
        391, 510, 539, 540, 589
Trichromatic theory, 562
Trivial game, 457, 526
Trivial strategy, 526
Troubleshooting, 183–184
Trunk strength, 280
Trust, in automation, 5, 7, 21–23, 312,
        436, 468, 533, 575
TTS, see Text-to-speech conversions
Turing test, 455, 456
Typewriter, 27f, 28, 31, 397, 409, 425,
        453

Ultimate tensile strength, 320
Unblocking effect, 507
Uniform resource locator (URL), 429
URL, see Uniform resource locator
Usability analysis and testing, 19,
        28–31
U.S. Department of Defense (DOD), 9,
        102, 103, 107–111, 144, 160,
        251f, 271, 425, 475f, 476, 531,
        574
U.S. Department of Transportation,
        132, 251f
USENET, 425
User friendliness, 425
User population, 15, 17, 26, 30, 33,
        273, 285, 397, 426, 432, 437

Validity (of AI products), 174,
        480–486
Vanishing point, 102
Variable capacity theory, 185, 188
Verbalization, role of, 224, 225, 482,
        519, 521
Verbal signs, 130
VERONICA, 425, 429
Vestibular sense, 585
Vibration, see Tactile displays
Videoconferencing, 508
Vigilance, 5, 81, 82, 87, 88, 149, 234,
        235, 298, 300, 304, 306, 309,
        327–329t, 334t, 336–339, 365,
        367
Virtual reality (VR), 3, 7, 102, 117,
        159, 161, 241, 259, 260f, 294,
        336, 365, 438–445, 585
Viscous resistance, 249
Visibility, 107–110f, 134, 246, 248,
        547, 557, 558, 563, 564
Vision, 193, 291; see also Field of
        vision
    color vision, 93–96
    eye, structure of, 91, 92
    night, 24, 91
    space environmental design and,
        585
    synthetic, 102, 122
Visual acuity, 92, 92
Visual displays, 102, 105, 106, 121,
        143, 304, 415, 416, 441;
        see also Heads-up displays;
        Visual icons
Visual icons, 123–127

Voice control, 251–253, 269; *see also* Controls
Volley theory, 137
Volumetric display, 117–119, 136
Von Neumann, John, 60, 451, 455, 457, 519
VR, *see* Virtual reality (VR)

Wait signals, 422
Walking, 2, 227, 267, 287, 288, 304, 407, 408, 438, 550, 552, 585
  motor control and, 259–260
  surfaces, *see* Workspace design
Wall-mounted displays, 93, 240f, 440–442
Weber's law, 66, 67
Web pages, *see* Internet
Windows
  architectural, 505, 548, 553, 562
  computer interface, 30, 184, 410, 418, 423, 437, 441, 448, 454–469, 555f
Wizard of Oz, 456, 469, 534
Word-processing, 292, 398, 409, 410, 422, 501
Word rhythm, 152
Work curve, 315, 316, 319, 330, 332
Working memory, *see* Memory
Workload, *see also* Limited capacity theory; Multitasking; Variable capacity theory; Yerkes-Dodson law
  cognitive, 6, 7, 9, 19, 20, 22–24, 28, 32, 49f, 50, 81, 122, 150, 170, 185, 191, 210, 214, 222, 223, 226, 251, 253, 285, 305, 311, 312, 319–323, 327, 329t, 332f, 342, 374, 380, 445, 446, 474, 502, 503, 506, 522, 523, 531, 534, 535, 537, 543
  group (or team), 326t, 327, 330, 493, 496, 535
  physical, 9, 285, 298, 320–322, 326, 329t, 535
  speed and, 303, 304
Work schedules, stress and, 297, 306, 307
  circadian rhythm, 307–309f
  dysregulation, 309–310
Workspace design
  confined spaces, 61, 277, 298
  computer workstations, 291–293
  emergency switch, 276
  health care workstations, 293, 294
  machine guards, 275–277f, 379, 388
  overcrowding, 277, 278f
  walking surfaces, 287–288
Workstations, *see* Workspace design
World Wide Web (WWW), *see* Internet
Wrist-finger speed, 236
WWW, *see* Internet; World Wide Web (WWW)
Www.webpagesthatsuck.com, 433

Yerkes-Dodson law, 311–312f

Zero-order control, 264
Zero-sum game, 457

Printed in the United States
by Baker & Taylor Publisher Services